ADVANCES IN ORGANIC CHEMISTRY:
Methods and Results
VOLUME 7

The Chemistry of Cyclic Enaminonitriles and *o*-Aminonitriles

ADVANCES IN ORGANIC CHEMISTRY: Methods and Results
E. C. TAYLOR, *Editor*

The Chemistry of Cyclic Enaminonitriles and *o*-Aminonitriles

EDWARD C. TAYLOR
Department of Chemistry
Princeton University
Princeton, New Jersey

ALEXANDER McKILLOP
School of Chemical Sciences
University of East Anglia
Norwich, England

1970
Interscience Publishers
A division of John Wiley & Sons, New York · London · Sydney · Toronto

The paper used in this book has pH of 6.5 or higher. It has been used because the best information now available indicates that this will contribute to its longevity.

1 2 3 4 5 6 7 8 9 10

LIBRARY OF CONGRESS CATALOG CARD NUMBER 79-79146
SBN 471 84661 9
PRINTED IN THE UNITED STATES OF AMERICA

PREFACE TO VOLUME 7

This book is intended primarily as a research monograph, and should be of particular interest to organic chemists concerned with the synthesis of condensed nitrogen heterocycles. Our own research interests have dealt for some years with the synthetic utilization of *o*-aminonitriles and cyclic enaminonitriles; we have as a result become aware both of the fascinating chemistry associated with these deceptively simple compounds, and with the general lack of appreciation of their potential as intermediates. No review has ever been written in this field, perhaps in part because of the difficulty inherent in searching the literature for the chemistry of compounds which have not been systematically indexed. We have, however, made every effort to be as complete as possible in covering the literature through August 1968; we apologize for the inevitable omissions.

The first part of this monograph is concerned with the synthesis of cyclic enaminonitriles (the immediate products of the Thorpe–Ziegler condensation) and their conversion to cyclic α-cyanoketones and cyclic ketones. The second part discusses the many diverse methods available for the preparation of *o*-aminonitriles, both aromatic and heterocyclic, and their utilization in the synthesis of condensed nitrogen heterocycles. In our treatment of both classes of compounds, we have concentrated upon the principles involved, and have made liberal use of illustrative examples; complete coverage of most reaction types has been achieved through the summary Tables. Table XXX lists all known cyclic enaminonitriles and *o*-aminonitriles.

It is a pleasure to acknowledge the secretarial prowess of Mrs. Evelyn Cantu, and the helpful criticisms of the complete manuscript made by Dr. A. J. Boulton of the School of Chemical Sciences, University of East Anglia.

EDWARD C. TAYLOR
ALEXANDER MCKILLOP

January 1969

EDITOR'S NOTE

Although most volumes in the Advances in Organic Chemistry Series will continue to be multi-authored works presenting authoritative, critical, and timely discussions of new developments in synthetic and instrumental methodology, in line with the general objectives of the series as set forth in the Preface to previous volumes, the present book marks an expansion of the concept of Advances to include single-authored research monographs. Many synthetic methods and research techniques, at the time publication is first contemplated, have already been developed beyond the point where adequate presentation can be made in chapter form. Research monographs then become the logical vehicle of publication. We hope that the rapidity of publication of such monographs in the Advances series will be attractive both to readers and to authors, and that the series as a whole will continue to present in a challenging, provocative, and stimulating manner new ideas, new techniques, and new methods which will become part of the classical repertoire of the practicing organic chemist.

EDWARD C. TAYLOR
Editor
Advances in Organic Chemistry

CONTENTS

Chapter I. Cyclic Enaminonitriles 1

I. Introduction. The Thorpe–Ziegler Reaction 1

II. Structure of Cyclic Enaminonitriles 3

A. Infrared Evidence 4

Table I: Infrared Spectra of Enaminonitriles 4

B. Ultraviolet Evidence 8

Table II: Ultraviolet Spectra of Enaminonitriles . . . 8

C. NMR Evidence 11

III. Scope and Limitations of the Thorpe–Ziegler Reaction 11

A. Five-Membered Rings 11

B. Six-Membered Rings 17

C. Seven-Membered Rings 20

D. Eight-Membered and Larger Rings 20

E. Heterocyclic Rings 21

Table III: Cyclic Enaminonitriles Prepared by the Thorpe–Ziegler Cyclization 30

F. Dimerization of Alkylidenemalononitriles 57

Table IIIA: Dimerization of Alkylidenemalononitriles . 58

IV. Hydrolysis of Cyclic Enaminonitriles. Synthesis of Cyclic α-Cyanoketones and Cyclic Ketones 60

Table IV: Cyclic α-Cyanoketones and Ketones from Acid Hydrolysis of Cyclic Enaminonitriles 62

Chapter II. *o*-Aminonitriles 79

I. Introduction 79

II. Syntheses from Malononitrile and Its Derivatives . . 79

A. Synthesis of Pyrazole *o*-Aminonitriles 80

Table V: Pyrazole *o*-Aminonitriles 82

B. Synthesis of Pyrimidine *o*-Aminonitriles 103

Table VI: Pyrimidine *o*-Aminonitriles 108

C. Synthesis of Furan *o*-Aminonitriles 126

Table VII: Furan *o*-Aminonitriles 127

D. Synthesis of Pyrrole *o*-Aminonitriles 129
Table VIII: Pyrrole *o*-Aminonitriles 130
E. Synthesis of Thiophene *o*-Aminonitriles 137
Table IX: Thiophene *o*-Aminonitriles 138
F. Synthesis of Isoxazole *o*-Aminonitriles 143
Table X: Isoxazole *o*-Aminonitriles 143
G. Synthesis of Imidazole and Oxazole *o*-Aminonitriles . . 144
Table XIA: Imidazole *o*-Aminonitriles 145
Table XIB: Oxazole *o*-Aminonitriles 145
H. Synthesis of Thiazole *o*-Aminonitriles 147
Table XII: Thiazole *o*-Aminonitriles 148
I. Synthesis of Isothiazole *o*-Aminonitriles 149
Table XIII: Isothiazole *o*-Aminonitriles 150
J. Synthesis of 1,2,3-Triazole *o*-Aminonitriles 150
Table XIV: 1,2,3-Triazole *o*-Aminonitriles 151
K. Synthesis of Miscellaneous *S*-Heterocyclic *o*-Aminonitriles . 152
Table XV: Miscellaneous *S*-Heterocyclic *o*-Aminonitriles . 154
L. Synthesis of Pyridine *o*-Aminonitriles 157
Table XVI: Pyridine *o*-Aminonitriles 158
M. Synthesis of Quinoline and Isoquinoline *o*-Aminonitriles . 165
Table XVII: Quinoline and Isoquinoline *o*-Aminonitriles . 166
Table XVIII: Pyrido(2,3-*b*)pyridine and Pyrido(2,3-*d*)-pyrimidine *o*-Aminonitriles 169
N. Synthesis of Pyrazine and Pteridine *o*-Aminonitriles . . 171
Table XIX: Pyrazine *o*-Aminonitriles 172
Table XX: Pteridine *o*-Aminonitriles 175
O. Acid-Catalyzed Cyclization of Alkylidenemalononitriles . 177
Table XXI: *o*-Aminonitriles by Acid-Catalyzed Cyclization of Alkylidenemalononitriles 178

III. Synthesis of *o*-Aminonitriles by Dehydration of Amides and Related Transformations 180

IV. Synthesis of *o*-Aminonitriles by Nucleophilic Displacement of Halogen, Oxygen, or Sulfur 180

Table XXII: *o*-Aminonitriles by Dehydration of Amides and Related Transformations 181

Table XXIII: *o*-Aminonitriles by Replacement of —Hal or —O by —NH_2 or —CN 188

V. Synthesis of *o*-Aminonitriles by Reduction of *o*-Nitronitriles 196

Table XXIV: *o*-Aminonitriles by Reduction of *o*-Nitronitriles 197

VI. Synthesis of *o*-Aminonitriles by the Bedford-Partridge Reaction 203

Table XXV: *o*-Aminobenzonitriles from the Bedford-Partridge Reaction 204

VII. Synthesis of *o*-Aminonitriles by Ring-Cleavage Reactions 207

A. Of Thiophenes 207
B. Of *s*-Triazine 208
C. Of Pteridines 209
D. Of Diels-Alder Adducts 213

Table XXVI: *o*-Aminonitriles by Rearrangement or Ring Cleavage Reactions 214

E. Miscellaneous 219

VIII. Reactions of *o*-Aminonitriles 219

A. Hydrolysis and Related Reactions. 219
B. Reduction. 224
C. Alkylation 225
D. Diazotization 225
E. Synthesis of Fused Pyrimidines: Acylation of *o*-Aminonitriles. 226
F. Synthesis of Fused Pyridines: Acylation of *o*-Aminonitriles 231
G. Synthesis of Fused 4-Aminopyrimidines 233

1. By Dimerization of *o*-Aminonitriles. Reaction with Nitriles 233
2. By Reaction of *o*-Aminonitriles with Orthoformate Esters and Amines. 238
3. By Reaction of *o*-Aminonitriles with Amidines . . . 243

Table XXVII: Condensed 4-Aminopyrimidines from *o*-Aminonitriles and $HC(OC_2H_5)_3/NH_3$ (or RNH_2) . . 244

4. By Reaction of *o*-Aminonitriles with Formamide and Other Amides 270
5. By Reaction of *o*-Aminonitriles with Guanidine, Cyanamide and Dicyandiamide 272

Table XXVIII: Condensed 4-Aminopyrimidines from *o*-Aminonitriles and Formamide 273

Table XXIX: Condensed 4-Aminopyrimidines from *o*-Aminonitriles and Amidines (Including Guanidine, Cyanamide, and Dicyandiamide) 286

6. By Reaction of *o*-Aminonitriles with Urea and Thiourea . . . 294
7. By Reaction of *o*-Aminonitriles with Isothiocyanates and Isocyanates . . . 295

H. Synthesis of Fused 4(3*H*)-Pyrimidinethiones (4-Mercaptopyrimidines) . . . 299
1. By Reaction of *o*-Aminonitriles with Orthoformate Esters and Hydrogen Sulfide . . . 299
2. By Reaction of *o*-Aminonitriles with Thioamides and Acid . . . 301
3. By Reaction of *o*-Aminonitriles with Isothiocyanates and Acid . . . 304

I. Synthesis of Fused 2,4(1*H*,3*H*)-Pyrimidinedithiones (2,4-Dimercaptopyrimidines) by Reaction of *o*-Aminonitriles with Carbon Disulfide . . . 306

J. Michael Additions of *o*-Aminonitriles . . . 306
Table XXX: Cyclic Enaminonitriles and *o*-Aminonitriles . . . 308

References . . . 375

Author Index . . . 391

Subject Index . . . 405

Chapter I

CYCLIC ENAMINONITRILES

I. Introduction. The Thorpe-Ziegler Reaction

In 1908, Moore and Thorpe (191) described the base-catalyzed conversion of 1,2-bis(cyanomethyl)benzene to "1-cyano-2-iminoindane," and a year later, Thorpe (195) discovered the sodium ethoxide-catalyzed intramolecular cyclization of adiponitrile to give what he described as "1-imino-2-cyanocyclopentane," and extended the reaction to the preparation of a number of other mono- and bi-cyclic five-membered "iminonitriles" (**1**) (now known to exist as the tautomeric enaminonitriles **2**). This work apparently attracted no attention, for it was not until 1933 that Ziegler, during his studies on the synthesis and properties of many-membered rings, adapted Rueggli's high dilution technique to the base-catalyzed intramolecular cyclization of α,ω-dinitriles (187). In a careful examination of Thorpe's reaction, as applied to the preparation of many-membered rings (188–190,193,200–202,235,243,448), Ziegler found that the most effective catalyst system was an alkali metal salt of an aralkyl amine (which has since become known as Ziegler's catalyst). The condensation reaction itself is now generally termed the Thorpe-Ziegler cyclization. Since this early work, the Thorpe-Ziegler cyclization has been found applicable to the preparation of cyclic enaminonitriles varying from 5 to 33 members; it has been used for the construction of *meta*-cyclophanes, catenanes, and of nitrogen, phosphorus, arsenic, and oxygen heterocyclic enaminonitriles and constitutes the most generally applicable and versatile of the available routes to these and derived compounds. The base-catalyzed intermolecular condensation of nitriles of the type R—CH_2CN is one of the oldest known methods of forming the carbon–carbon bond, and in the case of simple self-condensation, leads to aliphatic analogs of the cyclic enaminonitriles formed in the classical Thorpe-Ziegler cyclization. For example, the base-catalyzed dimerization of acetonitrile gives β-aminocrotononitrile (524), and the base-catalyzed dimerization of malononitrile gives 1,1,3-tricyano-2-aminopropene (26,84). Although these reactions are obviously closely related to their cyclic counterparts, they will not be discussed here except insofar as studies on their chemistry or properties relate directly to their cyclic counterparts.

Several reviews of the Thorpe-Ziegler reaction have appeared, the first by Ziegler in 1934 (448); its applications to the synthesis of large ring compounds were discussed, again by Ziegler, in 1955 (517), and a survey of the reaction was included in a recent *Organic Reactions* chapter on the Dieckmann condensation (518). This latter review, although the most recent, was surprisingly incomplete. We have attempted here, therefore, to present a complete coverage of the Thorpe-Ziegler reaction through August, 1968, and although some of the material has been summarized before, more than half of all examples discussed have not been covered in any previous review.

Although no studies on the mechanism of the Thorpe-Ziegler cyclization have been reported (see, however, ref. 495), it seems reasonable to assume that it is, in principle, only a modification of the familiar Dieckmann cyclization (replacement of COOR by CN). The reaction course can be depicted by Scheme 1. The ease and direction of cyclization are affected

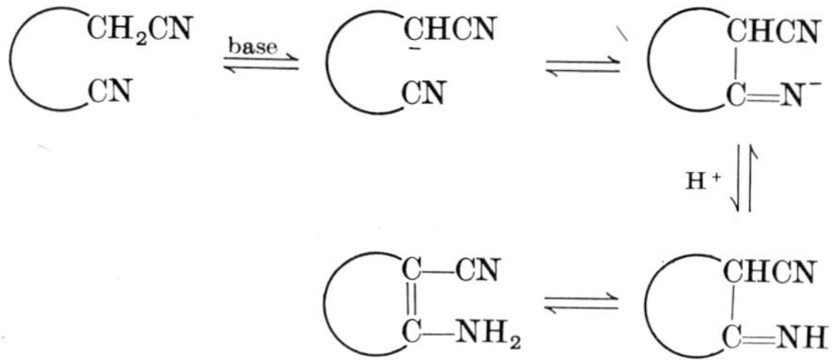

(2) (1)

Scheme 1

by the relative acidities of the methylene groups alpha to each nitrile, and by the relative electrophilic character of the two nitrile groups. Like the Dieckmann reaction, the Thorpe-Ziegler cyclization is particularly effective in the formation of five- and six-membered rings, but differs from the former in the relative ease with which seven- or eight-membered and larger rings are formed. Furthermore, five- and seven-membered rings are often formed with remarkable ease by the Thorpe-Ziegler cyclization and a variety of bases and conditions, some of them extremely mild, appear to be effective. For example, in the industrial preparation and purification of adiponitrile and of its reduction product, hexamethylene-diamine, one of the major by-products is 1-amino-2-cyanocyclopentene. Exposure of adiponitrile to ethyl magnesium bromide, catalytic reduction, HCN, or sodium ethoxide, and even reductive amination conditions upon adipic acid, effect intramolecular condensation. Conditions for effecting the Thorpe-Ziegler cyclization to seven-membered rings are even milder. Thus, 2,2′-bis(bromomethyl)biphenyl (**3**) is converted to **4** simply upon heating with potassium cyanide in aqueous ethanol (223,224). In contrast to the Dieckmann reaction, which is reversible and depends for its success

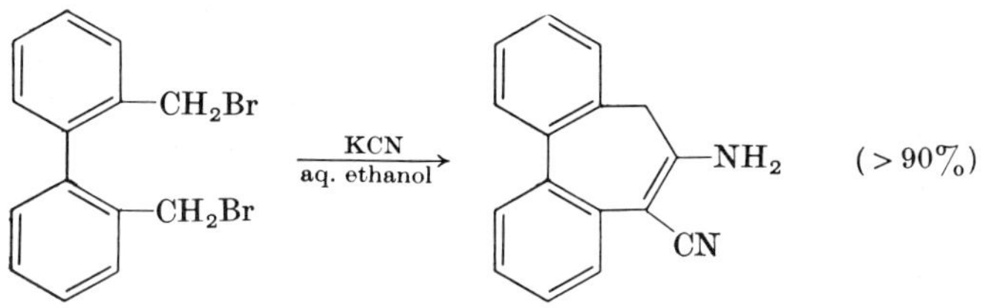

(3) (4)

upon stabilization of the immediate cyclization product as the enolate anion, the Thorpe-Ziegler cyclization leads to a stable enaminonitrile. In fact, the Thorpe-Ziegler cyclization appears to be irreversible even in the absence of a base sufficiently strong to convert the enaminonitrile to its anion.

A wide range of bases has been utilized for the Thorpe-Ziegler cyclization; sodium ethoxide, sodium/potassium sand in hydrocarbon solvents such as toluene or xylene, sodium (or lithium) ethyl (or methyl) anilide (the original Ziegler catalyst), sodium or potassium *t*-butoxide, sodamide, sodium hydride in toluene or dimethylsulfoxide, and even diethylamine have been used. The choice of base and reaction conditions appears in most cases to be arbitrary.

Sodium bis(trimethylsilyl)amide ($NaN(SiMe_3)_2$) has recently been advanced as a superior reagent for effecting Thorpe-Ziegler cyclizations, since it is not air sensitive, as are lithium ethyl anilide and sodium methyl anilide. Using this reagent, and without an inert atmosphere, suberonitrile was cyclized to 1-amino-2-cyanocyclohexene in 97.5% yield (500).

II. Structure of Cyclic Enaminonitriles

The product arising from the Thorpe-Ziegler cyclization of an α,ω-dinitrile can be formulated either as an enaminonitrile (**2**) or its α-iminonitrile tautomer (**1**). Based on the observed facile hydrolysis of the Thorpe-Ziegler cyclization products to α-cyanoketones, earlier workers favored the α-iminonitrile formulation, and subsequent workers have apparently seldom questioned this structural assignment, despite overwhelming physical evidence to the contrary. In fact, the advancement of the imino tautomer represents a classic example of the fallacy of basing structures upon chemical rather than physical evidence (although the failure of the "imino" group to undergo reduction was cited (135) as chemical evidence favoring the enaminonitrile structure). Since even recent workers have persisted in formulating these cyclization products incorrectly, we have summarized below the spectral evidence which clearly establishes the enaminonitrile structure **2** as the predominant form for the compounds. For the reasons to be discussed below, we shall describe all Thorpe-Ziegler cyclization products as enaminonitriles even though they may be described in the literature as their tautomers.

A. INFRARED EVIDENCE

Baldwin (66) and Karle (495) have compiled infrared data on a variety of cyclic and acyclic enaminonitriles, which are summarized in Table I.

TABLE I

Infrared Spectra of Enaminonitriles[a]

Compound	Phase	NH_2	CN	C=C
$CH_3C(NH_2){=}CHCN$	—[b]	3450, 3350, 1645	2180	1600
$C_2H_5OC(NH_2){=}C(CN)_2$	—[b]	3279, 3135, 1653	—[c]	1543
$H_2NC(NH_2){=}C(CN)_2$	—[b]	3300, 3175, 1656	—[c]	1555
$NCCH_2C(NH_2){=}C(CN)_2$	—[d]	3356, 3226, 1664	2370, 2217, 2198	1555
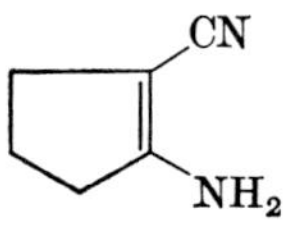	—[e]	3512, 3420, 1664	2189	1605
	—[f]	3497, 3401, 1647	2198	1608
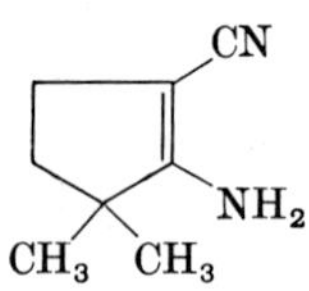	—[f]	3521, 3436, 1661	2212	1613

(*continued*)

TABLE I (*continued*)

Compound	Phase	NH_2	CN	C=C
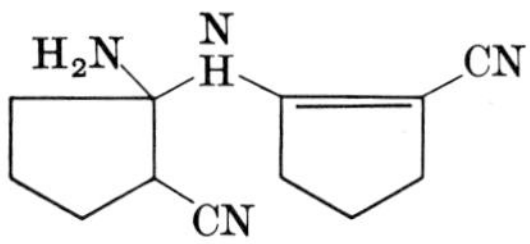	—[e]	3520, 3440, 1640	2180, 2240	1608, 1575
	—[f]	3448, 3226, 1672	2212, 2283	1616, 1587
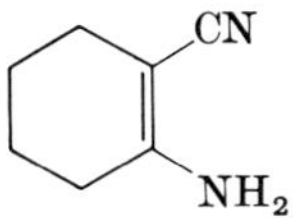	—[f]	3509, 3413, 1647	2212	1616
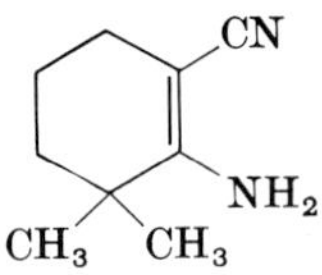	—[f]	3521, 3401, 1637	2193	1592
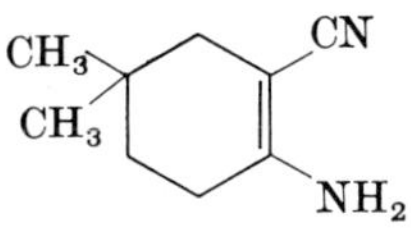	—[f]	3497, 3413, 1656	2198	1618
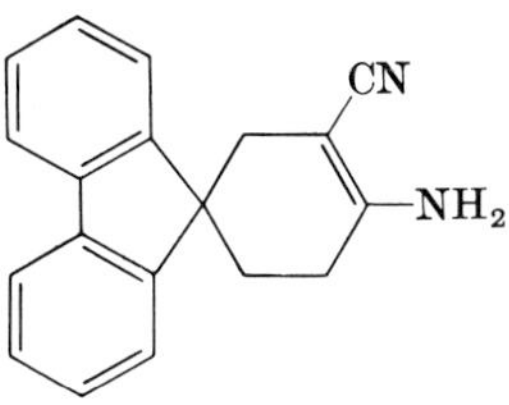	—[d]	3448, 3356, 1645	2169	1608
	—[f]	3378, 3279, 1645	2165	1616
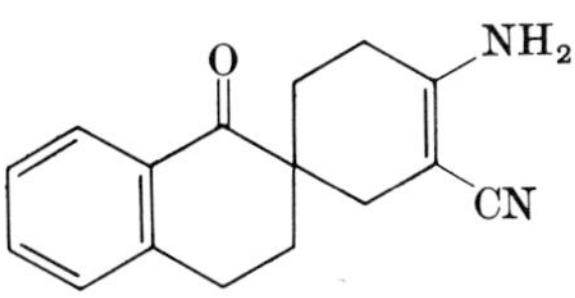	—[d]	3436, 3333, 1637	2179	1616
	—[e]	3460, 3390, 1642	2188	1610
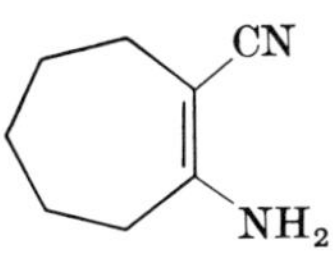	—[f]	3559, 3448, 1667	2217	1621

(*continued*)

TABLE I (*continued*)

Compound	Phase	NH_2	CN	C═C
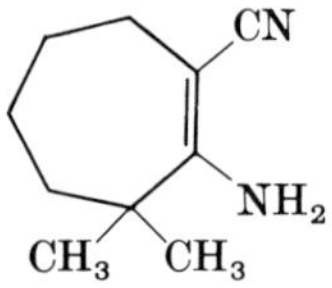	—[f]	3509, 3425, 1637	2193	1567
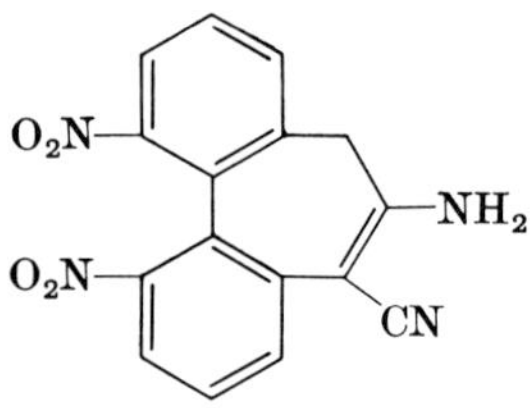 (±)	—[f]	3325, 1635	2190	—[c]
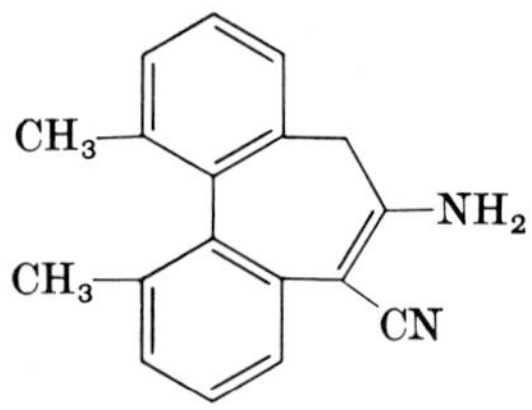 (+)	—[f]	3333, 1635	2174	—[c]
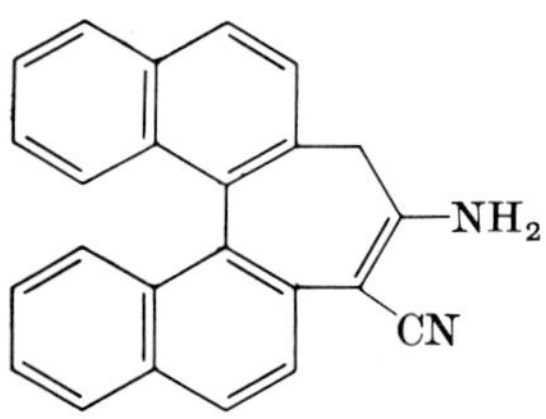 (±)	—[f]	3266, 3170, 1646	2167	—[c]
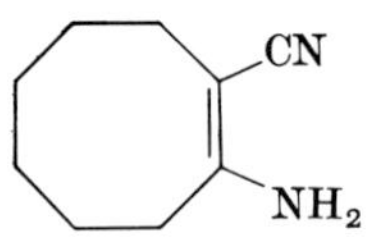	—[f]	3509, 3436, 1639	2183	1600
$CH_3C(N(CH_3)_2)$═CHCN	—[b]	None	2190	1590

(*continued*)

TABLE I (*continued*)

Compound	Phase	NH_2	CN	C=C
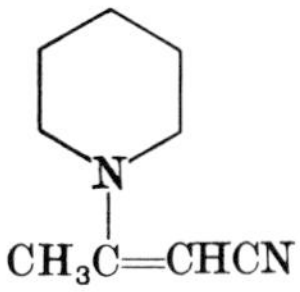	—[b]	None	2185	1570
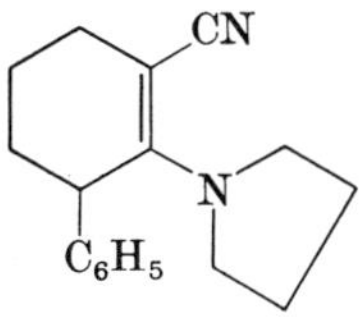	—[b]	None	2170–2180	1570–1580
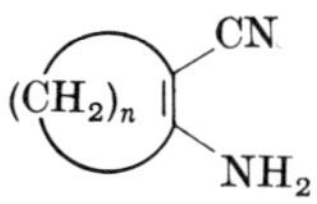 (n = 6,7)	—[b]	None		

[a] Taken from refs. 66 and 495.
[b] Not specified.
[c] No value reported.
[d] Nujol mull.
[e] Chloroform solution.
[f] Potassium bromide pellet.

The reader is referred to the original sources for a detailed discussion of these data; in summary, the most compelling infrared spectral evidence in favor of the enaminonitrile tautomer **2** is the unusually low stretching frequency for the nitrile group observed in all Thorpe-Ziegler products capable of tautomerization. A normal, unconjugated aliphatic nitrile grouping absorbs about 2250 cm^{-1} while aliphatic α,β-unsaturated nitriles absorb about 2225 cm^{-1}, a lowering of some 25 cm^{-1} due to conjugation. The enaminonitriles exhibit their nitrile stretching frequency at 2165–2190 cm^{-1}, a dramatic lowering which has been attributed to conjugation of the primary amino group through the carbon–carbon double bond with the nitrile. Analogous p–π conjugation is responsible for the lowering of the nitrile frequency in β-alkoxy-α,β-unsaturated nitriles, although the decrease in stretching frequency is somewhat less in the latter case, perhaps because of the greater electronegativity of oxygen compared to nitrogen. Almost all enaminonitriles show two bands in the

NH-stretching region compatible only with a primary amino group and incompatible with the imino formulation **1**. Furthermore, these observations exclude a tautomeric equilibrium in which the imino tautomer **1** is present to any detectable (by spectral observation) amount, since only one nitrile band is observed. This conclusion is supported by the observation that 2-cyanocyclanones exhibit two nitrile stretching frequencies, one for the unconjugated nitrile in the keto tautomer and one for the conjugated nitrile in the enol tautomer.

B. ULTRAVIOLET EVIDENCE

Baldwin (66) and Karle (495) have conveniently summarized available data on the ultraviolet absorption spectra of a variety of cyclic and acyclic enaminonitriles which are reproduced in Table II. The high

TABLE II

Ultraviolet Spectra of Enaminonitriles[a]

Compound	Solvent[b]	λ_{max}, nm	ϵ_{max}
$CH_3C(NH_2)=CHCN$	AE	258	13,600
	E	258	18,500
	M	258	19,800
$CH_3C(N(CH_3)_2)=CHCN$	M	267	23,200
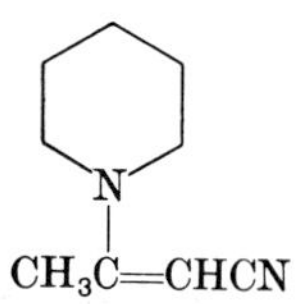	M	270	23,800
$C_2H_5OC(NH_2)=C(CN)_2$	NS	254	"strong"
$H_2NC(NH_2)=C(CN)_2$	NS	251	"strong"
$NCCH_2C(NH_2)=C(CN)_2$	E	276	14,000
	W	273	15,700

(continued)

TABLE II (*continued*)

Compound	Solvent[b]	λ_{max}, nm	ϵ_{max}
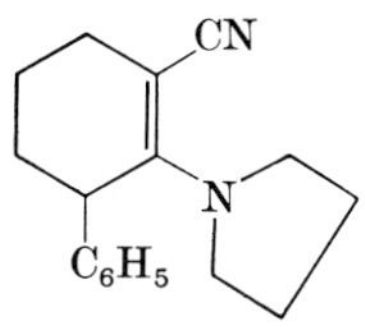	E	289	14,470
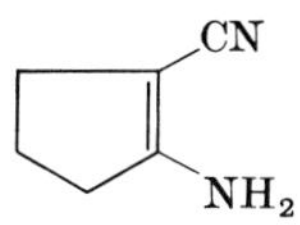	NS E	263 264	13,000 (ref. 66) 13,400 (ref. 465)
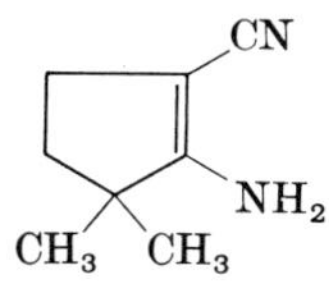	E	264	13,500
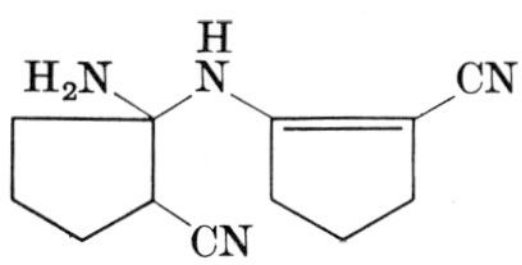	E	263	14,400
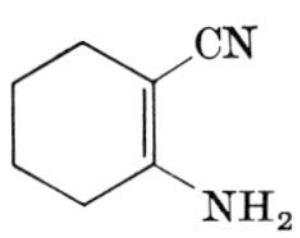	E	264	13,500
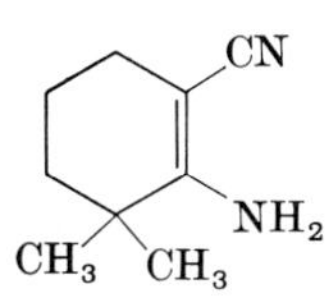	E	266	12,500
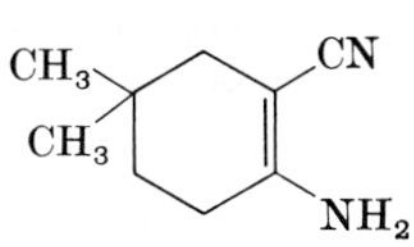	E	265	11,600

(*continued*)

TABLE II (*continued*)

Compound	Solvent[b]	λ_{max}, nm	ϵ_{max}
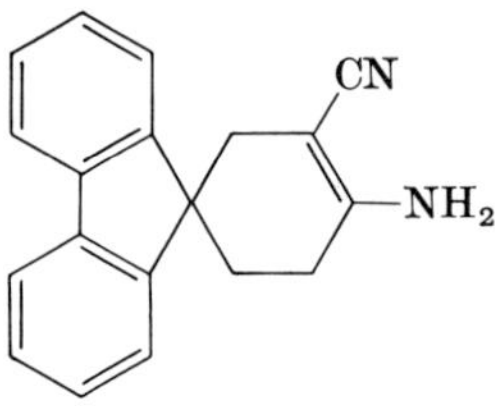	E	267.5	12,700
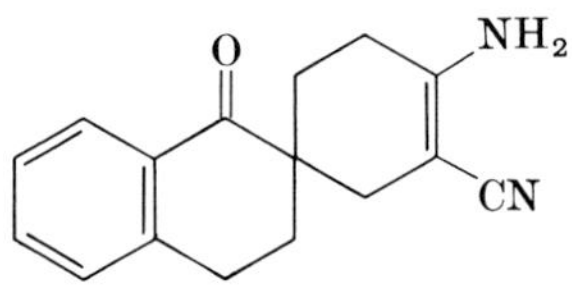	E	261.5	12,200
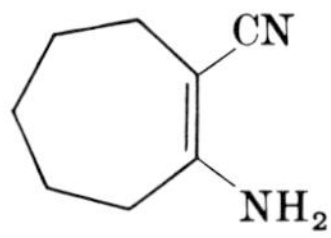	E	271	12,000
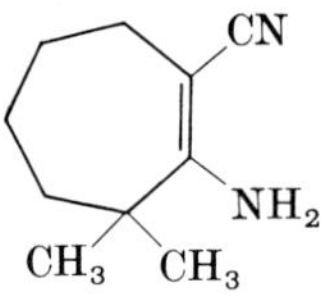	E	273	—
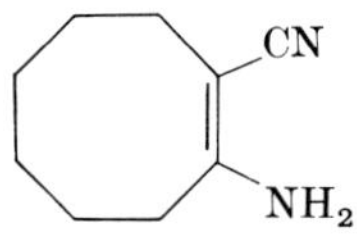	E	269	12,300

[a] Taken from refs. 66 and 495.
[b] AE, absolute ethanol; E, ethanol; M, methanol; NS, not specified; W, water.

position of maximum absorption and the intense extinction coefficients observed are totally incompatible with the imino formulation **1**, and again are consistent only with p–π overlap in the conjugated enaminonitrile formulation **2**. The two products formed upon dimerization of acetonitrile with base, formerly thought to be the amino and imino tautomers of β-aminocrotononitrile, were shown by examination of their UV spectra to be *cis–trans* isomers (498). Both hypsochromic and hyperchromic shifts are observed in acid solution as a result of protonation, while a small hyperchromic effect is noted in alkaline solution, probably due to conversion of the enaminonitrile to its anion.

C. NMR EVIDENCE

The NMR data which have been published, although not extensive, are again compatible only with the enamino tautomer **2** (66,181,273,495). In the compounds which have been examined, the primary amino signal appears at about δ 4.43, a position typical for conjugated primary amino groups. For compound 12, Table I (compound 15, Table II), for example, the integration for aromatic, amino, and methylene protons (4:2:10) is consistent with the aminonitrile formulation (e.g., **2**), but inconsistent with the imino formulation (e.g., **1**) (which requires a ratio of 4:1:11). Most convincing is the absence of fine structure in the methylene signal adjacent to the amino grouping, which would have been a consequence of spin–spin coupling with the C—H proton in the imino tautomer.

Although the study by Baldwin (66) (see, however, ref. 495) remains the most definitive and comprehensive summary of spectral data on Thorpe-Ziegler cyclization products, a number of authors have commented, usually without presentation of complete data, that their products possessed the enamino rather than the iminonitrile structure on the basis of raman spectra (135), molar refractions and dispersions (496,497), and UV, IR, and NMR evidence (7,12,34,81,89,99,111,231,273,281,450,500,542, 549). The data thus appear compelling, even for five-membered rings (which thus contradict Brown's generalizations on the relative preferences for *exo* vs. *endo* bonds in five- and six-membered rings) (516).

III. Scope and Limitations of the Thorpe-Ziegler Reaction

A. FIVE-MEMBERED RINGS

The base-catalyzed intramolecular cyclization of adiponitrile and substituted derivatives of adiponitrile proceeds with remarkable ease (see Table III) and constitutes a superior method (via hydrolysis of the resulting enaminonitriles) for the synthesis of 2-cyanocyclopentanone and related cyclopentene and cyclopentanone derivatives. A full discussion of the utility of Thorpe-Ziegler cyclization products for the preparation of cyclic ketones will be deferred until Section IV. The yields in the cyclization of adiponitrile itself to 1-amino-2-cyanocyclopentene vary from 49 to

100%, apparently depending upon the catalyst and solvent system employed. The reaction appears to be equally applicable to α,α-, α,α,β,-, and α,α,γ-substituted adiponitriles. Yields in almost all cases vary from 65 to 100%, with the apparent single exception of a very low yield (8.8%) reported for the sodamide cyclization of α,α-diphenyl-β-methyladiponitrile (220). The low yield in this case may well be due, however, to the fact that the adiponitrile was formed *in situ* from diphenylacetonitrile and γ-chlorovaleronitrile; assuming that a substantial amount of dehydrohalogenation accompanied alkylation in this reaction, the low yield of the final cyclization product may simply be a reflection of the low yield of direct alkylation.

It is interesting to note that in a number of instances attempted alkylation of diphenylacetonitrile with γ-bromobutyronitrile was claimed to give α,α-diphenyladiponitrile (**5**) whose subsequent chemistry became the subject of some puzzled discussion (287). Later work showed that the actual product formed in the above alkylation was 5,5-diphenyl-1-amino-2-cyanocyclopentene (**6**) (285). In fact, it is now recognized that

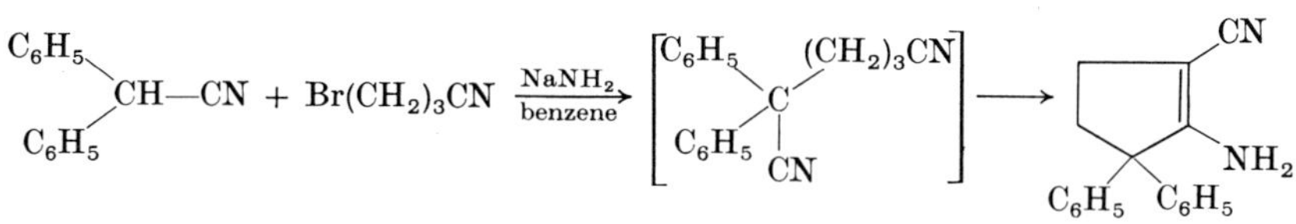

(**5**) (**6**)

the most satisfactory route to the latter type of substituted 1-amino-2-cyanocyclopentene derivatives lies in the direct alkylation of disubstituted acetonitriles with a γ-halobutyronitrile; the intermediate adiponitrile need not even be isolated, since it reacts intramolecularly under the alkaline alkylation conditions and is, in fact, difficult to isolate because of its propensity for intramolecular cyclization.

An interesting example of what appears to be yet another example of a substituted adiponitrile undergoing Thorpe-Ziegler cyclization under alkaline conditions is the conversion of a mixture of 1-carboethoxy-1-cyanocyclopropane and ethyl cyanoacetate in the presence of sodium ethoxide to 5-carboethoxy-1-amino-2-cyanocyclopentene (**7**). The latter compound is also formed as a by-product in the initial synthesis of 1-carboethoxy-1-cyanocyclopropane from ethyl cyanoacetate and 1,2-dibromoethane; its structure was recognized only later to be the cyclic enaminonitrile (194). The course of this unusual reaction may be reasonably depicted by Scheme 2.

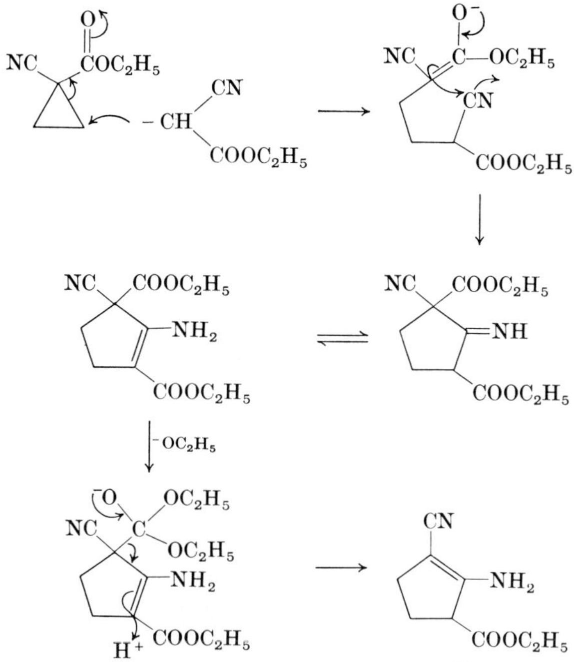

(7)

Scheme 2

Fused 1-amino-2-cyanocyclopentene derivatives are also readily formed by a Thorpe-Ziegler cyclization of what may be formally considered as adiponitrile derivatives of type **8**. For example, *cis*-2-cyano-3-amino-

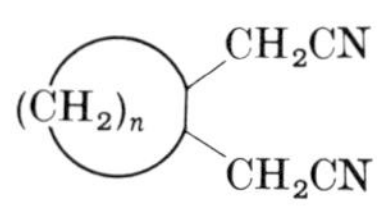

(**8**)

bicyclo[3.2.0]heptene-2 (**10**) is formed in excellent yield by treatment of *cis*-1,2-bis(bromomethyl)cyclobutane (**9**) with sodium cyanide in dimethylsulfoxide, followed by addition of sodium hydride (273). In a

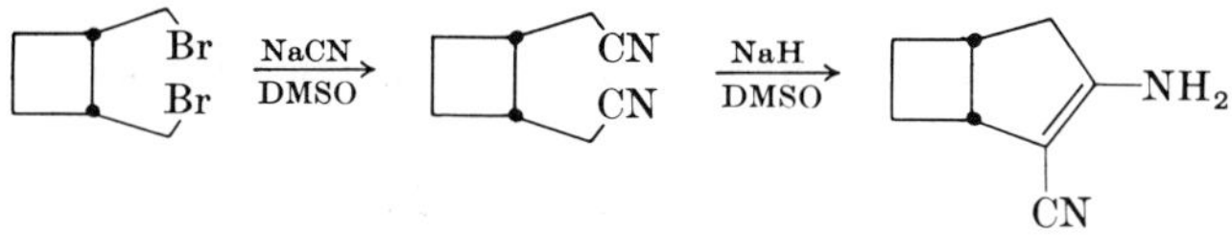

(**9**) (**10**)

variant of this procedure, *trans*-2-amino-3-cyano-4,7,8,9-tetrahydroindene (**12**) was prepared in quantitative yield from **11** and sodium cyanide in dimethylsulfoxide, followed again by addition of sodium hydride (273). This reaction sequence from intermediates readily

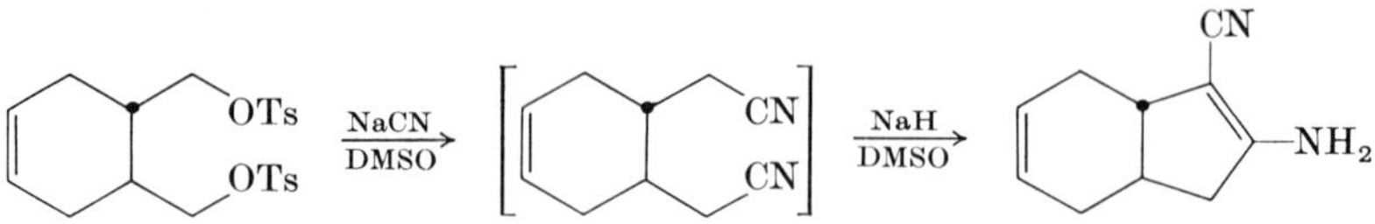

(**11**) (**12**)

accessible by the Diels-Alder reaction of 1,3-butadiene with maleic anhydride, followed by lithium aluminum hydride reduction, has obvious utility in the formation of bicyclic cyclopentanone derivatives of known ring fusion stereochemistry.

A recent synthesis of [4.4.3]propellane (**13**) was achieved by utilization of the Thorpe-Ziegler cyclization for construction of the five-membered ring (7). Thus, treatment of the dibromides **14** and **15a** with sodium cyanide in dimethylsulfoxide gave the corresponding dinitriles, which underwent cyclization *in situ* under the influence of sodium cyanide as the base to give the enaminonitriles **16** and **17**. Hydrolysis to the keto

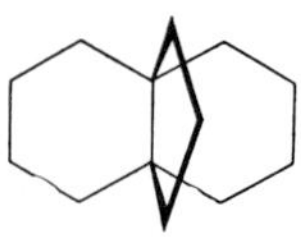

(**13**)

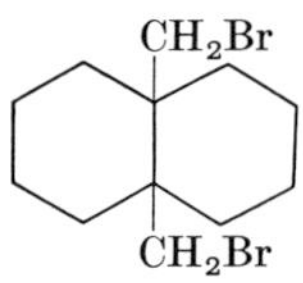

(**14**)

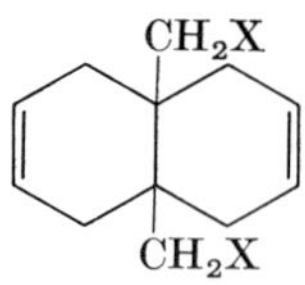

(**15a**) X = Br
(**15b**) X = OTs
(**15c**) X = OMs
(**15d**) X = CN

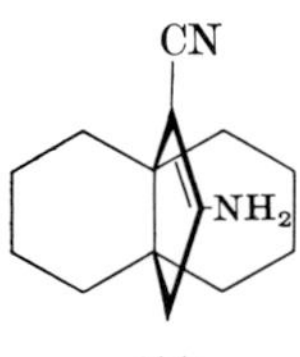

(**16**)

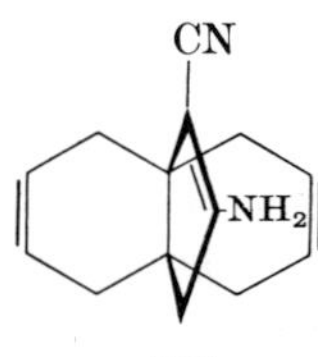

(**17**)

nitriles proceeded in quantitative yield; further hydrolysis of the keto nitrile derived from **16** to the cyclopentanone, followed by reduction, then gave [4.4.3]propellane (**13**). Although the ditosylate corresponding to **15** was unreactive toward sodium cyanide in dimethylsulfoxide, it did react

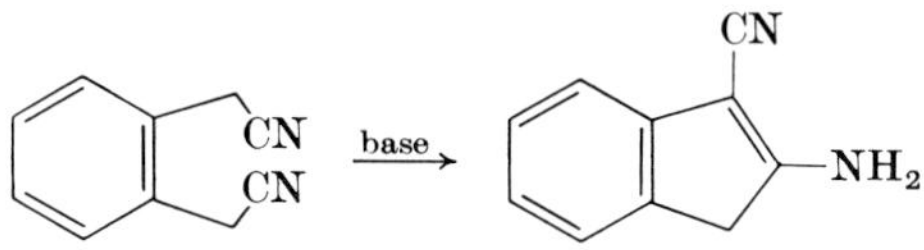

(18) **(19)**

in the presence of two equivalents of sodium iodide to give the Thorpe-Ziegler product **17** in 25% yield (12). A deliberate attempt to prepare the intermediate dinitrile **15d** from the more reactive dimesylate **15c** was frustrated by facile Thorpe-Ziegler cyclization to **17**; the use of *N*-methyl-2-pyrrolidinone rather than DMSO, DMF or pyridine as solvent gave the smallest amount of the undesired enaminonitrile (534).

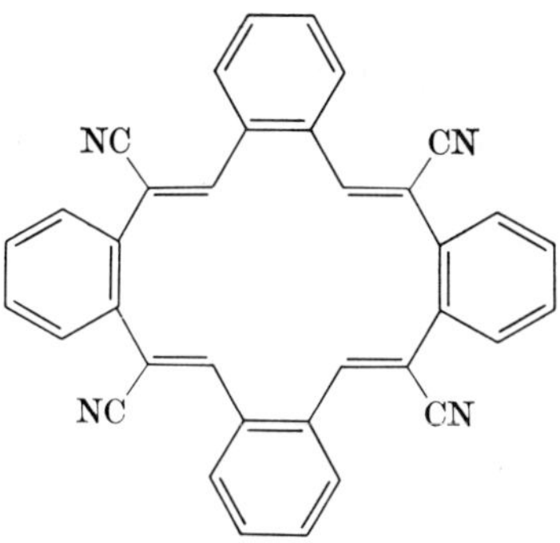

(20)

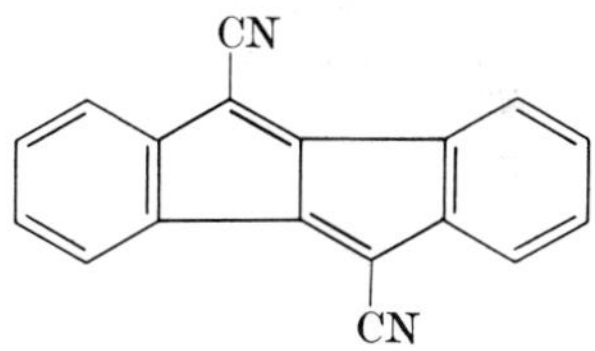

(21)

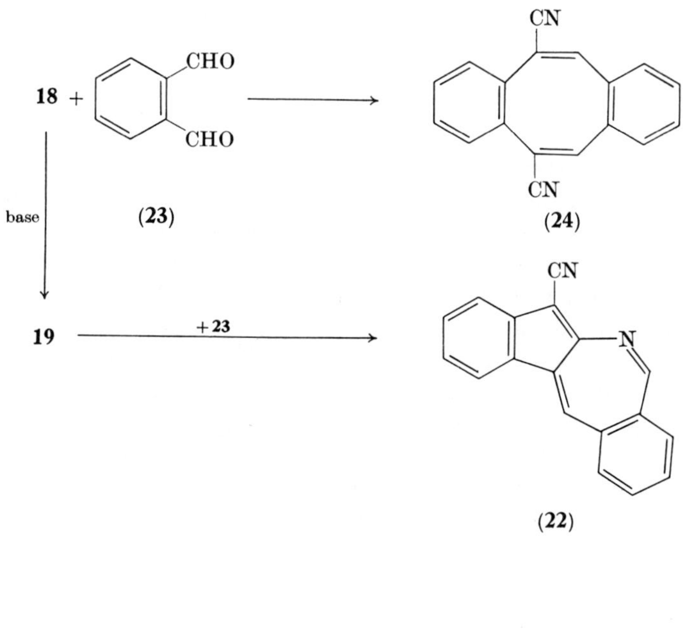

(23) **(24)** **(22)**

2-Amino-3-cyanoindene (**19**) is readily prepared by treatment of 1,2-bis(cyanomethyl)benzene (**18**) with bases such as sodium methoxide (191) and, in fact, occurs with such readiness that **19** has been isolated from a number of attempts to reduce **18** under alkaline conditions (454,455).

A by-product formed in the dibenzocyclooctatetraene (**24**) synthesis of Fieser and Pechet (60) which was first thought to be **20** and later revised to dicyanodibenzpentalene (**21**) (90) has been shown to be, in fact, 3-cyano-1,2,6,7-dibenzo-4-azazulene (**22**) (453). The origin of **22** lies in the facile Thorpe-Ziegler condensation undergone by 1,2-bis(cyanomethyl)-benzene (**18**) in the presence of sodium methoxide, the catalyst employed for the condensation of **18** with *o*-phthalaldehyde (**23**). The resulting 2-amino-3-cyanoindene (**19**) reacts competitively with **23** to give **22**. Several heterocyclic systems isoelectronic with dibenzazulene (e.g., **25** and **26**) were prepared analogously by condensation of **19** with salicylaldehyde and *o*-aminobenzaldehyde, followed by cyclization of the intermediate benzylidene derivatives with acid (450–452).

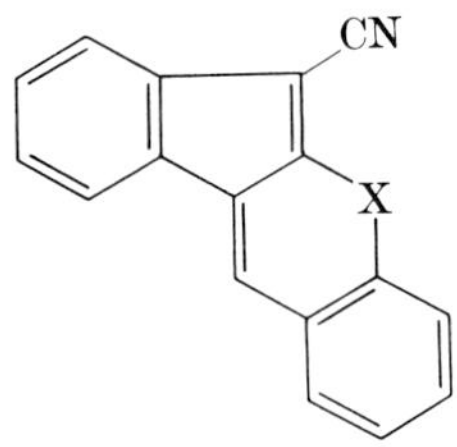

(**25**) X = NH
(**26**) X = O

Acid-catalyzed Thorpe-Ziegler condensations are also possible under optimum circumstances (for brief reviews see refs. 437 and 506). For example, dilute aqueous acid cyclizes disodium hexacyanobutenediide (**27**) to 1-amino-2,3,4,5,5-pentacyanocyclopentadiene (**28**), which is a bright yellow, unstable solid. It reverts to **27** with aqueous base, and is converted with concentrated hydrochloric acid to 1-amino-2,3,4,5-tetracyanocyclopentadienide (**29**), isolated as its tetraethylammonium salt. This latter compound is of some interest because upon diazotization it gives a diazo compound (**30**) whose properties resemble those of normal aromatic diazonium compounds (483).

A projected synthesis of the dinitrile **30a** by a modified Wittig reaction between 3-benzoyl-2-phenylpropionitrile and the sodium salt of diethyl cyanomethylphosphonate was frustrated by a facile Thorpe-Ziegler

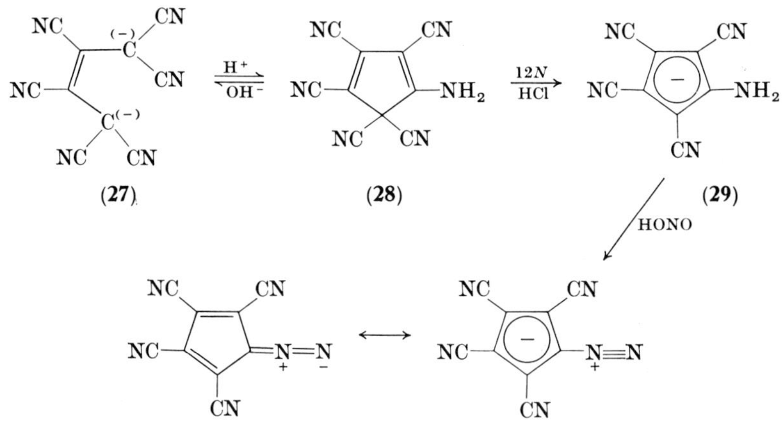

(27) (28) (29) (30)

cyclization (promoted by the phosphonate anion acting as the base) to give the iminonitrile **30b** (533). This latter compound represents an example of a "frozen" iminonitrile which cannot tautomerize to an enaminonitrile.

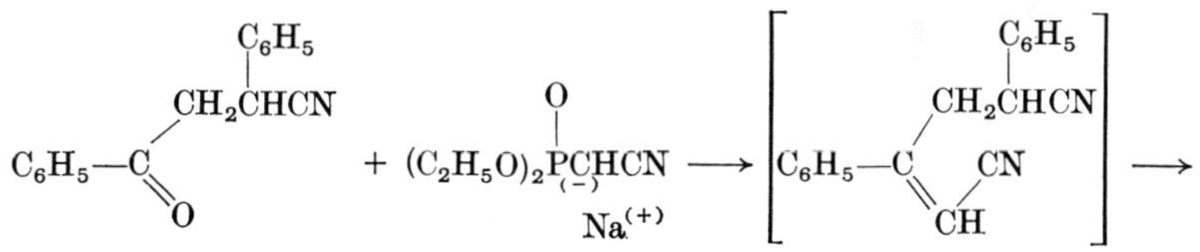

(30a)

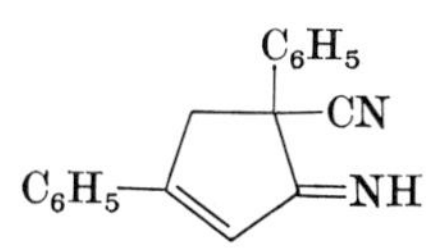

(30b)

B. SIX-MEMBERED RINGS

Pimelonitrile is readily cyclized with sodium hydride to 1-amino-2-cyanocyclohexene, and substituted pimelonitriles react similarly (89). The yields run consistently 10–15% lower than in the corresponding Thorpe-Ziegler cyclizations of analogous adiponitriles. The reduced facility of cyclization to six- as contrasted with five-membered enaminonitriles is underscored by the fact that it is possible to isolate α,α-disubstituted pimelonitriles by the alkylation of α-substituted phenylacetonitriles with δ-bromovaleronitrile; these intermediates may then be subjected to Thorpe-Ziegler cyclization by use of stronger bases such as sodamide. Bis-cyanoethylation of 3,4-methylenedioxyphenylacetonitrile to give **31** followed by

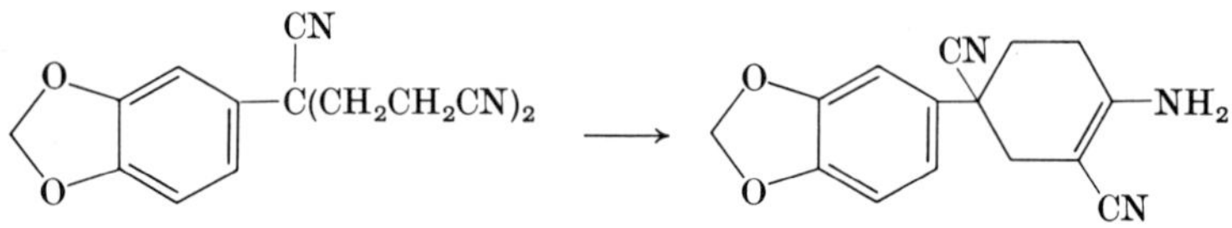

(**31**) (**32**)

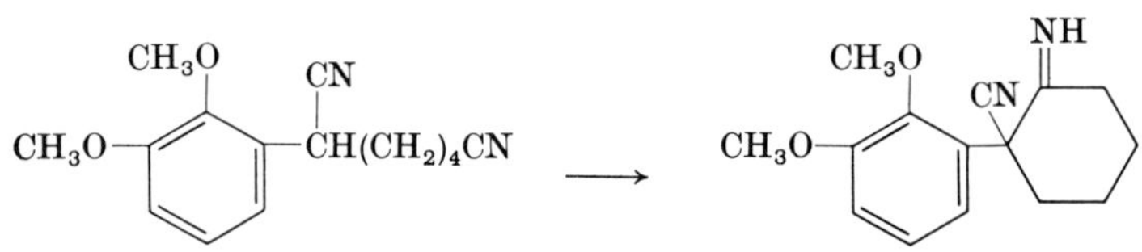

(**33**) (**34**)

sodamide cyclization yields **32** in 80% yield (230). Only when an α-aryl substituted pimelonitrile (e.g., **33**) is cyclized can a true iminonitrile be formed, since only in this instance is tautomerization precluded by the absence of the requisite proton (e.g., **34**) (222). These and related Thorpe-Ziegler cyclizations have been investigated in some detail because of the ease with which the resulting products are hydrolyzed by aqueous acid to α-arylcyclohexanones, of interest in the synthesis of products related to morphine.

The Thorpe-Ziegler cyclization is equally applicable to the synthesis of spiro derivatives of 1-amino-2-cyanocyclohexene. Thus, the bis-cyanoethylation product of fluorene (**35**) is cyclized in approximately 80%

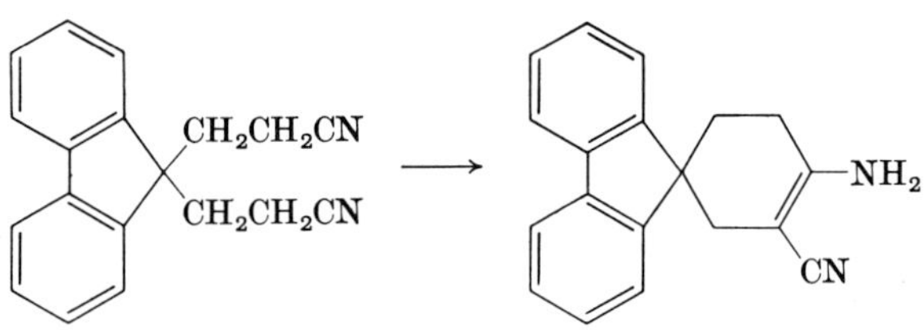

(**35**) (**36**)

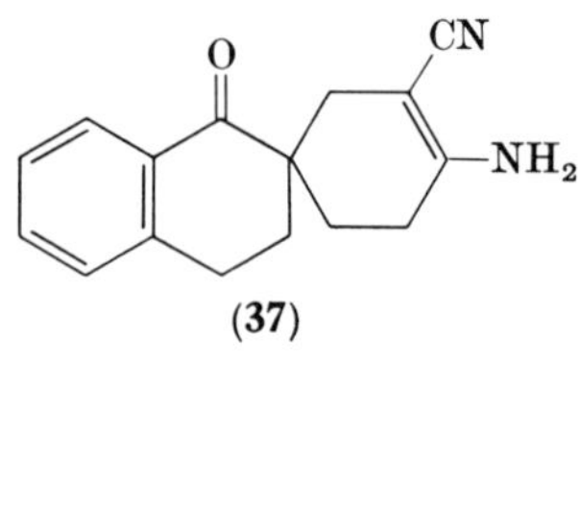

(**37**)

yield to **36** (210). Similarly, the bis-cyanoethylation product of 1-tetralone is reported to give **37** (81). In view of the ease with which these latter two reactions take place, the report (218) that **38** failed to undergo Thorpe-Ziegler cyclization, even in the presence of lithium ethyl anilide, is indeed surprising, and would appear to deserve reinvestigation.

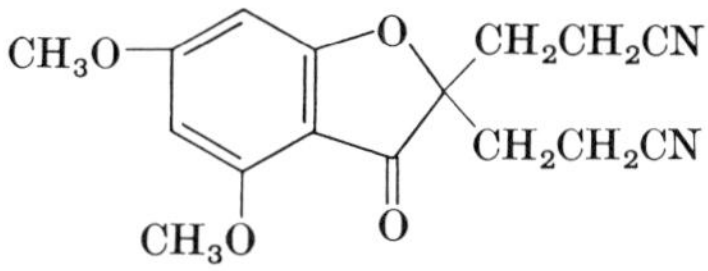

(38)

There is one report of the direct formation of an *aromatic* *o*-aminonitrile by a Thorpe-Ziegler-type reaction; this promising synthesis of functionalized aromatic rings should see further applications. Thus, the

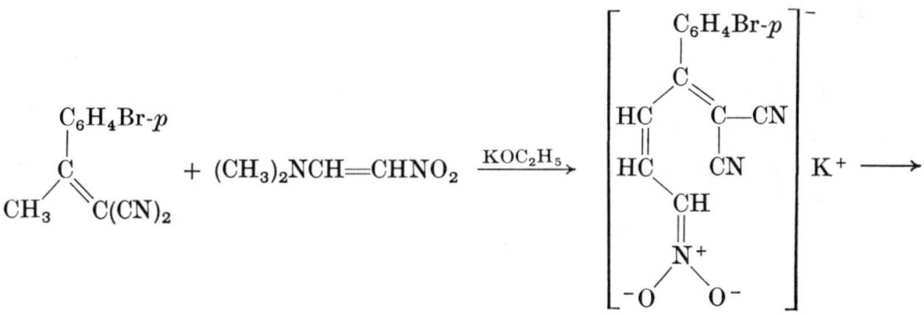

(39) **(40)**

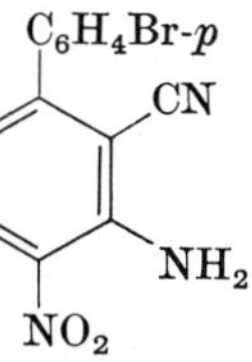

(41)

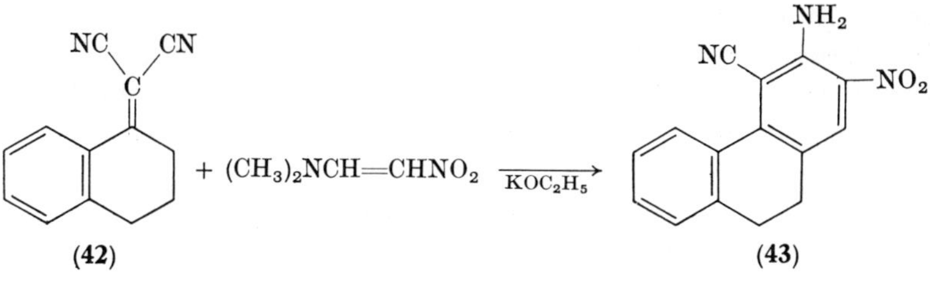

(42) **(43)**

condensation of **39** with 1-nitro-2-dimethylaminoethylene gives 2-amino-3-nitro-6-(*p*-bromophenyl)benzonitrile (**41**) directly; the intermediate salt (**40**) could not be isolated. Similarly, the product of condensation of 1-tetralone with malononitrile (**42**) reacts with 1-nitro-2-dimethylamino-ethylene to give the dihydrophenanthrene aminonitrile **43** (478). An indirect aromatic *o*-aminonitrile synthesis has been achieved from 1-(2-cyanoethyl)-2-cyanomethylbenzene by Thorpe-Ziegler cyclization to 1-cyano-2-amino-3,4-dihydronaphthalene (94% yield), followed by aromatization with selenium dioxide (549).

C. SEVEN-MEMBERED RINGS

The only reported examples of the formation of monocyclic seven-membered rings by the Thorpe-Ziegler cyclization are the conversions of suberonitrile to 1-amino-2-cyanocycloheptene by alkali alkyl anilides (190,495), lithium dicyclohexamide (200,202), or sodium bis(trimethyl-silyl)amide (500) (this latter reagent effects the conversion in 97.5% yield), and of α,α-dimethylsuberonitrile to the corresponding dimethyl derivative (495). It may be worthy of note that Ziegler, who first carried out this reaction (190), employed high dilution techniques, and it is perhaps significant that the yield observed (95–97%) is substantially better than that observed in the formation of six-membered rings without the use of high dilution techniques.

Several dibenz(*a*,*c*)[1.3]cycloheptatrienes (e.g., **4**, p. 3) have been prepared by reaction of bases on 2,2′-bis(cyanomethyl)biphenyl (29,192, 205,206,223,265,476), or even by treatment of 2,2′-bis(bromomethyl)-biphenyls with potassium cyanide in ethanol (223,224). The fact that these latter conditions lead directly to the cycloheptatriene is indeed remarkable; they represent the mildest conditions yet reported for an effective Thorpe-Ziegler cyclization.

D. EIGHT-MEMBERED AND LARGER RINGS

Ziegler has also reported on the sodium ethyl anilide cyclization of azeleonitrile to 1-amino-2-cyanocyclooctene, again using high dilution techniques (190). The 5-phenyl and 5-*t*-butyl derivatives have been prepared in analogous fashion (167,181,559). A mixture of the benzcyclo-octadiene enaminonitriles **45** and **46** was formed by the action of lithium methyl anilide in ether on the dinitrile **44**. The fact that a mixture of products was obtained was of little consequence, however, since the ultimate objective of this reaction, the benzcyclooctanone **47**, was the common hydrolysis product of both isomers (262).

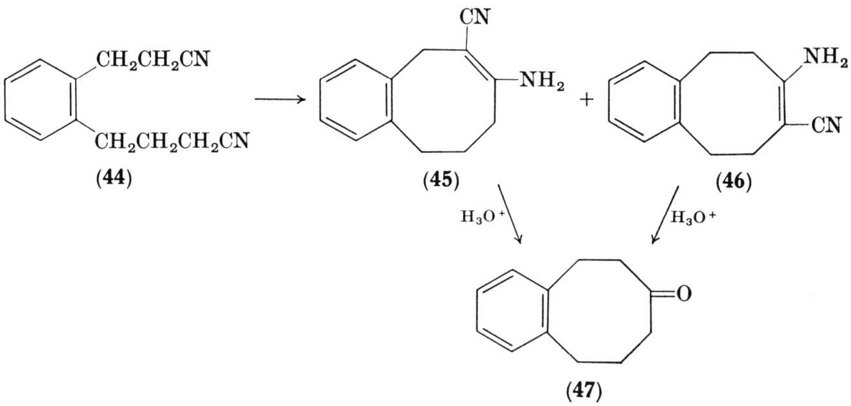

No 9- or 11-membered cyclic enaminonitriles have been reported as products of the Thorpe-Ziegler cyclization. In fact, the only 10-membered ring is the bis-enaminonitrile **48**, formed as a dimeric by-product in the reaction of α,α-diphenyladiponitrile with sodium hydride in dioxane (282). True intramolecular Thorpe-Ziegler cyclization products from α,ω-dinitriles have been reported which possess 13-, 14-, 15-, 16-, and 17-membered rings, while dimeric 10-, 18-, 20-, 22-, and 30-membered cyclic bis-enaminonitriles (from intermolecular condensation) have been reported (see Table III). The *meta*-cyclophane **49** results from the action of sodium ethyl anilide on 1,3-bis(2-cyanoethyl)benzene; no intramolecular product was observed (193).

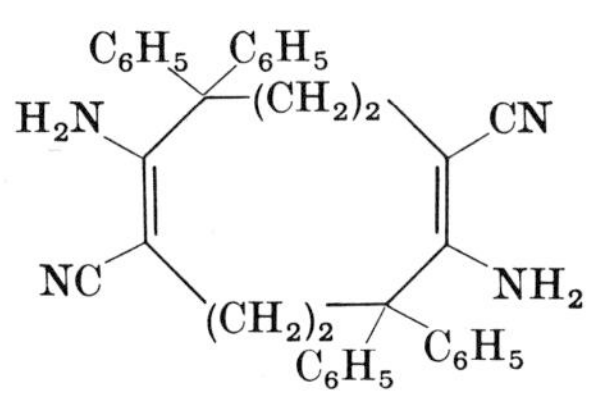

(48)

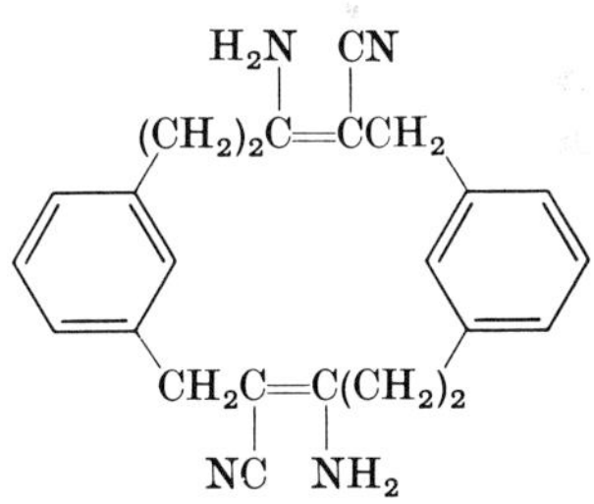

(49)

E. HETEROCYCLIC RINGS

The Thorpe-Ziegler cyclization has thus far found only limited application in the synthesis of heterocyclic enaminonitriles. Aldehyde or ketone cyanohydrins react readily with β-alkylaminopropionitriles to give

N-alkyl-*N*-(cyanoethyl)-*N*-(cyanomethyl)amines (**50**), which upon treatment with sodium *t*-butoxide cyclize to *N*-alkyl-3-amino-4-cyano-3-pyrrolines (**51**) in moderate to good yield (13,88). The low yield observed in the preparation of 1,2,2-trimethyl-3-amino-4-cyano-3-pyrroline from **50** ($R_1 = R_2 = CH_3$) may be attributed to steric hindrance at the *tert*-nitrile grouping.

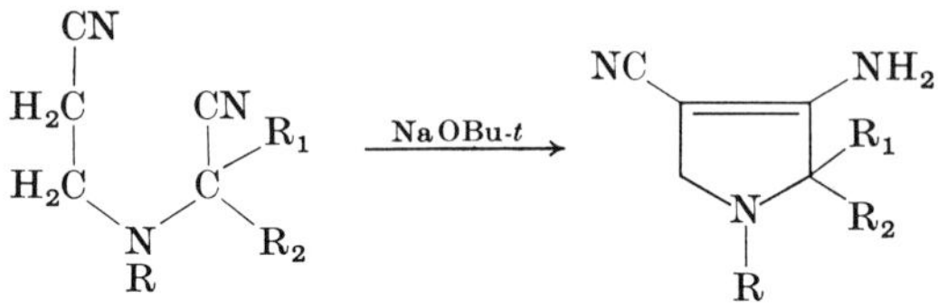

(50) **(51)**

3-Cyano-4-amino-3-piperideines (1,2,5,6-tetrahydropyridines) (**53a**) are readily prepared in high yield by Thorpe-Ziegler cyclization of the bis-cyanoethylation products of ammonia, alkyl, and aryl primary amines (**52a**) (88,95,96,111,112,117,219,441,536). Further members of the eutropic series of cyclic enaminonitriles (**53b,c**) have been prepared analogously by base-catalyzed cyclization of the bis-cyanoethylation products of monosubstituted phosphines (**52b**) and arsines (**52c**) (111,231).

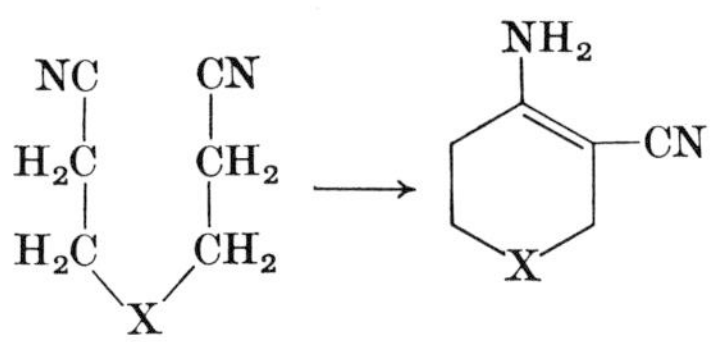

(52a) X = —NR
(52b) X = —PR
(52c) X = —AsC_6H_5

(53)

The ten-membered heterocyclic enaminonitrile **55** was prepared, albeit in low (17%) yield, by the action of lithium ethyl anilide on the bis-cyanoalkylether **54** with the use of high dilution techniques (189). In view of the success of this medium-ring synthesis, one must view with some skepticism the reported failure to effect intramolecular cyclization of bis-cyanoethyl ether and bis-cyanoethyl sulfide to the corresponding six-membered cyclic enaminonitriles (117), since in the corresponding carbocyclic series propensity for Thorpe-Ziegler cyclization is demonstrably higher for the formation of six-membered than of medium-sized rings. The only examples of large-membered cyclic enaminonitriles containing heteroatoms were provided by Ziegler (193), who prepared the *meta*- and *para*-cyclophanes **57** and **59** by sodium ethyl anilide-catalyzed cyclization of the corresponding cyanoalkyl resorcinol and hydroquinone ethers **56** and **58**, respectively, and by Allen and VanAllen (445), who converted the dinitrile **60** to the 4,15-dioxacyclopentadecene enaminonitrile **61**.

An attempt to cyclize 1,16-dicyano-3,14-dioxahexadecane under the same conditions led only to β-cleavage to give decamethylene glycol.

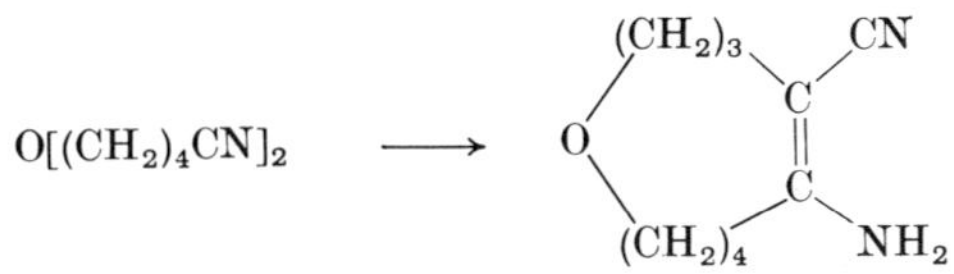

(54) (55)

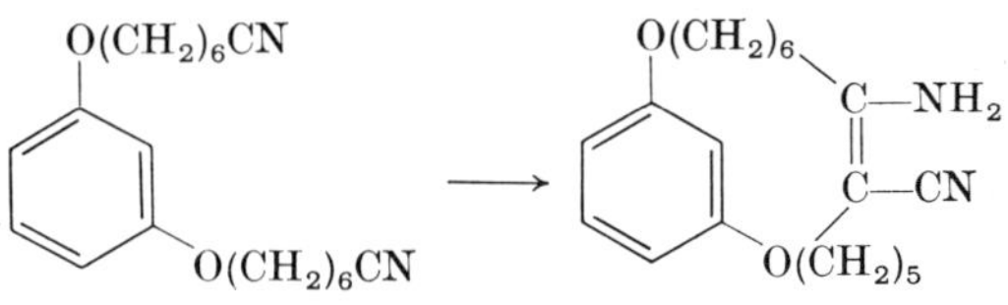

(56) (57)

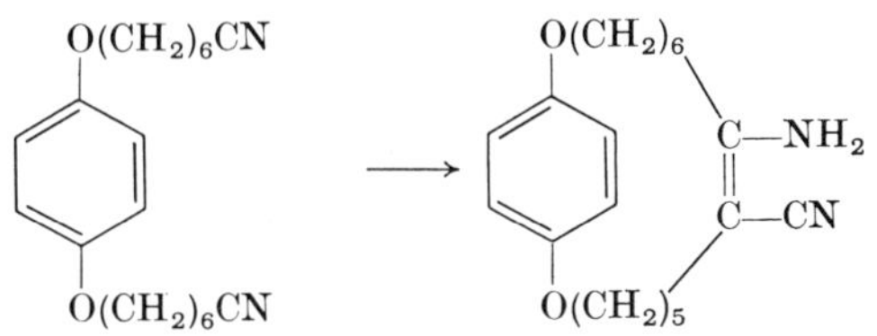

(58) (59)

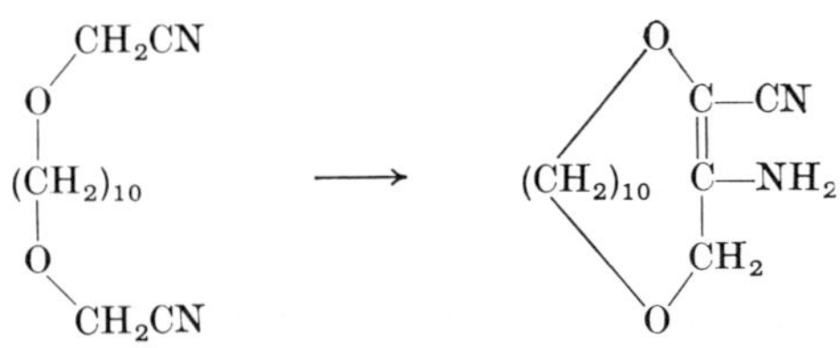

(60) (61)

The seven-membered cyclic enaminonitrile **64** was prepared from *N*-phenylpiperidine (**62**) by ring cleavage with cyanogen bromide, replacement of the primary bromide with cyanide to give the dinitrile **63**, and finally treatment with the "Thorpe-Ziegler catalyst" (431,447). This latter cyclization is one of the few reported examples of the participation

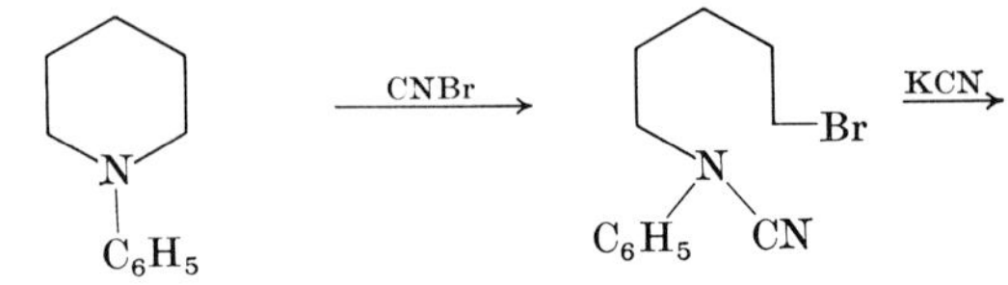

(62)

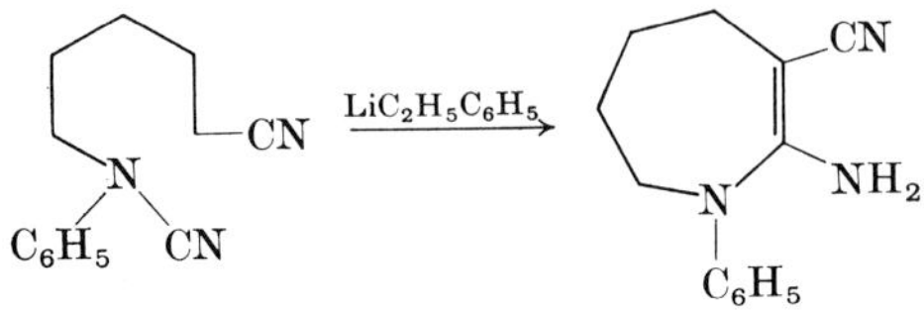

(63) **(64)**

of the N—CN grouping in this reaction. An attempt to prepare the homologous eight-membered ring **65** (using sodium methyl anilide rather than lithium ethyl anilide) inexplicably led to the cyclic amidine **66** with loss of the nitrile grouping (446).

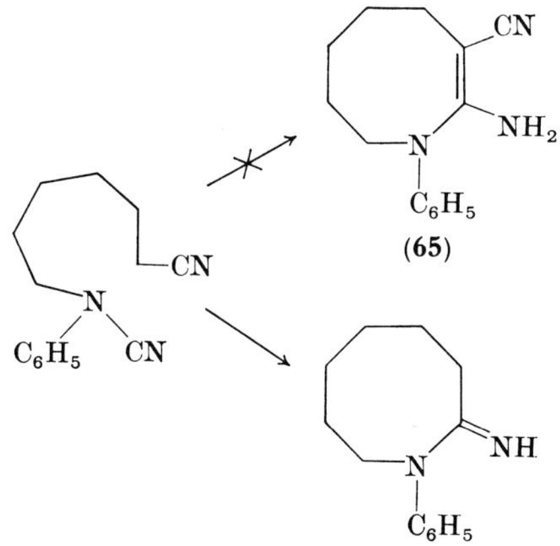

(65)

(66)

The synthetic potential of the Thorpe-Ziegler cyclization appears to be virtually unexplored for the preparation of other and more useful heterocyclic intermediates. One striking example of the potential utility of this method is found in a recent synthesis by Gompper (339) of 1-methyl-2-methylthio-4-amino-5-cyanoimidazole (**67**) by the route shown below; the aminonitrile thus formed provided a versatile intermediate for purine synthesis by methods which will be discussed in detail in Chapter II, Section VIII.

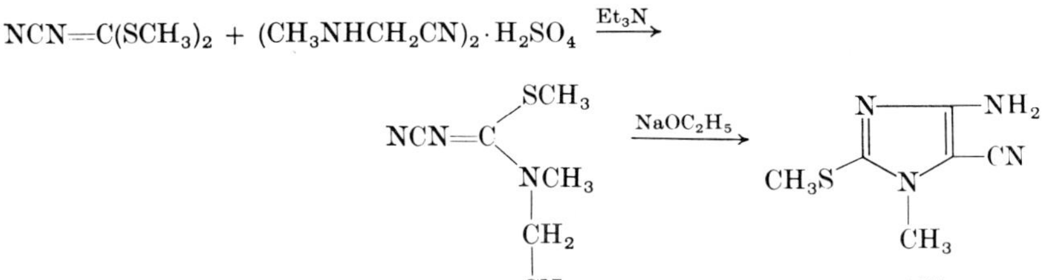

(67)

A second example, also due to Gompper, involves the utilization of ketene dithioacetals as intermediates. Condensation of phenyl acetonitrile with carbon disulfide in the presence of sodamide yields the disodium salt of the ketene dithioacetal **68**, which reacts with 2 moles of chloroacetonitrile to give **69**. This compound undergoes a Thorpe-

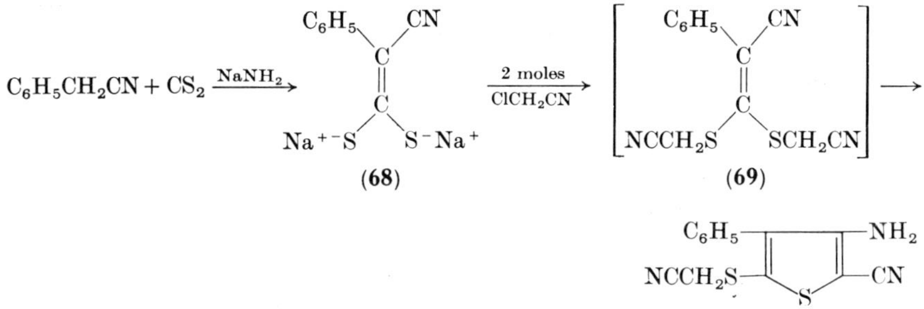

(68) (69)

(70)

Ziegler cyclization under the reaction conditions to give 2-cyano-3-amino-4-phenyl-5-cyanomethylthiophene (**70**) in 81% yield (171). Similarly, treatment of malononitrile with carbon disulfide, alkali, and chloroacetonitrile produces the ketene dithioacetal **71** *in situ* which undergoes a double Thorpe-Ziegler cyclization to give the bis-aminonitrile **72**

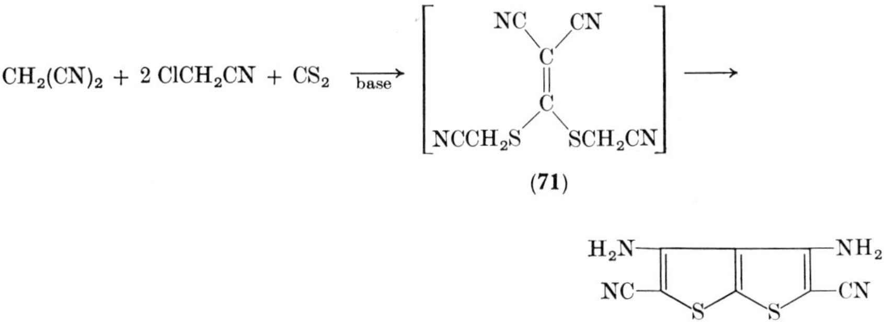

(71)

(72)

in 52% yield (171). It is interesting to note that the related ketene dithioacetal **73** upon treatment with sodium ethoxide does not undergo an analogous double Thorpe-Ziegler-type cyclization, but produces exclusively 2-carboethoxy-3-amino-4-cyano-5-carboethoxymethylthiophene (**74**) in 87% yield (171). This latter compound is one of the few

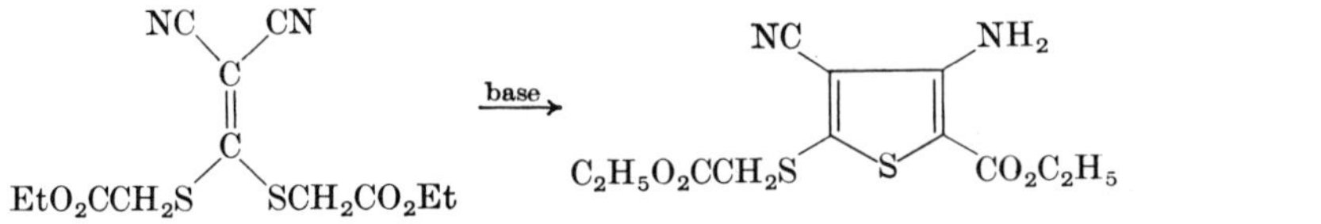

(73) (74)

aromatic or heterocyclic *o*-aminonitriles thus far prepared which cannot possess a normal carbon–carbon double bond between the amino and cyano groups; it might thus be of considerable interest to compare the spectral and chemical properties of this and similar aminonitriles (507,543) with the normal type (exemplified by **70**) in which the amino and nitrile functions are in the "enaminonitrile" relationship.

Dipotassium cyanimidodithiocarbonate (**75**) reacts exothermically with chloroacetonitrile in methanol to give 2-(cyanomethylmercapto)-4-amino-5-cyanothiazole (**77**); the intermediate dialkylated dithiocarbonate **76** could not even be isolated (107).

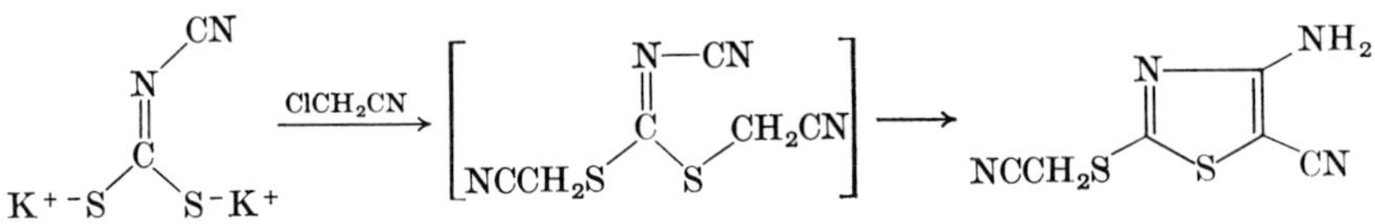

(75) (76) (77)

An interesting, and extremely unusual, heterocyclic synthesis via a Thorpe-Ziegler reaction was observed upon pyrolysis of the adduct **78** between indene and tetracyanoethylene. The product was shown to be 3-amino-2,5-dicyanobenzo(*f*)quinoline (**79**); it was suggested that it arose by thermal cleavage of the cyclobutane ring, Thorpe-Ziegler condensation of the resulting homocinnamonitrile anion (presumably formed by a trace of cyanide ion acting as the requisite base) with the benzylidenemalonitrile grouping, followed by a further cyclization by an amine–nitrile addition. Since the final product is itself basic, the pyrolysis may be autocatalytic (469).

In one instance dimerization of the product of a Thorpe-Ziegler condensation under the conditions of its formation has been reported (99).

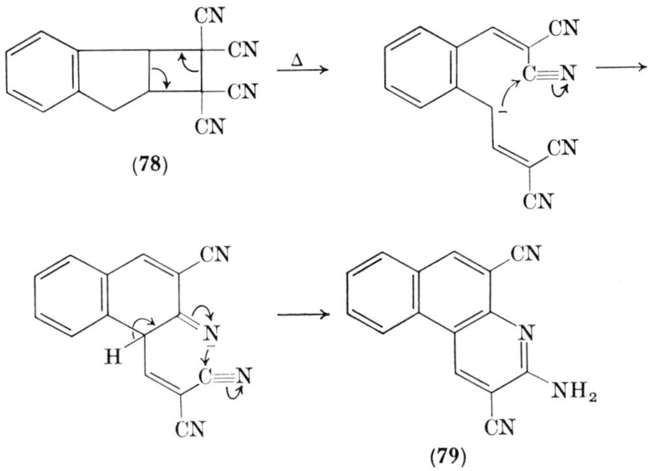

Treatment of adiponitrile with a molar equivalent of sodium *t*-butoxide in refluxing toluene gave the expected 1-amino-2-cyanocyclopentene (**80**), but the use of only a catalytic amount of potassium *t*-butoxide gave a dimeric compound which was transformed into a second, isomeric dimer

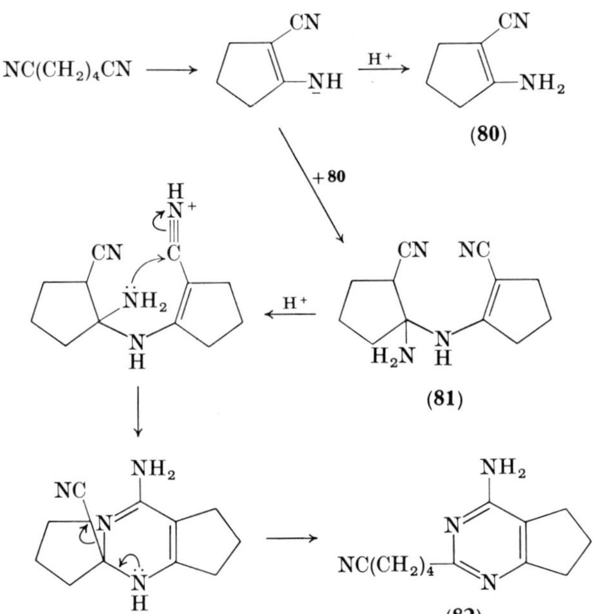

Scheme 3

upon work-up in the presence of acid. The initially formed dimer has been shown to be 2-amino-2-(2-cyano-1-penten-1-ylamino)-cyclopentacarbonitrile (**81**), formed by addition of the anion of **80** to **80**. This Michael addition is prevented when a full equivalent of *t*-butoxide is present, since not only is the sodium salt of **80** precipitated from the toluene solution and thus removed from reaction, but no free enaminonitrile is present to undergo Michael addition. The second, isomeric dimer formed upon treatment of **81** with acid is 2-(4-cyanobutyl)-4-amino-5,6-trimethylenepyrimidine (**82**); its formation is outlined in Scheme 3.

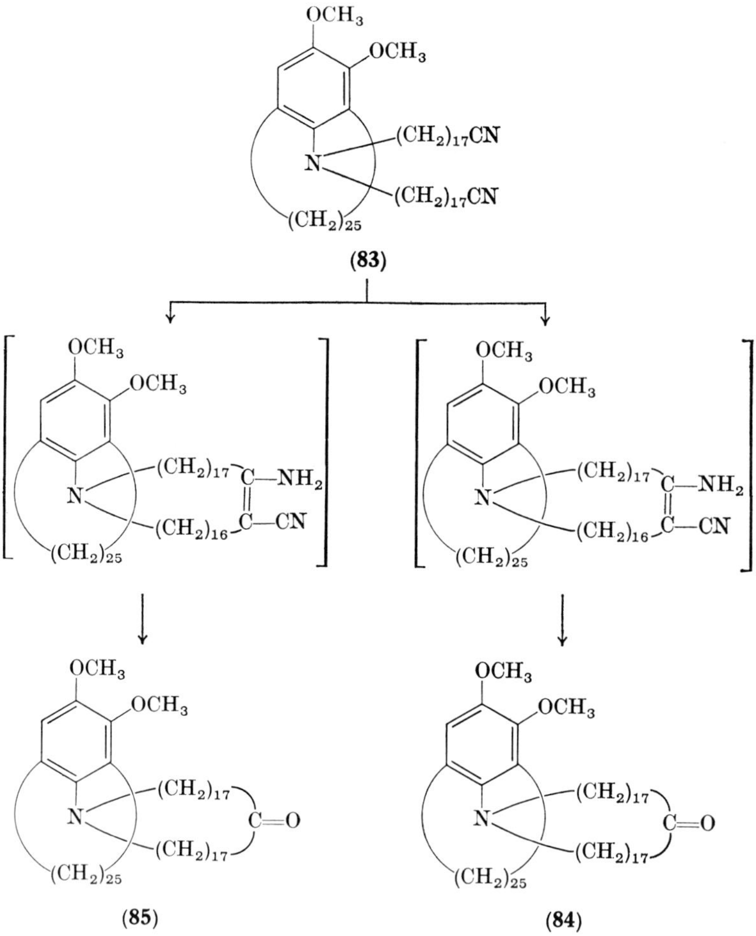

Scheme 4

In a synthesis of a catenane by what has been termed the "semi-statistical principle" (as contrasted with the "planned" synthesis of Schill (522)), Thorpe-Ziegler cyclization of the dinitrile **83** followed by hydrolysis gave a mixture of the intraannular (ca. 6%) and the extra-annular (ca. 94%) macrocycles **84** and **85** (Scheme 4). The former compound was cleaved to the catenane **86** (530).

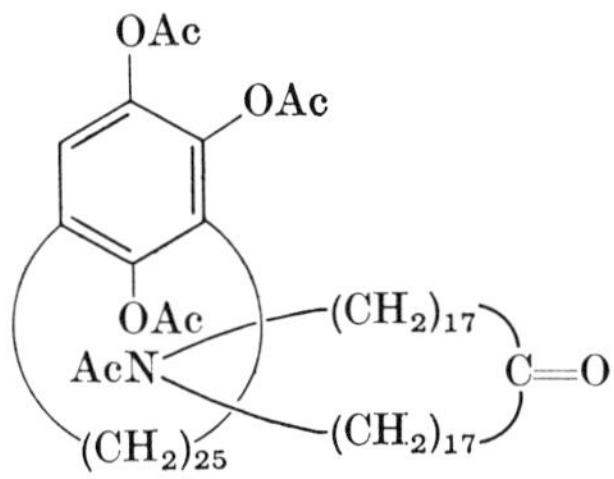

(86)

Finally, a key step in an ingenious total synthesis of the iboga alkaloid *d,l*-epiibogamine (**90**) was a Thorpe-Ziegler cyclization of the trinitrile **87** to the seven-membered enaminonitrile **88**, which was subsequently converted to the ketone **89** and then by further transformations to **90** (558).

Despite these examples, the potential versatility of the Thorpe-Ziegler cyclization for the construction of multifunctional intermediates suitable for further synthetic exploitation has been largely ignored; this would appear to be a fruitful area for further exploration.

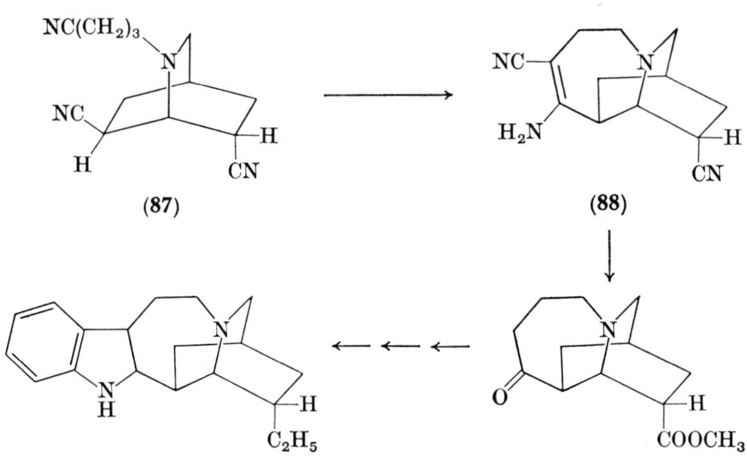

TABLE III

Cyclic Enaminonitriles Prepared by the Thorpe-Ziegler Cyclization

Starting dinitrile	Base	Solvent	Yield, %	Product	Ref.
1. Cyclic Enaminonitriles Containing Five-Membered Rings					
$NC(CH_2)_4CN$	Na	Toluene	64	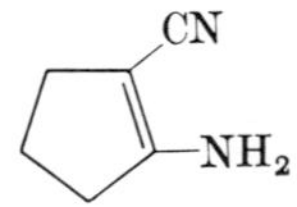	59
	NaOBu-*t*	Toluene	85		99
	NaOBu-*t*	Xylene	70		495
	t-BuNHMgBr	Ether	49		99
	$(C_2H_5)_2NMgBr$	Ether	—		200, 202
	$NaOC_2H_5$	Ethanol	80		195
	—	—	—		233
	C_2H_5MgBr	Ether	10–11		236
	H_2/catalyst	—	—		284
	H_2/catalyst	—	—		237
	NH_3/H_2/catalyst	—	—		238
	HCN	—	—		239
	KOBu-*t*	HOBu-*t*	76	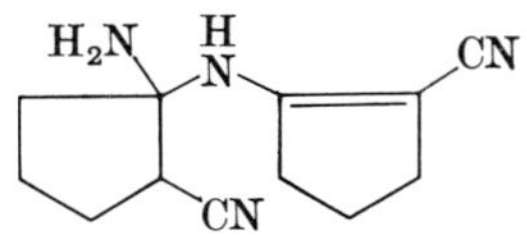	99

	$NaOC_2H_5$	Ethanol	—	Recovered starting material or 25% dimer + 39% enaminonitrile (see pp. 27–28)	99
	$NaOC_2H_5$ + $(C_2H_5O)_2CO$	Ethanol	53		512
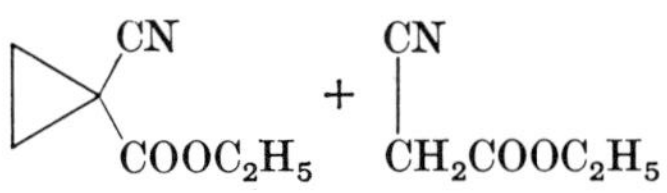	$NaOC_2H_5$	Ethanol	75	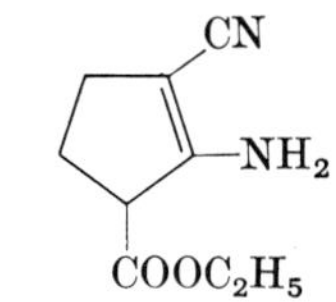	194
$C_2H_5OOCCH(CN)CH_2CH_2CH(CN)COOC_2H_5$	$NaCH(CN)COOC_2H_5$	Ethanol	—	Same as above	194
(prepared *in situ*)	$NaOC_2H_5$	Ethanol	—	Same as above	514
			35	Same as above	512
$NC—C(CH_3)_2(CH_2)_3CN$	NaH	Dioxane	90	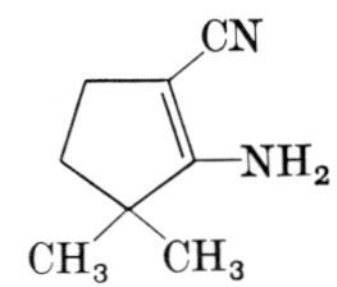	281
	NaOBu-*t*	Xylene	80		495

(*continued*)

TABLE III (*continued*)

Starting dinitrile	Base	Solvent	Yield, %	Product	Ref.
$NC-\overset{C_2H_5}{\underset{C_2H_5}{C}}(CH_2)_3CN$	NaH	Dioxane	70	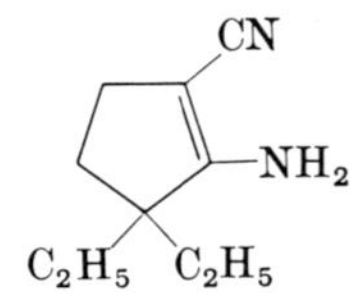	281
$NC-\underset{C_6H_5}{CH}(CH_2)_3CN$	NaH	Dioxane	98	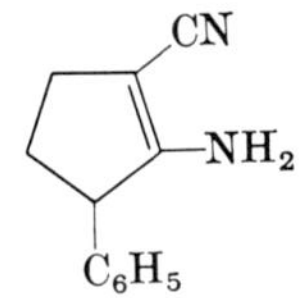	281
$NC-\overset{C_6H_5}{\underset{CH_3}{C}}(CH_2)_3CN$	$NaNH_2$	Benzene	85	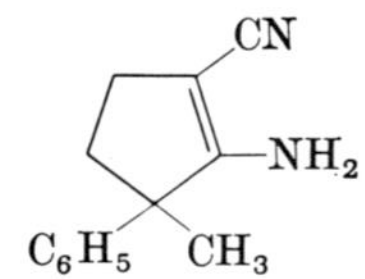	182
$NC-\overset{C_6H_5}{\underset{C_2H_5}{C}}(CH_2)_3CN$	— $NaNH_2$ $NaNH_2$	— Benzene Benzene	85 80 70–80	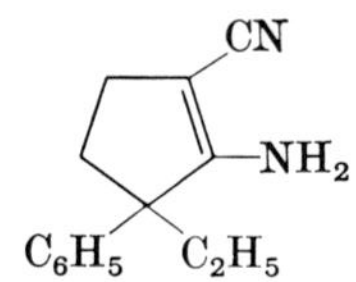	281 269 486

$NC—C(C_6H_5)(C_3H_7\text{-}n)(CH_2)_3CN$	$NaNH_2$ $NaNH_2$	Benzene Benzene	Excellent 70–80	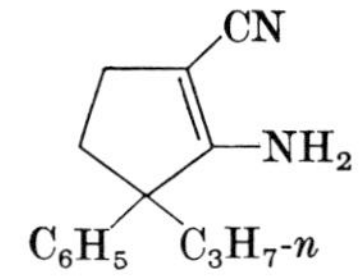	138 486
$NC—C(C_6H_5)(C_4H_9\text{-}n)(CH_2)_3CN$	$NaNH_2$ $NaNH_2$	Benzene Benzene	Excellent 70–80	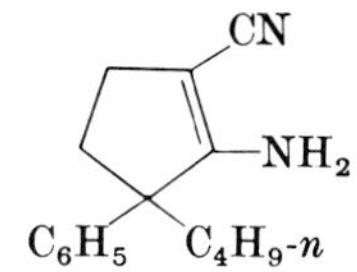	138 486
$NC—C(C_6H_5)(C_5H_{11}\text{-}n)(CH_2)_3CN$	$NaNH_2$ $NaNH_2$	Benzene Benzene	Excellent 70–80	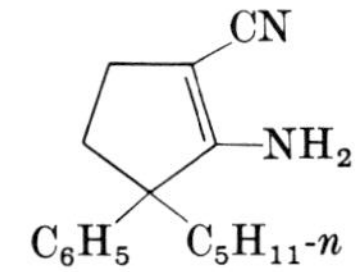	138 486
$NC—C(C_6H_5)(C_6H_{11}\text{-cyclo})(CH_2)_3CN$	$NaNH_2$ $NaNH_2$	Benzene Benzene	Excellent 70–80	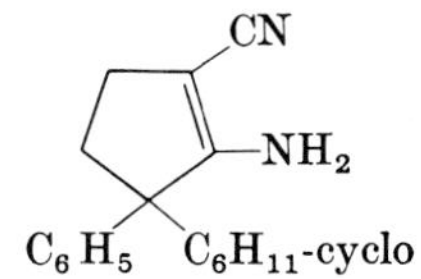	138 486
$NC—C(C_6H_5)(CH_2C_6H_5)(CH_2)_3CN$	$NaNH_2$ $NaNH_2$	Benzene Benzene	Excellent 70–80	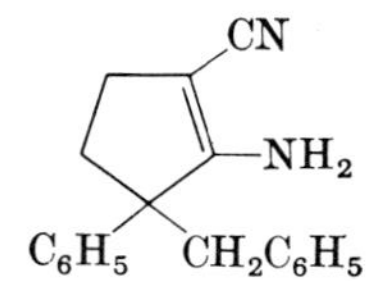	138 486

(*continued*)

TABLE III (*continued*)

Starting dinitrile	Base	Solvent	Yield, %	Product	Ref.
$NC-C(C_6H_5)(CH_2CH_2C_6H_5)(CH_2)_3CN$	$NaNH_2$	Benzene	Excellent	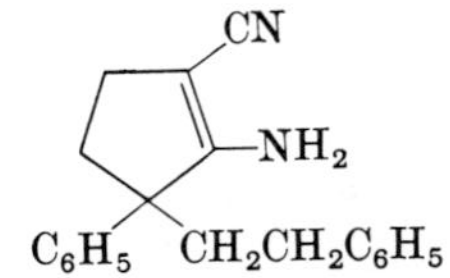	138
	$NaNH_2$	Benzene	70–80		486
$NC-C(C_6H_5)_2(CH_2)_3CN$	—	—	80	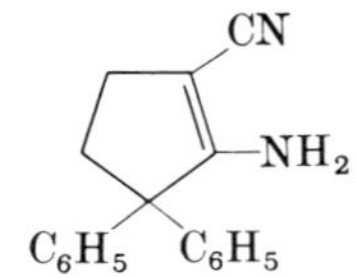	282
	$NaNH_2$	Benzene	—		285
	$NaNH_2$	Benzene	70–80		486
	$NaNH_2$	Benzene	—		277
	NaOBu-*t*	HOBu-*t*	100		221
$(C_6H_5)_2CHCN + Br(CH_2)_3CN$	$NaNH_2$	Benzene	65		268
	$NaNH_2$	Benzene	—		287
$(C_6H_5)_2CHCN + CH_3CH(Cl)CH_2CH_2CN$	$NaNH_2$	Benzene	8.8	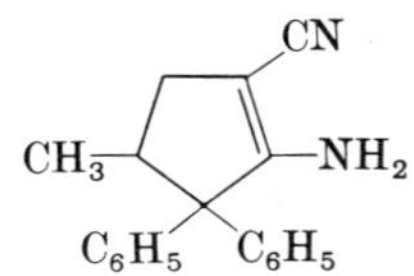	220
	NaOBu-*t*	HOBu-*t*	—		220
$NC-C(C_6H_5)_2-CH(CH_3)CH_2CH_2CN$	NaOBu-*t*	HOBu-*t*	88		485
	$NaNH_2$	Benzene	95		485

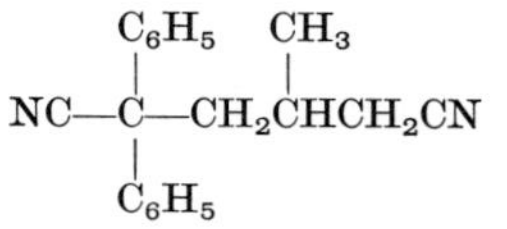	NaOBu-*t*	HOBu-*t*	—	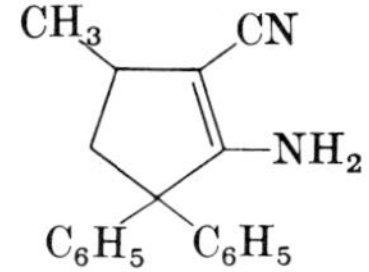	220
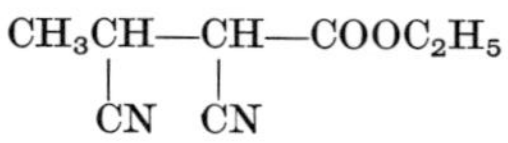	NH_3	—	8 —	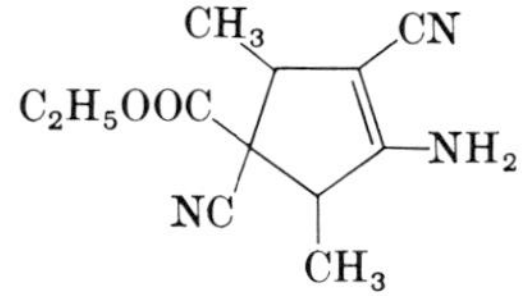	491 515
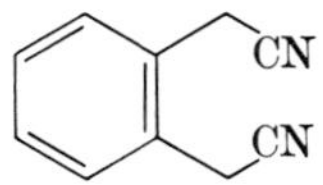	$NaOC_2H_5$ $NaOC_2H_5$	Ethanol Ethanol	91 100	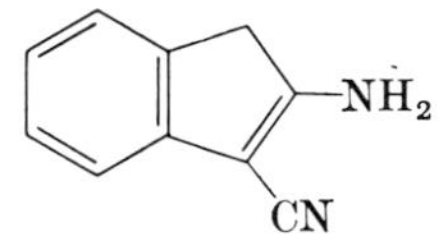	64 191
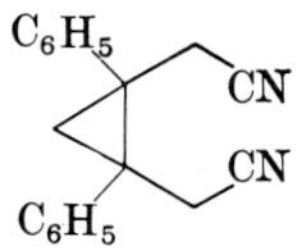	$LiCH_3NC_6H_5$	Ether	80–90	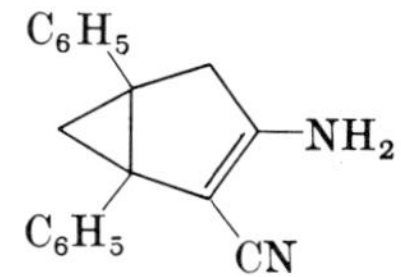	545
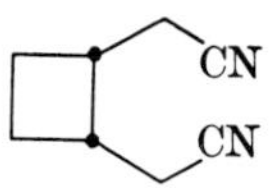	NaH	DMSO	78	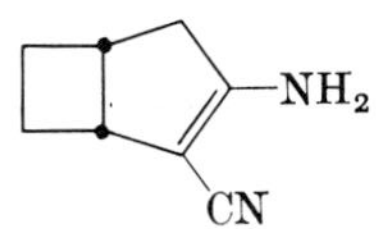	273

(continued)

TABLE III (*continued*)

Starting dinitrile	Base	Solvent	Yield, %	Product	Ref.
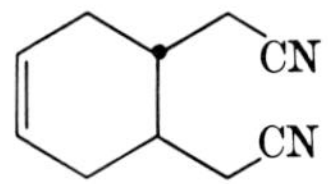	NaH	DMSO	100	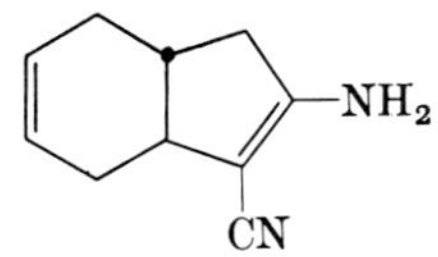	273
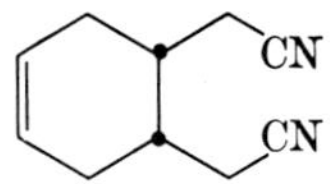	NaH	DMSO	74	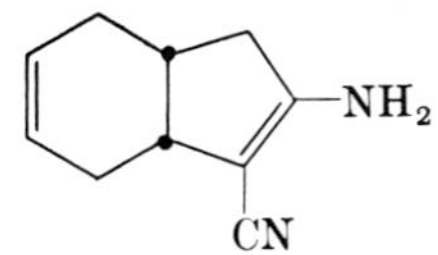	273
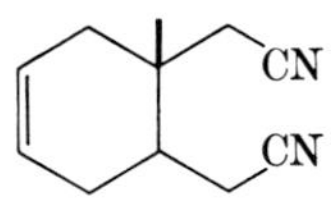	NaH	DMSO	85	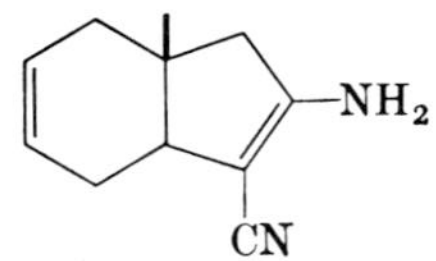	273
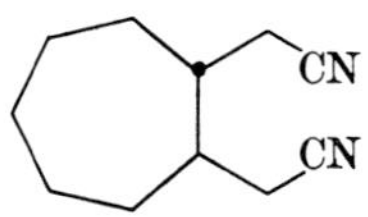	$NaOCH_2CH_2OH$	$HOCH_2CH_2OH$	75	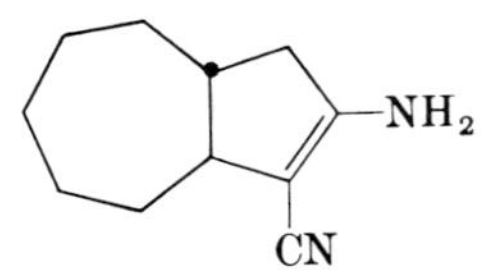	511

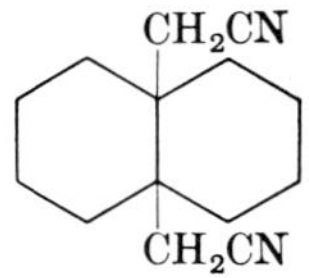 (prepared *in situ* from the dibromide)	NaCN	DMSO/N_2	15	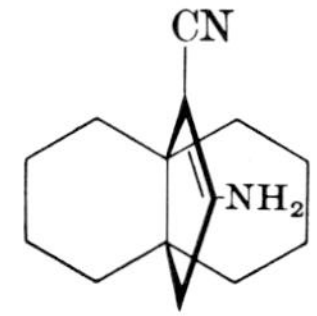	7
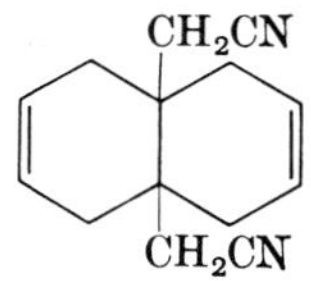 (prepared *in situ* from the dibromide)	NaCN	DMSO	78	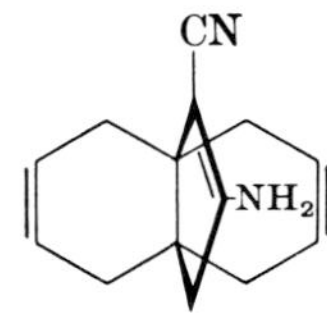	7
(prepared *in situ* from the ditosylate)	NaCN, NaI	DMSO	25	Same as above	12
(prepared *in situ* from the dimesylate)	NaCN	DMSO or DMF or pyridine or *N*-methyl-2-pyrrolidinone	Little	Same as above	534
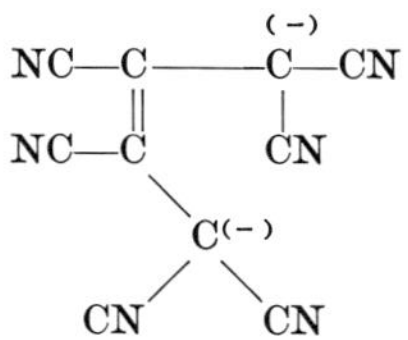	—[a]	H_2O	82	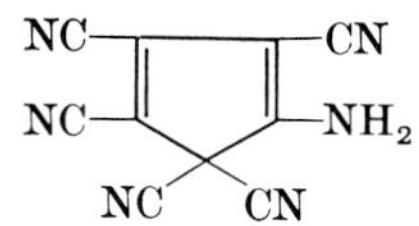	483

(*continued*)

TABLE III (*continued*)

Starting dinitrile	Base	Solvent	Yield, %	Product	Ref.
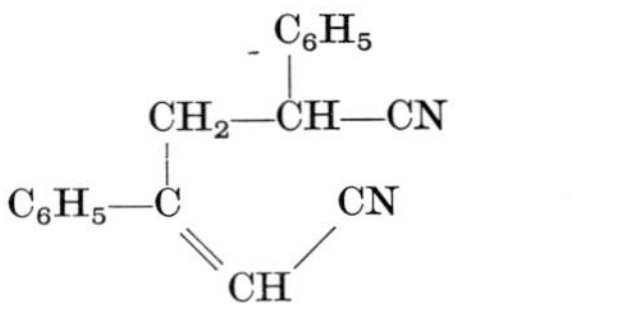 (not isolated: formed *in situ*)	$(C_2H_5O)_2P(O)\bar{C}HCN\ Na^+$	$CH_3OCH_2CH_2OCH_3$	—	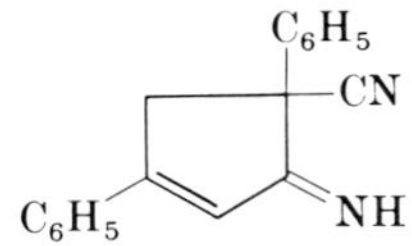	533
2. Cyclic Enaminonitriles Containing Six-Membered Rings					
$NC(CH_2)_5CN$	NaH NaOBu-*t* $Li(C_2H_5)_2N$	Toluene Xylene Ether	67 — 92	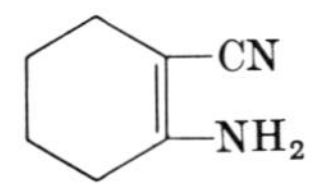	89 495 542
$NC—C(CH_3)_2(CH_2)_4CN$	$LiCH_3NC_6H_5$ $Li(C_2H_5)_2N$	Ether Ether	48 87	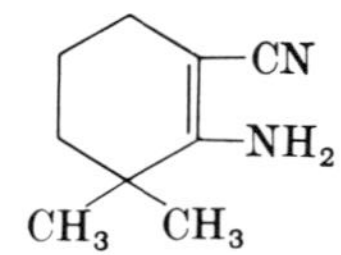	495 542
$NCCH_2CH_2C(CH_3)_2CH_2CH_2CN$	NaOBu-*t*	Xylene	48	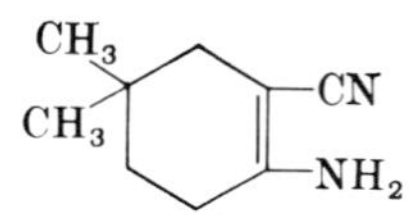	495
$NC—C(C_2H_5)_2(CH_2)_4CN$	$Li(C_2H_5)_2N$	Ether	51	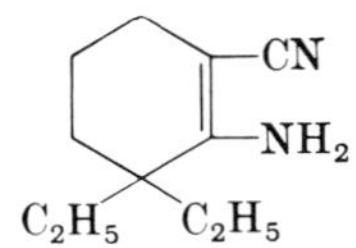	542

$NC\text{—}\overset{C_6H_5}{\underset{C_2H_5}{C}}(CH_2)_4CN$	$Li(C_2H_5)_2N$	Ether	96	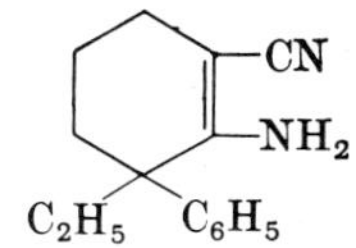	542
$NC\text{—}\overset{C_6H_5}{\underset{C_6H_5}{C}}(CH_2)_4CN$	$Li(C_2H_5)_2N$	Ether	91	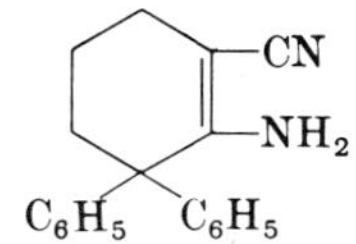	542
$NCCH_2CH_2\overset{C_6H_5}{\underset{CN}{C}}CH_2CH_2CN$	$NaOC_2H_5$	Ethanol	54	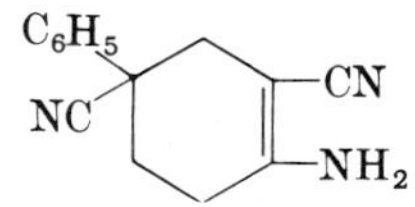	488
	$NaNH_2$	Toluene	40	Same as above	230
$C_6H_5CH_2CN + BrCH_2CH_2CN$	$NaNH_2$	Toluene	—	Same as above	230
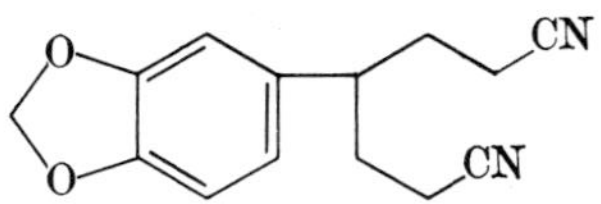	$NaNH_2$	Toluene	80	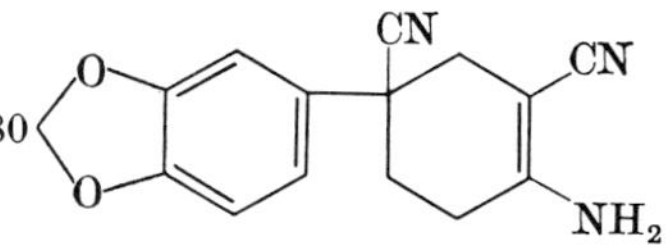	230
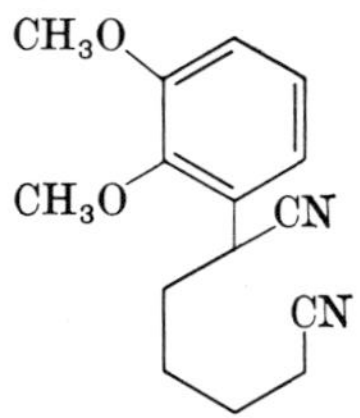	Na/K	Toluene	50	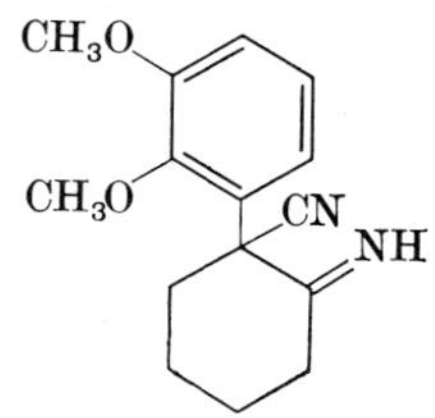	222

(continued)

TABLE III (*continued*)

Starting dinitrile	Base	Solvent	Yield, %	Product	Ref.
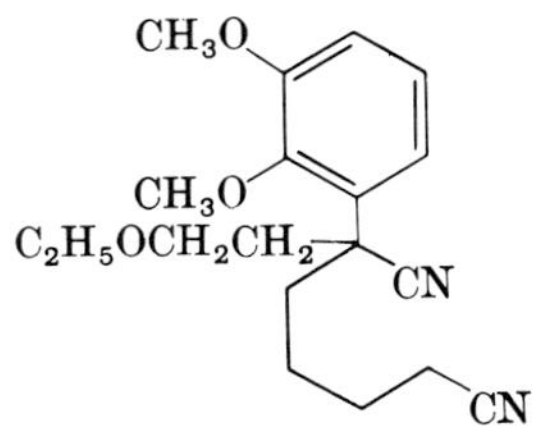	$NaNH_2$	Benzene	80	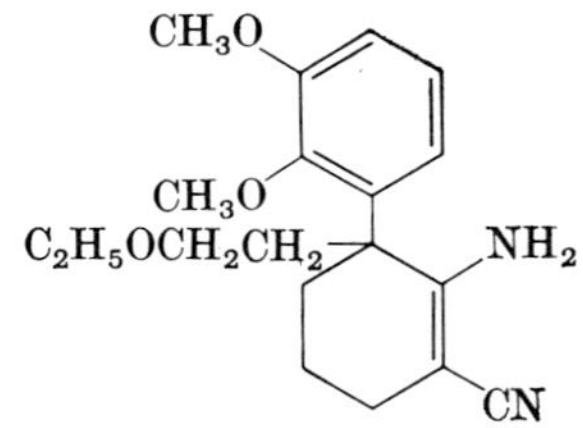	225
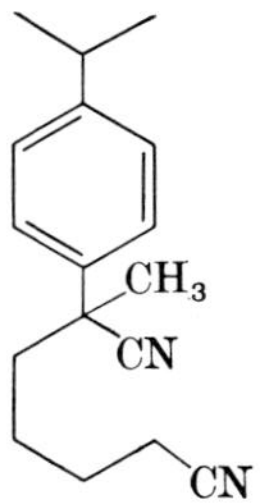	$NaNH_2$	Benzene	70	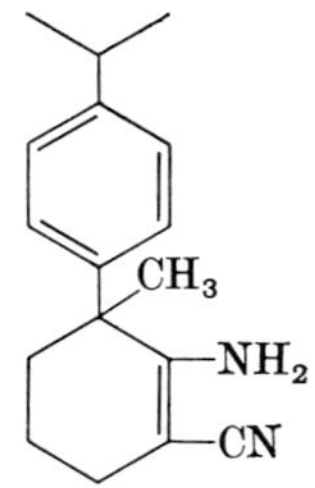	226
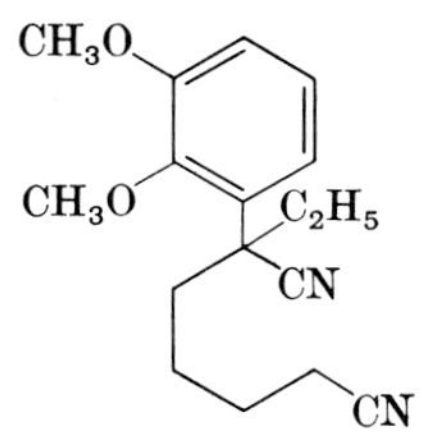	$NaNH_2$ $NaNH_2$	Benzene Benzene	53 58	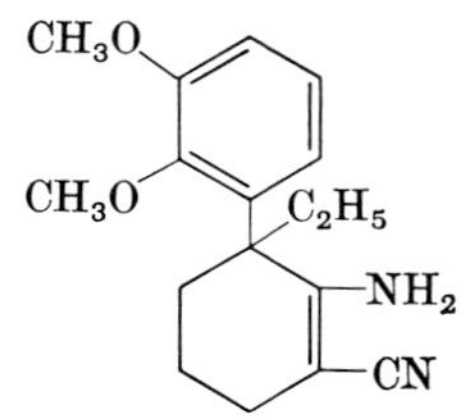	227 228

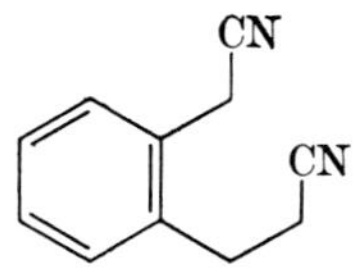	$NaOC_2H_5$	Ethanol	94	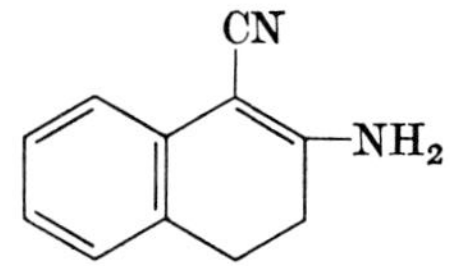	549
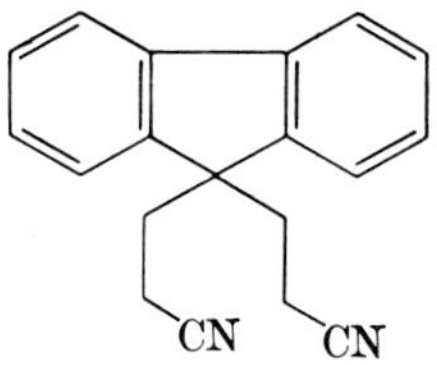	KOBu-*t* Na/K	HOBu-*t* Toluene	77 80	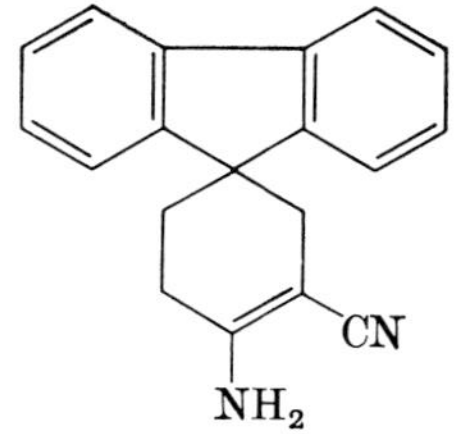	81 210
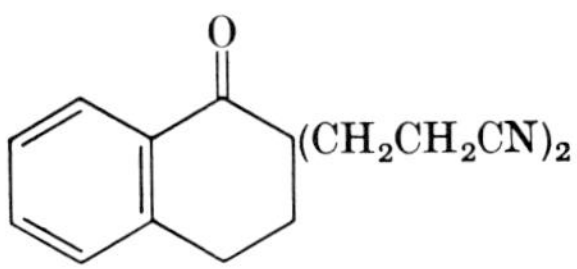	—	—	—	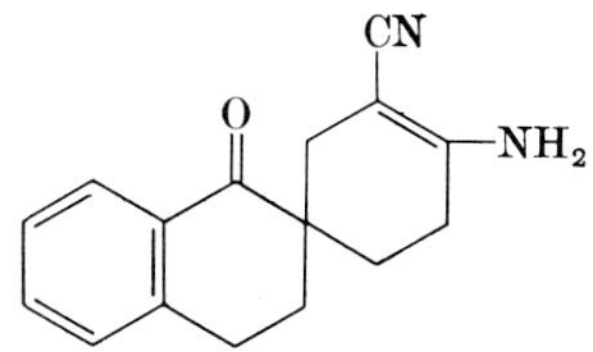	81
p-$BrC_6H_4C(CH_3){=}C(CN)_2$ + $(CH_3)_2NCH{=}CH{-}NO_2$	KOC_2H_5	Ethanol	40	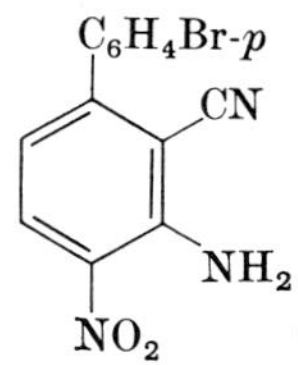	478

(continued)

TABLE III (*continued*)

Starting dinitrile	Base	Solvent	Yield, %	Product	Ref.
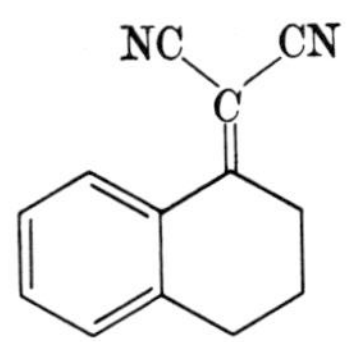 + $(CH_3)_2NCH{=}CH{-}NO_2$	KOC_2H_5	Ethanol	45	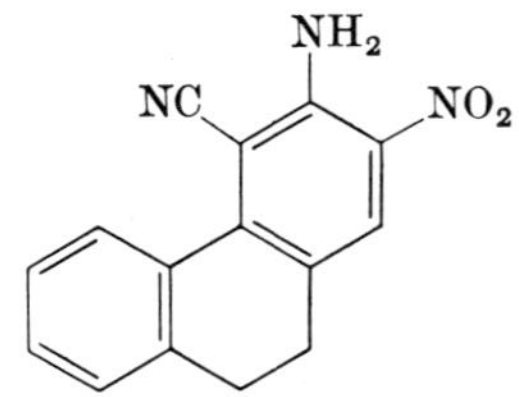	478
	3. Cyclic Enaminonitriles Containing Seven-Membered Rings				
$NC(CH_2)_6CN$	$NaC_2H_5NC_6H_5$	Ether	95–97	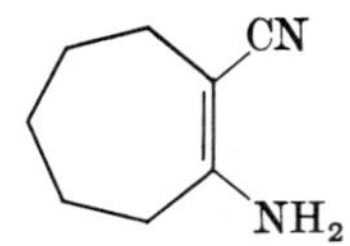	190
	$LiCH_3NC_6H_5$	Ether	26		495
	$Li(C_6H_{11})_2N$	Ether	—		200, 202
	$NaN[Si(CH_3)_3]_2$	Ether	97.5		500
$NCC(CH_3)_2(CH_2)_5CN$	$LiCH_3NC_6H_5$	Ether	—	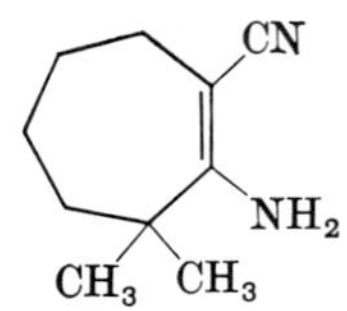	495

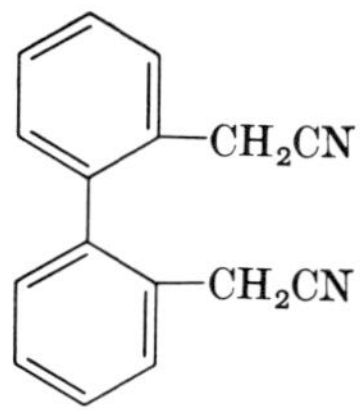	KCN on dibromide	Aq. ethanol	96	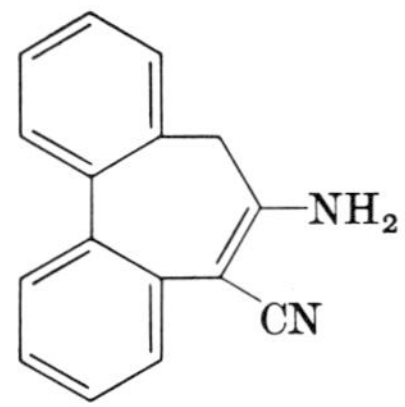	224
	KCN on dibromide	Aq. ethanol	90		223
	$NaOC_2H_5$	Ethanol	—		265
	$NaOC_2H_5$	Ethanol	80		192
	$NaOC_2H_5$	Ethanol	75		223
	KOH	Aq. ethanol	83		223
	KOH	Aq. ethanol	97		29
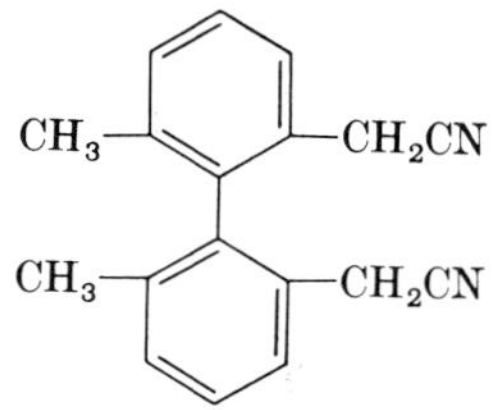	$NaOC_2H_5$ (N_2)	Ethanol	66	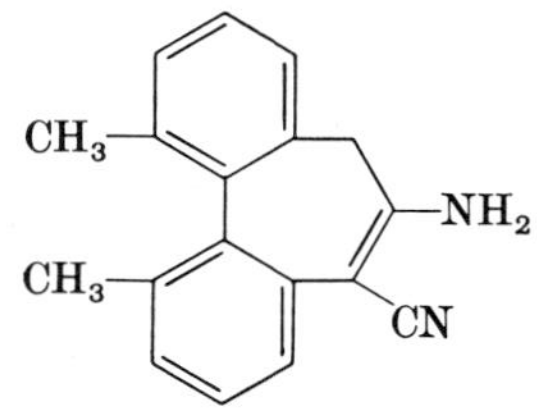	205
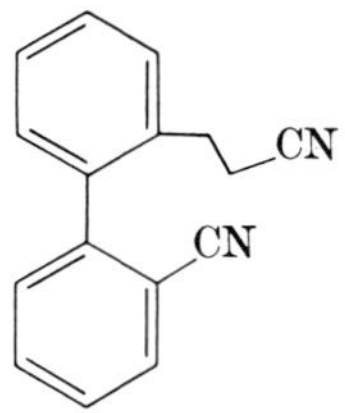	$LiCH_3NC_6H_5$	Ether	98	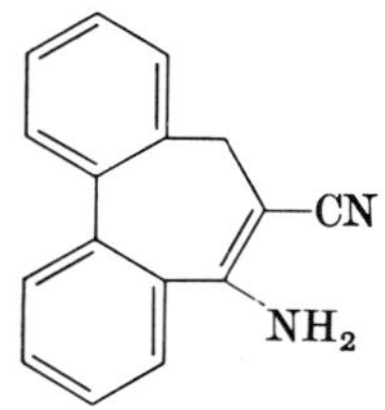	229 (also 224)
	$LiCH_3NC_6H_5$	Ether	—		29

(continued)

TABLE III (*continued*)

Starting dinitrile	Base	Solvent	Yield, %	Product	Ref.
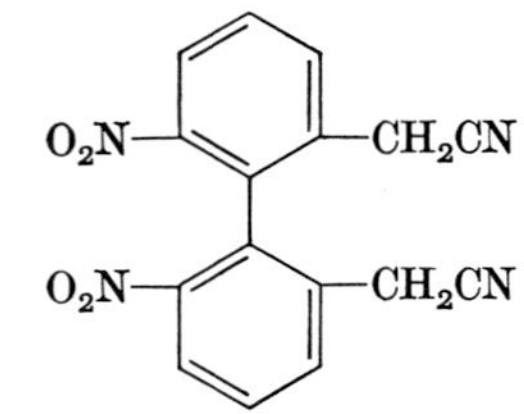	KCN on dibromide; $NaOC_2H_5$	Aq. ethanol	61	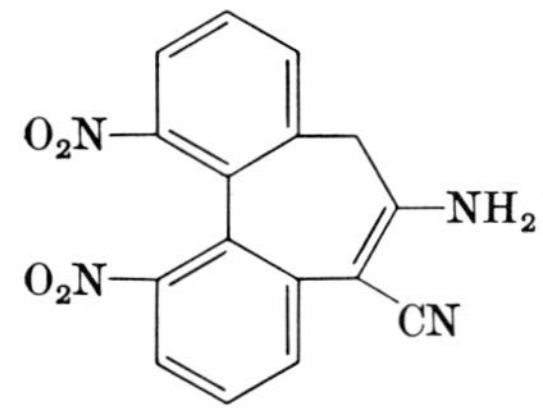	205 (also 206)
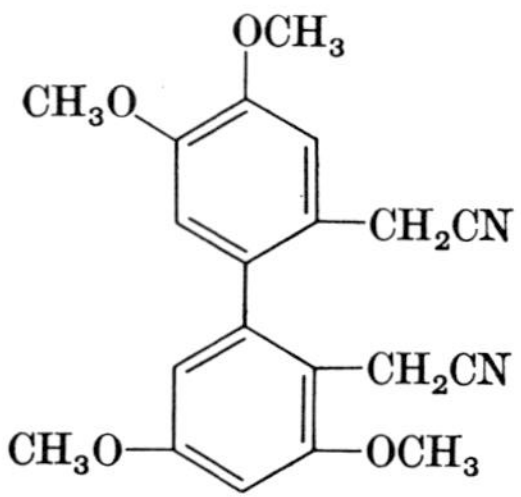	$NaOC_2H_5$	Ethanol	100	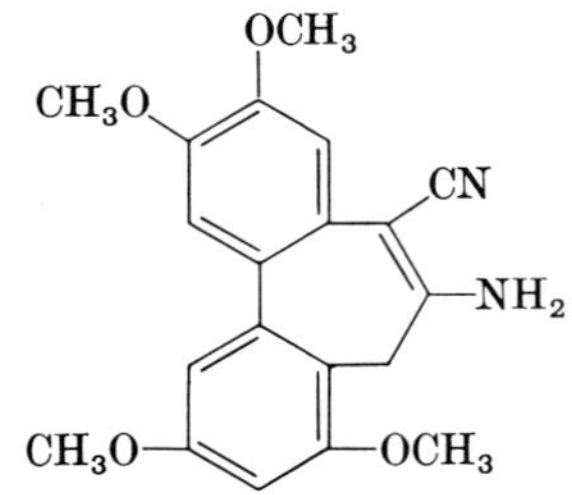 *or* 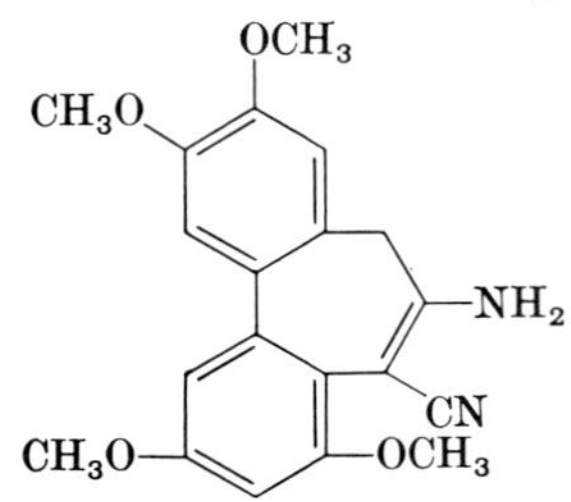	476

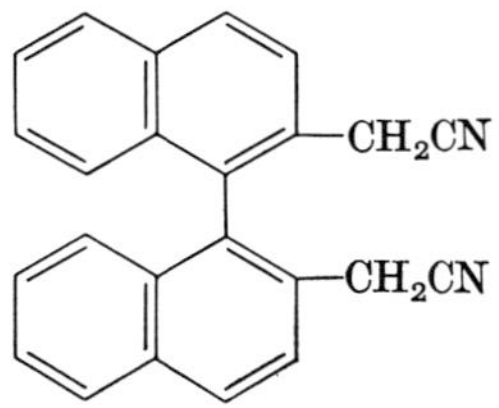	KCN on dibromide; $NaOC_2H_5$	Aq. ethanol	68	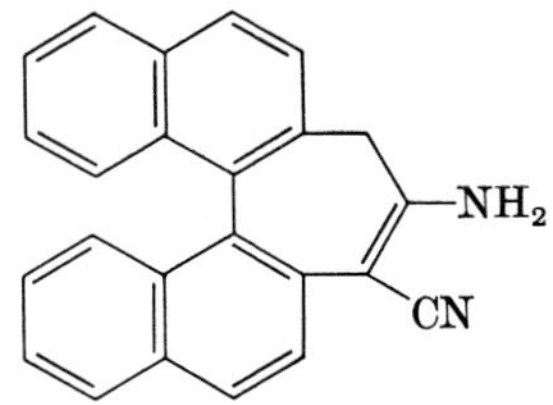	209

4. Cyclic Enaminonitriles Containing Eight-Membered and Larger Rings

$NC(CH_2)_7CN$	$NaC_2H_5NC_6H_5$ $LiCH_3NC_6H_5$	Ether Ether	89 18	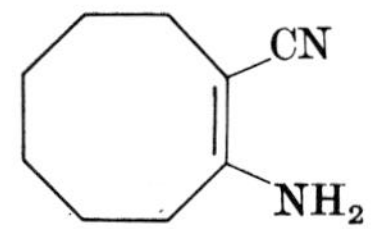	190 495
$C_6H_5CH[(CH_2)_3CN]_2$	$NaCH_3NC_6H_5$ $LiC_2H_5NC_6H_5$	Ether Ether/naphthalene	54 89	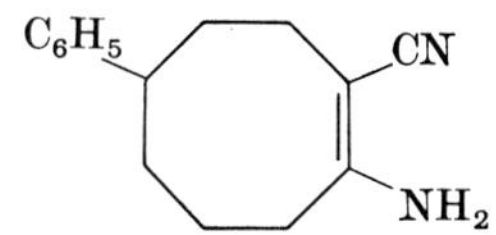	181 167
$(CH_3)_3CCH[(CH_2)_3CN]_2$	$LiCH_3NC_6H_5$	Ether	—	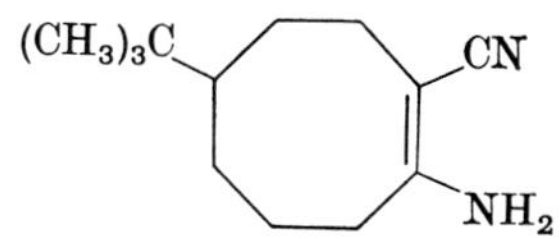	559

(continued)

TABLE III (*continued*)

Starting dinitrile	Base	Solvent	Yield, %	Product	Ref.
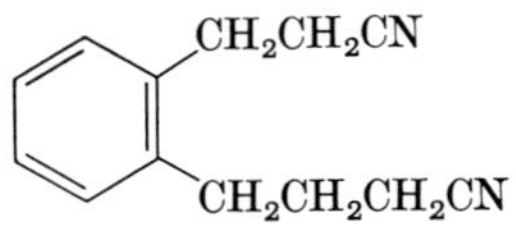	$NaCH_3NC_6H_5$	Ether	51*A* 20*B*	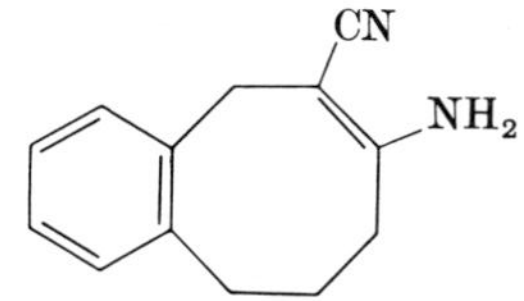 *A* 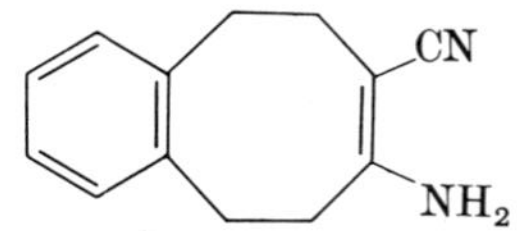 *B*	262
$NC(CH_2)_{13}CN$	—	—	62	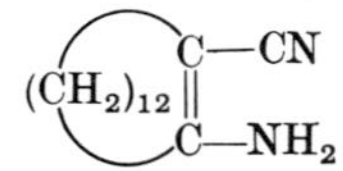	190
	$NaCH_3NC_6H_5$	Ether	—		201, 202
$NC(CH_2)_{14}CN$	—	—	—	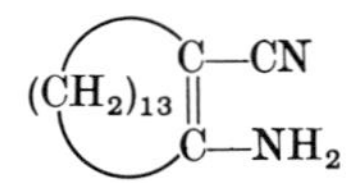	200, 202, 243
	$LiC_2H_5NC_6H_5$	Ether	30–40	(dimer also formed)	187
$NC(CH_2)_{15}CN$	$NaC_5H_{11}NC_6H_5$	Ether	—	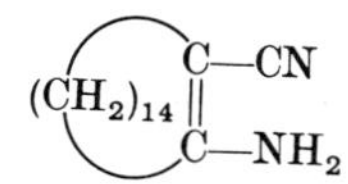	201, 202

$NC(CH_2)_{16}CN$	$LiC_2H_5NC_6H_5$	Ether	80	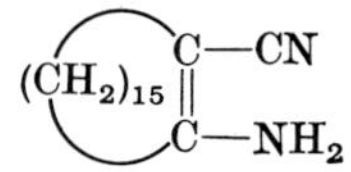	188
	$LiC_2H_5NC_6H_5$	Ether	Low		187
$NC-C(C_6H_5)_2(CH_2)_3CN$	NaH	Dioxane	12	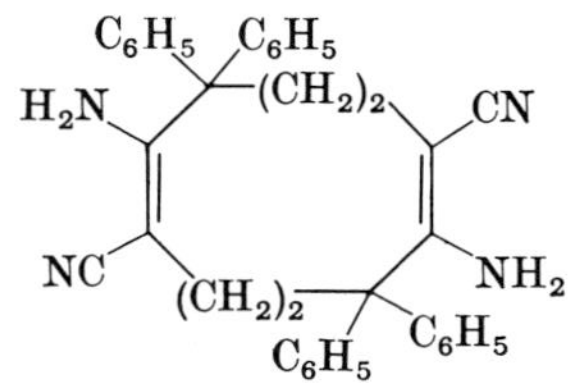	282
$NC(CH_2)_8CN$	$NaC_2H_5NC_6H_5$	Ether	33.6	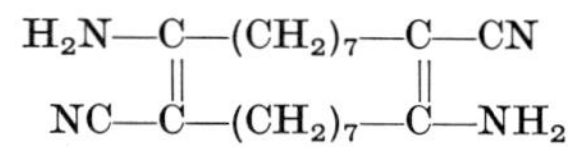	190
$NC(CH_2)_9CN$	$NaC_2H_5NC_6H_5$	Ether	45	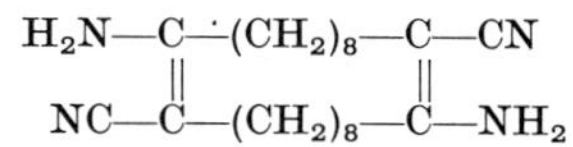	190
$NC(CH_2)_{10}CN$	$NaC_2H_5NC_6H_5$	Ether	60	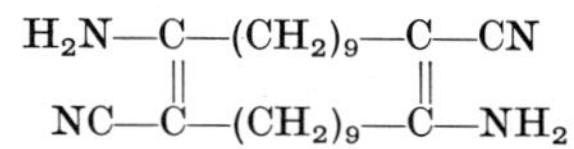	190
$NC(CH_2)_{14}CN$	$LiC_2H_5NC_6H_5$	Ether	28	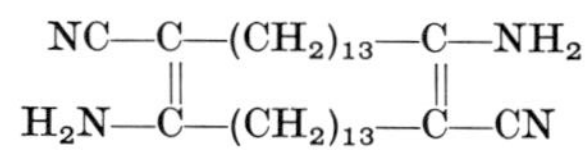	187
	—	—	—		200, 202

(continued)

TABLE III (*continued*)

Starting dinitrile	Base	Solvent	Yield, %	Product	Ref.
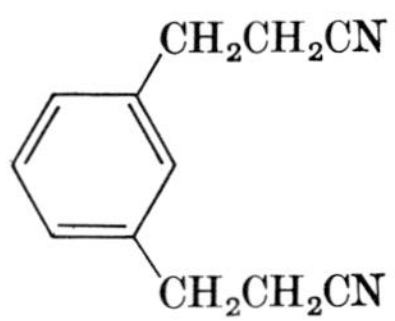	$NaC_2H_5NC_6H_5$	Ether	22	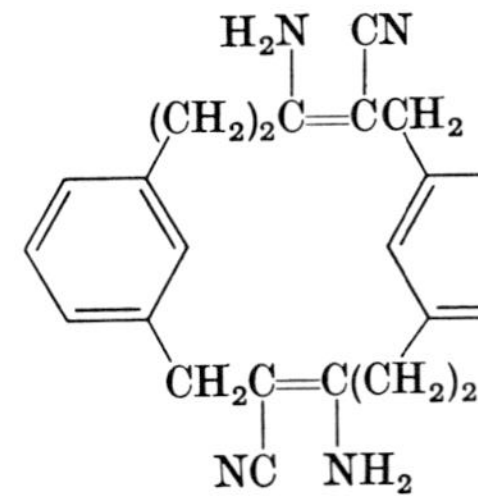	193
5. Heterocyclic Enaminonitriles					
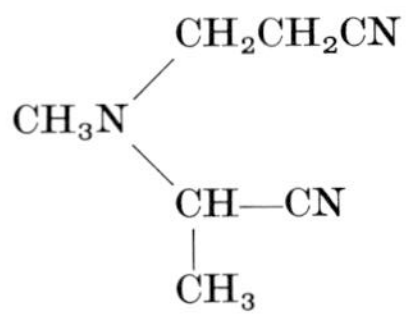	NaOBu-*t*	HOBu-*t*	85	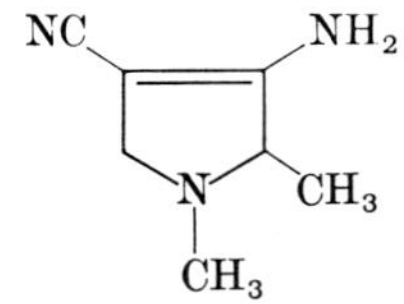	88
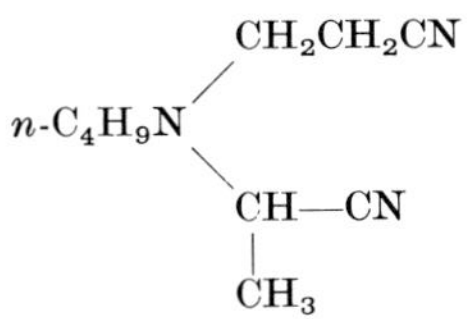	NaOBu-*t*	HOBu-*t*	55	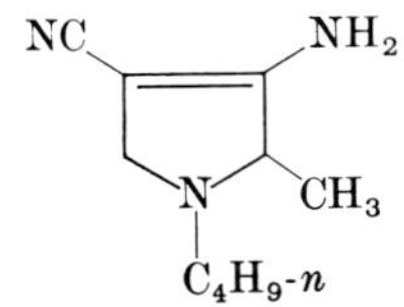	88

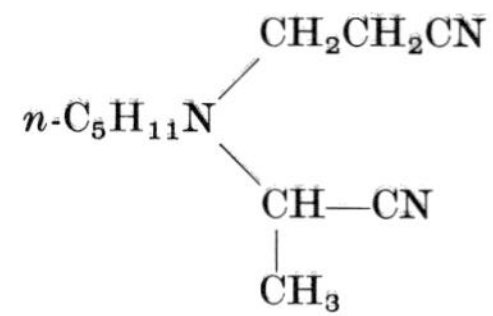	NaOBu-*t*	HOBu-*t*	60	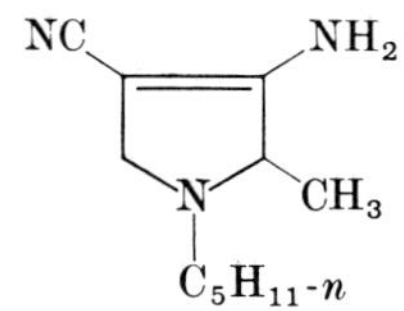	88
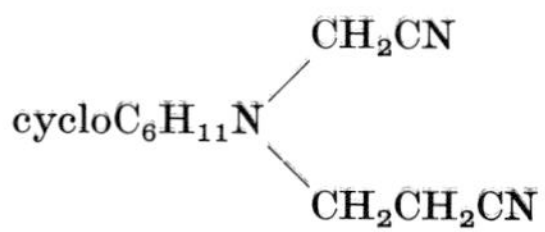	NaOH	Toluene	72	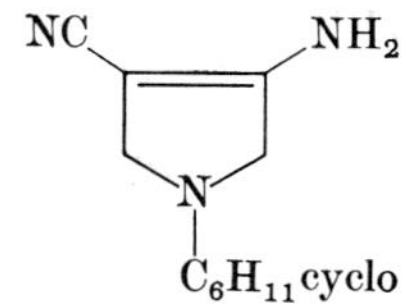	100
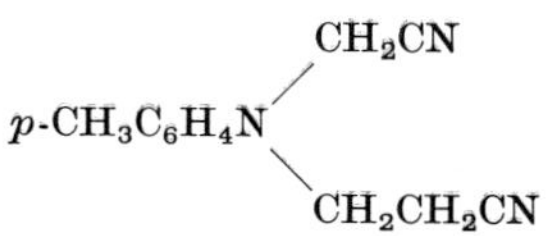	NaOH KOBu-*t*	Toluene HOBu-*t*	73–80 47	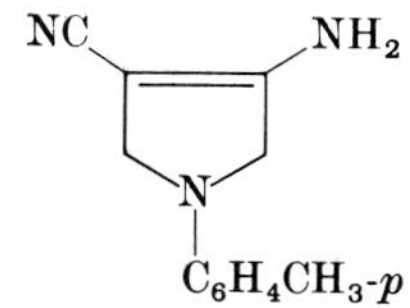	100
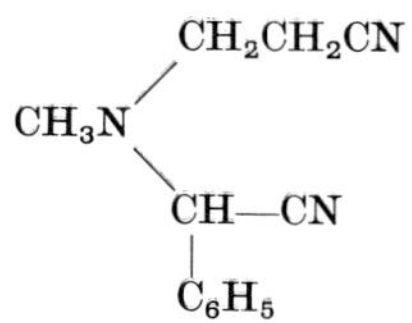	NaOBu-*t*	HOBu-*t*	47	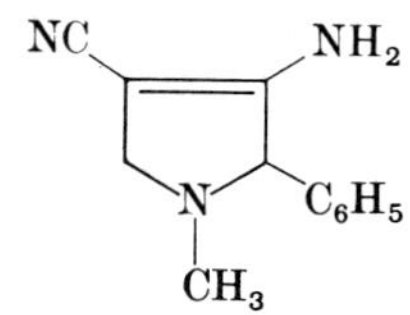	13
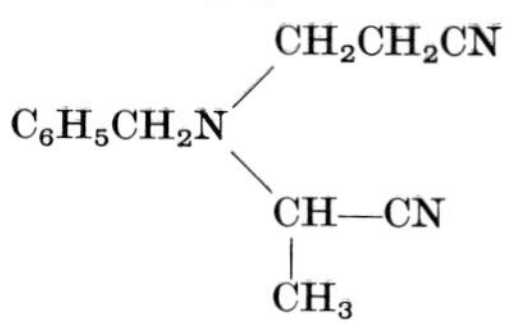	NaOBu-*t*	HOBu-*t*	73	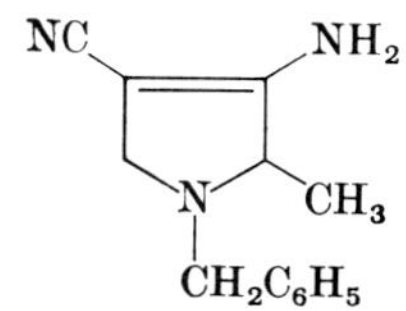	88

(continued)

TABLE III (*continued*)

Starting dinitrile	Base	Solvent	Yield, %	Product	Ref.
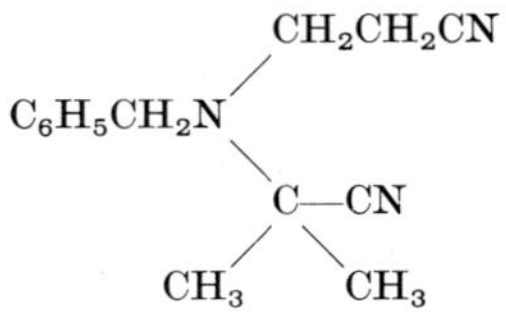	NaOBu-*t*	HOBu-*t*	60	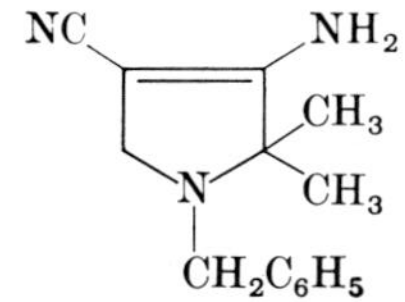	28
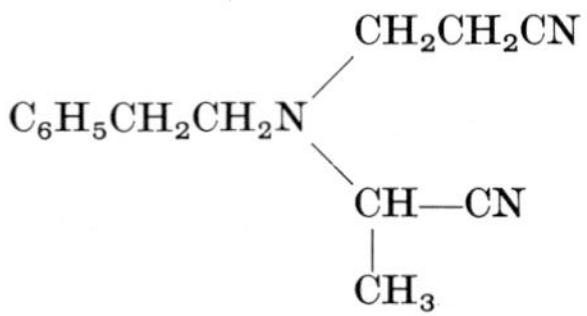	NaOBu-*t*	HOBu-*t*	77	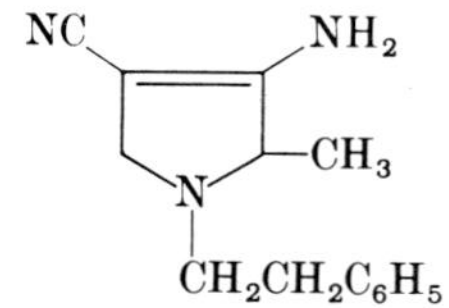	88
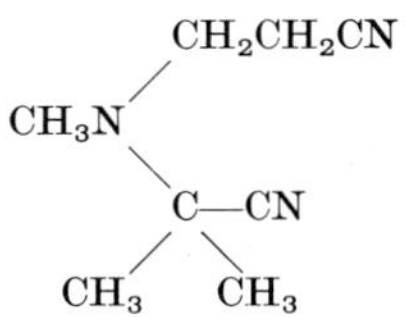	NaOBu-*t*	HOBu-*t*	32	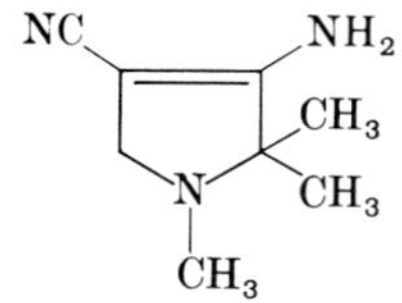	88
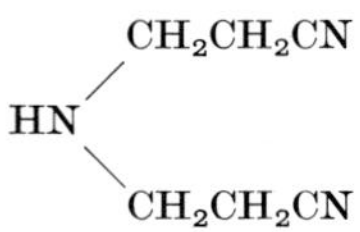	Na	Dioxane/ naphthalene	85	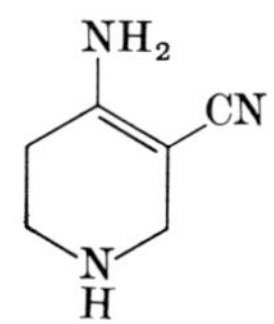	219
	KOBu-*t*	Toluene	85–90		536

$CH_3N(CH_2CH_2CN)_2$	NaOBu-*t*	HOBu-*t*	70	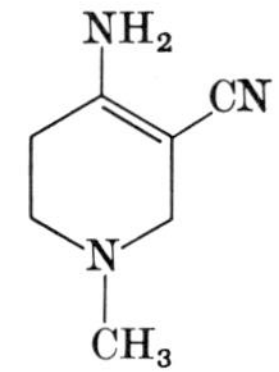	88
	Na	Toluene	—		112
	$NaCH_3NC_6H_5$	Ether	87		117
	$NaCH_3NC_6H_5$	Ether	—		441
$C_2H_5N(CH_2CH_2CN)_2$	$NaCH_3NC_6H_5$	Ether	85	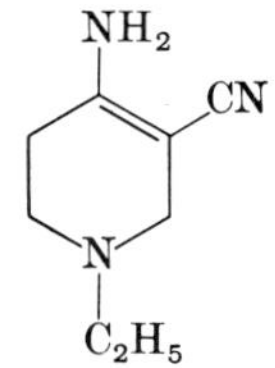	117
	$NaCH_3NC_6H_5$	Ether	—		441
$C_6H_5N(CH_2CH_2CN)_2$	—	—	72	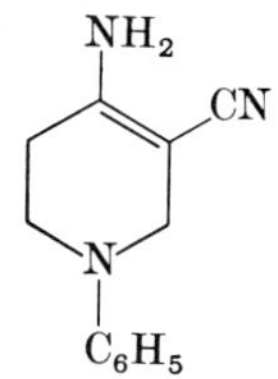	117
	NaOBu-*t*	Xylene	68		111
	$NaCH_3NC_6H_5$	Ether	—		441
$C_6H_5CH_2N(CH_2CH_2CN)_2$	KOBu-*t*	HOBu-*t*	99	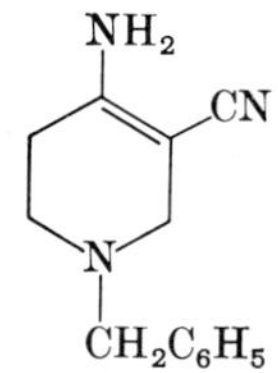	95, 96

(*continued*)

TABLE III (*continued*)

Starting dinitrile	Base	Solvent	Yield, %	Product	Ref.
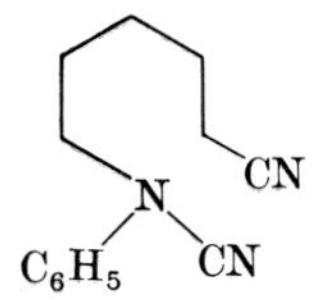	$LiC_2H_5NC_6H_5$ $LiCH_3NC_6H_5$	Ether Ether	— 93	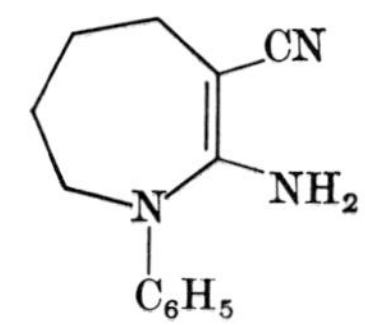	431 447
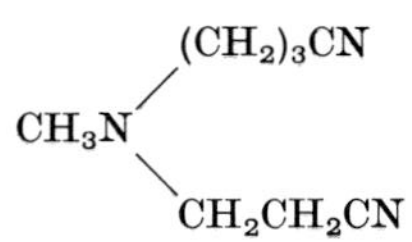	$LiC_2H_5NC_6H_5$	Ether	25.3 (combined)	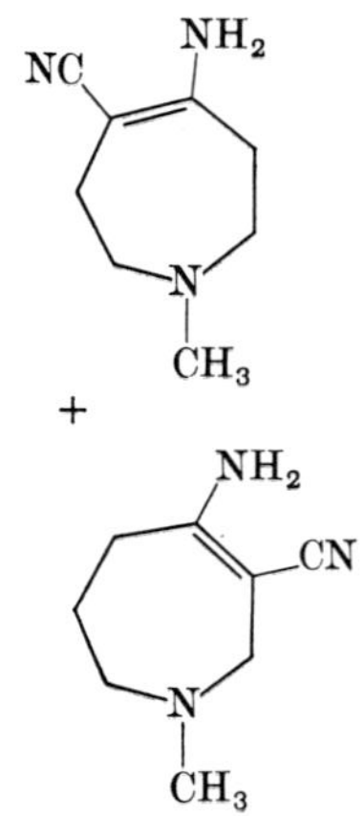	432

$C_2H_5P(CH_2CH_2CN)_2$	NaOBu-*t*	Toluene	83	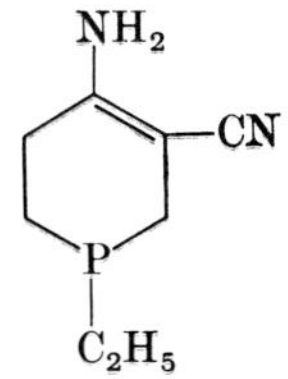	231, 417
n-$C_8H_{17}P(CH_2CH_2CN)_2$	NaOBu-*t*	Toluene	—	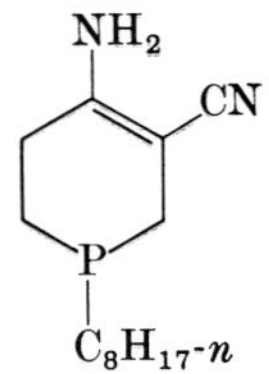	417
$C_6H_5P(CH_2CH_2CN)_2$	NaOBu-*t* NaOBu-*t* —	Toluene Toluene —	80 — —	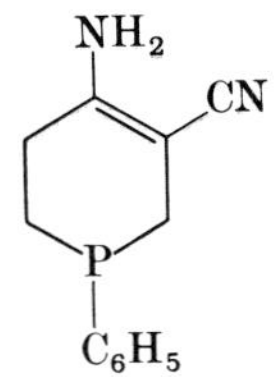	231 417 111
$C_6H_5As(CH_2CH_2CN)_2$	NaOBu-*t* NaOBu-*t*	Toluene Toluene	— —	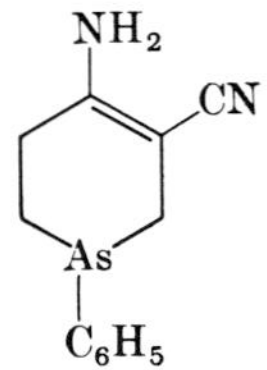	231, 417 111

(*continued*)

TABLE III (*continued*)

Starting dinitrile	Base	Solvent	Yield, %	Product	Ref.
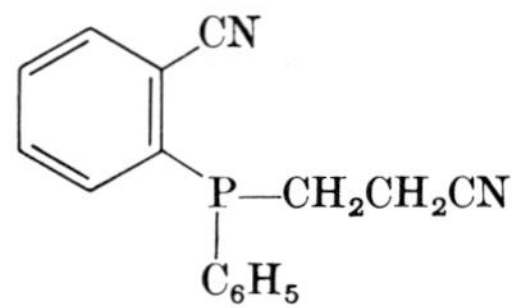	NaOBu-*t*	Xylene	28	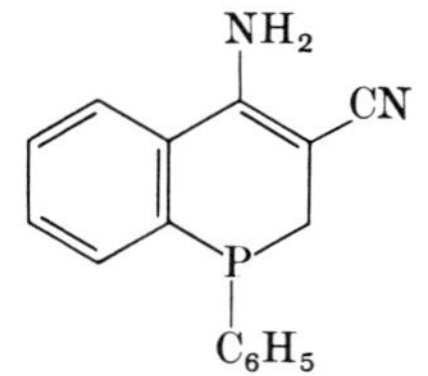	110
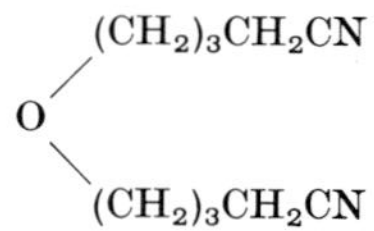	$NaCH_3NC_6H_5$	Ether	17	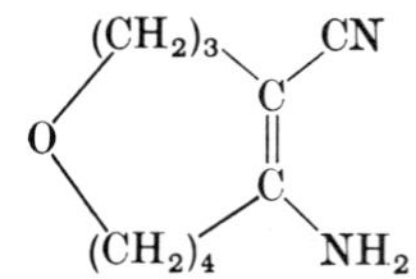	189
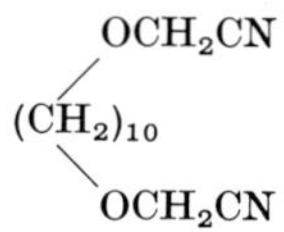	$NaCH_3NC_6H_5$	Ether	37	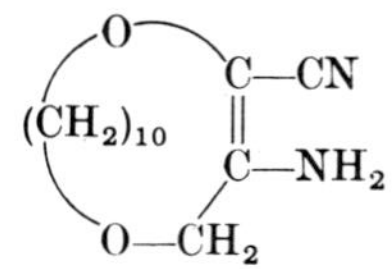	445
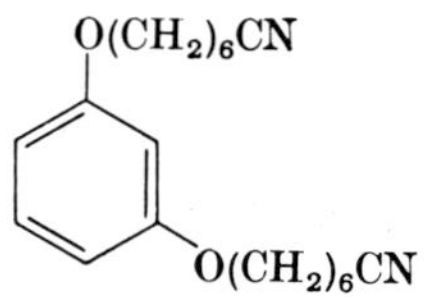	$NaC_2H_5NC_6H_5$	Ether	53.5	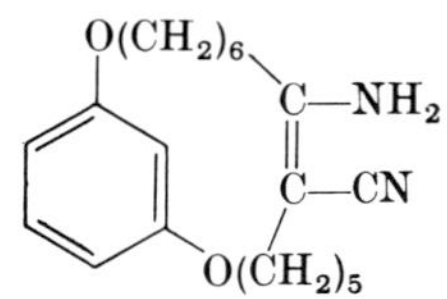	193

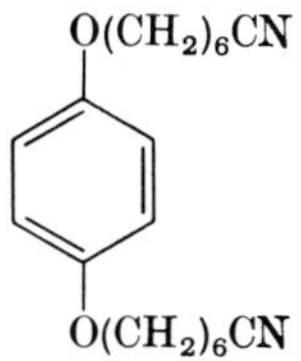	$NaC_2H_5NC_6H_5$	Ether	52	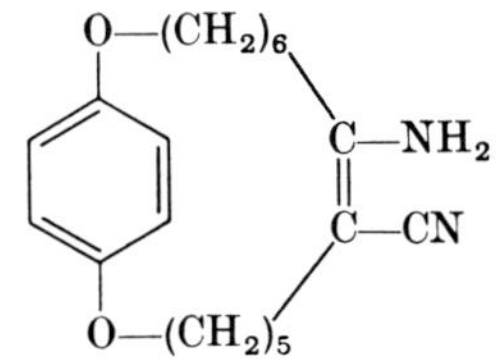	193
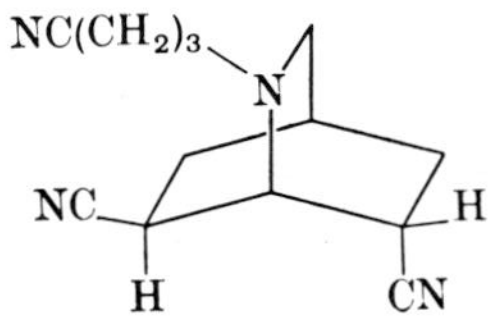	$LiCH_3NC_6H_5$	—	—	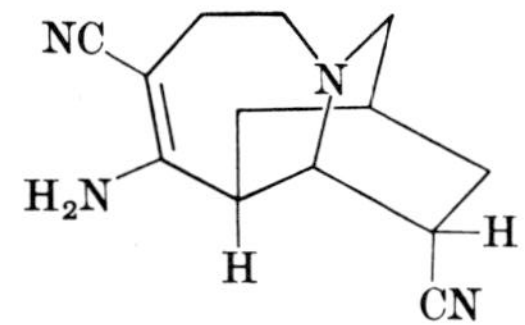	558
		6. Heterocyclic *o*-Aminonitriles			
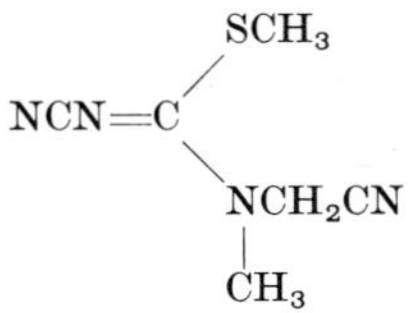	$NaOC_2H_5$	Ethanol	100	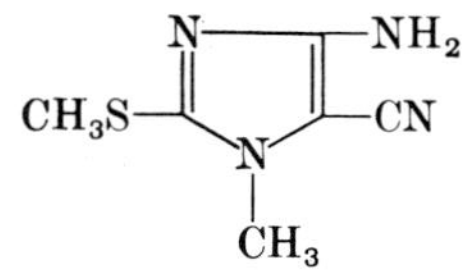	339
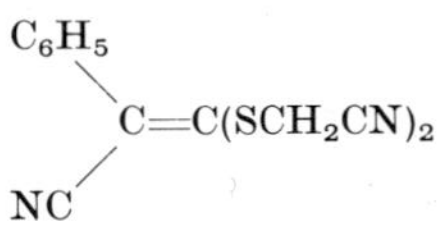 (formed *in situ*)	$NaNH_2$	Ether	81	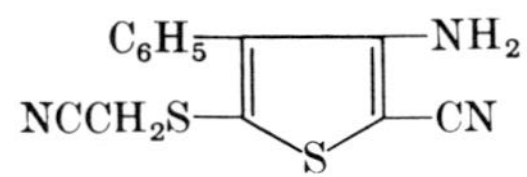	171

(continued)

TABLE III (*continued*)

Starting dinitrile	Base	Solvent	Yield, %	Product	Ref.
$(NC)_2C{=}C(SCH_2COOC_2H_5)_2$	$NaOC_2H_5$	Ethanol	87	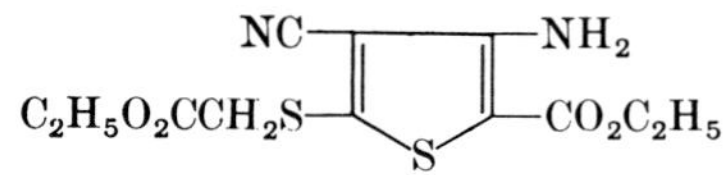	171
$(NC)_2C{=}C(SCH_2CN)_2$ (formed *in situ*)	$NaOCH_3$	Methanol	52	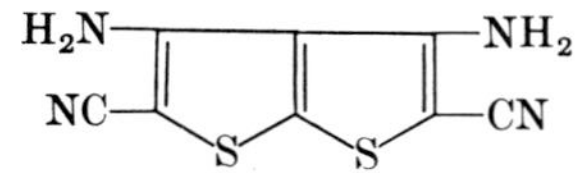	171
$NCN{=}C(SCH_2CN)_2$ (formed *in situ*)	—	Methanol	82	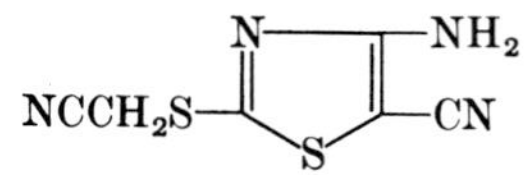	107

[a] This cyclization was effected in acidic medium.

F. DIMERIZATION OF ALKYLIDENEMALONONITRILES

In the preparation of α,β-unsaturated *gem*-dinitriles (alkylidenemalononitriles) of type **91** by base-catalyzed condensation of aldehydes and ketones with malononitrile (69,89,204), by-products have often been observed whose structures have been a source of some confusion for many years. The structures of these products, which are formally dimeric with the expected alkylidenemalononitriles, have been shown by modern

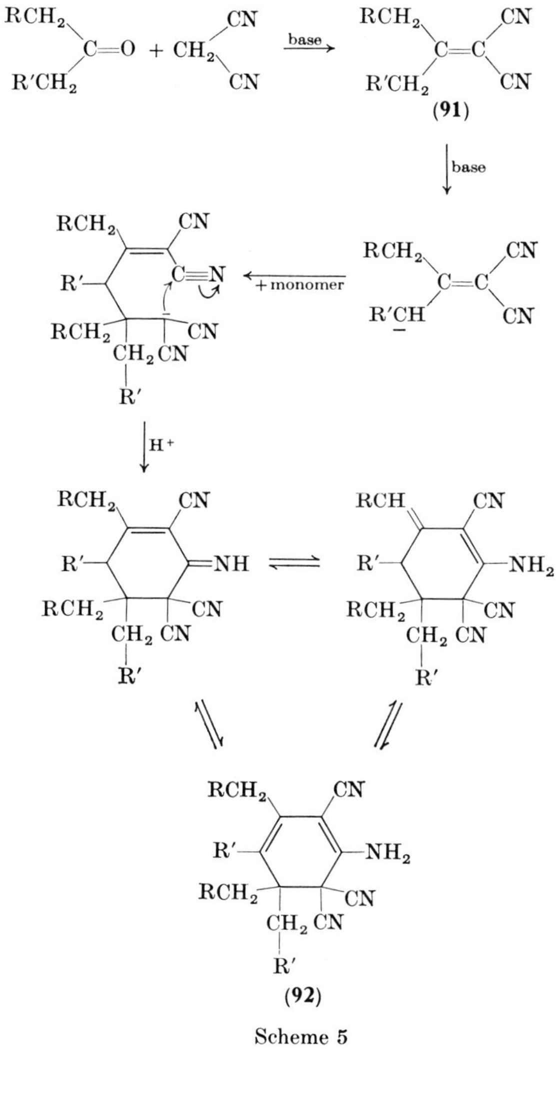

(92)

Scheme 5

TABLE IIIA

Dimerization of Alkylidenemalononitriles

Starting dinitrile	Base	Solvent	Yield, %	Product	Ref.
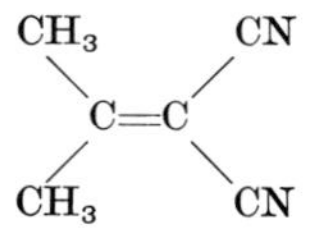	—	—	—	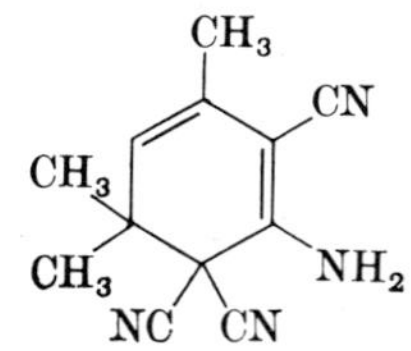	89
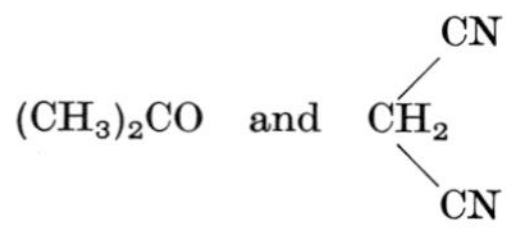	$(C_2H_5)_2NH$	Acetone	—		242
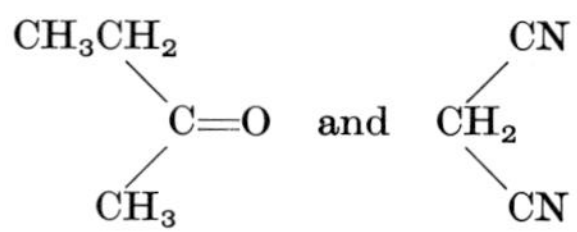	$(C_2H_5)_2NH$	—	—	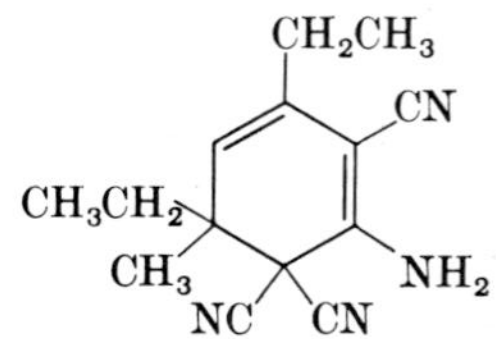	242
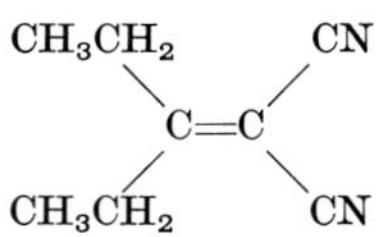	$(C_2H_5)_2NH$	—	—	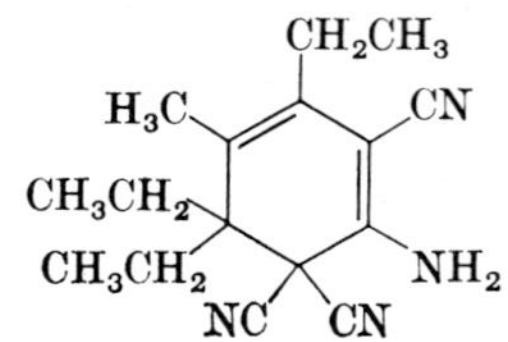	242

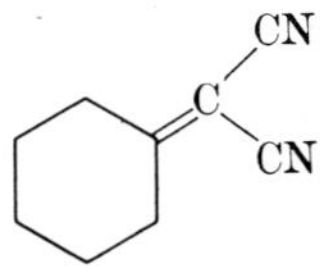	Piperidine	—	—	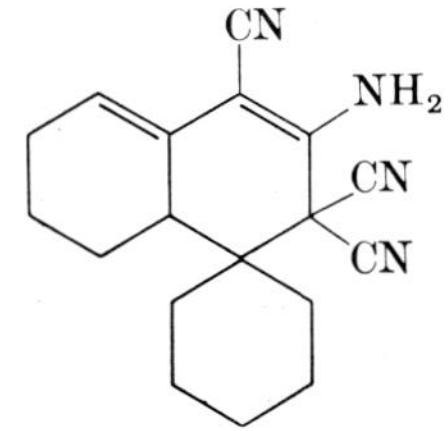	69
	—	—	—		242
HCHO + $CH_2(CN)_2$	Piperidine	Ethanol	20	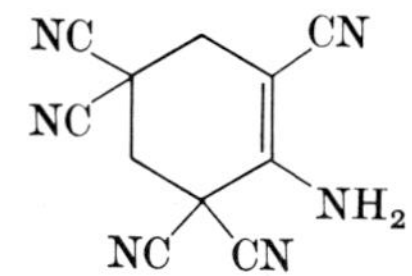	204
	—	—	—		89
(*1*) $(CH_3)_2C{=}O + CH_2(CN)_2$	NaOH	Ethanol	80	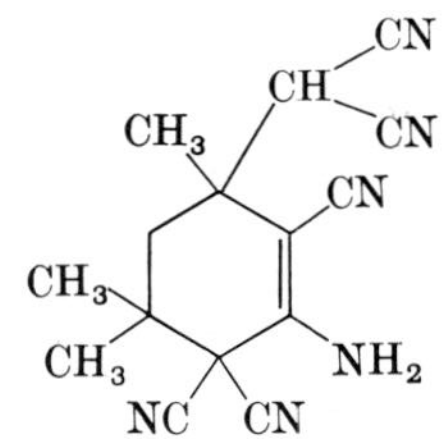	89
(*2*) $(CH_3)_2C{=}C(CN)_2 + CH_2(CN)_2$	NaOH	Ethanol	22		89

spectroscopic means to be cyclic enaminonitriles (**92**) which apparently arise by an initial Michael addition of the anion of an alkylidenemalononitrile to the carbon–carbon double bond of a second molecule of alkylidenemalononitrile, followed by an intramolecular Thorpe-Ziegler cyclization. This reaction sequence is outlined in Scheme 5 (242); it has been observed upon condensation of malononitrile with formaldehyde, acetone, methyl ethyl ketone, diethyl ketone, and cyclohexanone, among others, and is probably general for this type of aldol condensation with malononitrile (see Table IIIA).

IV. Hydrolysis of Cyclic Enaminonitriles. Synthesis of Cyclic α-Cyanoketones and Cyclic Ketones

Cyclic enaminonitriles, the vast majority of which have been prepared by the Thorpe-Ziegler cyclization, have in most instances not been considered as end products but rather as intermediates for the synthesis, by hydrolysis, of cyclic α-cyanoketones and cyclic ketones. Nevertheless, the synthetic utilization and exploitation of this simple route to these two classes of versatile and much-studied compounds appears to have been both neglected and misjudged.

Relatively few satisfactory synthetic methods (aside from that under discussion here) for the preparation of cyclic α-cyanoketones are available. All other methods start with a preformed cyclic ketone which may be (*a*) condensed with a formic ester, followed by reaction with hydroxylamine to give a condensed isoxazole which is then ring-opened with base (101), (*b*) chlorinated to the α-chloroketone, followed by displacement of halogen by cyanide ion (499), or (*c*) converted to the enamine, which is then treated with cyanogen chloride (443). Although in individual cases these methods are satisfactory for the preparation of α-cyanoketones, they do require a preformed ketone, and may be incompatible with some structural features in the latter intermediate (i.e., groupings which interfere with chlorination or can react with hydroxylamine). Complementary to these methods, and far superior to them in the case of many-membered rings, is the hydrolysis of cyclic enaminonitriles formed by the Thorpe-Ziegler cyclization of α,ω-dinitriles. For example, 2-cyanocyclopentanone (**2**) is formed in quantitative yield by mild acid hydrolysis of 1-amino-2-cyanocyclopentene (**1**) (233) and the reaction appears to be equally applicable to the preparation of substituted 2-cyanocyclopentanones (see Table IV). This reaction sequence is uniquely applicable to the preparation of such *gem*-disubstituted α-cyanocyclohexanones as 2-cyano-6-ethyl-6-(2,3-dimethoxyphenyl)cyclohexanone (**4**) which is formed in quantitative yield by acid hydrolysis of the corresponding enaminonitrile **3** (228). Similarly, 2-cyanocycloheptanone is formed in 90% yield from 1-amino-2-cyanocycloheptene (187), and 2-cyanocyclooctanone in 30% yield from 1-amino-2-cyanocyclooctene (187). The reaction is also applicable to the preparation of heterocyclic α-cyanoketones such as 1-methyl-3-cyano-4-piperidone (**7**) which is formed by acid hydrolysis of the Thorpe-Ziegler cyclization product **6** of the bis-cyanoethylation derivative of methylamine **5** (112).

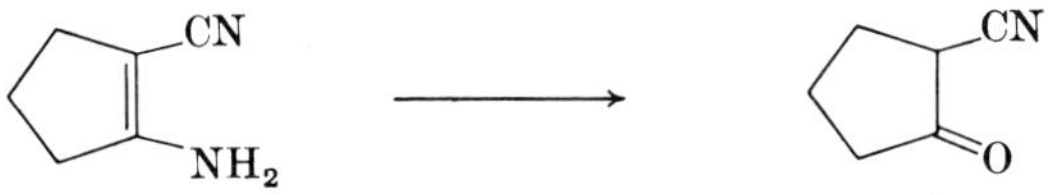

(**1**) (**2**)

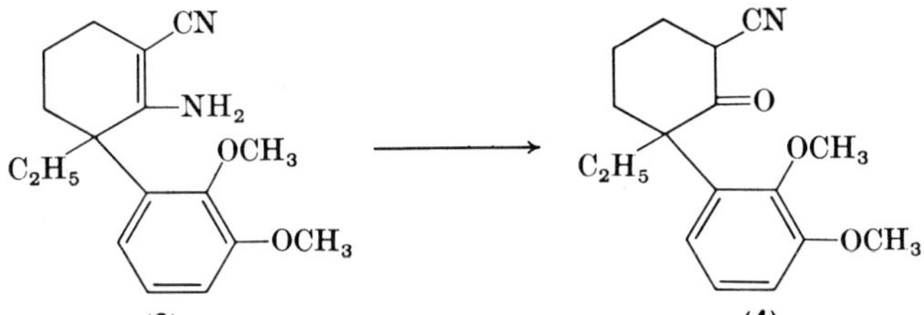

(**3**) (**4**)

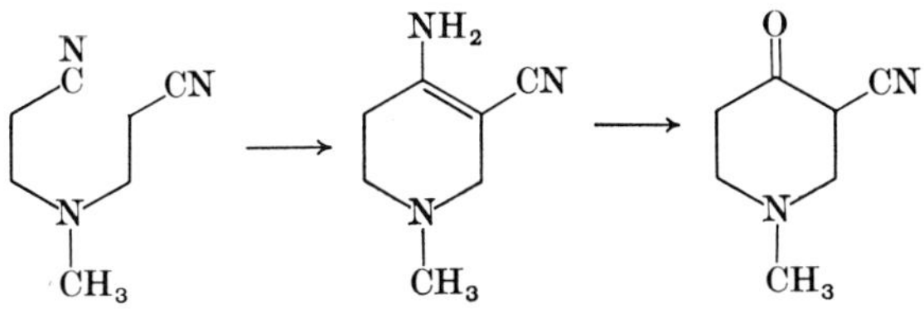

(**5**) (**6**) (**7**)

Acid hydrolysis of cyclic enaminonitriles can readily be carried beyond the α-cyanoketone stage to give cyclic ketones, and this reaction has seen rather wide application. It constitutes, in fact, a superior route to many cyclic ketones, particularly those with many-membered rings. As can be seen from Table IV, cyclopentanones, cyclohexanones, cycloheptanones, cyclooctanones, and, indeed, cyclic ketones up to cyclotritriacontanone can be prepared, generally in excellent yields.

Attempts to prepare cyclic ketones of medium-sized rings (C_9–C_{12}) from α,ω-dinitriles lead instead to cyclic diketones. For example, a projected Thorpe-Ziegler cyclization of 1,10-dicyanodecane gave instead

TABLE IV

Cyclic α-Cyanoketones and Ketones from Acid Hydrolysis of Cyclic Enaminonitriles

Starting material[a]	α-Cyanoketone (% yield)	Ketone (% yield)	Ref.
	Cyclopentanones		
A	2-Cyanocyclopentanone (100) (—)		233, 512
A		Cyclopentanone (—)	195
A	5-Carbethoxy-2-cyanocyclopentanone (—)		194
A	5,5-Dimethyl-2-cyanocyclopentanone (70)		281
B		2,2-Dimethylcyclopentanone (67)	513
A	5,5-Diethyl-2-cyanocyclopentanone (66)		281
A	5-Phenyl-2-cyanocyclopentanone (90)		281
A	5-Ethyl-5-phenyl-2-cyanocyclopentanone (82)		281
A		2-Methyl-2-phenylcyclopentanone (86)	182
A		2-Ethyl-2-phenylcyclopentanone (70) (—)	269, 486
A		2-*n*-Propyl-2-phenylcyclopentanone (excellent) (—)	138, 486
A		2-*n*-Butyl-2-phenylcyclopentanone (excellent) (—)	138, 486
A		2-*n*-Pentyl-2-phenylcyclopentanone (excellent) (—)	138, 486
A		2-Cyclohexyl-2-phenylcyclopentanone (excellent) (—)	138, 486
A		2,2-Diphenylcyclopentanone (81) (—)	221, 486
A		2-Phenyl-2-benzylcyclopentanone (excellent) (—)	138, 486

A		2-Phenyl-2-phenethylcyclopentanone (excellent) (—)	138, 486
A		2,2-Diphenyl-3-methylcyclopentanone (78) ("100")	220, 485
A		2,2-Diphenyl-4-methylcyclopentanone (66)	220
A		1,5-Diphenylbicyclo[3.1.0]hexanone-3 (90)	545
A	1-Cyano-2-indanone (—)		191
A		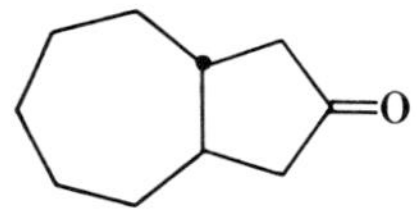 (51)	511
A	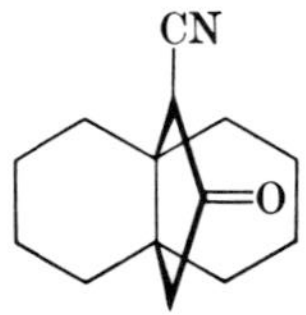 ("100")		7
A	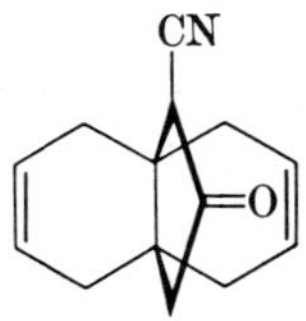 ("100")		7, 12
A		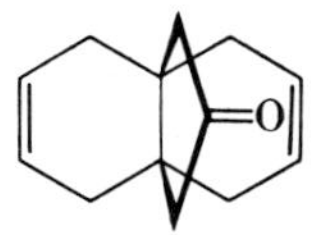 (45)	7

(continued)

TABLE IV (*continued*)

Starting material[a]	α-Cyanoketone (% yield)	Ketone (% yield)	Ref.
	Cyclohexanones		
A	2-Cyanocyclohexanone (—) (60)		101, 542
A		Cyclohexanone (—)	542
A	2-Cyano-6-methylcyclohexanone (—)		101
A	2-Cyano-6,6-dimethylcyclohexanone (80)		542
B		2,2-Dimethylcyclohexanone (—)	513
A		2,2-Dimethylcyclohexanone (—)	542
A	2-Cyano-6,6-diethylcyclohexanone (84)		542
A	2-Cyano-6-ethyl-6-phenylcyclohexanone (95)		542
A		2-Ethyl-2-phenylcyclohexanone (—)	542
A	2-Cyano-6,6-diphenylcyclohexanone (90)		542
A		2,2-Diphenylcyclohexanone (—)	542
A	2,6,6-Tricyano-3,5,5-trimethyl-2-cyclohexen-1-one (72)		89
A	1-Carbomethoxy-1,3-dicyano-4,6,6-trimethyl-2-oxo-3-cyclohexene (90)		89
A	1-Carbethoxy-1,3-dicyano-4,6,6-trimethyl-2-oxo-3-cyclohexene (93)		89
A	2-Cyano-6-methyl-6-(*p*-isopropylphenyl)-cyclohexanone ("100")		226
A		2-(2,3-Dimethoxyphenyl)cyclohexanone (57)	222
A	2-Cyano-6-ethyl-6-(2,3-dimethoxyphenyl)-cyclohexanone ("100")		228
A	2-Cyano-6-(2-ethoxyethyl)-6-(2,3-dimethoxyphenyl)cyclohexanone ("100")		225

A	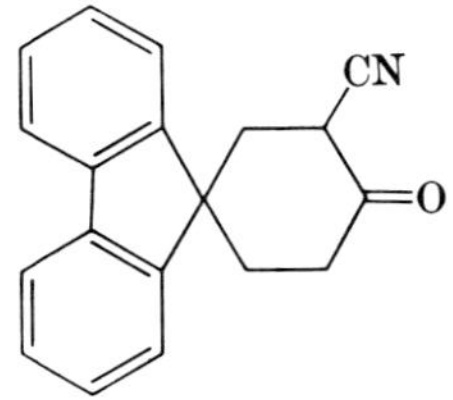 (96)	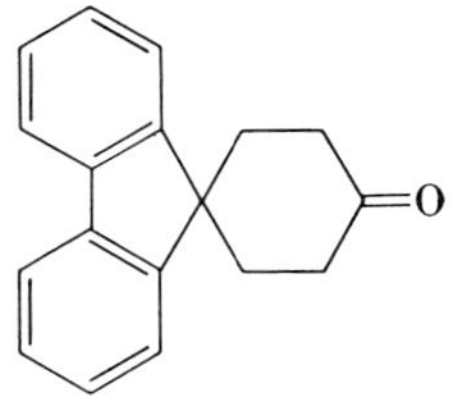 (22)(88)	210, 81
A	2,4-Dicyano-4-phenylcyclohexanone (6)		230
A	2,4-Dicyano-4-(2′,3′-dimethoxyphenyl)-cyclohexanone (3)		230
A	1-Cyano-2-hydroxy-3,4-dihydro-naphthalene (79)[b]		549
	Cycloheptanones		
A		Cycloheptanone (85–95)	190
B	2-Cyanocycloheptanone (90)		187
A	2-Cyanocycloheptanone (—)		200, 202
B	2-Cyano-4,5-*trans*-dimethoxycyclo-heptanone (88)	4,5-*trans*-Dimethoxycycloheptanone (57)	428
B		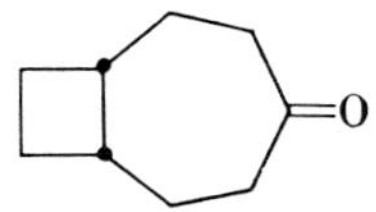 (72)	427
B		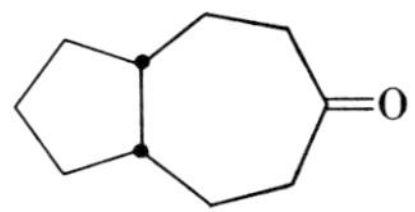 (58)	510

(*continued*)

TABLE IV (*continued*)

Starting Material[a]	α-Cyanoketone (% yield)	Ketone (% yield)		Ref.
		Cycloheptanones (*continued*)		
B		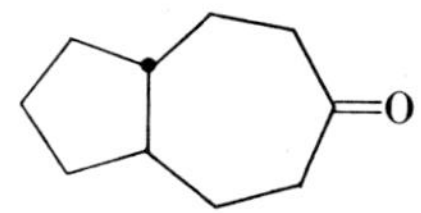	(48)	510
A		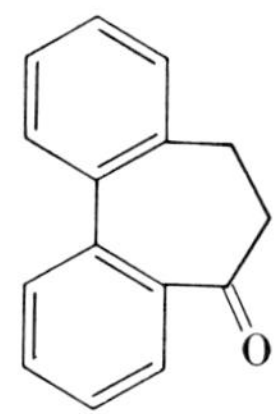	(76)	229
A			(80)	29
A		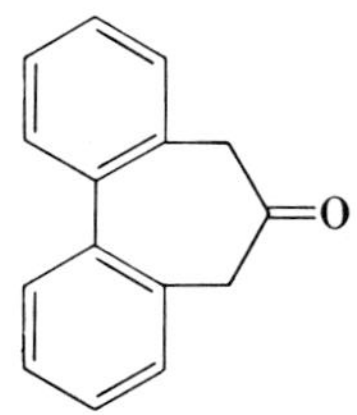	(74)	224
A			(85)	29

A	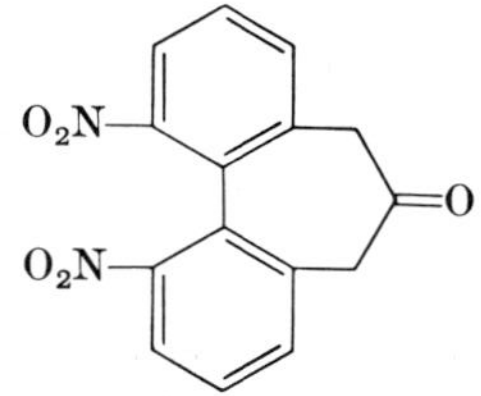	(50)	205
A	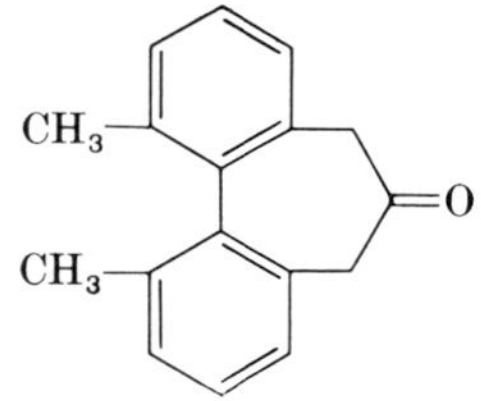	(?)	205
A	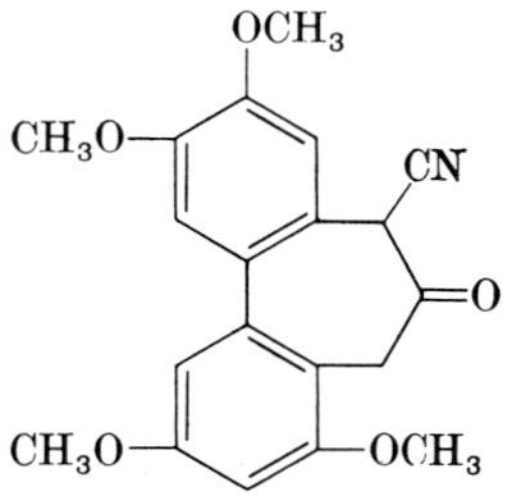 *or* 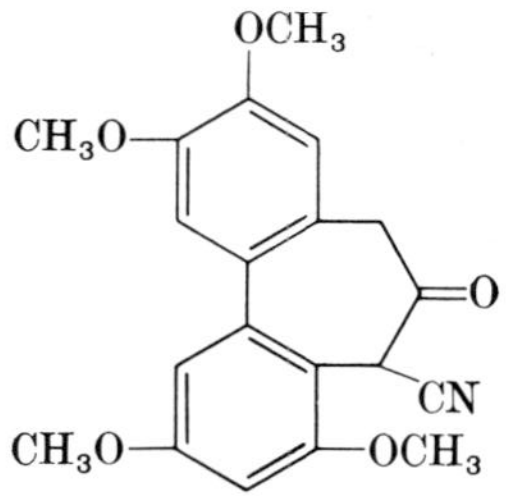 (mixture of keto- and enol-tautomers)	("100")	476

(*continued*)

TABLE IV (*continued*)

Starting material[a]	α-Cyanoketone (% yield)	Ketone (% yield)	Ref.
A		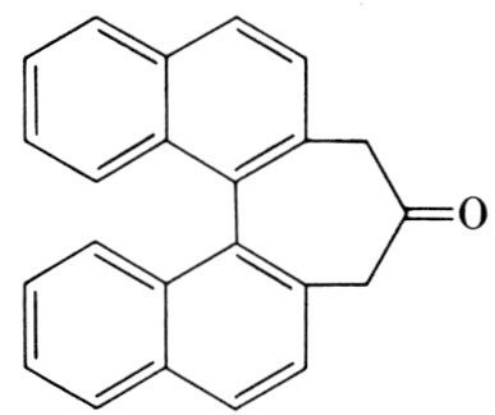 (50)	209
		Cyclooctanones	
B	2-Cyanocyclooctanone (30)	Cyclooctanone (—)	187, 200, 202
A		Cyclooctanone (85)	190
B		4,6-Dimethylcyclooctanone ("100")	426
B		5-*t*-Butylcyclooctanone (89) (73)	424, 559
A		5-Phenylcyclooctanone (80)	167, 181
B		5-(*p*-Chlorophenyl)cyclooctanone (4)	425
A	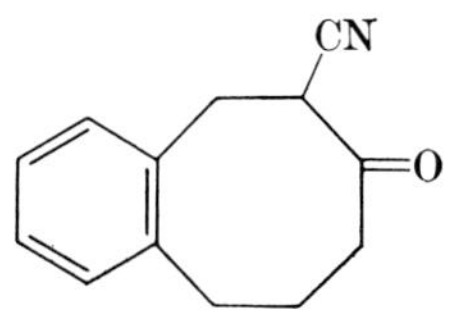 (87)	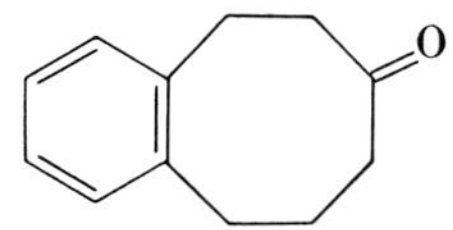 (95)	262

A	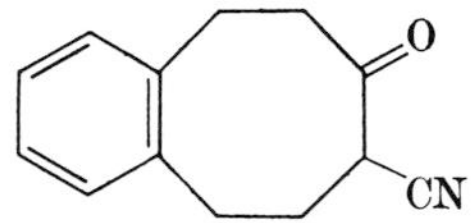 (—)		262
	Medium and Large-Membered Cyclic Ketones		
B		Cyclononanone (2.75)	190
B		Cyclodecanone (0.37)	190
B		Cycloundecanone (1.35)	190
B		Cyclododecanone (8)	190
B		Cyclotridecanone (14–15)	190
		Cyclotridecanone (4.1)	430
A	2-Cyanocyclotetradecanone (—)		200, 202
B		Cyclotetradecanone (62)	190
A		Cyclotetradecanone (—)	201, 202
A		Cyclopentadecanone (37)	187
B		Cyclopentadecanone (60) (67)	190, 430
B		Muscone (54)	235
B		Cyclohexadecanone (77)	190
A		Cyclohexadecanone (—)	201, 202
B		Cycloheptadecanone (98) (82)	188, 190
B		Cyclooctadecanone (82)	190
B		Cyclononadecanone (73)	190
B		Cycloeicosanone (85)	188
B		Cycloheneicosanone (70)	188
B		Cyclodocosanone (83)	188
B		Cyclotricosanone (65)	188
B		Cyclotetracosanone (74)	188

(*continued*)

TABLE IV (*continued*)

Starting material[a]	α-Cyanoketone (% yield)	Ketone (% yield)	Ref.
		Medium and Large-Membered Cyclic Ketones (*continued*)	
B		Cyclopentacosanone (64)	188
B		Cyclohexacosanone (68)	188
B		Cyclooctacosanone (86.5)	188
B		Cyclononacosanone (76)	188
B		Cyclotritriacontanone (68)	188
B		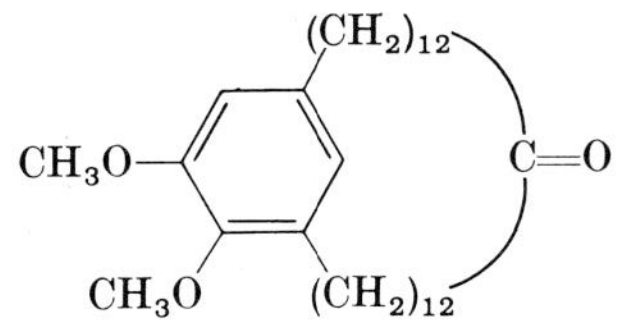 (—)	522
A		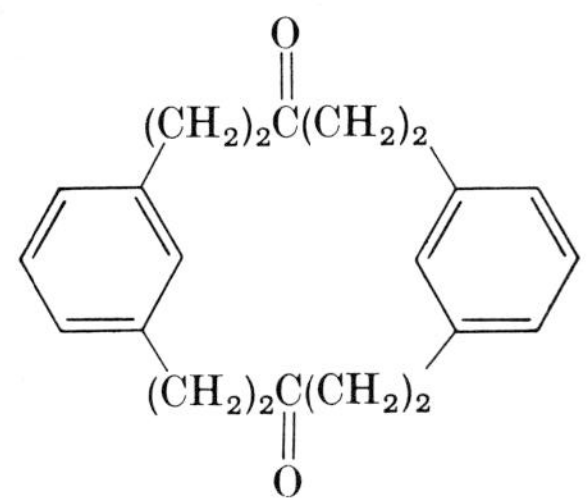 (—)	193

B	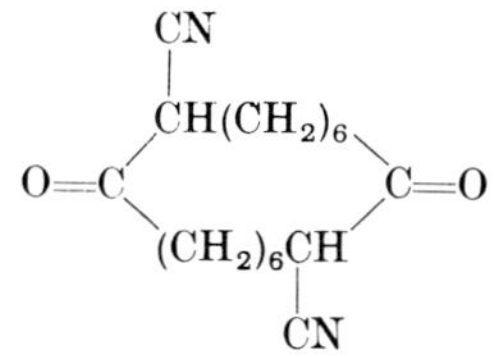 (—)		187, 200, 202
A		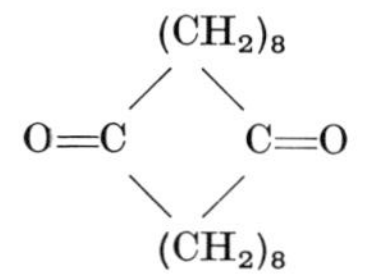 (62)	190
A		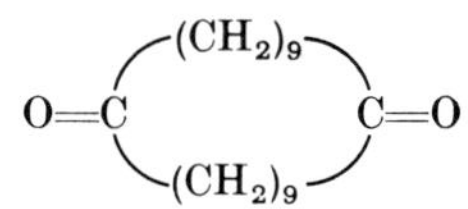 (85–95)	190
A		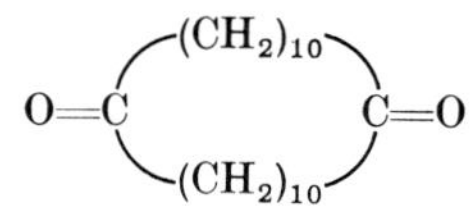 (60)	190
B		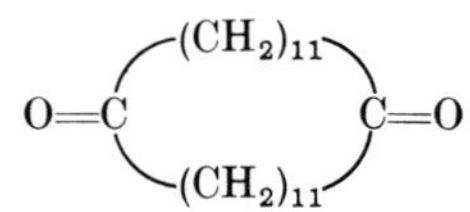 (—)	190

(continued)

TABLE IV (*continued*)

Starting material[a]	α-Cyanoketone (% yield)	Ketone (% yield)	Ref.
Medium and Large-Membered Cyclic Ketones (*continued*)			
A	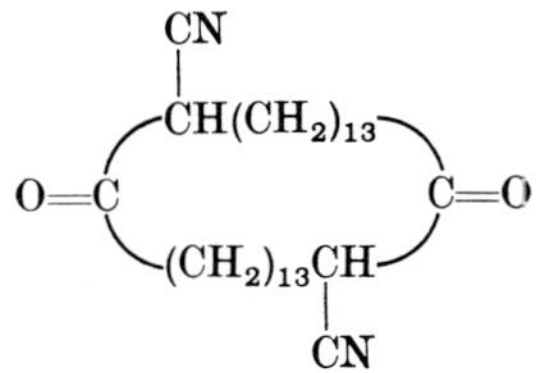 (—)		200, 202
B		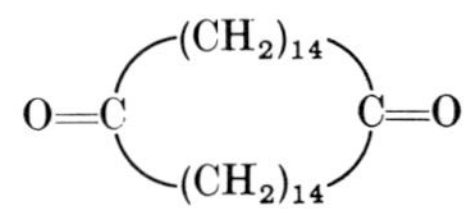 (18)	187
B		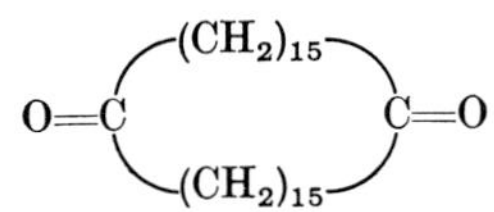 (16)	187
B		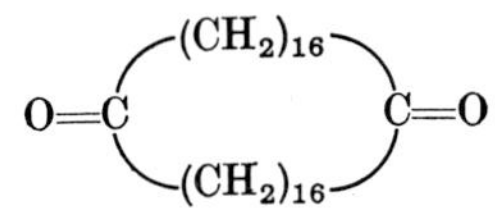 (~20)	187

Heterocyclic Ketones

A	3-Hydroxy-4-cyano-1-cyclohexyl-Δ^3-pyrroline betaine (—)		100
A	1-(*p*-Tolyl)-4-cyano-3-pyrrolidone (60)		100
A		1,2-Dimethyl-3-pyrrolidone (60)	88
A		1-(*n*-Butyl)-2-methyl-3-pyrrolidone (61)	88
A		1-(*n*-Pentyl)-2-methyl-3-pyrrolidone (50)	88
A		1-Benzyl-2-methyl-3-pyrrolidone (70)	88
A		1-Phenethyl-2-methyl-3-pyrrolidone (57)	88
A	1-Methyl-3-cyano-4-piperidone (—)		112
A	1-Acetyl-3-cyano-4-piperidone ("100")		536
A		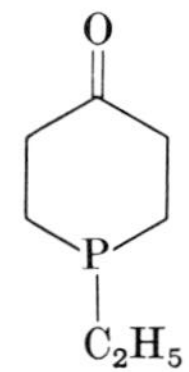 (23)	231
A		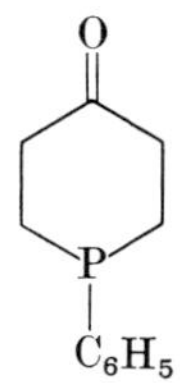 (70)(21)	111, 231

(*continued*)

TABLE IV (*continued*)

Starting Material[a]	α-Cyanoketone (% yield)	Ketone (% yield)	Ref.
		Heterocyclic Ketones (*continued*)	
A		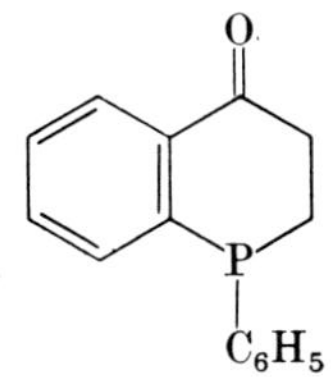 (55)	110
A		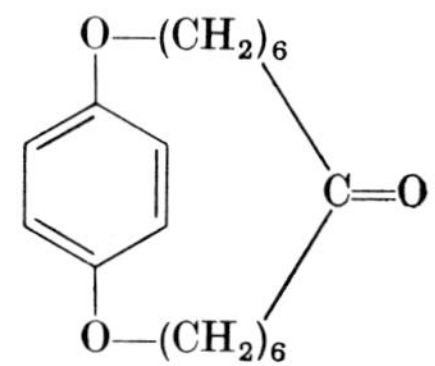 (—)	193
A	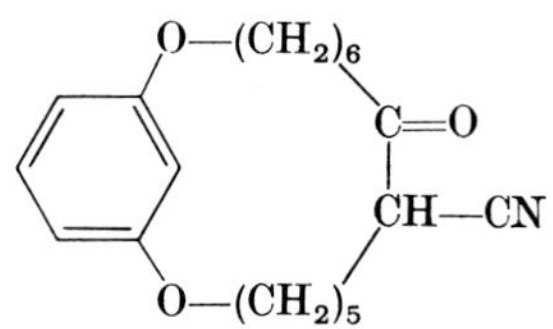 (—)		193
B	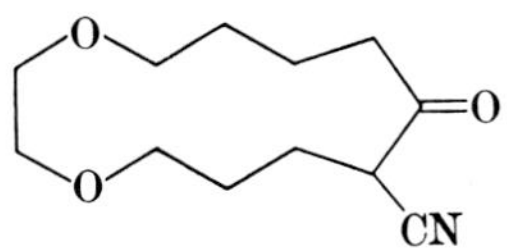 (70)		189

B	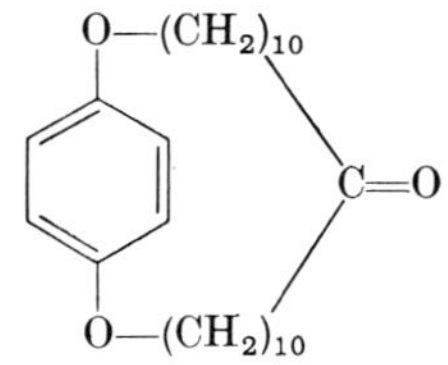	(54)	523
B	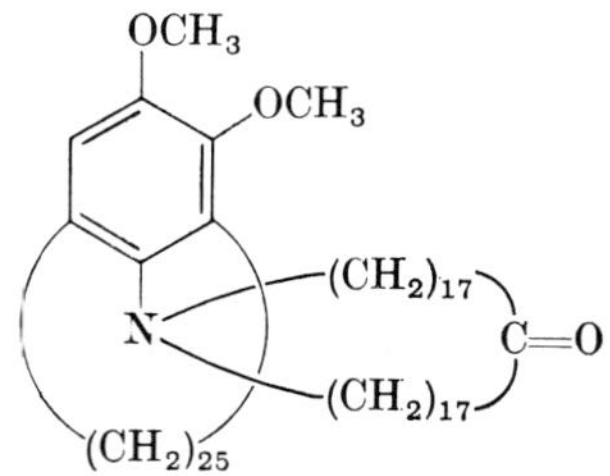	(6)	530
B	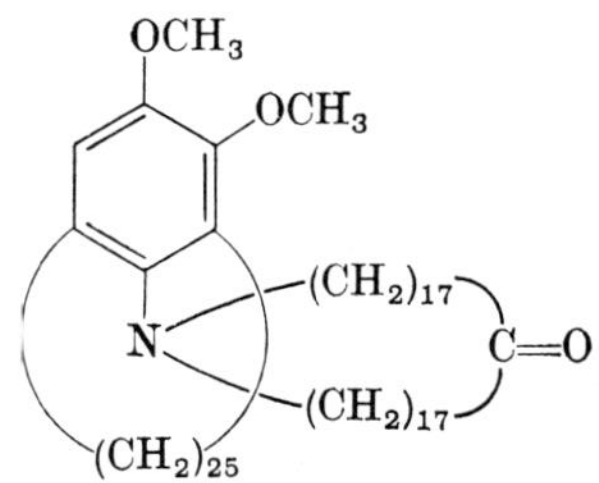	(94)	530

(continued)

TABLE IV (*continued*)

Starting Material[a]	α-Cyanoketone (% yield)	Ketone (% yield)	Ref.
	Heterocyclic Ketones (*continued*)		
B	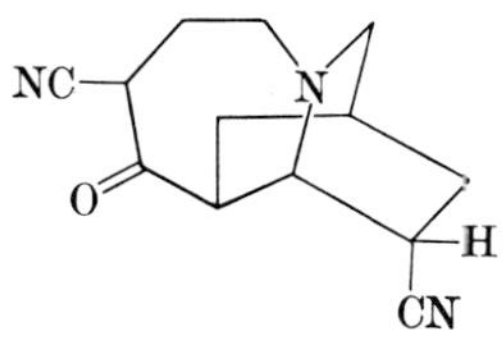 (20)		558

[a] A. The starting material was the isolated cyclic enaminonitrile, prepared as described in Table III.

B. The starting material was the appropriate α,ω-dinitrile; the initially formed cyclic enaminonitrile was not isolated, but hydrolyzed directly.

[b] IR and NMR studies showed conclusively that this compound possessed the enolic structure.

the bis-enaminonitrile **8**, which on acid hydrolysis gave 1,11-cyclodocosanedione (**9**). This route to cyclic diketones appears to have been little exploited despite the high yields and the ease with which the reaction takes place. High dilution techniques are not only unnecessary, but are to be avoided, since the dienaminonitrile precursors to the cyclic diketones are actually the result of a failure of the intramolecular Thorpe-Ziegler cyclization.

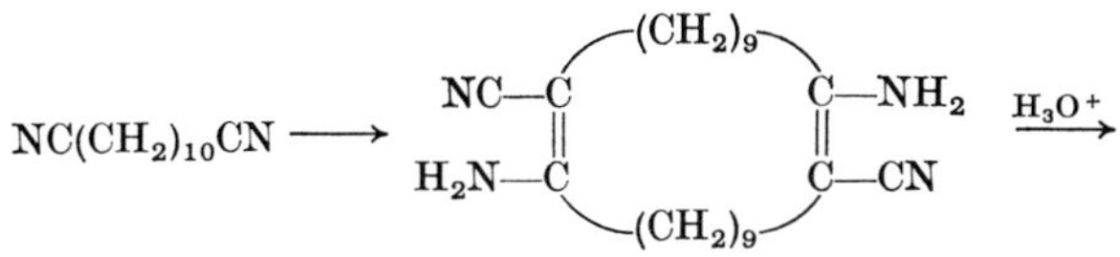

(**8**)

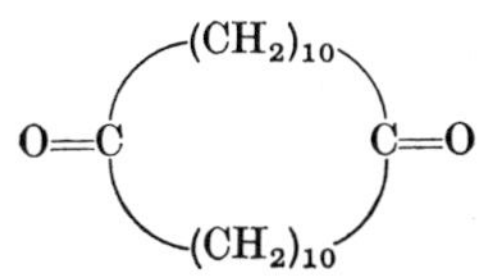

(**9**)

Aside from the acid hydrolysis of cyclic enaminonitriles to give α-cyanoketones and ketones, as discussed above, cyclic enaminonitriles appear to have received scant attention as synthetic intermediates. Typical of the trivial reactions to which these compounds have been subjected are reduction (i.e., of 1-amino-2-cyanocyclopentene to 1-amino-2-aminomethylcyclopentane) (237) and a variety of hydrolytic conditions which gave ring-cleaved products, *o*-aminoamides, etc. (265,268). In some instances it is possible to effect a direct replacement of the primary amino group of 2-amino-3-cyanoindene by an alkoxy group by treatment with an alcohol in sulfuric acid (191), or by an aryl amino group by direct treatment with an aryl amine in the presence of its hydrochloride (64).

Displacement of the amino group of enaminonitriles by sulfur has also occasionally been observed, although the results appear erratic and unpredictable. For example, although 3-amino-4-cyano-1,2-dimethyl-3-pyrroline (**10**) and hydrogen sulfide yield only 3-mercapto-1,2-dimethyl-4-thiocarbamoyl-3-pyrroline (**11**), by reaction at both the amino and nitrile groups, the closely related 2,2-dimethyl derivative **12** has been found to give three products under the same conditions—the expected thiol **13**, a thione **14** (postulated to arise by air oxidation), and 3-amino-1,2,2-trimethyl-4-thiocarbamoyl-3-pyrroline (**15**), arising from reaction of hydrogen sulfide only at the nitrile grouping. The same difference in reaction course was observed with 3-amino-1-benzyl-4-cyano-2-methyl-3-pyrroline and with its 2,2-dimethyl analog; the former gave only **16**, whereas the latter gave both **17** and **18**. There appears to be no explanation for these curious results (28).

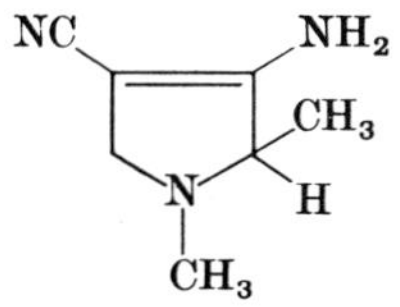

(10)

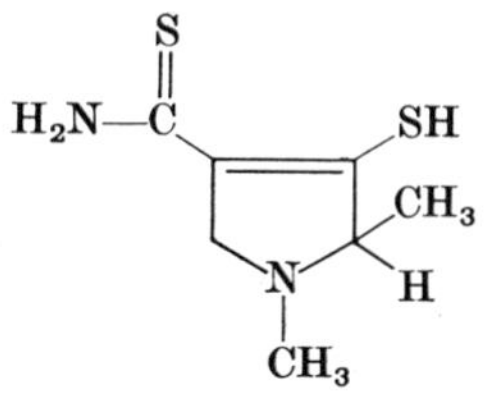

(11)

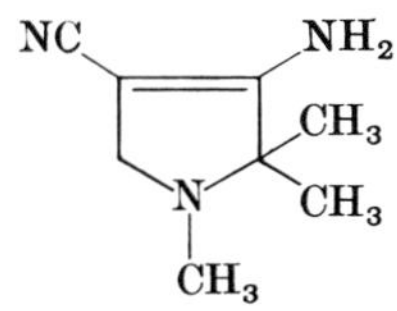

(12)

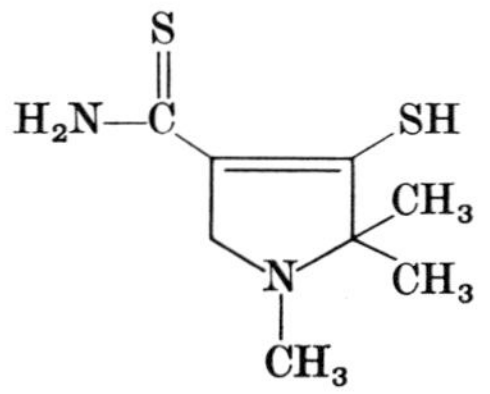

(13)

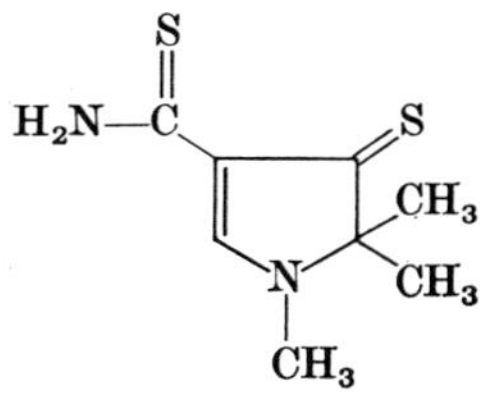

(14)

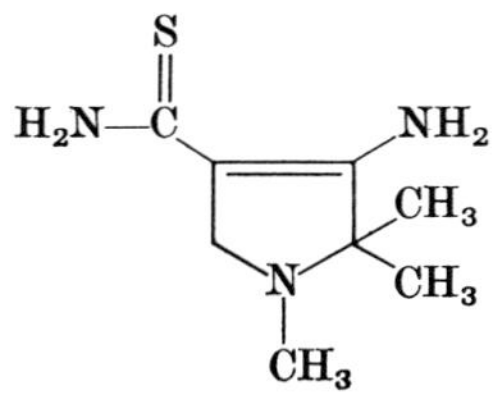

(15)

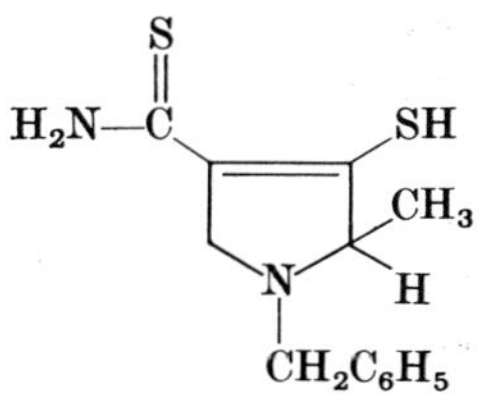

(16)

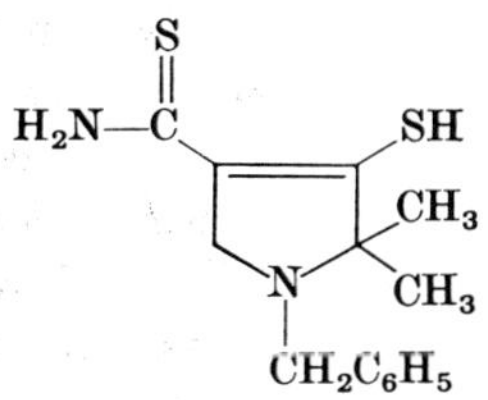

(17)

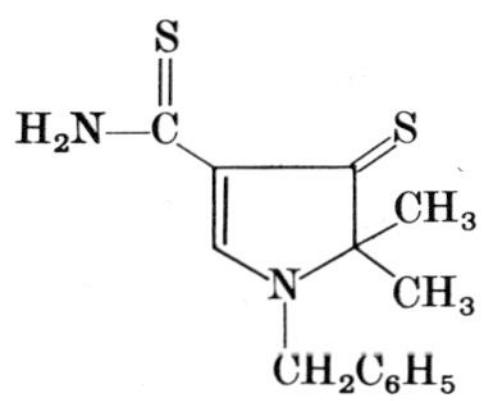

(18)

It has recently been shown that many cyclic enaminonitriles may be utilized as effective intermediates for the synthesis of condensed pyrimidines. These applications are discussed in detail in Chapter II, Section VIII. It is worth noting in this context, therefore, that an initial report that **10** did not undergo such cyclization reactions (88) has since been retracted (13,116) and it would appear desirable to reinvestigate other reports of anomalous reactions attributed to such cyclic enaminonitriles.

Chapter II

o-AMINONITRILES

I. Introduction

o-Aminonitriles are readily accessible derivatives of almost every known important heterocycle, as well as the simple aromatic systems. In many instances, they may be prepared in a single step by condensation of an appropriate reagent with a malononitrile derivative. They may also be prepared by ring rearrangement, isomerization or cleavage reactions, from *o*-halo-, *o*-mercapto-, and *o*-alkoxy-nitriles by nucleophilic displacement with ammonia, from *o*-haloamines by displacement with cyanide ion, and by the more prosaic processes of reduction of *o*-nitronitriles or dehydration of *o*-aminoamides. Their versatility as synthetic intermediates lies principally in the ease with which they may be transformed into a variety of condensed heterocyclic systems, often in efficient, one-step reactions.

The following sections discuss both the synthesis of *o*-aminonitriles and their remarkable versatility and utility in synthesis.

II. Syntheses from Malononitrile and its Derivatives

The great majority of heterocyclic *o*-aminonitriles known to date have been prepared by condensation reactions involving the participation of malononitrile or one of its derivatives (ethoxymethylenemalononitrile, aminomethylenemalononitrile, tetracyanoethane, tetracyanoethylene) as a key component. In contrast to the enaminonitriles prepared by Thorpe-Ziegler cyclization of α,ω-dinitriles, which have been utilized almost entirely as intermediates for the preparation of cyclic ketones (see Chapter I, Section IV), the many heterocyclic *o*-aminonitriles prepared as described below from malononitrile have in most cases been considered as end products and their utility as synthetic intermediates has been largely ignored.

The preparation of malononitrile and its various derivatives utilized as intermediates will not be described here; readers are referred to the original papers describing their use for descriptions of their preparation. Malononitrile, ethoxymethylenemalononitrile, and tetracyanoethylene are all commercially available, and the other intermediates derived from malononitrile are readily preparable.

The terminal step in the preparation of all of the heterocyclic *o*-aminonitriles described below is an intramolecular nucleophilic addition of an NH, OH, or SH group to one of the nitrile groups of the malononitrile-derived unit; subsequent tautomerization of the resulting imine gives the final aminonitrile. A generalized depiction of this synthetic sequence is given in Scheme 6.

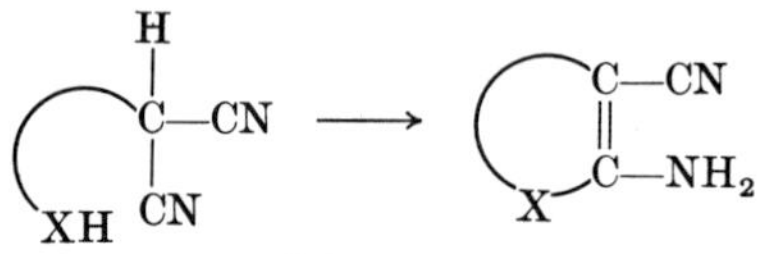

Scheme 6

A. SYNTHESIS OF PYRAZOLE *o*-AMINONITRILES (Table V)

The condensation of ethoxymethylenemalononitrile or its C-alkyl or C-aryl derivatives with hydrazine and substituted hydrazines constitutes the most general and versatile synthesis of pyrazole *o*-aminonitriles (**1**). Although no detailed study of the mechanism of this condensation has

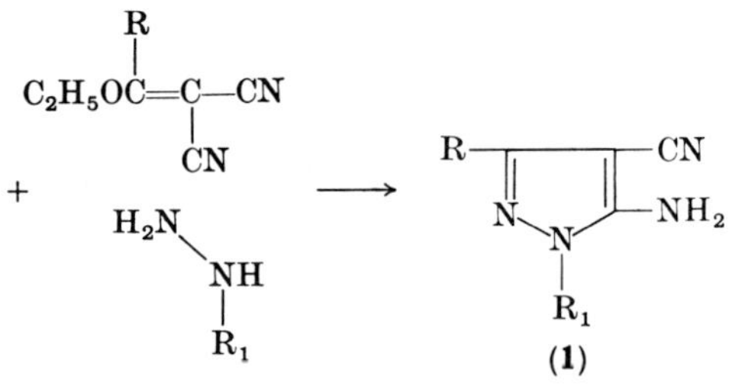

(**1**)

been made, the first step is undoubtedly Michael addition of the hydrazine to the α,β-unsaturated *gem*-dinitrile, followed by elimination of ethanol, intramolecular cyclization by addition of the hydrazine NH grouping to one of the nitriles and final tautomerization. It is noteworthy that both alkyl and aryl hydrazines yield 1-substituted-4-cyano-5-aminopyrazoles; in both cases the primary NH_2 group of the substituted hydrazine undergoes the initial Michael addition. A choice of hydrazines offers the possibility of wide substitution at position 1, but the only substituents which can be introduced at position 3 are hydrogen, alkyl, or aryl. A variety of other substituents can be introduced at this position, however, by the use of other malononitrile intermediates. For example, the reaction of hydrazine with dicyanoketene acetals (**2**, X = O) leads to 3-alkoxy-4-cyano-5-aminopyrazoles (**3**, X = O, R′ = H) (105). A thioether grouping (**3**, X = S) is introduced analogously by the use of dicyanoketene dithioacetals (232,266). Condensation with 1,1-dicyano-2-amino-2-(2-ethoxyethoxy)ethylene (**4**) yields 3,5-diamino-4-cyanopyrazole (**5**)

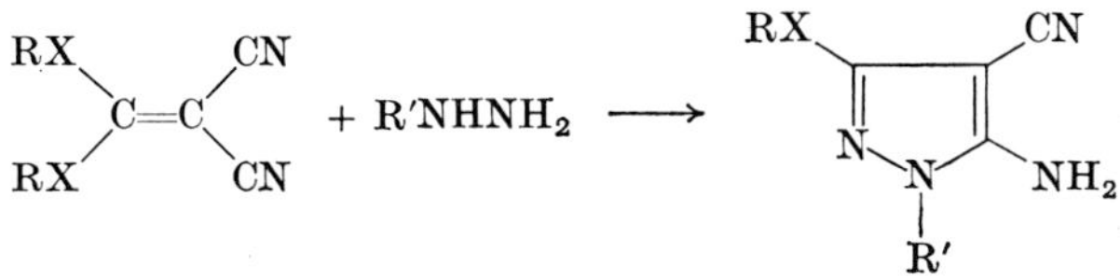

(**2**) (**3**)

(X = O, S)

(105); 1,1-dicyano-2,2-dichloroethylene in analogous fashion gives a 3-chloropyrazole (232). Further variations in this synthetic approach to pyrazole *o*-aminonitriles appear to be limited only by the availability of suitably substituted 1,1-dicyanoethylene intermediates.

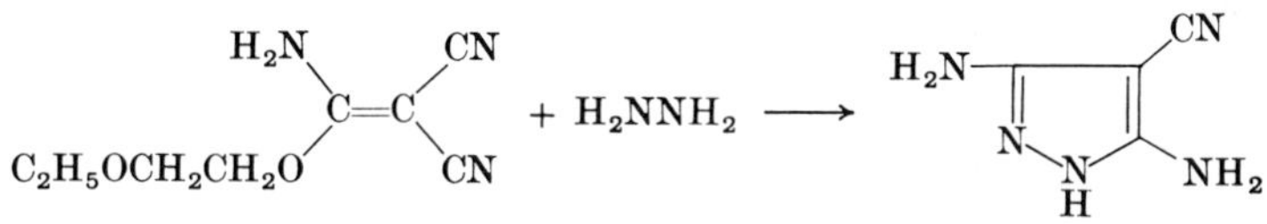

(**4**) (**5**)

Surprisingly, the reaction of malononitrile itself with hydrazine gives 3-cyanomethyl-4-cyano-5-aminopyrazole and not 3,5-diaminopyrazole as was once claimed. 2-Amino-1,1,3-tricyanopropene (26,84) (formed by base-catalyzed dimerization of malononitrile and henceforth referred to as malononitrile dimer), has been shown to be an intermediate in this conversion; its subsequent reaction with hydrazine (26) and substituted hydrazines (25,84,540) involves Michael addition followed by elimination of ammonia and final intramolecular cyclization. Both alkyl- and arylhydrazines give 1-substituted 3-cyanomethyl-4-cyano-5-aminopyrazoles; the claim that phenylhydrazine gives the 2-phenyl derivative (84) has been shown to be incorrect (25).

A nitrile group can be introduced into position 3 of the pyrazole ring by taking advantage of the propensity of tetracyanoethylene for Michael additions. Thus, the reaction of tetracyanoethylene with hydrazines leads to 3,4-dicyano-5-aminopyrazoles (**6**) by initial addition to the carbon–carbon double bond, elimination of hydrogen cyanide and terminal intramolecular cyclization (232). Aryl- and alkyl-hydrazines, as well as hydrazides, semicarbazides, and thiosemicarbazides all react similarly to give 1-substituted derivatives (see Table V). A closely related pyrazole

TABLE V

Pyrazole *o*-Aminonitriles

Reactants				
Malononitrile derivative	Hydrazine derivative	Product	Yield, %	Ref.
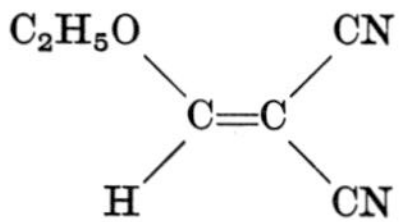	H_2NNH_2	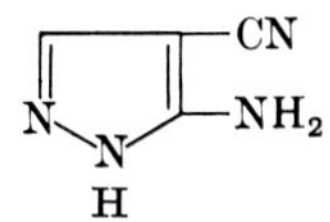	85	4
			—	18
			—	395
			—	154
			—	168
	H_2NNHCH_3	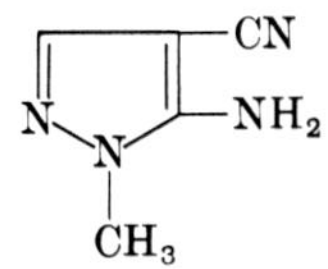	86.4	3
			—	169
			—	405
			—	168
			—	20
			—	532
			—	149

$H_2NNHC_2H_5$	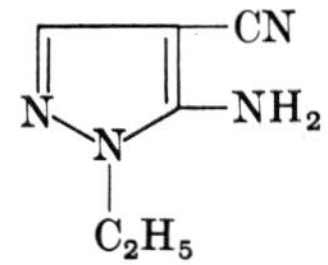	85	532
$H_2NNHCH_2CH_2OH$	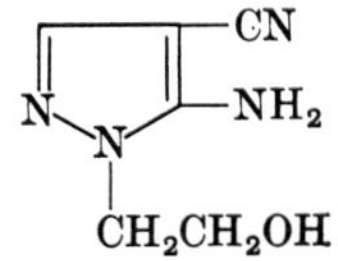	83.5 — — —	3 150 168 407
$H_2NNHCH_2CH_2OCH_3$	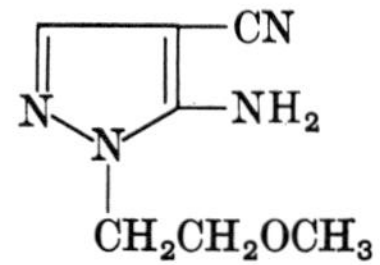	64	532
$H_2NNHCH_2CH_2CH_3$	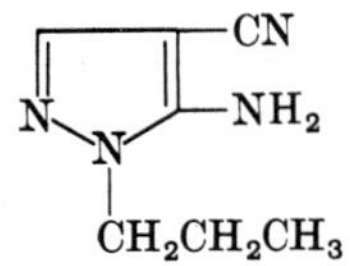	36	532

(continued)

TABLE V (*continued*)

Reactants		Product	Yield, %	Ref.
Malononitrile derivative	Hydrazine derivative			
		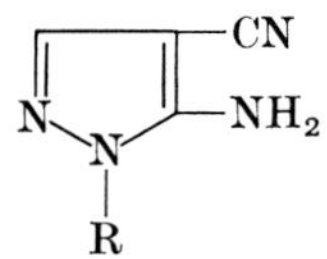		
		R		
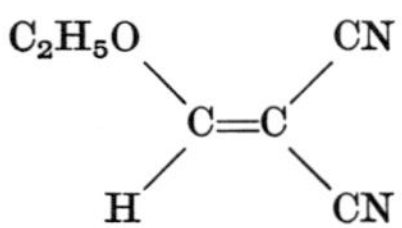 (*cont.*)	$H_2NNHCH(CH_3)CH(CH_3)_2$	$CH(CH_3)CH(CH_3)_2$	—	144
			—	149
			—	408
	$H_2NNHCH(CH_2CH_3)_2$	$CH(CH_2CH_3)_2$	—	159
			—	155
			—	408
	$H_2NNHCH(CH_3)_2$	$CH(CH_3)_2$	—	148
			—	149
			—	373
			—	403
	$H_2NNHCH_2CH_2CH_2CH_3$	$CH_2CH_2CH_2CH_3$	60	170
	$H_2NNHCH(CH_3)CH_2CH_3$	$CH(CH_3)CH_2CH_3$	—	144
			—	155
			—	408

H_2NNH— 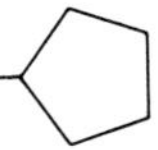		—	155
		—	159
		—	408
H_2NNH— 	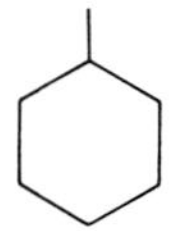	—	144
		—	149
		—	155
		50.4	170
		—	408
$H_2NNHC_6H_{13}$-*n*	C_6H_{13}-*n*	47.6	170
$H_2NNHCH(CH_3)C_5H_{11}$	$CH(CH_3)C_5H_{11}$	—	408
$H_2NNHC_6H_5$	C_6H_5	80	3
		—	145
		—	157
		—	168
		—	169
H_2NNH— 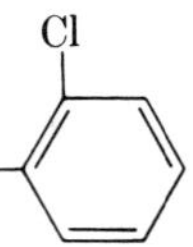	*o*-ClC_6H_4	61	3
		—	168
H_2NNH— 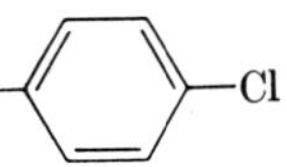	*p*-ClC_6H_4	77.5	3
		—	157
		—	168
		—	169

(continued)

TABLE V (*continued*)

Reactants		Product	Yield, %	Ref.
Malononitrile derivative	Hydrazine derivative			
		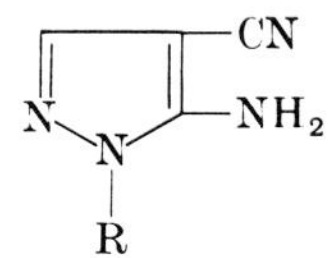 R		
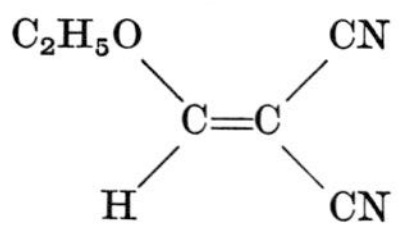 (*cont.*)	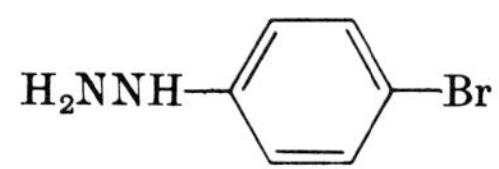	p-BrC_6H_4	40	3
			—	168
	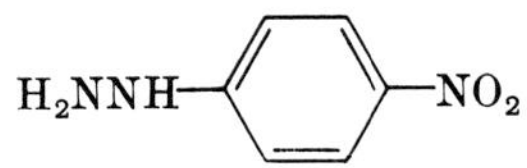	p-$O_2NC_6H_4$	81	3
			—	168
	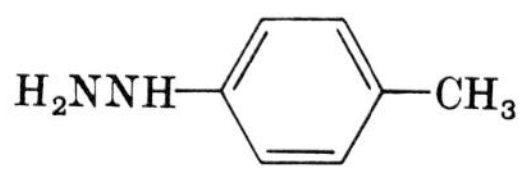	p-$CH_3C_6H_4$	84	3
			—	168
	$H_2NNHCH_2C_6H_5$	$CH_2C_6H_5$	57.6	170
			74	371
	$H_2NNHCONH_2$	$CONH_2$	20	232
	$C_6H_5CH{=}NNHCH_3$	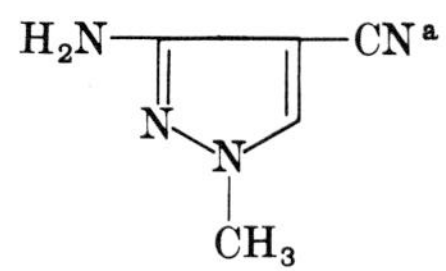 [a]	93	97

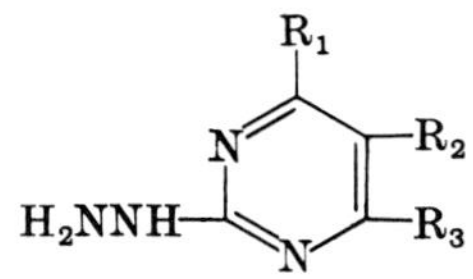

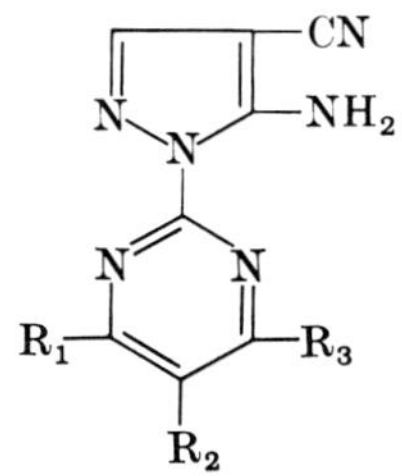

R_1	R_2	R_3		
H	H	H	—	419
CH_3	—	—	—	400
			—	419
H	CN	NH_2	—	411
			—	419
H	$COOC_2H_5$	CH_3	—	411
			—	419
H	$COOC_2H_5$	OH	—	411
			—	419
H	$COOC_2H_5$	NH_2	—	400
CH_3	H	CH_3	—	400
			—	419
CH_3	H	OH	—	400
			—	411
			—	419
OH	—$(CH_2)_3$—		—	400
			—	419

(continued)

TABLE V (*continued*)

Reactants: Malononitrile derivative	Reactants: Hydrazine derivative	Product			Yield, %	Ref.
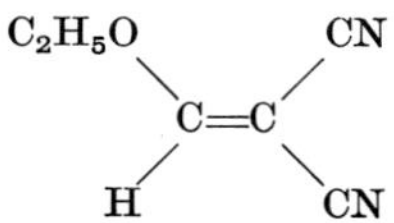 (*cont.*)	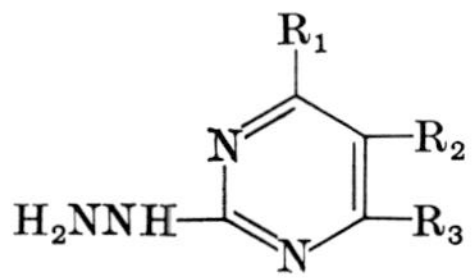	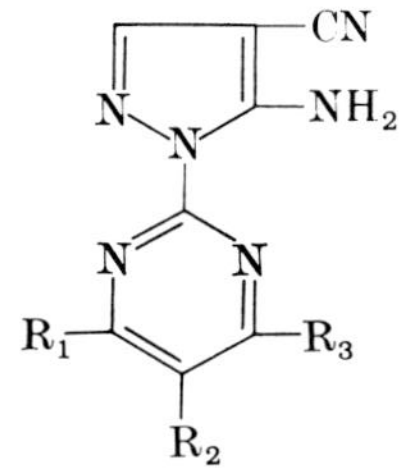				
		R_1	R_2	R_3		
		OH	—$(CH_2)_4$—		—	411
					—	419
		OH	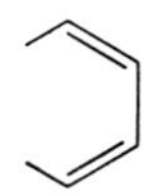		—	419
					—	423
		OH	H	C_6H_5	—	400
					—	411
					—	419
		OH	CH_3	CH_3	—	411

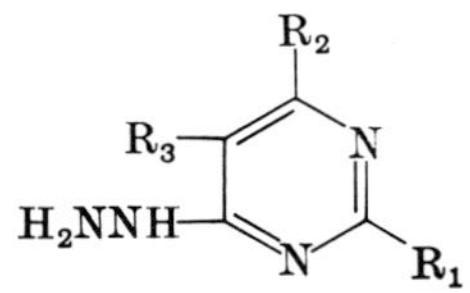

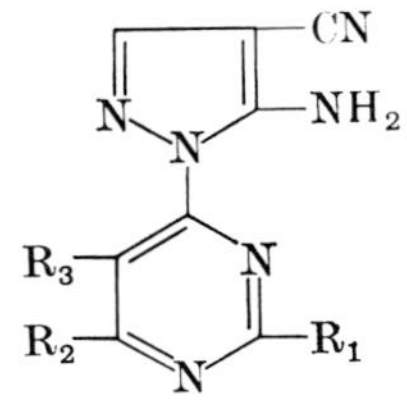

R_1	R_2	R_3		
OH	CH_3	H	—	419
NH_2	CH_3	H	—	419
C_6H_5NH	CH_3	H	—	419
$C_6H_5CH_2NH$	CH_3	H	—	419
CH_3S	CH_3	H	—	419
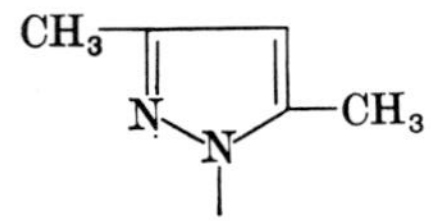	CH_3	H	—	419
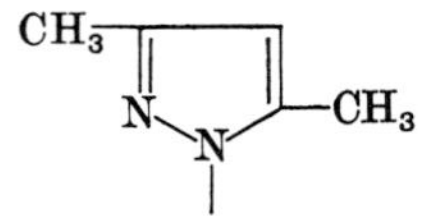	—$(CH_2)_4$—		—	419
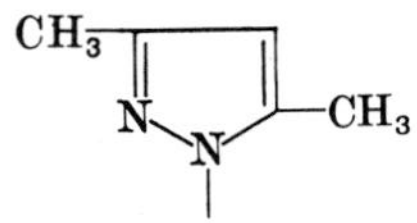	C_6H_5	H	—	419

(continued)

TABLE V (*continued*)

Reactants		Product	Yield, %	Ref.
Malononitrile derivative	Hydrazine derivative			
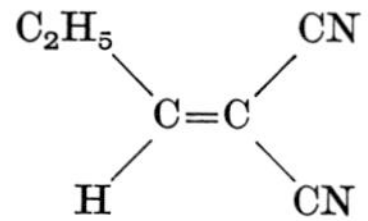 (*cont.*)		R_1 R_2 R_3 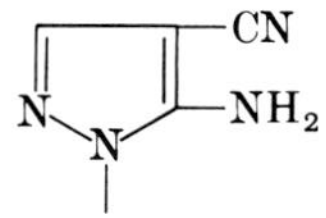 —$(CH_2)_3$—	—	419
	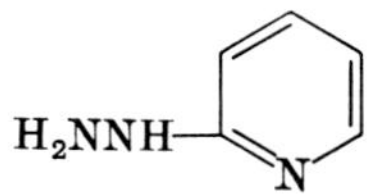	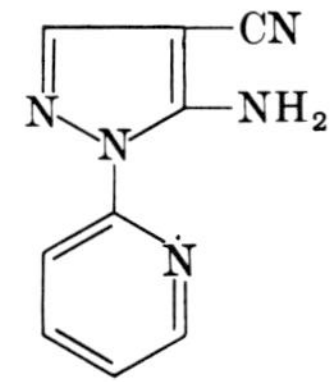	—	419
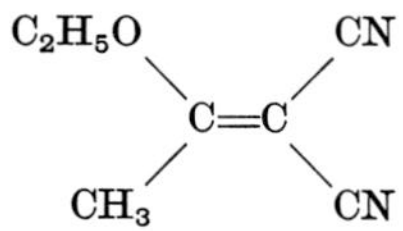	H_2NNH_2	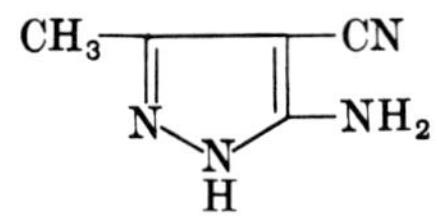	96 —	3 124
	H_2NNHCH_3	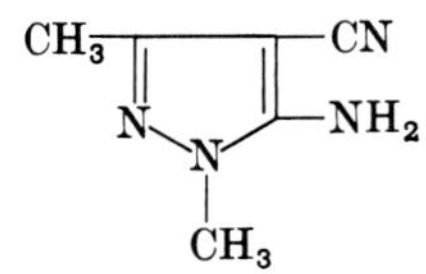	87 —	3 25

	$H_2NNHC_6H_5$	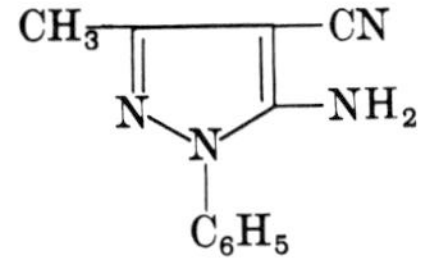	80 —	3 25
	$H_2NNHCONH_2$	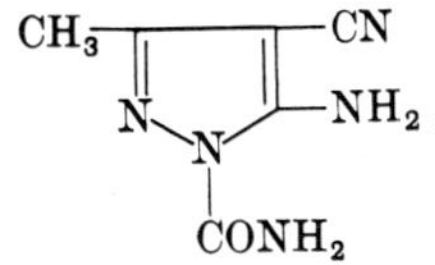	30	232
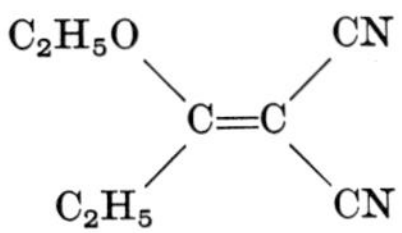	H_2NNHCH_3	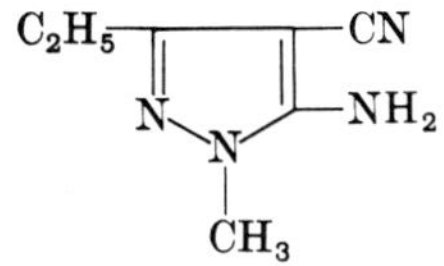	94	532
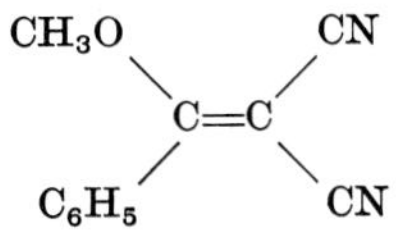	H_2NNH_2	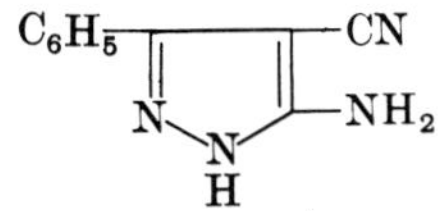	78	33
	$H_2NNHC_6H_5$	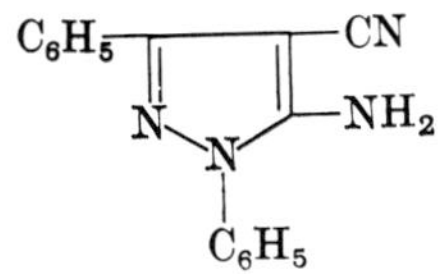	80	232

(*continued*)

TABLE V (*continued*)

Reactants				
Malononitrile derivative	Hydrazine derivative	Product	Yield, %	Ref.
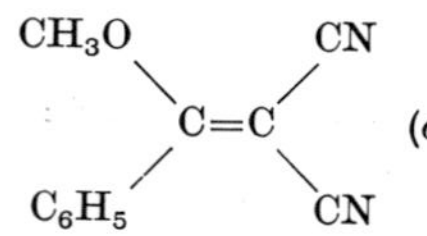 (*cont.*)	$H_2NNHCONH_2$	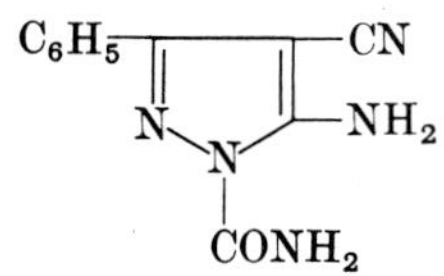	70	232
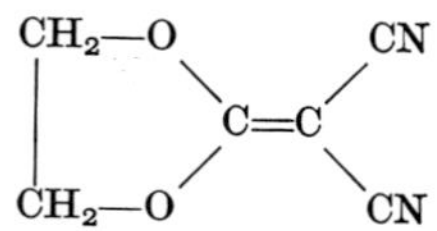	H_2NNH_2	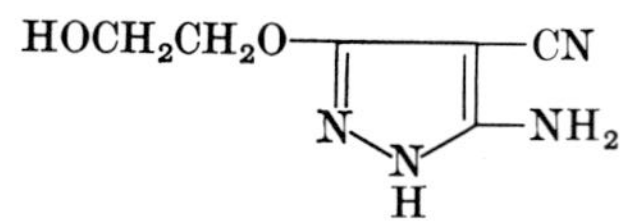	72	105
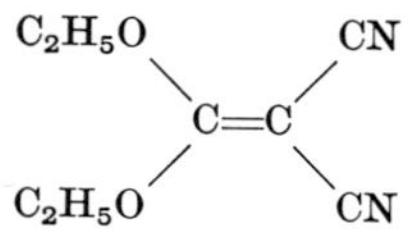	H_2NNH_2	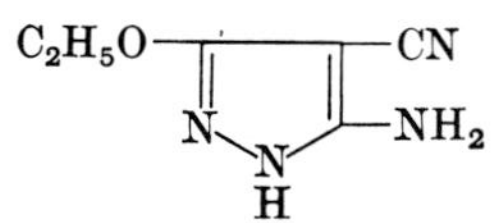	72	105
	$H_2NNHCONH_2$	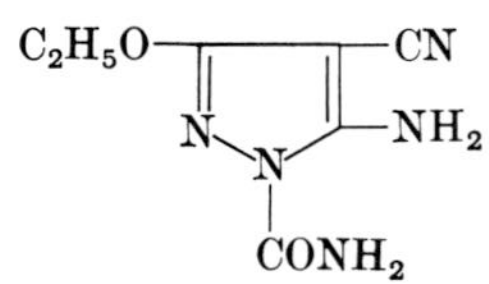	30	232

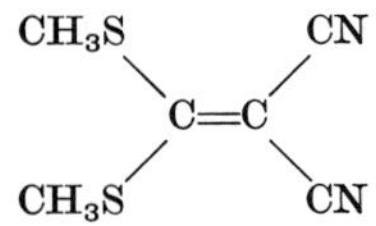	H_2NNH_2	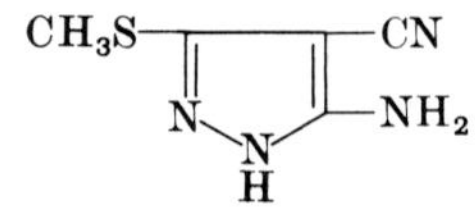	97	266
	$H_2NNHC_6H_5$	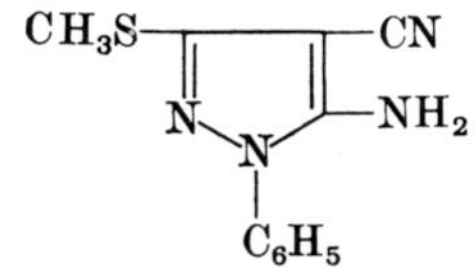	98	266
	$H_2NNHCONH_2$	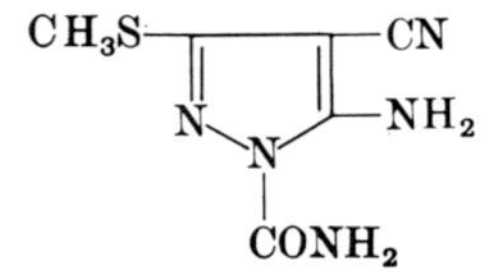	68 —	232 406
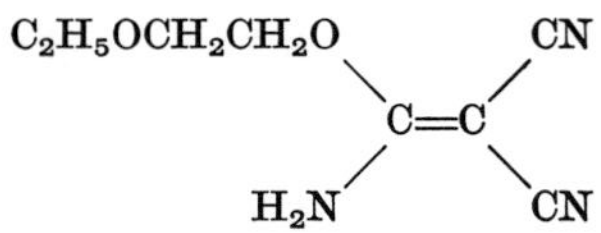	H_2NNH_2	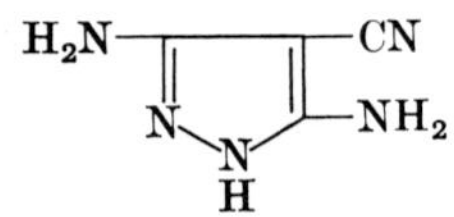	60	105
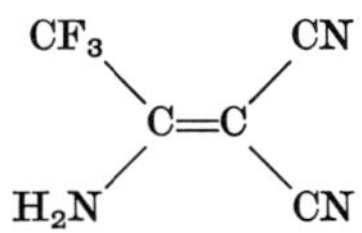	H_2NNH_2	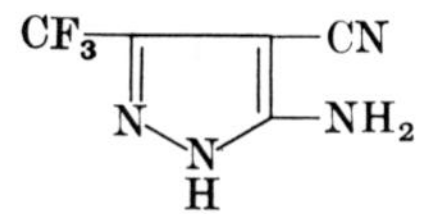	37	349

(continued)

TABLE V (*continued*)

Reactants				
Malononitrile derivative	Hydrazine derivative	Product	Yield, %	Ref.
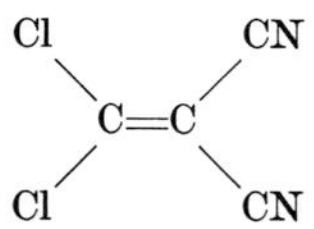	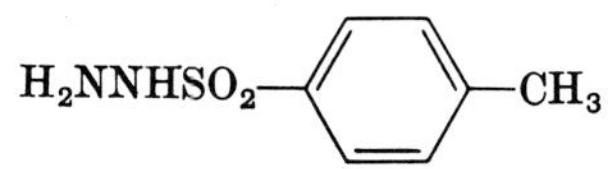	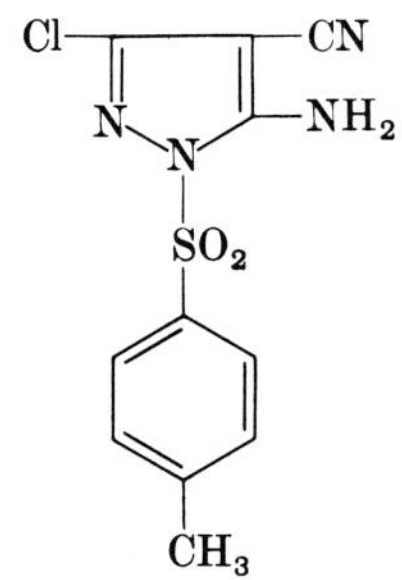	90 —	232 406
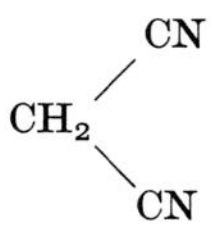	H_2NNH_2	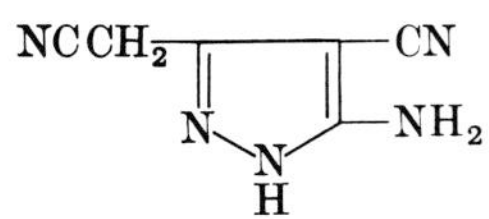	40	26
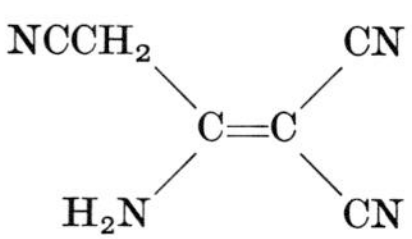	H_2NNH_2		71.5	26

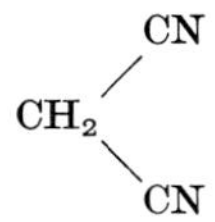	H_2NNHCH_3	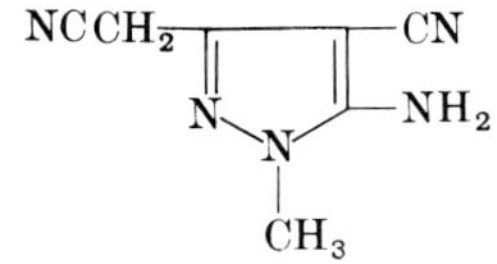	30	25
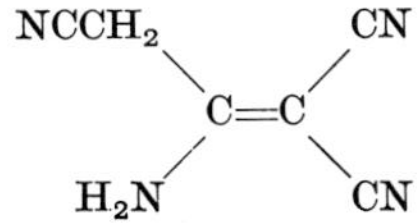	H_2NNHCH_3		59	25
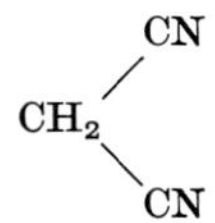	$H_2NNHC_6H_5$	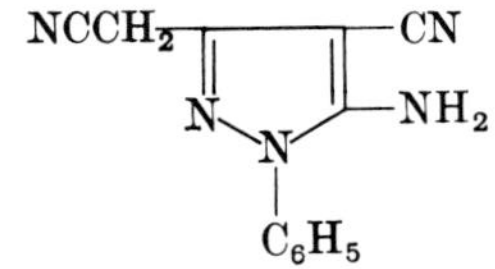	38	25
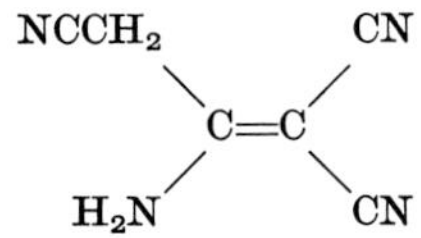	$H_2NNHC_6H_5$		58 48	25 84
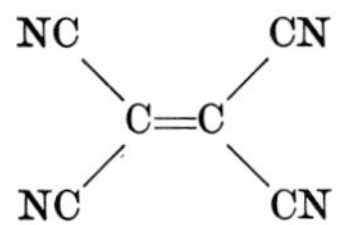	H_2NNHCH_3	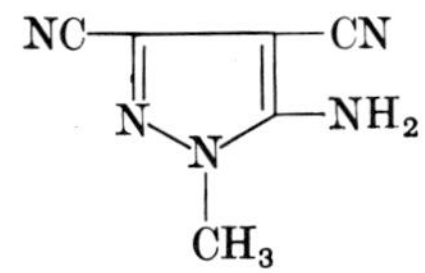	25	232
	$H_2NNHC_6H_5$	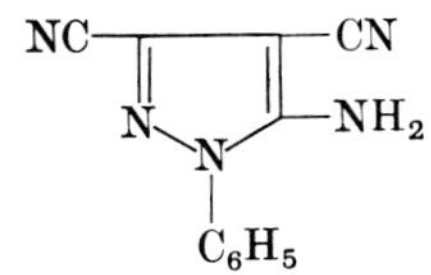	58	232

(continued)

TABLE V (*continued*)

Reactants				
Malononitrile derivative	Hydrazine derivative	Product	Yield, %	Ref.
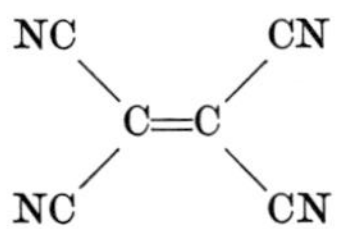 (*cont.*)	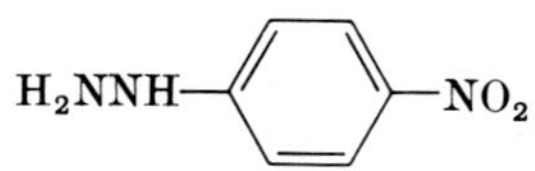	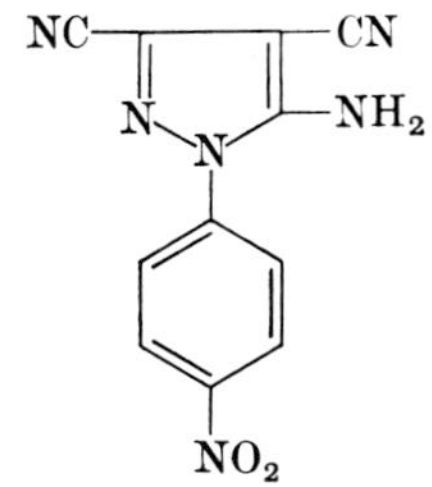	82	232
	$H_2NNHCONH_2$	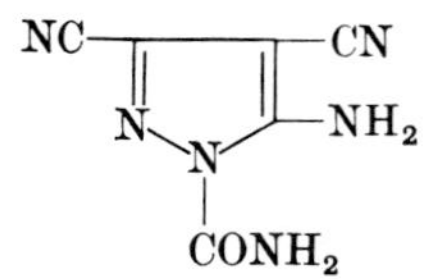	35	232
	$H_2NNHCSNH_2$	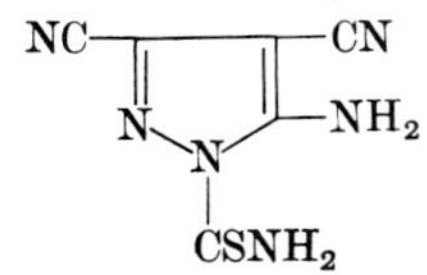	35	232
	$H_2NNHCON(CH_3)_2$	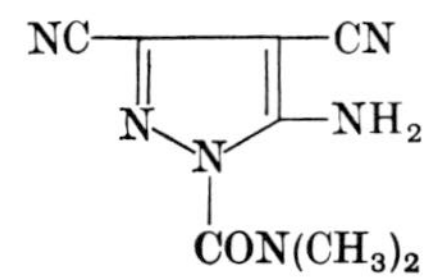	80	232

$H_2NNHCOCH_3$	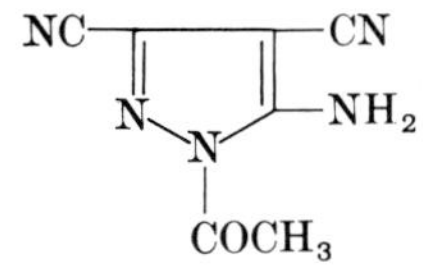	68	232
$H_2NNHCOCH_2Br$	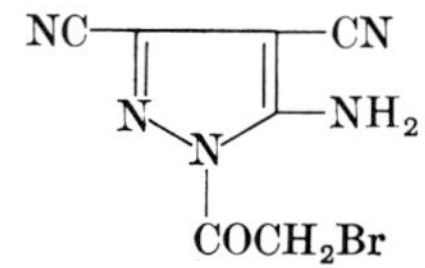	35	232
$H_2NNHCOC_6H_5$	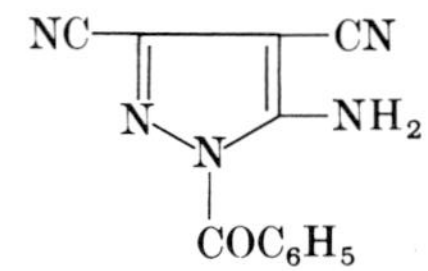	68	232
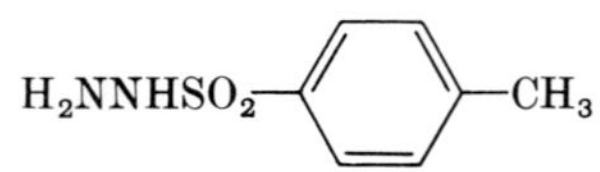	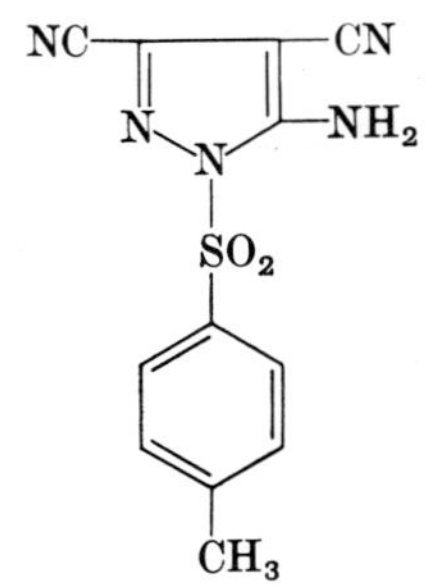	87	232

(continued)

TABLE V (*continued*)

Reactants				
Malononitrile derivative	Hydrazine derivative	Product	Yield, %	Ref.
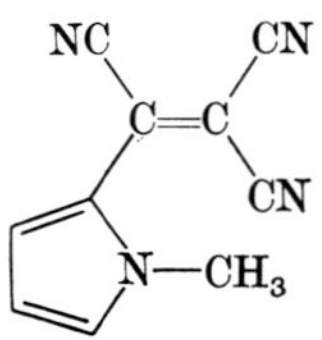	H_2NNH_2	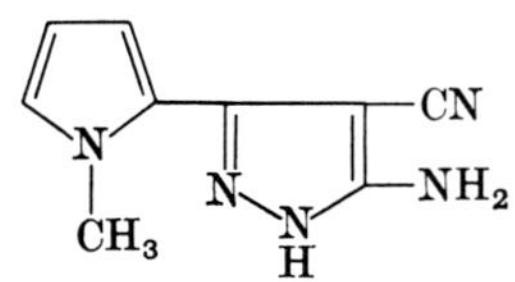	46	232
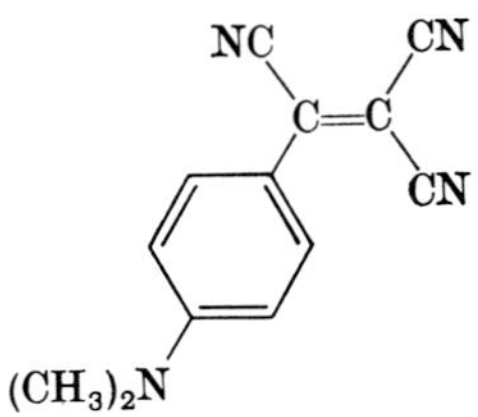	H_2NNH_2	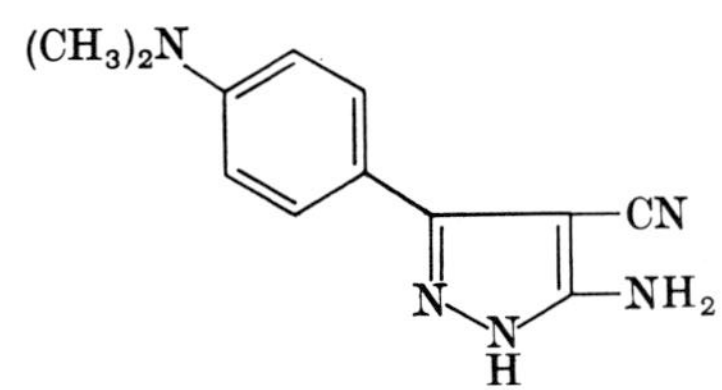	67	232
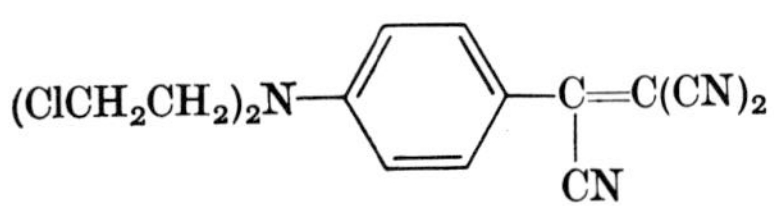	H_2NNH_2	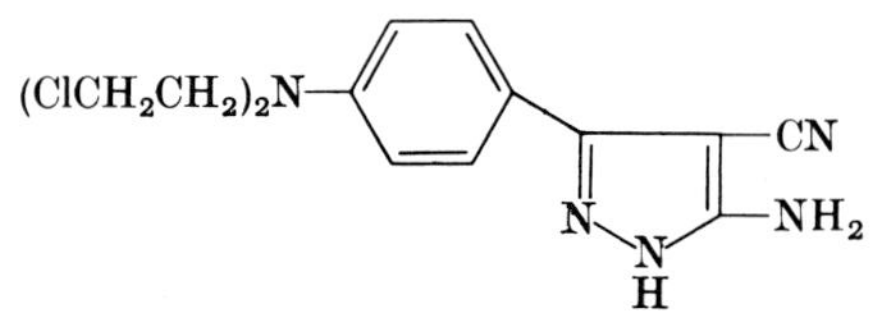	85	477

	H_2NNHCH_3	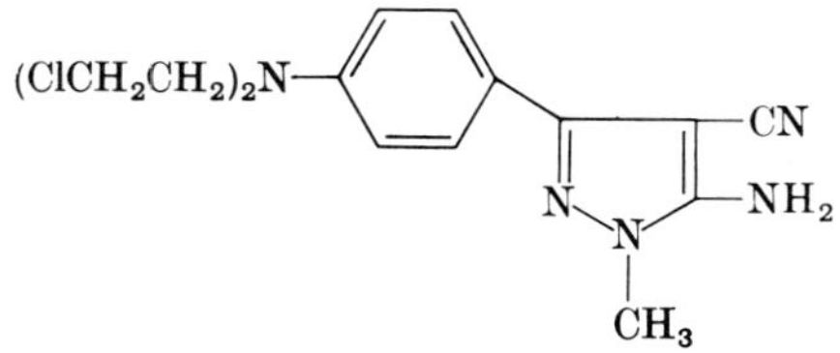	95	477
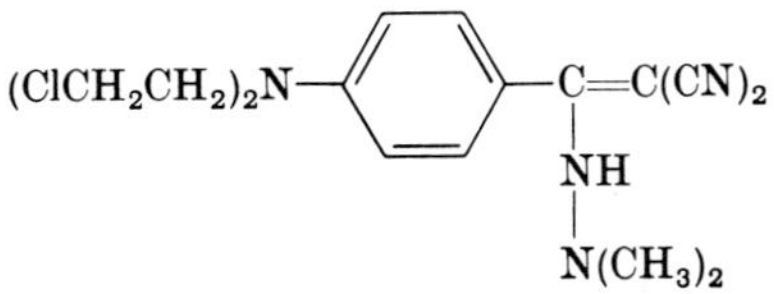	(HCl)	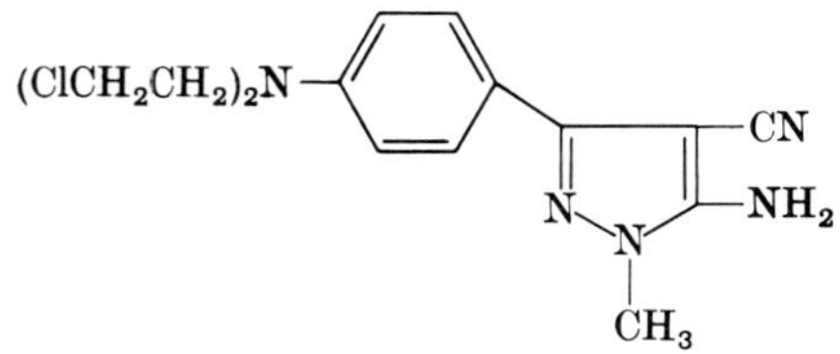	90	477
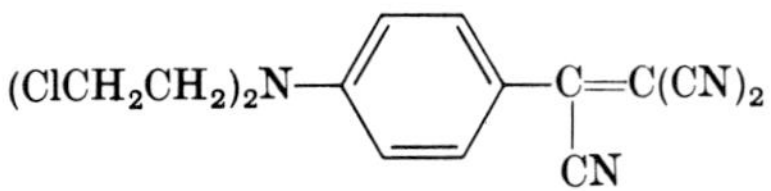	$H_2NNHC(=NH)NH_2$	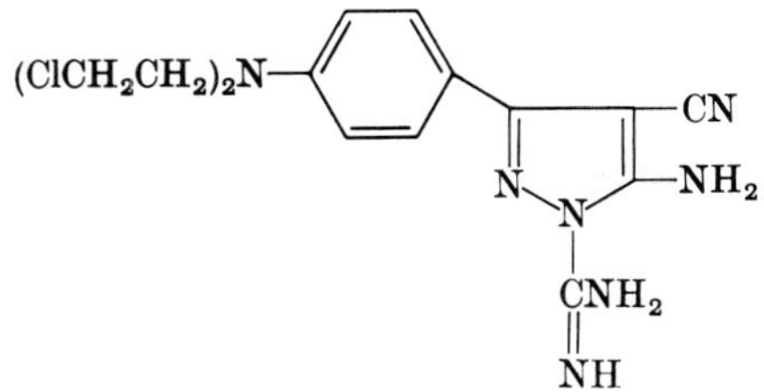	60	477
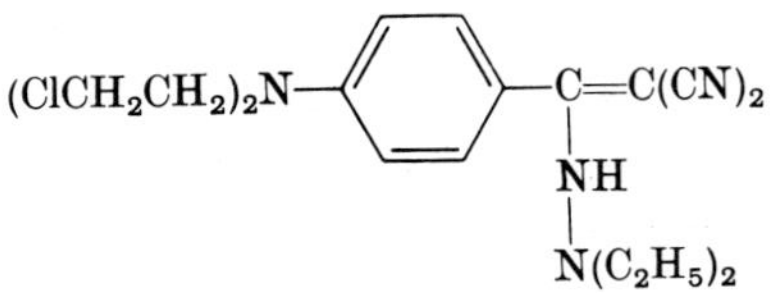	(HCl)	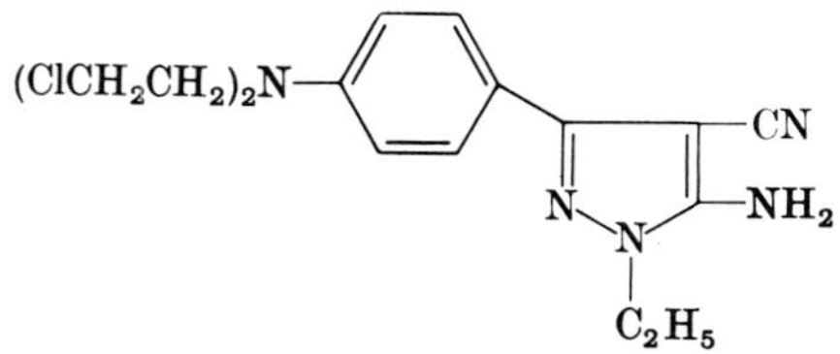	75	477

(*continued*)

TABLE V (*continued*)

Reactants				
Malononitrile derivative	Hydrazine derivative	Product	Yield, %	Ref.
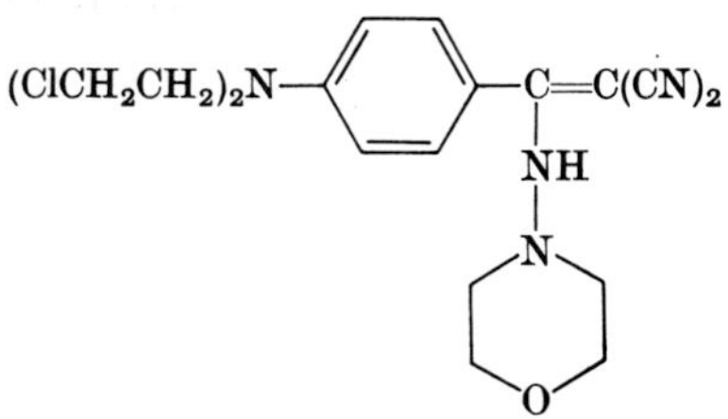	(HCl)	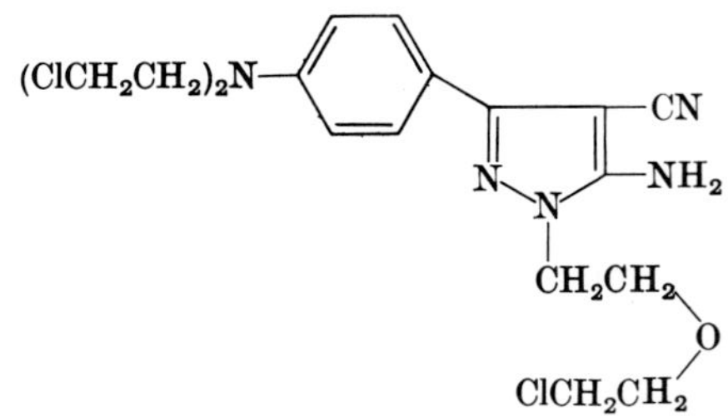	85	477
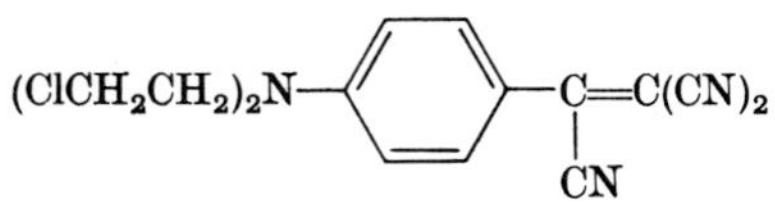	$H_2NNHC_6H_5$	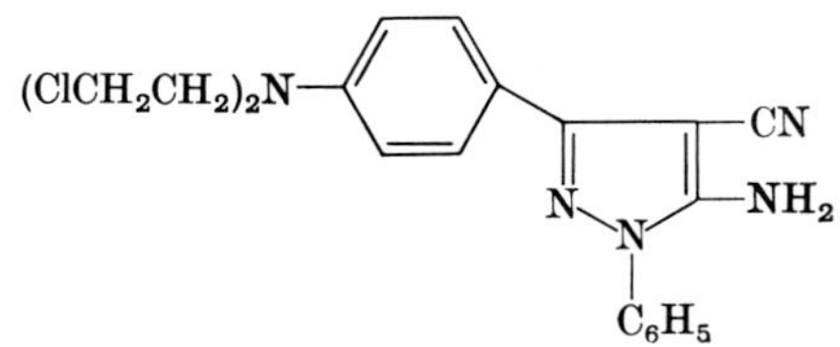	75	477
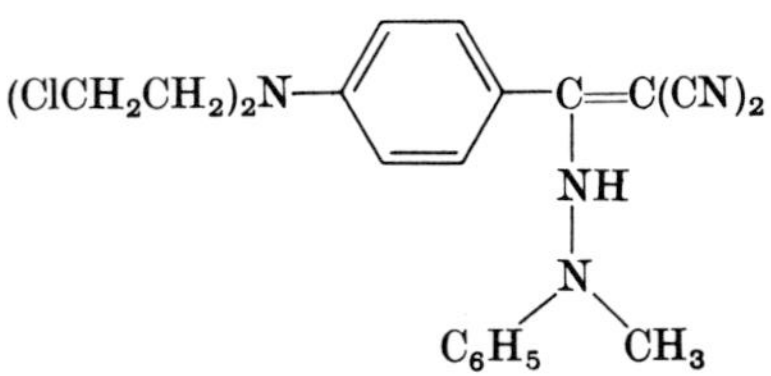	(HCl)	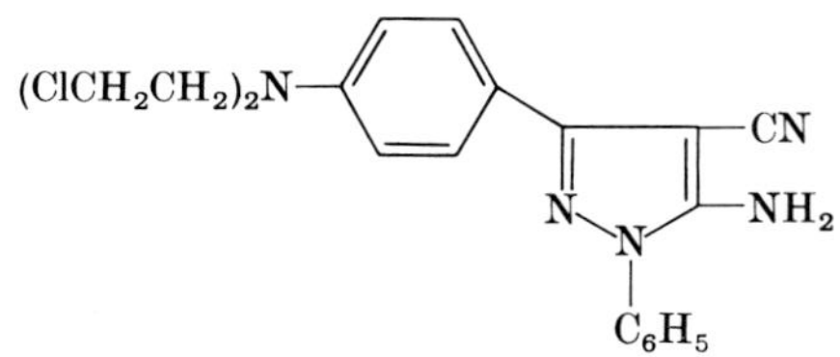	80	477

$(ClCH_2CH_2)_2N{-}C(CN){=}C(CN)_2$	H_2NNH_2	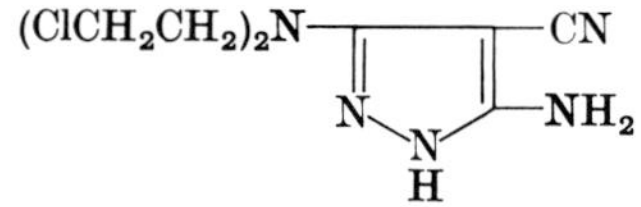	85	543
	H_2NNH_2	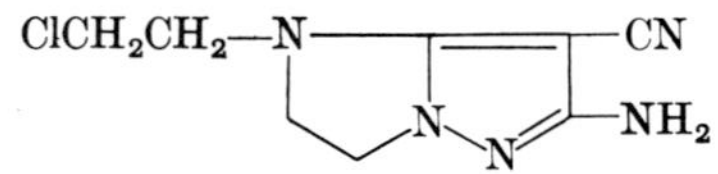	30	543
$(CH_3O)(C_2H_5OC(=O))C{=}C(CN)_2$	H_2NNH_2	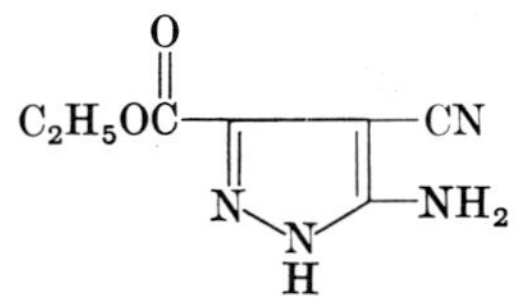	58	457 568
	$H_2NNHC_6H_5$	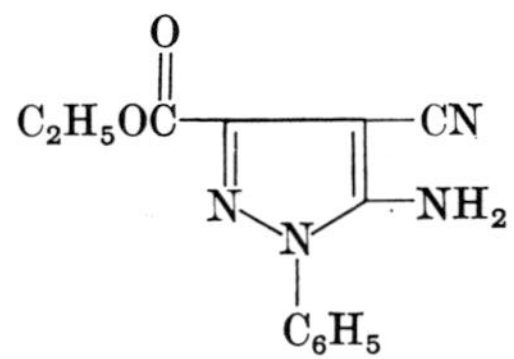	66	457
$CH_2(CN)_2$	$C_6H_5C(Cl){=}NNHC_6H_5$	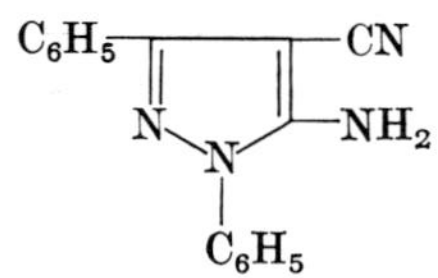	90	85

(continued)

TABLE V (*continued*)

Reactants				
Malononitrile derivative	Hydrazine derivative	Product	Yield, %	Ref.
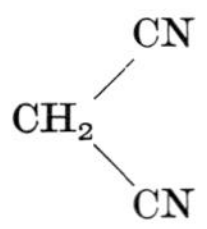 (*cont.*)	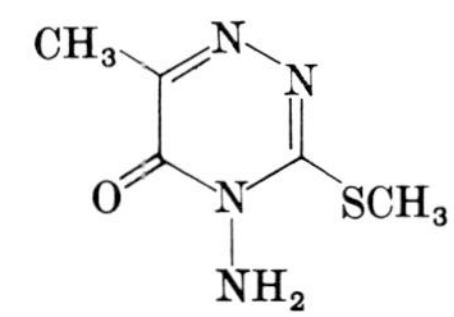	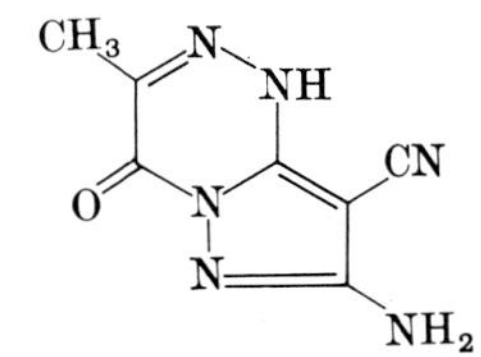	63	507
	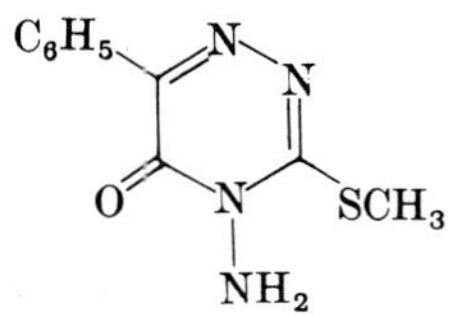	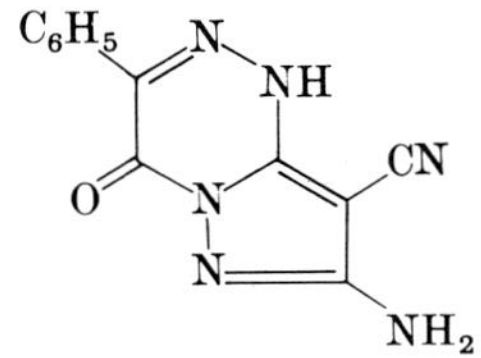	37	507

[a] This is a two-step reaction. Initial condensation gives

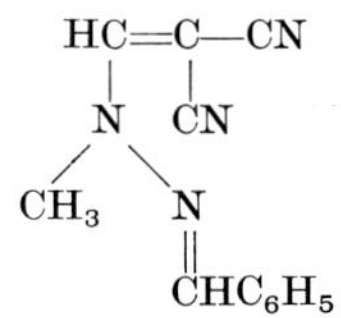

in 81% yield, and this is then cyclized in acid to the aminonitrile.

synthesis is the reaction of 1,1,2-tricyanostyrenes with hydrazines (232, 447).

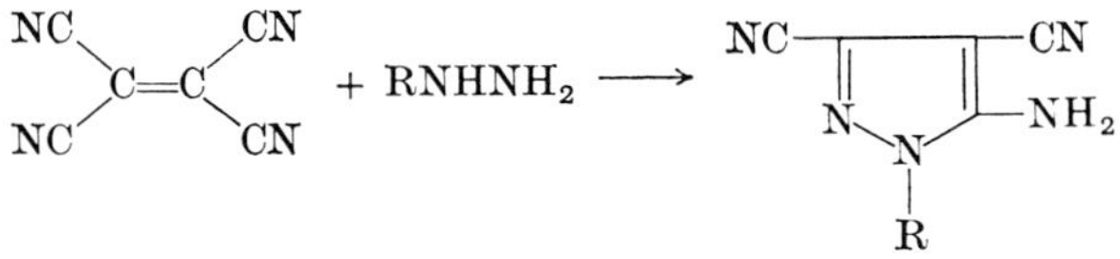

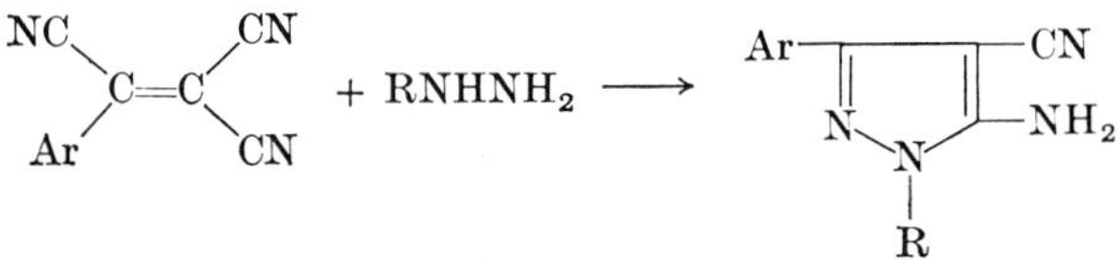

The propensity for pyrazole formation from the intermediate 1,1-dicyano-2-hydrazinostyrenes is strikingly demonstrated by the formation of **10** in 90% yield by treatment of **8** (formed from *N,N*-dimethylhydrazine and *p*-[*N,N*-bis(β-chloroethyl)amino]tricyanostyrene (**9**)) with hydrochloric acid. The methyl group is lost as methyl chloride. The product **10** is identical with the pyrazole formed from **9** and methylhydrazine (477) (Scheme 7).

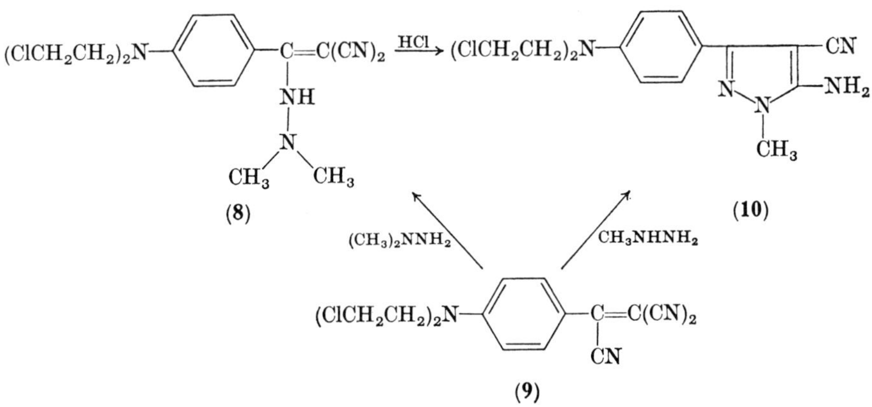

Scheme 7

B. SYNTHESIS OF PYRIMIDINE o-AMINONITRILES (Table VI)

The base-catalyzed condensation of guanidine, urea, and thiourea with malononitrile to give 4,6-diaminopyrimidines (**1**) is well documented and much exploited (215), and it was thus to be anticipated that the apparently closely related condensation of amidines with malononitrile would lead in analogous fashion to 4,6-diaminopyrimidines. The condensation of formamidine with malononitrile does not, however, give 4,6-diaminopyrimidine (**2**, R = H) as anticipated, but instead leads in modest yield to

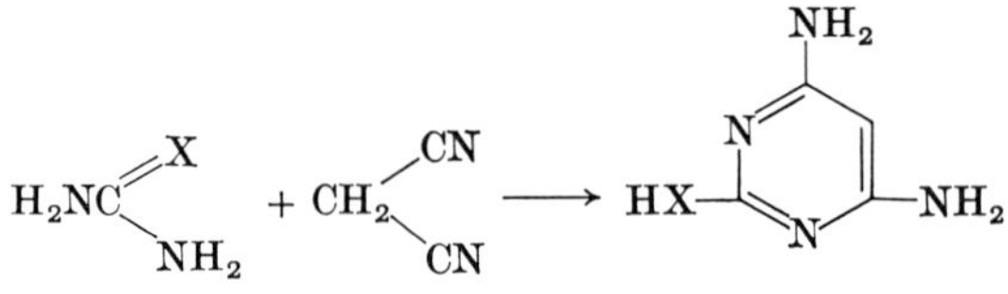

(X = O, S, NH) (**1**)

4-amino-5-cyanopyrimidine (**5**, R = H) (44). Acetamidine and benzamidine react similarly with malononitrile to give 2,4-dimethyl- and 2,4-diphenyl-5-cyano-6-aminopyrimidine (**5**, R = CH_3, C_6H_5) (36,367). The anticipated formation of a 4,6-diaminopyrimidine in this reaction is thwarted by an initial condensation between the active methylene group of malononitrile and the amidine to give an aminomethylenemalononitrile (**3**) which can be isolated. A further reaction of this intermediate with a second mole of amidine leads by amidine exchange to the intermediate **4**, which then undergoes spontaneous intramolecular cyclization (Scheme 8). Pyrimidines arising from this route must of necessity carry

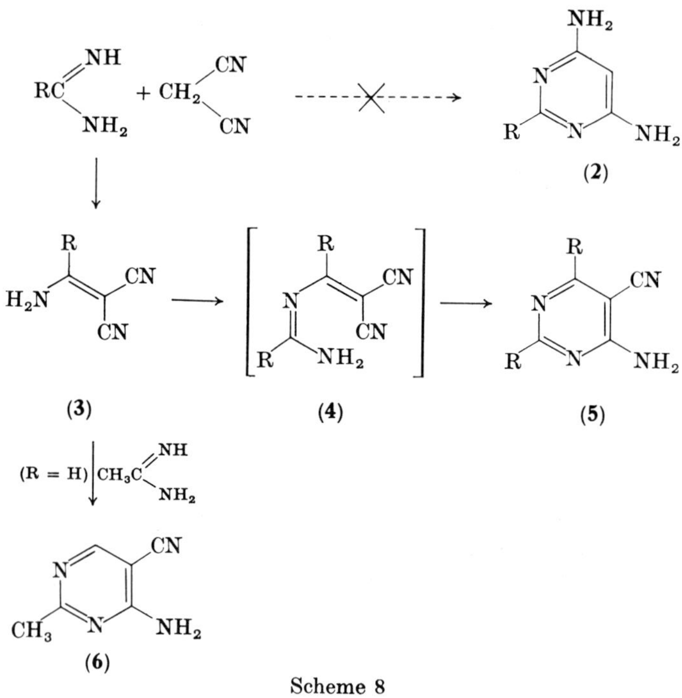

Scheme 8

the same substituent (alkyl or aryl) in positions 2 and 4, but the scheme is readily adaptable to the preparation of 2-substituted-4-amino-5-cyano-pyrimidines by utilization of aminomethylenemalononitrile (**3**, R = H) as a preformed intermediate, and its subsequent condensation with a second amidine. Thus, 2-methyl-4-amino-5-cyanopyrimidine (**6**), a key intermediate in the preparation of vitamin B_1, may be prepared by condensation of **3** (R = H) with acetamidine (prepared *in situ* from ammonia and triethyl orthoacetate (50) or acetonitrile ethyl imino ether (49)), with acetonitrile ethyl imino thioether (50), or with thioacetamidine (389).

An alternate, and considerably more convenient, route to the latter pyrimidine derivatives consists of the condensation of amidines, as well as ureas, thioureas, and guanidines, with ethoxymethylenemalononitrile. This reaction is thus directly analogous to the pyrazole syntheses discussed above. Similarly, the use of an aryl- or alkyl-ethoxymethylene-malononitrile leads to derivatives carrying an aryl or alkyl group at position 4 (e.g., **7**).

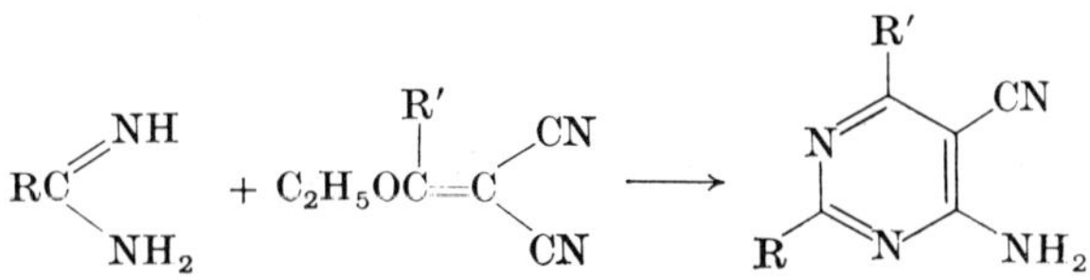

(**7**)

4-Amino-5-cyanopyrimidines carrying methyl groups in positions 2 and/or 4(6) have been prepared (467) in good yield by irradiation of the desmethyl compound in methanol containing 2% hydrogen chloride (w/w). For example, 4-amino-5-cyanopyrimidine (**8**) itself (readily available from formamidine and malononitrile or ethoxymethylenemalononitrile, as described above) is converted in 6 hr to 2,6-dimethyl-4-amino-5-cyanopyrimidine (**9**) in 60% yield. The same final product is formed in 86% yield by irradiation of 2-methyl-4-amino-5-cyanopyrimidine (**6**) under the same conditions. Similarly, 2,4-diamino-5-cyanopyrimidine is methylated to 2,4-diamino-5-cyano-6-methylpyrimidine in 57% yield. This extraordinary methylation procedure is not specific for pyrimidine aminonitriles (which appear to have been the pyrimidine substrates most readily accessible to the investigators) and may have general utility for methylations of other systems.

An earlier report (105) that the reaction of 1-amino-1-chloro-2,2-dicyanoethylene (**10**) with benzamidine gives 2-phenyl-4,6-diamino-5-cyanopyrimidine (**11**) has been shown to be incorrect; the product actually formed is benzamidinium tricyanomethanide (**12**) which, however, may be converted to **11** in poor (21%) yield upon prolonged heating (105,122).

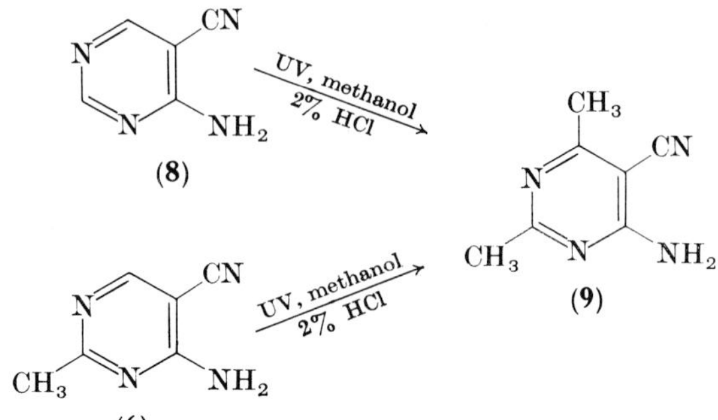

A preferable synthesis of the pyrimidine **11** is found in the condensation of 1,1-diamino-2,2-dicyanoethylene (**13**) with benzonitrile in the presence of alkali; an analogous condensation with acetonitrile gives the 2-methyl derivative **14** in moderate yield (366) (Scheme 9).

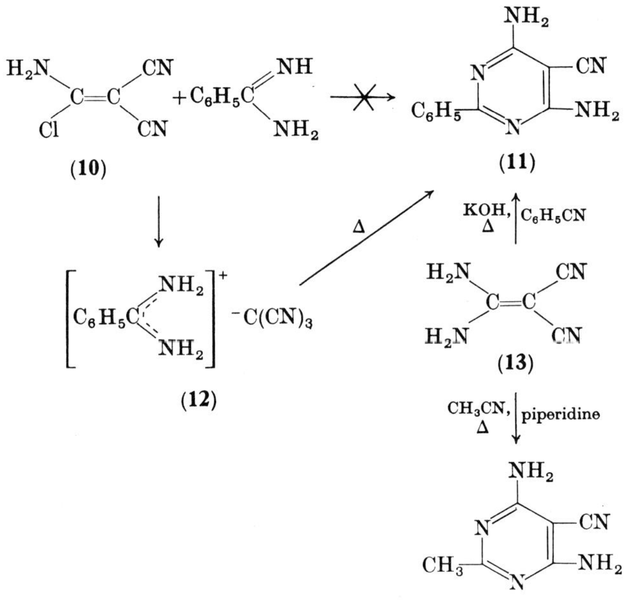

Scheme 9

Many 3-substituted 4-amino-5-cyano-2(3*H*)pyrimidinones (3-substituted 5-cyanocytosines) (**16**) have been prepared by modifications of the above general methods involving malononitrile or ethoxymethylenemalononitrile. Thus, the reaction of ethyl orthoformate, malononitrile, and *N*-alkyl- or *N*-cycloalkyl-ureas gives β-(3-alkylureido)-α-cyano-acrylonitriles (**15**) which either on heating or upon treatment with basic catalysts cyclize to **16**. The corresponding β-(3-arylureido)-α-cyano-acrylonitriles were prepared by condensation of *N*-arylureas with ethyl orthoformate, malononitrile, and excess acetic anhydride; this reaction mixture upon heating gave 3-aryl-5-cyanocytosines directly (43).

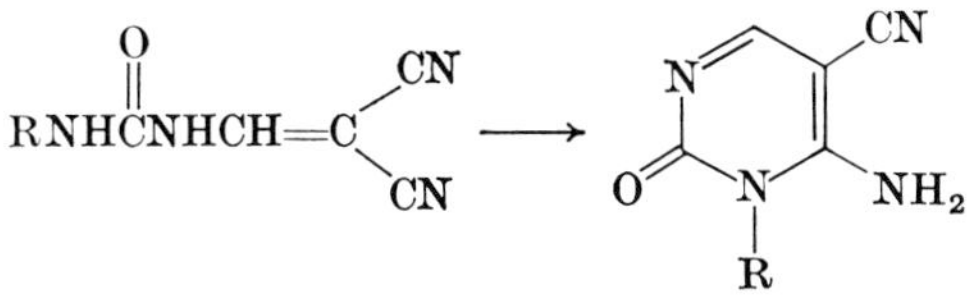

(**15**) (**16**)

A route to 2,4-diaryloxy-5-cyano-6-aminopyrimidines (**18**) has been described (72) in which malononitrile is treated with 1 mole of an aryl cyanate to give **17** which is then treated with a second mole of an aryl cyanate in the presence of triethylamine to give the pyrimidine **18**. With two different aryl cyanates, two isomeric pyrimidine aminonitriles may be obtained depending upon the order of use of the aryl cyanates in the above sequence of condensation reactions.

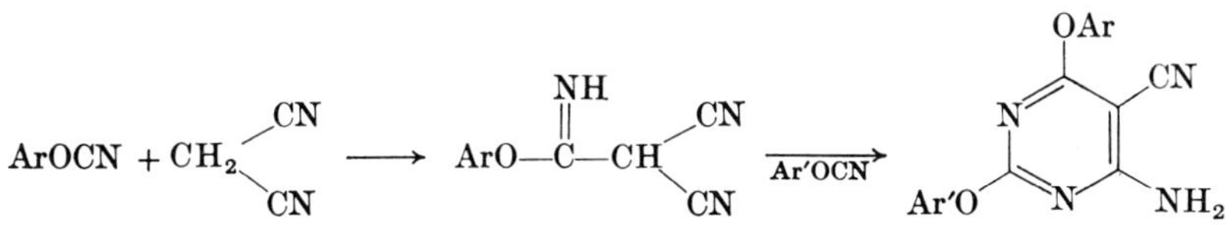

(**17**) (**18**)

The reaction of 1,1,2-tricyanostyrenes with hydrazines to give pyrazoles has been described on p. 103. 4-Amino-5-cyanopyrimidines are readily prepared from the same intermediates by reaction with amidines. For example, the tricyanostyrene **19** condenses with guanidine to give the pyrimidine *o*-aminonitrile **20** in 65% yield (477).

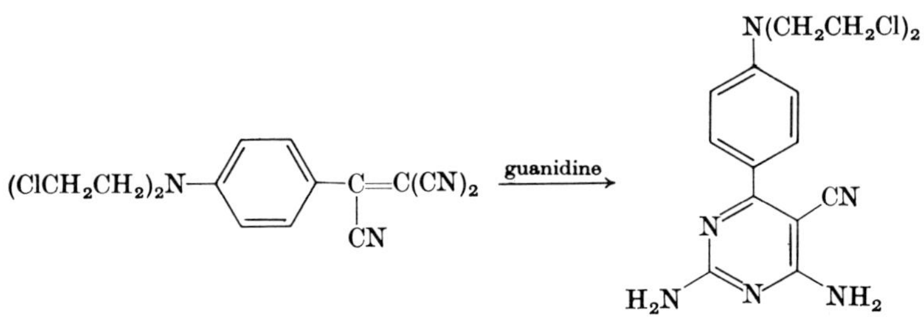

(**19**) (**20**)

TABLE VI
Pyrimidine *o*-Aminonitriles

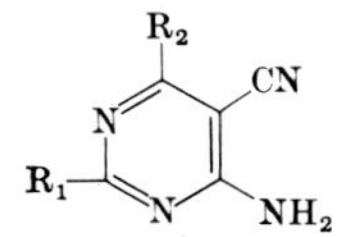

Reactants: Malononitrile derivative	Reactants: Other component	Product: R_1	Product: R_2	Yield, %	Ref.
$CH_2(CN)_2$	$HC(=NH)NH_2$	H	H	45	44
	$CH_3C(=NH)NH_2$	CH_3	CH_3	50 90	36 367
	$CF_3C(=NH)NH_2$	CF_3	CF_3	17	345
	$C_6H_5C(=NH)NH_2$	C_6H_5	C_6H_5	45 26	36 122
	CCl_3CH_2OCN	CCl_3CH_2O	CCl_3CH_2O	86	72
	CCl_3CH_2OCN followed by *p*-$CH_3OC_6H_4OCN$	CCl_3CH_2O	*p*-$CH_3OC_6H_4O$	45	72
	p-ClC_6H_4OCN followed by C_6H_5OCN	C_6H_5O	*p*-ClC_6H_4O	78	72
	p-$CH_3COC_6H_4OCN$ followed by C_6H_5OCN	C_6H_5O	*p*-$CH_3COC_6H_4O$	87	72

	C_6H_5OCN followed by p-$CH_3COC_6H_4OCN$	p-$CH_3COC_6H_4O$	C_6H_5O	52	72
	C_6H_5OCN followed by o-$CH_3OOCC_6H_4OCN$	o-$CH_3OOCC_6H_4O$	C_6H_5O	75	72
	2,4-di$CH_3C_6H_3OCN$	2,4-di$CH_3C_6H_3O$	2,4-di$CH_3C_6H_3O$	31	72
	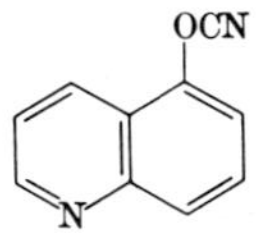 followed by C_6H_5OCN	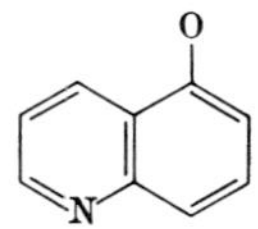	C_6H_5O	68	72
	C_6H_5OCN followed by 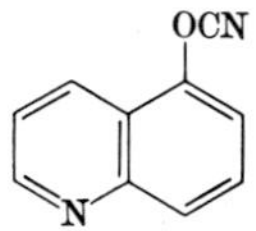	C_6H_5O	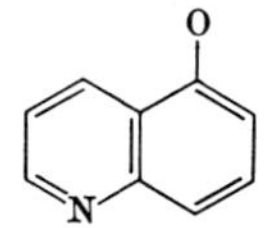	85	72
	$NH_3(g)$	$NCCH_2$	$NCCH_2$ (?)	—	342
	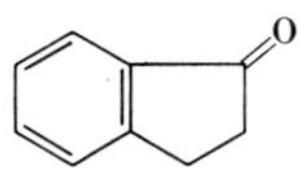	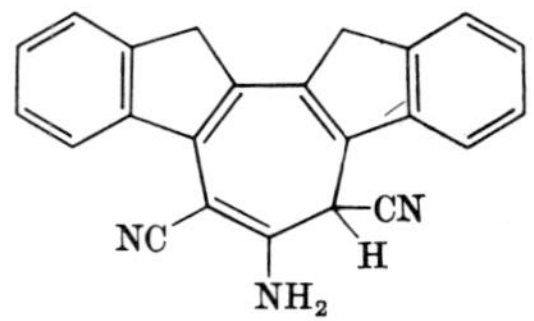	(?)	—	342
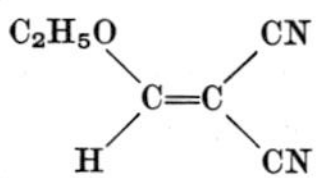	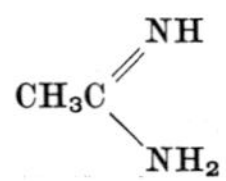	CH_3	H	— — 71 — —	384 387 161 203 42

(continued)

TABLE VI (*continued*)

Reactants		Product		Yield, %	Ref.
Malononitrile derivative	Other component				
		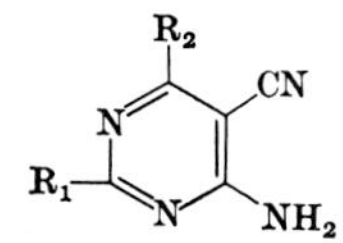			
		R_1	R_2		
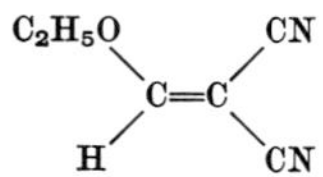 (*cont.*)	$CF_3C(=NH)NH_2$	CF_3	H	72	346
	$CH_3C(=NH)OC_2H_5$ + NH_3	CH_3	H	68 65	49 380
	$CH_3CH_2C(=NH)NH_2$	C_2H_5	H	80	75
	$CF_3CF_2C(=NH)NH_2$	CF_3CF_2	H	69	347
	$HOCH_2CH_2C(=NH)NH_2$	$HOCH_2CH_2$	H	—	51

$n\text{-}C_3H_7C(=NH)NH_2$	$n\text{-}C_3H_7$	H	55 — —	163 415 409
$(CH_3)_2CHC(=NH)NH_2$	$(CH_3)_2CH$	H	—	415
$n\text{-}C_4H_9C(=NH)NH_2$	$n\text{-}C_4H_9$	H	33 —	383 415
$n\text{-}C_5H_{11}C(=NH)NH_2$	$n\text{-}C_5H_{11}$	H	—	415
$CF_3CF_2CF_2C(=NH)NH_2$	$CF_3CF_2CF_2$	H	62	347
$C_6H_5C(=NH)NH_2$	C_6H_5	H	35	75
$H_2NC(=NH)NH_2$	NH_2	H	13 54	75 2
$CH_3NHC(=NH)NH_2$	CH_3NH	H	56	163

(continued)

TABLE VI (*continued*)

Reactants		Product			
Malononitrile derivative	Other component	Product 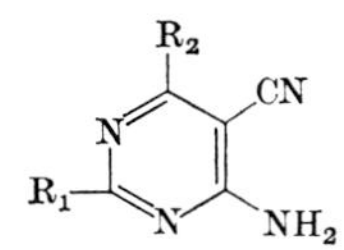		Yield, %	Ref.
		R_1	R_2		
$C_2H_5O(H)C{=}C(CN)_2$ (*cont.*)	$(CH_3)_2NC(=NH)NH_2$	$(CH_3)_2N$	H	51	163
				—	143
				—	414
	$CH_2{=}CHCH_2NHC(=NH)NH_2$	$CH_2{=}CHCH_2NH$	H	—	143
	$(C_2H_5)_2NC(=NH)NH_2$	$(C_2H_5)_2N$	H	—	414
	n-$C_4H_9NHC(=NH)NH_2$	n-C_4H_9NH	H	—	414
	$(CH_3)_2NCH_2CH_2NHC(=NH)NH_2$	$(CH_3)_2NCH_2CH_2NH$	H	—	414

$(C_2H_5)_2NCH_2CH_2NHC(=NH)NH_2$	$(C_2H_5)_2NCH_2CH_2NH$	H	—	414
$(C_2H_5)_2NCH_2CH_2CH_2NHC(=NH)NH_2$	$(C_2H_5)_2NCH_2CH_2CH_2NH$	H	—	414
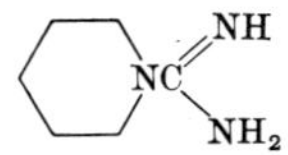		H	25	22
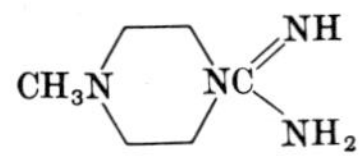	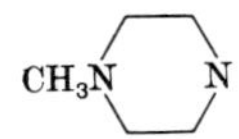	H	49	22
$H_2NC(=O)NH_2$	HO	H	—	53
$H_2NC(=NH)OCH_3$	CH_3O	H	36 40	53 22
$H_2NC(=S)NH_2$	HS	H	67 41 12	53 161 246
$H_2NC(=NH)SCH_3$	CH_3S	H	— — 65	147 404 246

(*continued*)

TABLE VI (*continued*)

Reactants		Product		Yield, %	Ref.
Malononitrile derivative	Other component				
		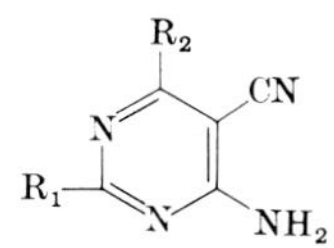			
		R_1	R_2		
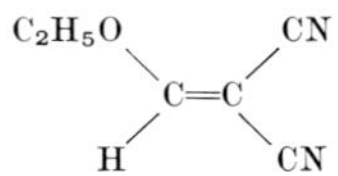 (*cont.*)	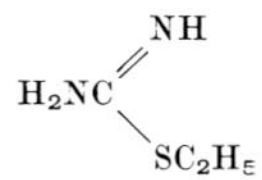	C_2H_5S	H	70	246
				—	161
				16.5	75
				56	53
	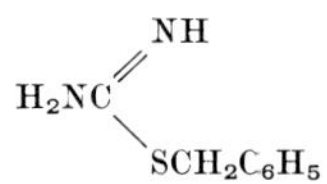	$C_6H_5CH_2S$	H	72	246
				85	53
		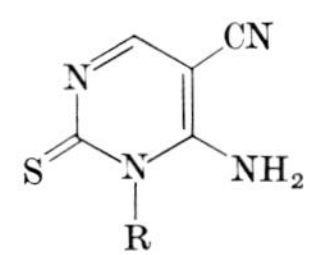			
		R			
	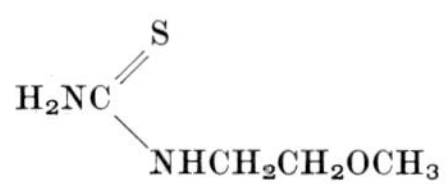	$CH_3OCH_2CH_2$		60	35

$H_2NC(=S)NHCH(CH_3)_2$	$(CH_3)_2CH$	90	35
$H_2NC(=S)NHC_4H_9$-*n*	*n*-C_4H_9	95	35
$H_2NC(=S)NHC_8H_{17}$-*n*	*n*-C_8H_{17}	95	35
$H_2NC(=S)NHC_6H_5$	C_6H_5	73	53
$H_2NC(=O)NHC_5H_{11}$-*n*	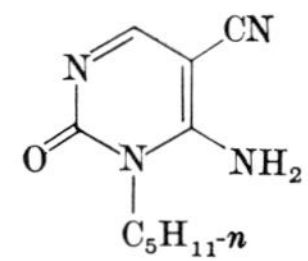	70	43
$H_2NC(=O)NHC_7H_{15}$-*n*	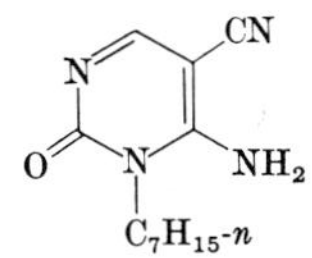	95	43

(continued)

TABLE VI (*continued*)

Reactants		Product			
Malononitrile derivative	Other component	Product 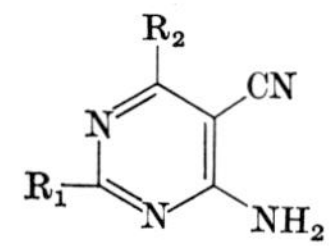 R_1	R_2	Yield, %	Ref.
$C_2H_5O(CH_3)C{=}C(CN)_2$	$HC(=NH)NH_2$	H	CH_3	63	22
	$CH_3C(=NH)NH_2$	CH_3	CH_3	58 76	75 122
	$H_2NC(=S)NH_2$ followed by CH_3I	CH_3S	CH_3	53	22
	$H_2NC(=NH)SCH_3$	CH_3S	CH_3	—	416
	$H_2NC(=NH)SC_2H_5$	C_2H_5S	CH_3	39	52
	$H_2NC(=NH)SCH_2C_6H_5$	$C_6H_5CH_2S$	CH_3	—	53
	$H_2NC(=NH)NH_2$	NH_2	CH_3	51	122

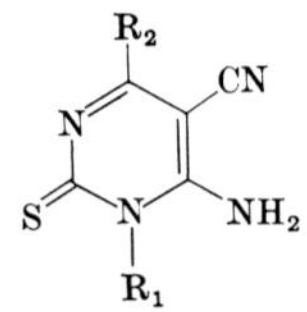

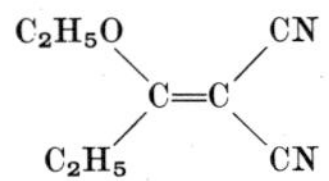	$H_2NC(=S)NHCH_2CH_2OCH_3$	R_1: $CH_3OCH_2CH_2$	R_2: CH_3	65	35
	$H_2NC(=S)NHC_4H_9$-*n*	*n*-C_4H_9	CH_3	54	35
	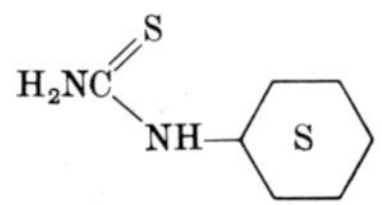		CH_3	85	35
	$H_2NC(=S)NHCH_2CH_2OCH_3$	$CH_3OCH_2CH_2$	C_2H_5	75	35
	$H_2NC(=S)NHCH(CH_3)_2$	$(CH_3)_2CH$	C_2H_5	79	35
	$H_2NC(=S)NHC_8H_{17}$-*n*	*n*-C_8H_{17}	C_2H_5	60	35

(continued)

TABLE VI (*continued*)

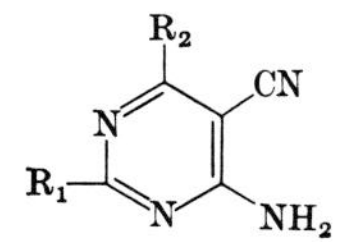

Reactants: Malononitrile derivative	Reactants: Other component	Product: R_1	Product: R_2	Yield, %	Ref.
$(C_2H_5O)(C_2H_5)C{=}C(CN)_2$	$H_2NC({=}NH)NH_2$	NH_2	C_2H_5	—	122
$(CH_3O)(C_6H_5)C{=}C(CN)_2$	$HC({=}NH)NH_2$	H	C_6H_5	51	122
	$CH_3C({=}NH)NH_2$	CH_3	C_6H_5	85	33
	$H_2NC({=}NH)NH_2$	NH_2	C_6H_5	85 80–85	33 122
	$H_2NC({=}S)NH_2$ followed by CH_3I	CH_3S	C_6H_5	—	141
	$H_2NC({=}NH)SC_2H_5$	C_2H_5S	C_6H_5	50	52

	$C_6H_5C(=NH)NH_2$	C_6H_5	C_6H_5	80	122
$CH_3(p\text{-}ClC_6H_4)C{=}C(CN)_2$	$CH_3C(=NH)NH_2$	CH_3	p-ClC_6H_4	16	33
	$H_2NC(=NH)NH_2$	NH_2	p-ClC_6H_4	16	33
$RNHC(=O)NHCH{=}C(CN)_2$	Base	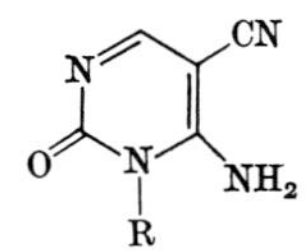			
R		R			
CH_3		CH_3		56	43
C_2H_5		C_2H_5		84	43
$CH_3OCH_2CH_2$		$CH_3OCH_2CH_2$		96	43
$CH_2{=}CHCH_2$		$CH_2{=}CHCH_2$		—	43
$CH_3CH_2CH_2$		$CH_3CH_2CH_2$		81	43
$(CH_3)_2CH$		$(CH_3)_2CH$		94	43
$CH_3CH_2CH_2CH_2$		$CH_3CH_2CH_2CH_2$		—	43
t-C_4H_9		t-C_4H_9		57	43
n-C_5H_{11}		n-C_5H_{11}		95 71	43 361
$(CH_3)_2CHCH_2CH_2$		$(CH_3)_2CHCH_2CH_2$		92	43
n-C_6H_{13}		n-C_6H_{13}		94	43

(*continued*)

TABLE VI (*continued*)

Reactants: Malononitrile derivative	Reactants: Other component	Product	Yield, %	Ref.
$RNHC(=O)NHCH{=}C(CN)_2$ (*cont.*)	Base	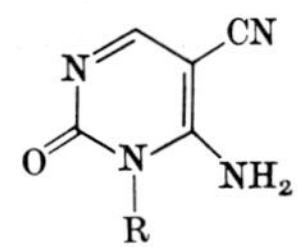		
R		R		
n-C_7H_{15}		n-C_7H_{15}	95	43
n-C_8H_{17}		n-C_8H_{17}	95	43
$C_6H_5CH_2$		$C_6H_5CH_2$	75	43
$C_6H_5CH(CH_3)$		$C_6H_5CH(CH_3)$	82	43
p-$CH_3C_6H_4$		p-$CH_3C_6H_4$	50	43
p-$CH_3OC_6H_4$		p-$CH_3OC_6H_4$	58	43
$CH_2(CN)_2$	$C_6H_5NHC(=O)N{=}CHNHC(=O)NHC_6H_5$	C_6H_5	61	43
	$H_2NC(=O)NH$-p-ClC_6H_4, $HC(OC_2H_5)_3$, $(CH_3CO)_2O$	p-ClC_6H_4	54	43
	$H_2NC(=O)NH$-2,6-$(CH_3)_2C_6H_3$, $HC(OC_2H_5)_3$, $(CH_3CO)_2O$	2,6-$(CH_3)_2C_6H_3$	31	43

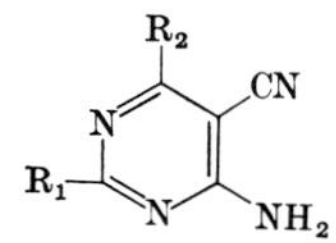

		R_1	R_2		
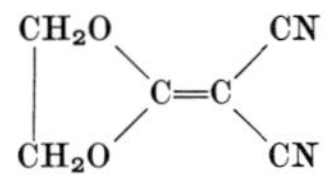	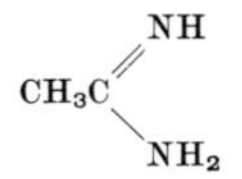	CH_3	$HOCH_2CH_2O$	53	105
	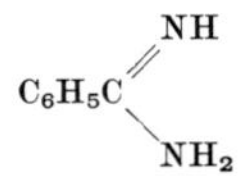	C_6H_5	$HOCH_2CH_2O$	63	105
	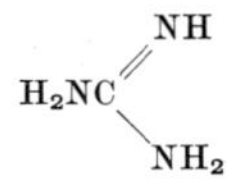	NH_2	$HOCH_2CH_2O$	70, 55	105
	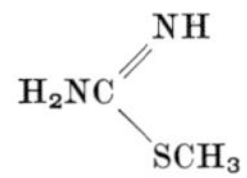	CH_3S	$HOCH_2CH_2O$	58	105
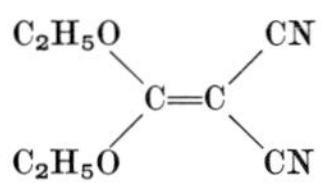	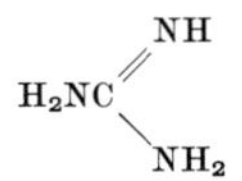	NH_2	C_2H_5O	61	105
	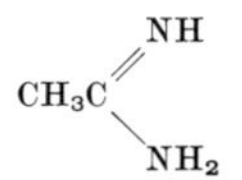	CH_3	C_2H_5O	82	164
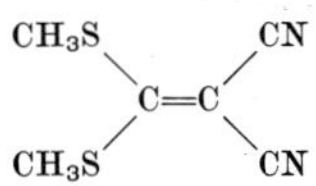	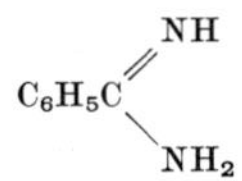	C_6H_5	CH_3S	87(?) 30	105 122

(continued)

TABLE VI (*continued*)

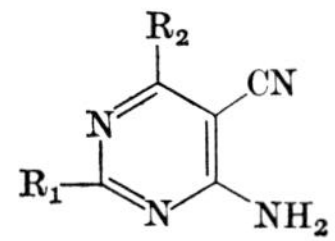

Reactants: Malononitrile derivative	Reactants: Other component	Product: R_1	Product: R_2	Yield, %	Ref.
$(CH_3S)_2C{=}C(CN)_2$ (*cont.*)	$H_2NC(=NH)SCH_3$	CH_3S	CH_3S	60	105
$H_2N(Cl)C{=}C(CN)_2$	$C_6H_5C(=NH)NH_2$	C_6H_5	NH_2	40	105
$H_2N(HOCH_2CH_2O)C{=}C(CN)_2$	$C_6H_5C(=NH)NH_2$	C_6H_5	$HOCH_2CH_2O$	70	105
	$H_2NC(=NH)NH_2$	NH_2	$HOCH_2CH_2O$	45	105
$C_2H_5O(CH_3C(Cl){=}N)C{=}C(CN)_2$	NH_3	CH_3	C_2H_5O	51	164

$H_2NCH{=}C(CN)_2$	$HC({=}NH)NH_2$	H	H	42	36
	$HC(OC_2H_5)_3$ and NH_3	H	H	—	50
	$CH_3C(OC_2H_5)_3$ and NH_3	CH_3	H	—	50
	$CH_3C({=}NH)OC_2H_5$	CH_3	H	"Good"	59
				—	376
				—	375
				—	378
	$CH_3C({=}S)NH_2$	CH_3	H	—	389
	$CH_3C({=}NH)SC_2H_5$	CH_3	H	—	50
	$C_6H_5CH_2C({=}NH)SC_2H_5$	$C_6H_5CH_2$	H	—	50
	$C_6H_5CH_2C({=}NH)OC_2H_5$	$C_6H_5CH_2$	H	—	378
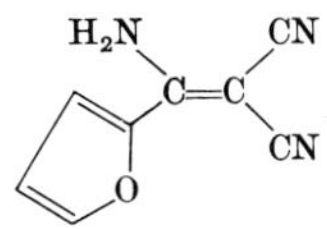	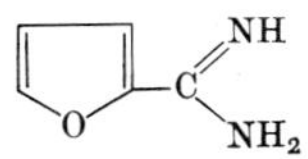	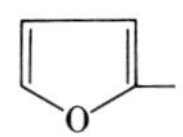	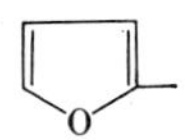	—	36

(*continued*)

TABLE VI (*continued*)

Reactants: Malononitrile derivative	Reactants: Other component	Product		Yield, %	Ref.
		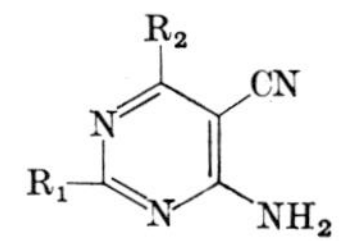			
		R_1	R_2		
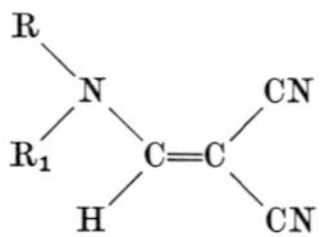 (R = R_1 = alkyl; R = alkyl, R_2 = C_6H_5)	$CH_3C(=NH)NH_2$	CH_3	H	90	411
$(CH_3)_2NCH_2C(OH)(CN)_2$	$HC(=NH)NH_2$	H	H	—	556
	$CH_3C(=NH)NH_2$	CH_3	H	84	556
	$H_2NC(=O)NH_2$	HO	H	—	556
	$H_2NC(=S)HN_2$	HS	H	—	556

	$H_2NC(=NH)NH_2$	H_2N	H	—	556
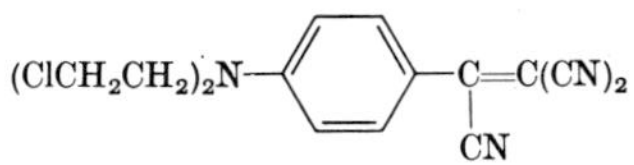	$H_2NC(=NH)NH_2$	NH_2	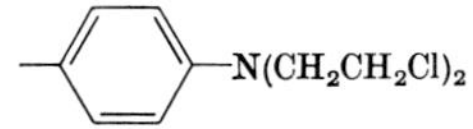	65	477
	$CH_3C(=NH)NH_2$	CH_3	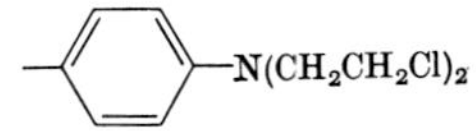	30	477
	$C_6H_5C(=NH)NH_2$	C_6H_5	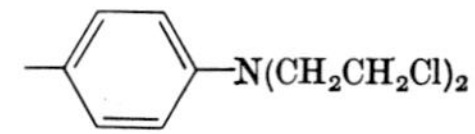	25	477
	$C_6H_5CH_2SC(=NH)NH_2$	$C_6H_5CH_2S$	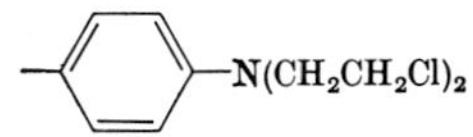	30	477
$NC—N—C(NH_2)—C(CN)_2$ $(-)$	(HCl)	Cl	NH_2	72	561
$(ClCH_2CH_2)_2NC(CN)=C(CN)_2$	$H_2NC(=NH)NH_2$	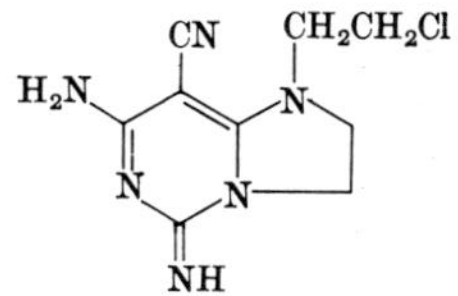		52	543
	$CH_3C(=NH)NH_2$	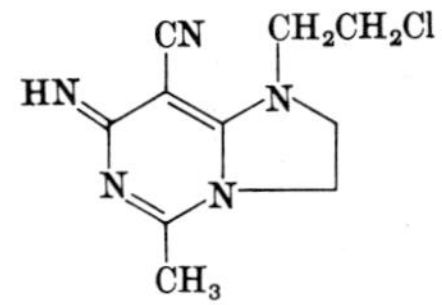		38	543

The reaction of malononitrile with ammonia in benzene solution has been reported, with skepticism, to give 2,4-bis(cyanomethyl)-5-cyano-6-aminopyrimidine (**21**), but this structural assignment is only tentative, and no evidence supporting it has been advanced (342). In the same

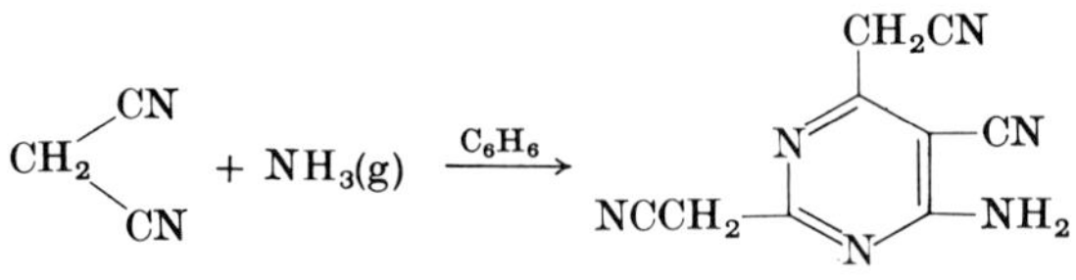

(?)

(21)

investigation, the reaction of 1-indanone and malononitrile in ethanol in the presence of potassium carbonate was reported to give **22**, but once again no support for the structure was presented, and it should be regarded as highly tentative.

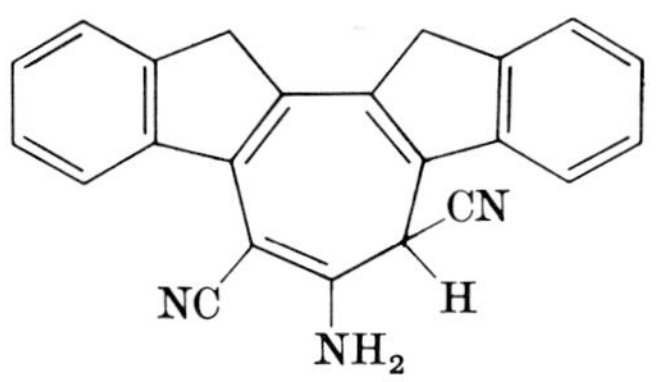

(?)

(22)

C. SYNTHESIS OF FURAN *o*-AMINONITRILES (Table VII)

An extrapolation of the *o*-aminonitrile synthesis outlined in principle on p. 80 to the construction of 2-amino-3-cyanofurans (**2**) would require the intermediacy of an alkylidenemalononitrile such as **1** in which the final ring closure would involve the addition of an OH group α to the alkylidene double bond across one of the two nitriles. As will be seen in the succeeding sections, analogous intermediates in which the α-substituent (OH in structure **1**) is replaced by NH_2 and SH lead to pyrrole and thiophene *o*-aminonitriles, respectively.

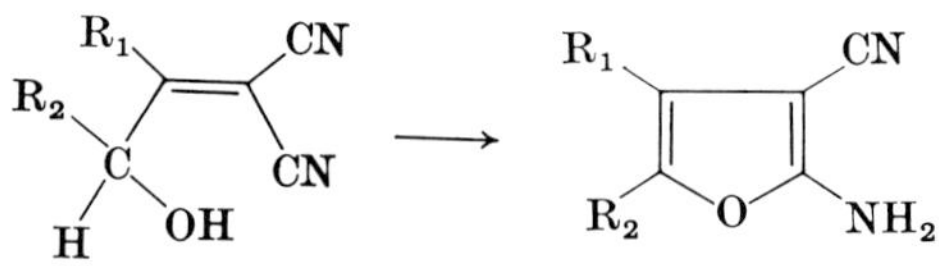

(1) **(2)**

TABLE VII

Furan o-Aminonitriles

Reactants		Product		Yield, %	Ref.
alononitrile	Other component	Product		Yield, %	Ref.
		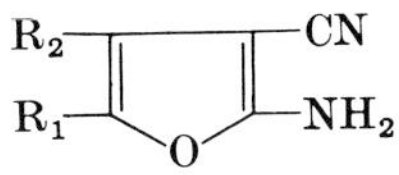			
		R_1	R_2		
CN 2 CN	$CH_3—C{=}O$ CH_2OH	H	CH_3	67	258
	$CH_3—C{=}O$ $CH_3—CHOH$	CH_3	CH_3	67	258
	$C_6H_5—C{=}O$ $C_6H_5—CH—OH$	C_6H_5	C_6H_5	75	258
	$C_6H_5—C{=}O$ $C_6H_5—CH—Hal$	C_6H_5	C_6H_5	74	492
	(Hal = Cl)			57.7	5
	$CH_3C_6H_4—C{=}O$ $C_6H_5—CH—Hal$	$CH_3C_6H_4$	C_6H_5	80	492
	$C_6H_5—C{=}O$ $CH_3C_6H_4—CH—Hal$	C_6H_5	$CH_3C_6H_4$	83	492
	$CH_3OC_6H_4—C{=}O$ $C_6H_5—CH—Hal$	$CH_3OC_6H_4$	C_6H_5	77	492
	$C_6H_5—C{=}O$ $CH_3OC_6H_4—CH—Hal$	C_6H_5	$CH_3OC_6H_4$	71	492
	$C_2H_5OC_6H_4—C{=}O$ $C_6H_5—CH—Hal$	$C_2H_5OC_6H_4$	C_6H_5	72	492
	$C_6H_5C_6H_4—C{=}O$ $C_6H_5—CH—Hal$	$C_6H_5C_6H_4$	C_6H_5	77	492
	$C_6H_5—C{=}O$ $C_6H_5C_6H_4—CH—Hal$	C_6H_5	$C_6H_5C_6H_4$	84	492

(continued)

TABLE VII (*continued*)

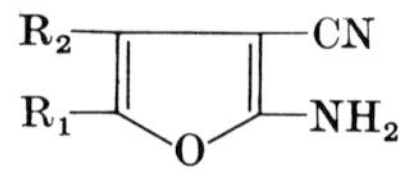

Reactants: Malononitrile	Reactants: Other component	Product: R_1	Product: R_2	Yield, %	Re
$CH_2(CN)_2$ (*cont.*)	C_6H_5—C=O / ClC_6H_4—CH—Hal	C_6H_5	ClC_6H_4	80	49
	C_6H_5—C=O / BrC_6H_4—CH—Hal	C_6H_5	BrC_6H_4	63	49
	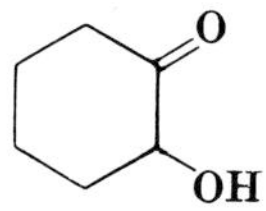	—$(CH_2)_4$—		51	25
	$(CH_3CO)_2C(Cl)^{(-)}Na^{(+)}$	CH_3	CH_3CO	93	27
	CH_3—C=O / C_2H_5OOC—CH—Cl	CH_3	$COOC_2H_5$	—	
	C_6H_5—C=O / C_2H_5OOC—CH—Cl	C_6H_5	$COOC_2H_5$	78.3	
	C_2H_5OOC—C=O / C_2H_5OOC—CH—Cl	$COOC_2H_5$	$COOC_2H_5$	43.6	
	C_6H_5—C=O / CH_3CO—CH—Cl	C_6H_5	CH_3CO	—	
	C_6H_5—C=O / C_6H_5CO—CH—Cl	C_6H_5	C_6H_5CO	71	

Many of the furan *o*-aminonitriles which have thus far been reported have indeed been prepared by condensation of α-hydroxyketones (acyloins, benzoin) with malononitrile in the presence of a base such as di- or triethylamine which promotes both the initial aldol condensation and the terminal ring closure (258). Yields are good and the reaction appears capable of wide extension, considering the accessibility of acyloins. The same key intermediate **1** may be prepared alternately by condensation of an α-haloketone with the sodium salt of malononitrile, except that the positions of R_1 and R_2 in the ensuing furan *o*-aminonitrile are now reversed (5,492). α-Chloro-β-keto esters and α-chloro-β-diketones react similarly (5,276).

D. SYNTHESIS OF PYRROLE *o*-AMINONITRILES (Table VIII)

The base-catalyzed condensation of α-aminoketones (**1**) with malononitrile yields 2-amino-3-cyanopyrroles (**2**) (175) in a manner analogous to the furan synthesis described above. Yields are moderate to good (40–

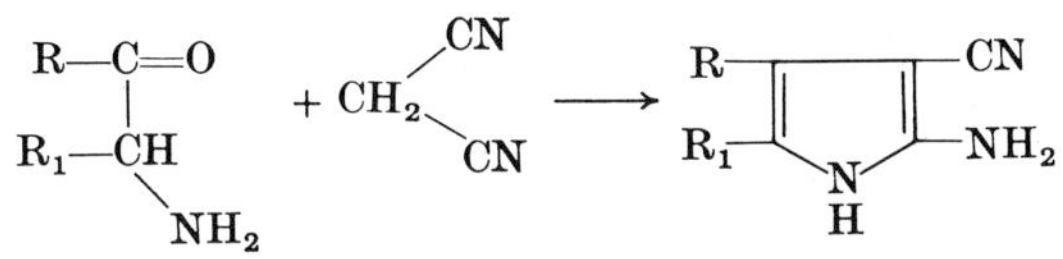

(**1**) (**2**)

70%), and the reaction appears limited only by the accessibility of the requisite α-aminoketones. These intermediates are not stable as their free bases (because of self-condensation to give dihydropyrazines) and thus must be generated *in situ* by neutralization of their salts. Since condensation in basic medium to give the desired pyrrole *o*-aminonitrile must therefore compete both with self-condensation of the α-aminonitrile and base-catalyzed dimerization of malononitrile, the observed yields are remarkably good.

2-Amino-5-bromo-3,4-dicyanopyrrole (**3**) is formed by the reaction of tetracyanoethylene with hydrogen bromide in acetone solution at −40° (104). This remarkable reaction appears to proceed by initial reduction of the tetracyanoethylene by hydrogen bromide to give tetracyanoethane

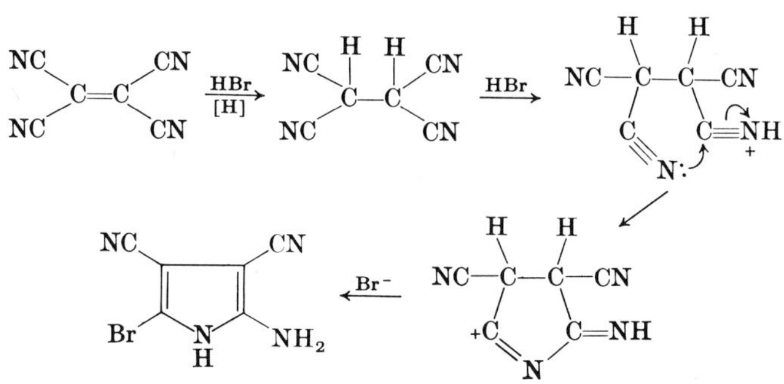

(**3**)

TABLE VIII

Pyrrole *o*-Aminonitriles

Reactants: Malononitrile derivative	Reactants: Other component	Product		Yield, %	Ref.
		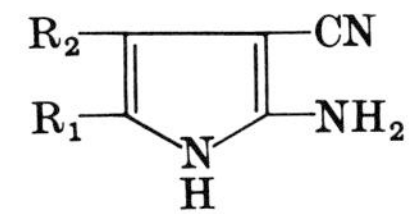			
		R_1	R_2		
$CH_2(CN)_2$	$CH_3—C(=O)—CH_2—NH_2$	H	CH_3	40–70	175
	$CH_3—C(=O)—CH(CH_3)—NH_2$	CH_3	CH_3	40–70	175
	$C_6H_5—C(=O)—CH_2—NH_2$	H	C_6H_5	40–70	175
	$C_6H_5—C(=O)—CH(CH_3)—NH_2$	CH_3	C_6H_5	40–70	175
	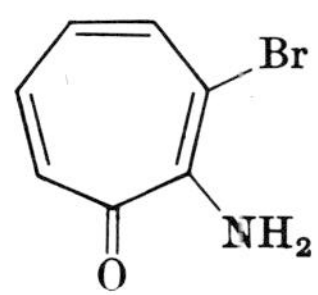	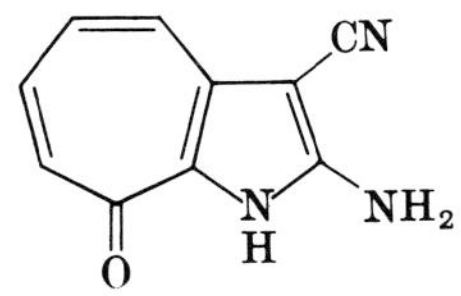		60	421

	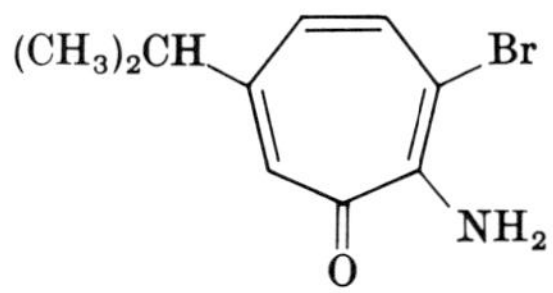	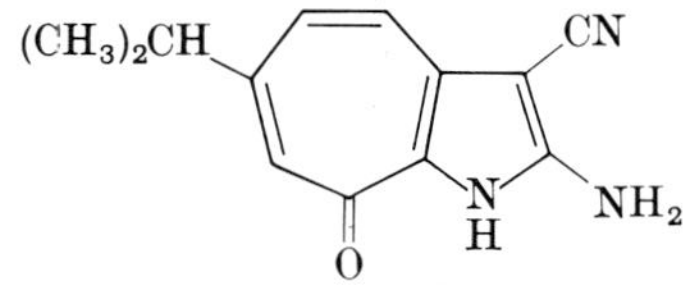		—	421
	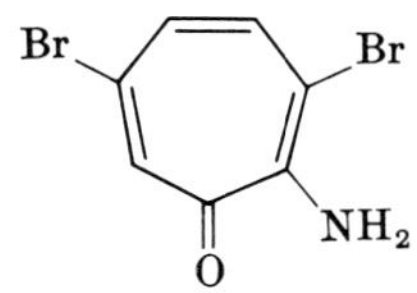	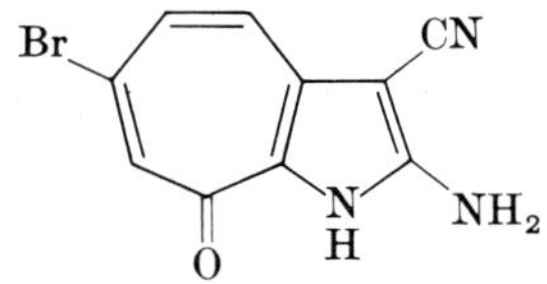		—	421
		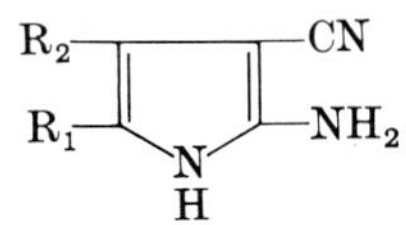			
		R_1	R_2		
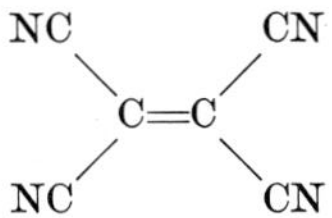	HBr gas (−40°)	Br	CN	33	104
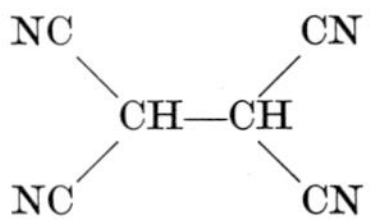	HBr gas (10°)	Br	CN	10	104

(continued)

TABLE VIII (*continued*)

Reactants: Malononitrile derivative	Reactants: Other component	Product	Yield, %	Ref.
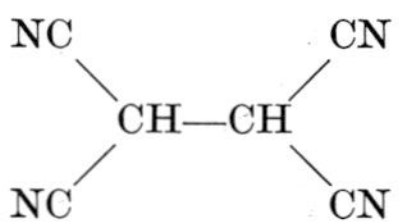 (*cont.*)	$NaHSO_3$ followed by 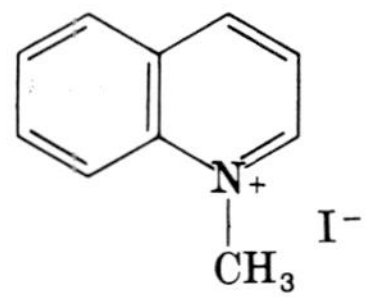	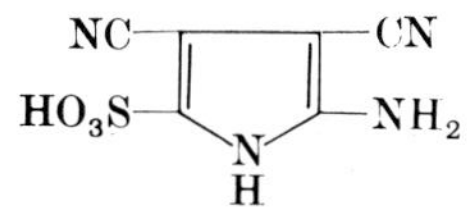 (as N-methylquinolinium salt)	27	104
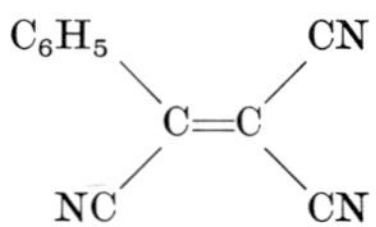	C_2H_5SH	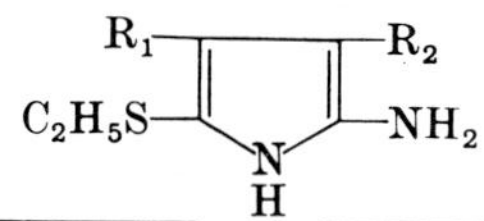 R_1: C_6H_5 or CN; R_2: CN or C_6H_5	43	103
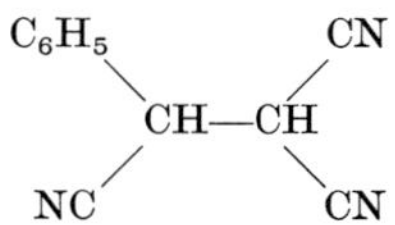	C_2H_5SH	R_1: C_6H_5 or CN; R_2: CN or C_6H_5	77	103
		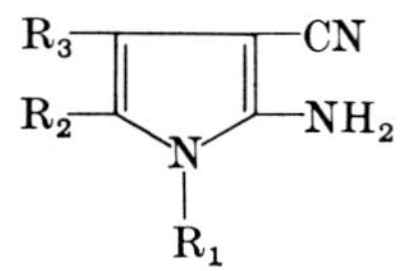		

R_1	R_2	R_3

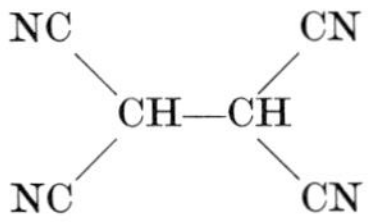	H_2NNH_2	NH_2	NH_2	CN	47	365
					—	402
	H_2NNHCH_3	CH_3NH	NH_2	CN	14	365
					—	402
		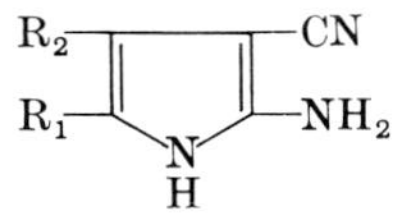				
		R_1	R_2			
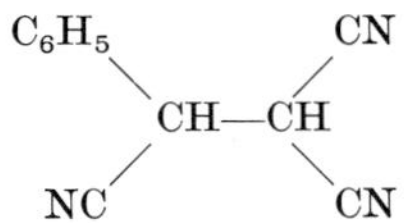	$HSCH_2CH_2OH$	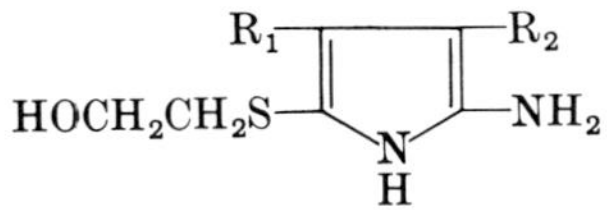			68	103
		C_6H_5 or CN	CN or C_6H_5			
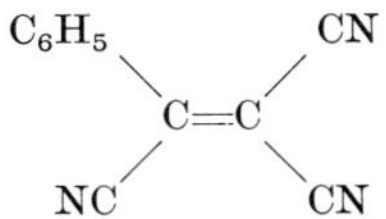	$HSCH_2CH_2OH$	Same as above			"Low"	103
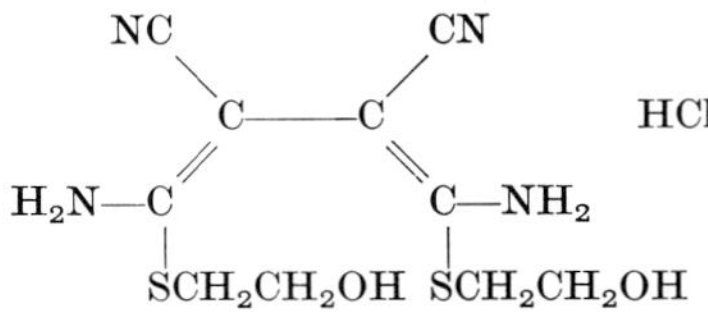	HCl	SCH_2CH_2OH	CN		78	104

(continued)

TABLE VIII (*continued*)

Reactants				
Malononitrile derivative	Other component	Product	Yield, %	Ref.
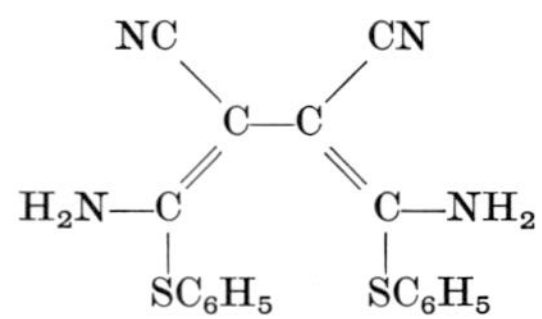	HCl and heat	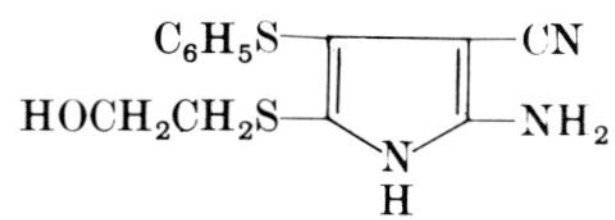	47	104
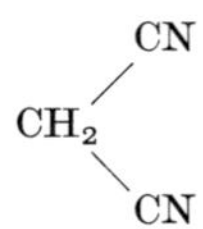	(NH_3, then heat)	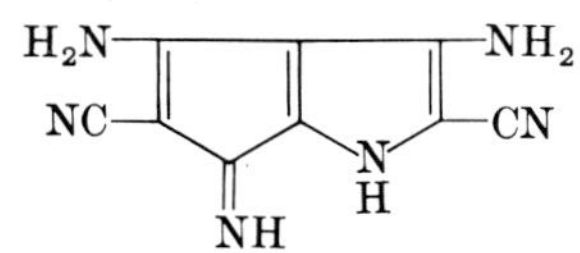	—	548
	(Piperidine/ethanol)	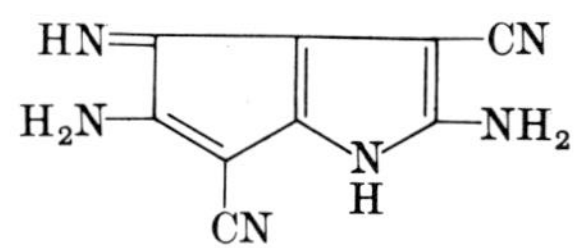	—	548

which is then converted to the pyrrole by hydrogen bromide, presumably via protonation of one of the nitrile groups, followed by nucleophilic addition by nitrogen of an adjacent nitrile group, addition of bromide ion and final tautomerization. Acetone is a requisite solvent for this reaction, since it reacts with the bromine formed in the reduction step to remove it from the reaction medium. No pyrrole was formed when methyl formate or tetrahydrofuran were substituted as solvents, apparently because the tetracyanoethane is readily reoxidized to tetracyanoethylene by free bromine. It should be noted, however, that a substantially lower yield (10% vs. 33%) of the pyrrole **3** was formed when tetracyanoethane, the presumed intermediate in the above reaction sequence, was treated with hydrogen bromide in acetone (104).

Reaction of tricyanostyrene (**4**) with ethylmercaptan or 2-mercaptoethanol does not lead to displacement of cyanide ion with the formation of an α-thio derivative as anticipated, but rather to a pyrrole aminonitrile (**7**), apparently by initial reduction of **4** to **5** with the mercaptan, followed by addition of the latter to one of the nitrile groups of **5** and final cyclization of the intermediate *S*-alkylthioiminoether (**6a** or **6b**) to the pyrrole (103) (Scheme 10). Although spectral data confirmed the fact that the product was indeed a pyrrole, it was not determined whether the latter was a

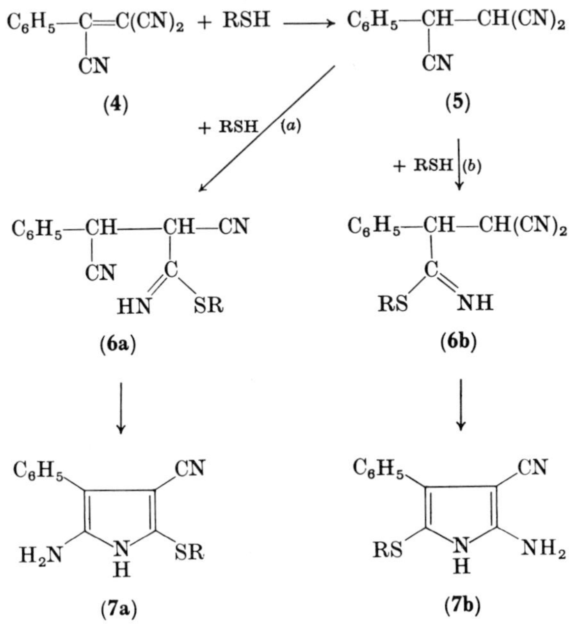

Scheme 10

2-amino-4-cyano (**7a**) or a 2-amino-3-cyano (**7b**) derivative because of ambiguity as to the nitrile group undergoing the initial nucleophilic addition of mercaptide ion. This point could be readily settled by application to the reaction product (**7a** or **7b**) of any of the numerous cyclization procedures which have been developed for the conversion of *o*-aminonitriles to condensed pyrimidines, as described in detail on pp. 226–306.

In contrast to the reaction of tricyanostyrene with mercaptans, the du Pont workers (103) found that the reaction of mercaptans with tetracyanoethylene gave 1,4-bis(alkylthio)-1,4-diamino-2,3-dicyanobutadienes (**8**), again via the intermediacy of tetracyanoethane. 2-Amino-3,4-dicyano-5-alkylthiopyrroles (**9**) are formed by acid-catalyzed elimination of alkyl mercaptans from **8**. It should be noted, however, that the only reported successful examples of this conversion involved elimination of ethanethiol or thiophenol from the intermediate **8**; all other alkylthio derivatives of **8** examined gave 2,5-bis(alkylthio)-3,4-dicyanopyrroles instead, although the reasons for the different reaction courses observed are not understood.

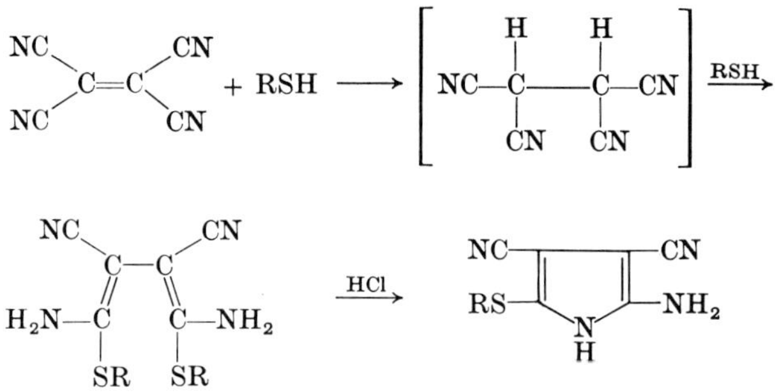

(**8**) (**9**)

(R = C_6H_5, $HOCH_2CH_2$)

1,2,5-Triamino-3,4-dicyanopyrrole (**11**, R = H) is formed in moderate (47%) yield by the reaction of hydrazine hydrate with tetracyanoethane (**10**); the use of methyl hydrazine gives the corresponding 1-methylamino derivative (365). These condensations are analogous to the formation of 2,5-diamino-3,4-dicyanothiophene by reaction of **10** with hydrogen sulfide (see p. 142).

In what appears at first glance to be a straightforward reaction, malononitrile reacts in the presence of sodium ethoxide, with 2-amino-3-bromotropone (**12**) to give **13** in 60% yield. However, 2-amino-3-cyano-8-hydroxyquinoline (**14**) is formed as a by-product in 14% yield. When

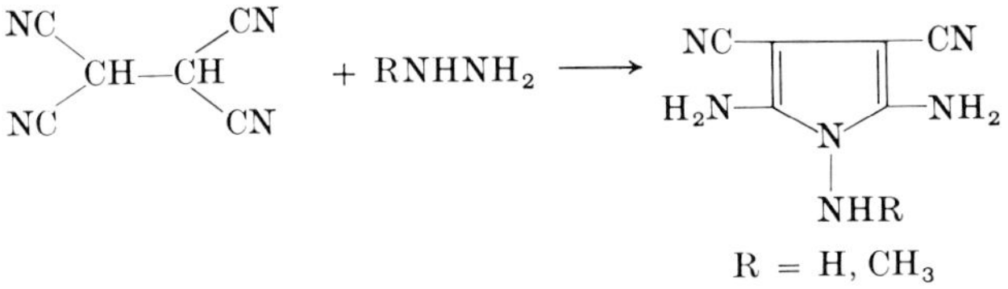

(10) (11)

potassium *t*-butoxide is employed as the base, **14** is the major product (420,421). Analogous results were obtained with a number of variously substituted 2-amino-3-bromotropones (418).

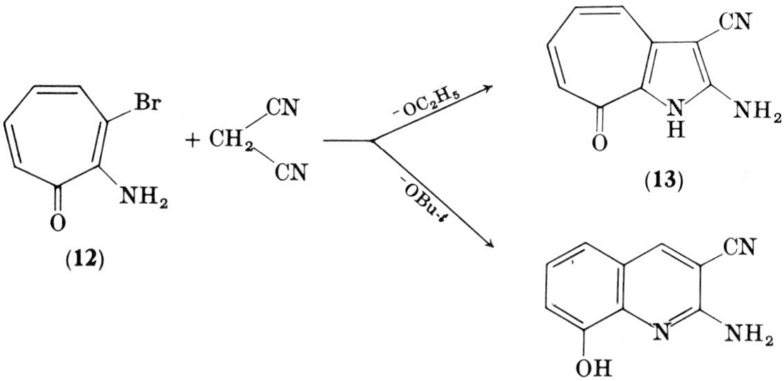

(12) (13) (14)

E. SYNTHESIS OF THIOPHENE *o*-AMINONITRILES (Table IX)

The base-catalyzed condensation of malononitrile with α-mercaptoaldehydes and ketones has been shown to give 2-amino-3-cyanothiophenes and appears to be the simplest general synthetic route to the latter compounds (74,217,334,341). The α-mercapto carbonyl compound need not be prepared in advance but may be generated *in situ*. Thus, 2-amino-3-cyano-4,5-dimethylthiophene (**3**) results from the reaction of ethyl methyl ketone with sulfur in the presence of triethylamine and malononitrile (113); the intermediate alkylidenemalononitrile **1** is apparently oxidized by sulfur to the α-mercapto alkylidene derivative **2**, which then undergoes intramolecular cyclization. In a variation of this procedure, the alkylidenemalononitrile **1** can be prepared initially and then treated with sulfur and triethylamine; yields appear to be substantially higher when this modification is employed (47,113,234), although it should be pointed out that the preparation of the requisite alkylidenemalononitriles may be complicated by dimerization (see pp. 57–60).

TABLE IX

Thiophene *o*-Aminonitriles

Reactants		Product		Yield, %	Ref.
Malononitrile derivative	Other component(s)	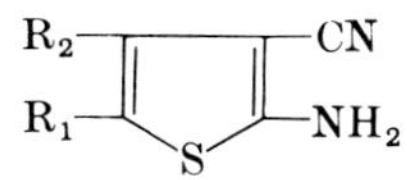			
		R_1	R_2		
$CH_2(CN)_2$	$HSCH_2CHO$	H	H	55	74
	CH_3—C(=O)—CH_2—SH	H	CH_3	73 87 67	74 334 341
	CH_2(SH)—C(=O)—CH_2—SH	H	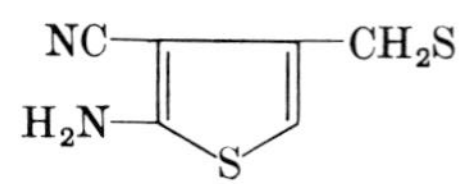	52	74
	CH_3—C(=O)—CH(CH_3)—SH	CH_3	CH_3	70 53 —	74 334 91
	CH_3—C(=O)—CH_2—CH_3 and S and $(C_2H_5)_3N$	CH_3	CH_3	42	113

$CH_3C(CH_2CH_3)=C(CN)_2$	S and $(C_2H_5)_3N$ (or morpholine)	CH_3	CH_3	41	113
$CH_2(CN)_2$	$CH_3CH_2—C(=O)—CH(CH_3)—SH$	CH_3	CH_3CH_2	51	74
	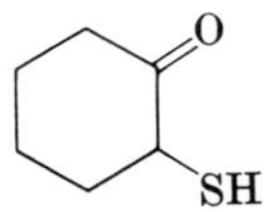	—$(CH_2)_4$—		70	74
	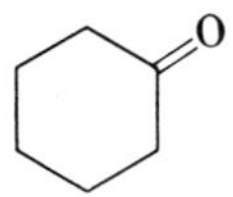 S, and $(C_2H_5)_3N$	—$(CH_2)_4$—		95 80 86 —	334 234 113 91
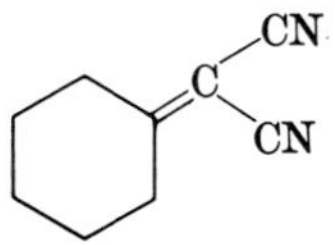	S and $(C_2H_5)_2NH$	—$(CH_2)_4$—		90 90	234 113
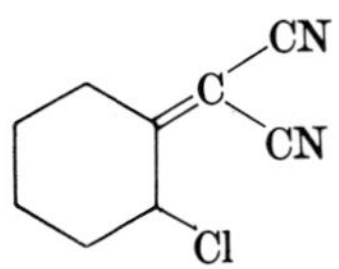	NaHS	—$(CH_2)_4$—		95	74

(continued)

TABLE IX (*continued*)

Reactants		Product			
Malononitrile derivative	Other component(s)	Product		Yield, %	Ref.
		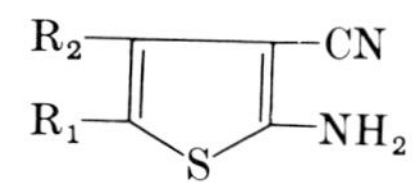			
		R_1	R_2		
$CH_2(CN)_2$	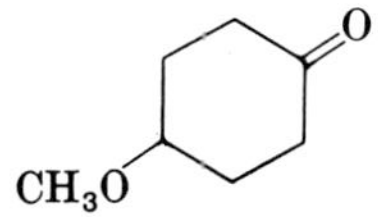 S, and morpholine	$—CH_2CH(OCH_3)CH_2CH_2—$		45	341
	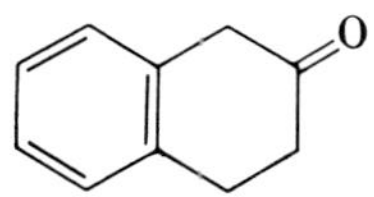 S, and morpholine	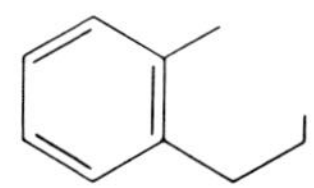		34	341
$(NC)_2C=C(CN)_2$	H_2S H_2S/CS_2/acetone/pyridine	NH_2	CN	92 79–85	104 374
$(NC)_2CH—CH(CN)_2$	Na_2S	NH_2	CN	25	104

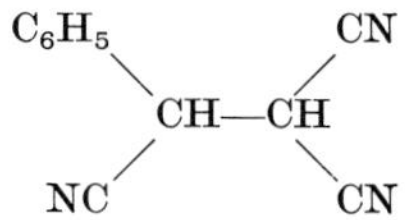	H_2S/pyridine	NH_2	C_6H_5	64	103
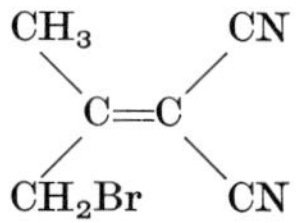	NaHS	H	CH_3	60	74
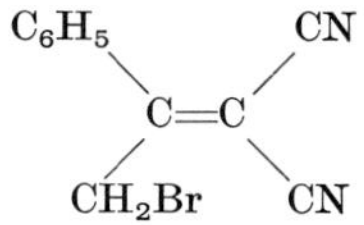	NaHS	H	C_6H_5	75	74
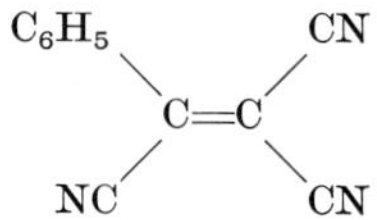	H_2S/pyridine	NH_2	C_6H_5	—	398
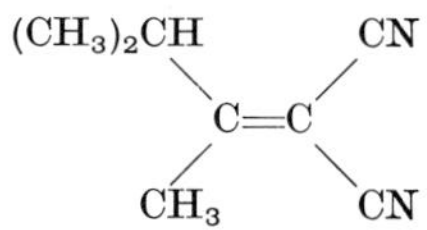	$S/(C_2H_5)_2NH$	H	$(CH_3)_2CH$	—	47
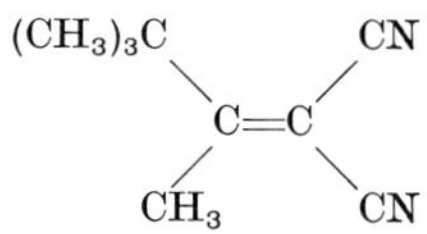	$S/(C_2H_5)_2NH$	H	$(CH_3)_3C$	40	47

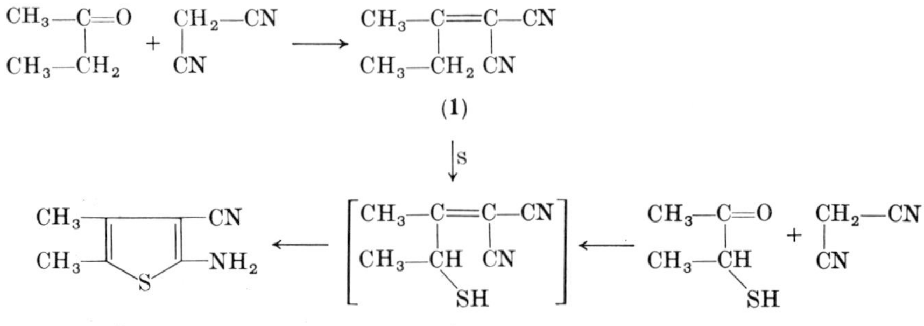

(3) (2)

As a further variation, α-chloroketones (e.g., **4**) may be condensed with malononitrile to give α-chloroalkylidenemalononitriles which upon treatment with sodium hydrosulfide give 2-amino-3-cyanothiophenes (e.g., **5**) (74).

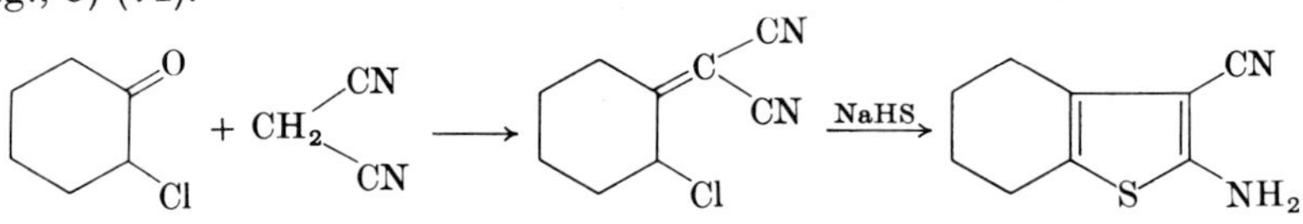

(4) (5)

2,5-Diamino-3,4-dicyanothiophene (**6**) is formed in 92% yield upon treatment of tetracyanoethylene with hydrogen sulfide in the presence of pyridine (102,104). The reaction course involves initial reduction of tetracyanoethylene to tetracyanoethane followed (presumably) by addition of mercaptide ion, generated *in situ* from the reaction of hydrogen sulfide with pyridine, to one of the nitrile groups and subsequent intramolecular cyclization. It is again noteworthy, however, that only poor yields (25%) of **6** were formed when preformed tetracyanoethane was treated with sodium hydrosulfide in an attempt to substantiate the reaction course suggested above. Hydrogen sulfide itself fails to convert tetracyanoethane to the thiophene **6** in the absence of a base.

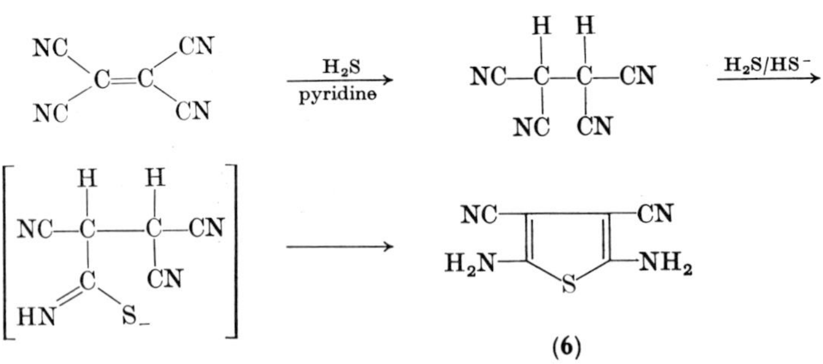

(6)

F. SYNTHESIS OF ISOXAZOLE *o*-AMINONITRILES (Table X)

Relatively few isoxazole aminonitriles are known; most have been prepared by the reaction of a suitable derivative of malononitrile with hydroxylamine. For example, ethoxymethylenemalononitrile and its alkyl and aryl derivatives (**1**) react with a stoichiometric amount of hydroxylamine to give 3-substituted 4-cyano-5-aminoisoxazoles (**2**) (11,176,177). It is important to avoid an excess of hydroxylamine, since amidoximes are formed in the presence of an excess of this reagent. 3-Alkoxy-4-cyano-5-aminoisoxazoles are readily formed by the reaction of one equivalent of hydroxylamine with dicyanoketene acetals (105).

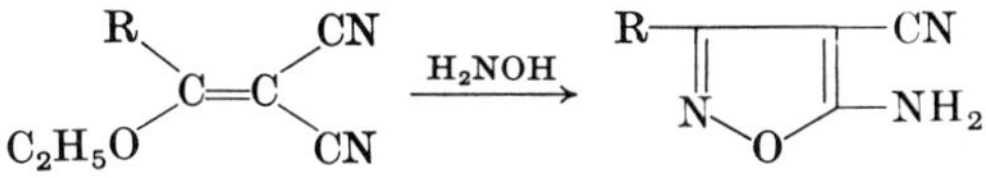

(**1**) (**2**)

TABLE X

Isoxazole *o*-Aminonitriles

Reactants: Malononitrile derivative	Reactants: Other component	Product	Yield, %	Ref.
		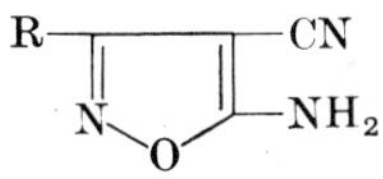 R		
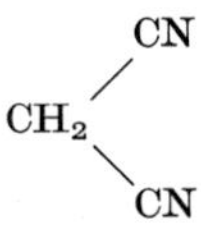	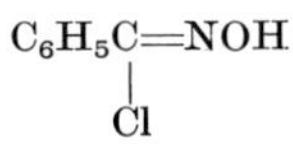	C_6H_5	81	6
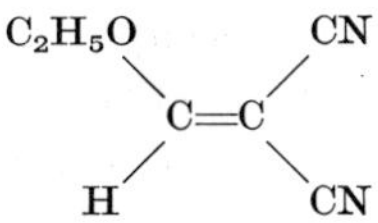	H_2NOH	H	60	176
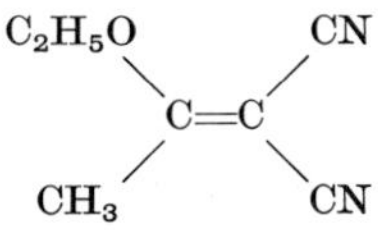	H_2NOH	CH_3	51	11

(*continued*)

TABLE X (*continued*)

Reactants				
Malononitrile derivative	Other component	Product	Yield, %	Ref.
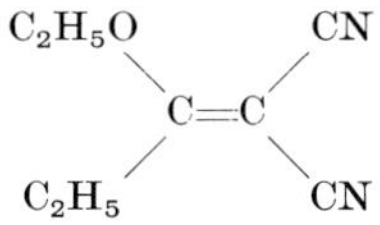	H_2NOH	C_2H_5	82	11
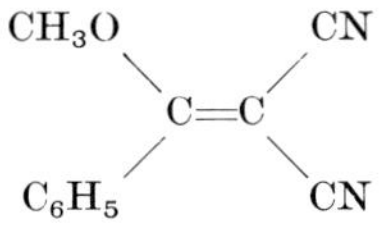	H_2NOH	C_6H_5	40	177
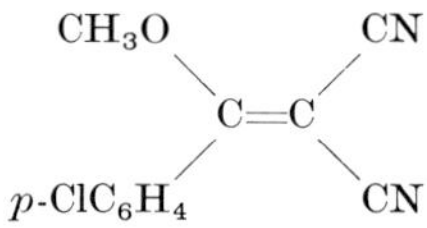	H_2NOH	*p*-ClC_6H_4	72	177
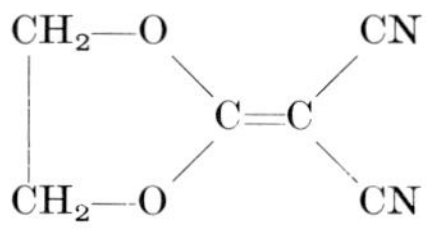	H_2NOH/NaOH	$HOCH_2CH_2O$	—	105
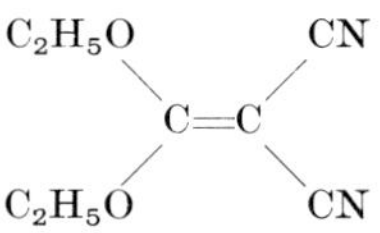	H_2NOH/NaOH	C_2H_5O	73	105
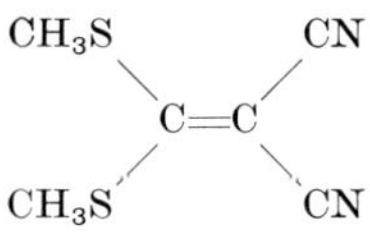	H_2NOH	CH_3S	32	266

G. SYNTHESIS OF IMIDAZOLE AND OXAZOLE *o*-AMINONITRILES (Tables XIA and XIB)

4-Amino-5-cyanoimidazole (**2**) is of considerable interest as a possible biogenetic precursor to purines (i.e., adenine), originating perhaps by tetramerization of hydrogen cyanide under primitive earth conditions. This new tetramer of hydrogen cyanide has recently been synthesized by irradiation at 350 nm of a $10^{-4}M$ aqueous solution of the HCN tetramer **1**; yields of 77–82% were estimated from the intensity of the 247 nm peak of **2**. In a remarkable photochemical isomerization, **2** is also formed upon irradiation at 253 nm of 1,1-diamino-2,2-dicyanoethylene (**3**) or 3-amino-4-cyanopyrazole (**4**) (336). This latter isomerization is analogous to the pyrazole → imidazole photochemical isomerization first

TABLE XIA

Imidazole *o*-Aminonitriles

Reactants				
Malononitrile derivative	Other component	Product	Yield, %	Ref.
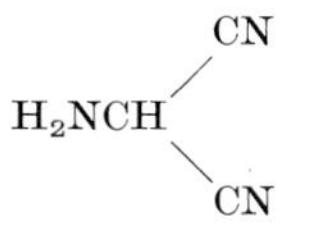	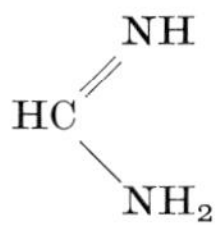	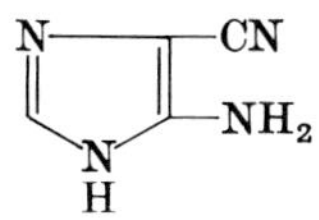	35 50	32, 335 336
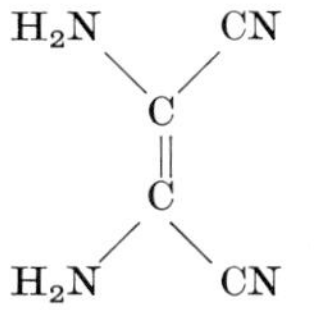 (HCN tetramer)	Irradiation 350 nm	Same as above	77–82	336
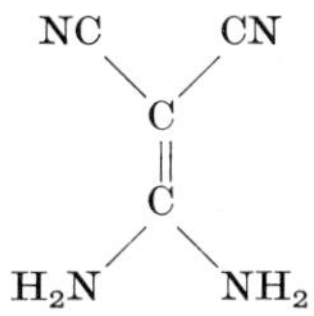	Irradiation 253 nm (in THF)	Same as above	—	336
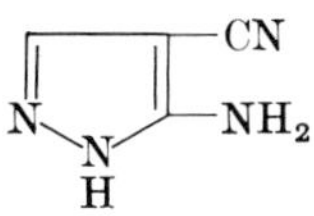	Irradiation 253 nm	Same as above	—	336

TABLE XIB

Oxazole *o*-Aminonitriles

Reactants				
Malononitrile derivative	Other component	Product	Yield, %	Ref.
		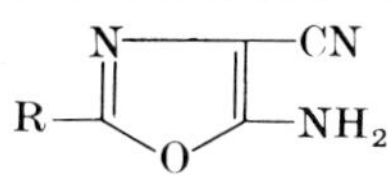		
		R		
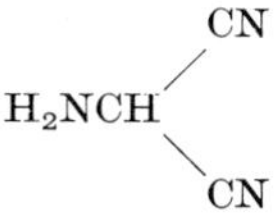	$(CH_3CO)_2O$ and HCOOH	H	— 43	335 32
	$(CH_3CO)_2O$	CH_3	—	32, 335
	$(CH_3CH_2CO)_2O$	CH_3CH_2	—	32, 335
	$(C_6H_5CO)_2O$	C_6H_5	—	32, 335

(*continued*)

TABLE XIB (*continued*)

Reactants				
Malononitrile derivative	Other component	Product	Yield, %	Ref.
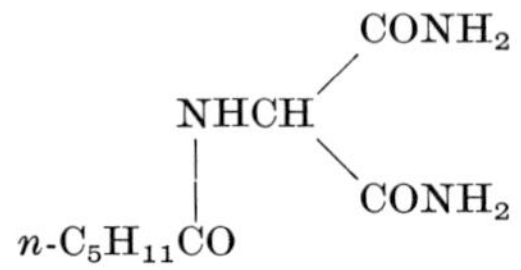	$POCl_3$	n-C_5H_{11}	73	337
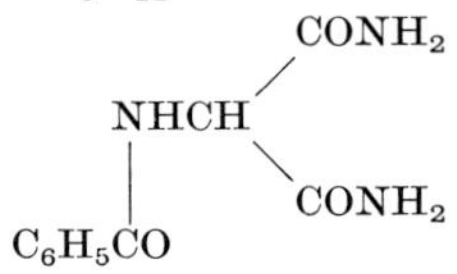	$POCl_3$	C_6H_5	66	337

reported by Schmidt and co-workers (429). These photochemical syntheses of **2** cannot, however, be used for its preparation in quantity, since very dilute solutions must be used. A practical synthesis of 4-amino-5-cyanoimidazole (**2**) has been found in the reaction of aminomalononitrile (**5**) with formamidine acetate (32).

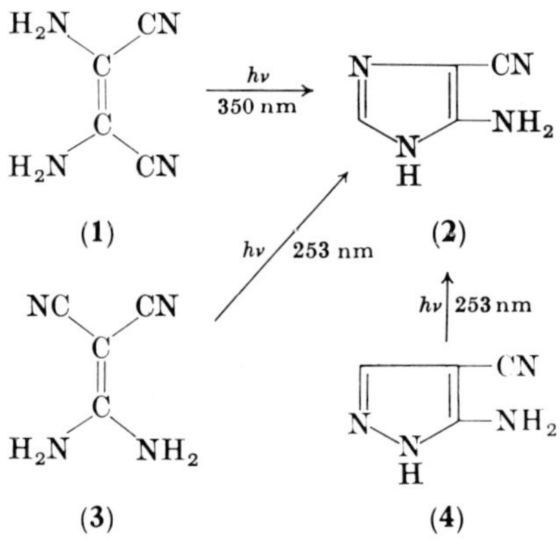

The reaction of aminomalononitrile (**5**) with mixed formic-acetic anhydride gives 4-cyano-5-aminooxazole (**6**, R = H) (32,335); other anhydrides react analogously to give 2-substituted derivatives of **6**. It is remarkable that cyclization of the intermediate acyl and aroyl aminomalononitriles appears to proceed spontaneously in the absence of basic catalysts. The product derived from benzoic anhydride, 2-phenyl-4-

cyano-5-aminooxazole (**6**, R = C_6H_5), had previously been prepared by dehydrative cyclization with phosphorus oxychloride of benzamido-malondiamide (**7**) (337).

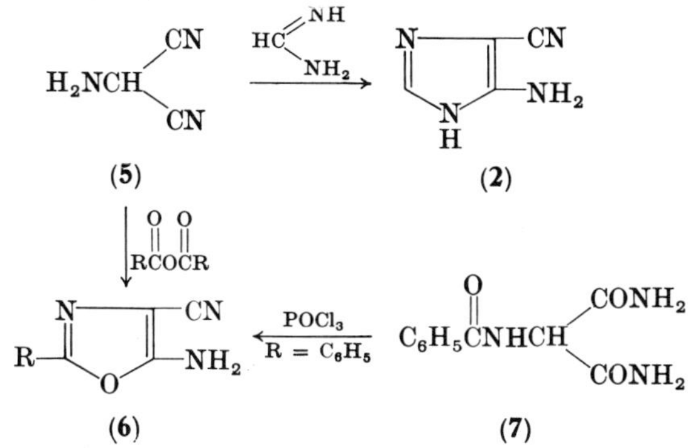

H. SYNTHESIS OF THIAZOLE *o*-AMINONITRILES (Table XII)

The condensation of malononitrile with isothiocyanates and sulfur in the presence of triethylamine has been reported to give 3-substituted 4-amino-5-cyanothiazoline-2-thiones (**1**), which can alternately be prepared by initial formation of a dithiocarbamate from a primary amine and carbon disulfide, followed by treatment with malononitrile and sulfur in the presence of triethylamine (38,73). The simplest derivative in the

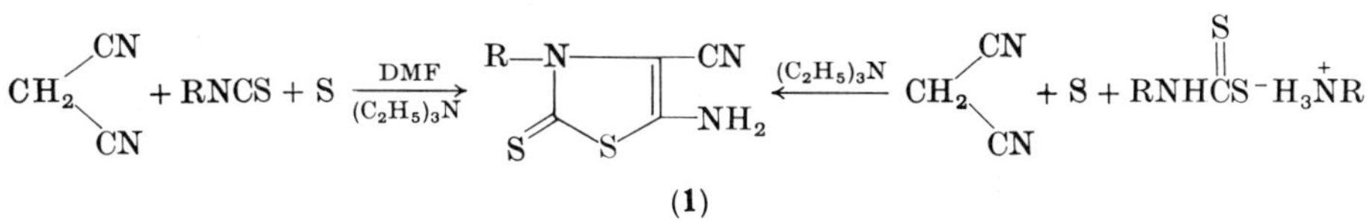

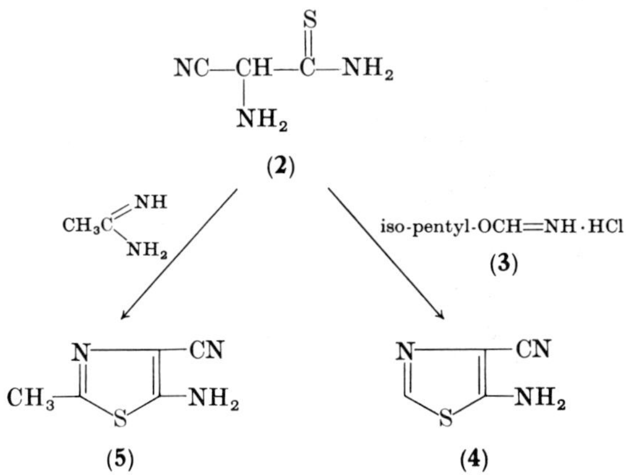

TABLE XII

Thiazole *o*-Aminonitriles

Reactants: Malononitrile derivative (or other nitrile intermediate)	Reactants: Other component	Product	Yield, %	Ref.
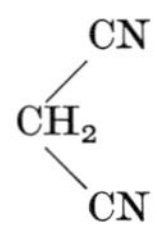	$C_6H_5NCS/S/(C_2H_5)_3N$	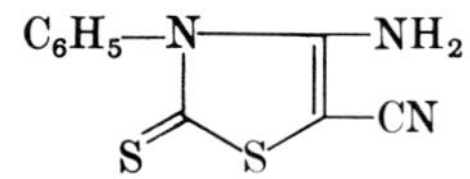	80	38
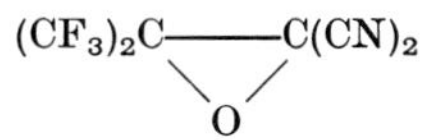	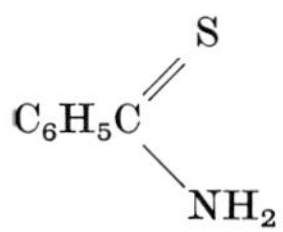	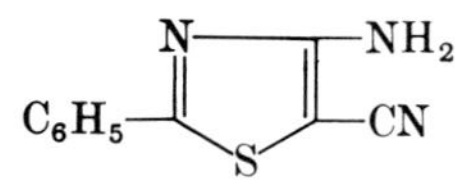	90	473
	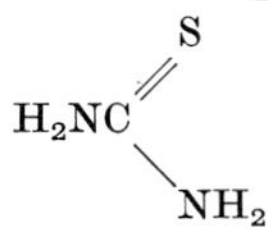	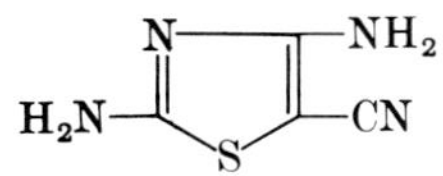	38	473
	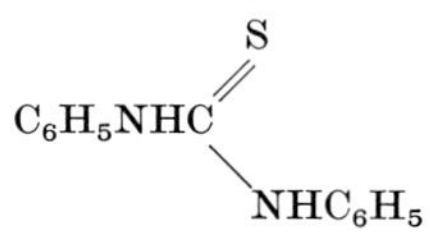	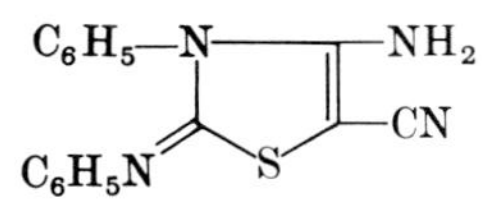	79	473

thiazole series, 4-cyano-5-aminothiazole (**4**), has been prepared in low (16%) yield by the condensation of aminocyanothioacetamide (**2**) with the iminoether **3**; the corresponding 2-methyl derivative (**5**) results from the reaction of **2** with acetamidine (114).

In a continuation of du Pont's study of the fascinating chemistry of tetracyanoethylene and its derivatives, Middleton (473) found that 1,1-dicyano-2,2-bis(trifluoromethyl)ethylene oxide (**6**) reacts rapidly and exothermically with thiourea in ethanol to give (diaminomethylene)-sulfonium dicyanomethylid which is best represented by a number of resonance structures (**7**). This compound isomerizes by intramolecular cyclization upon heating in water to give 2,4-diamino-5-cyanothiazole (**8**). The reaction of **6** with thiobenzamide gave 2-phenyl-4-amino-5-cyano-thiazole (**9**) directly; the intermediate ylid could not be isolated. An analogous reaction between **6** and *s*-diphenylthiourea led to 4-amino-5-cyano-3-phenyl-2-phenylimino-4-thiazoline.

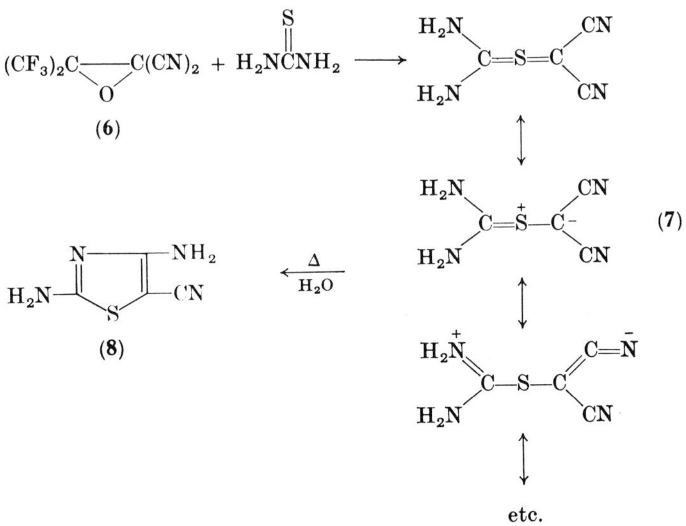

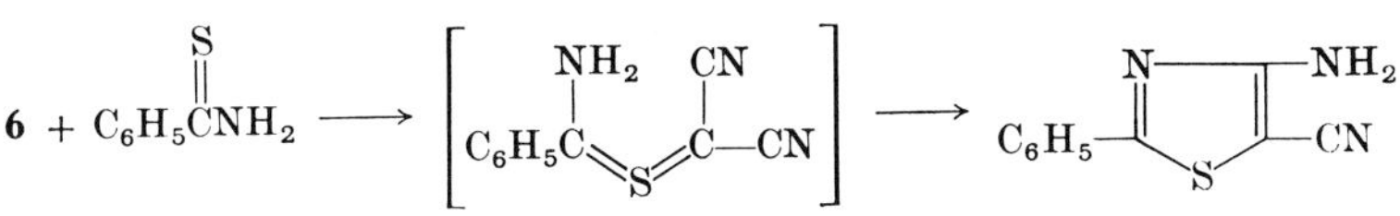

(**9**)

I. SYNTHESIS OF ISOTHIAZOLE *o*-AMINONITRILES (Table XIII)

Treatment of salts of type **1** with aqueous chloramine (or, less generally, hydroxylamine-O-sulfonic acid) gives 3-amino-4-cyanoisothiazoles (**3**);

TABLE XIII

Isothiazole *o*-Aminonitriles

Reactants				
Malononitrile derivative	Other component	Product	Yield, %	Ref.
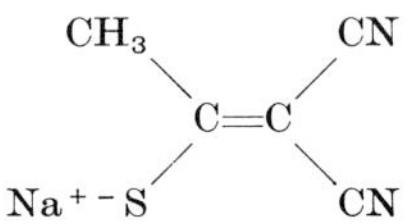	H_2NCl (or H_2NOSO_3H)	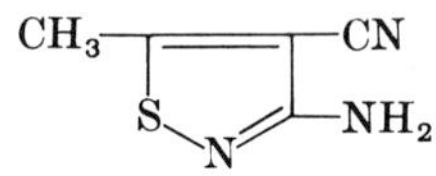	86	156
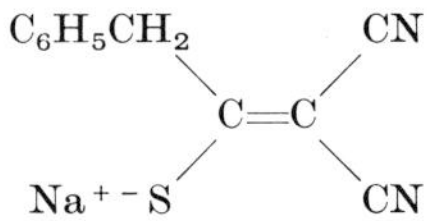	H_2NCl	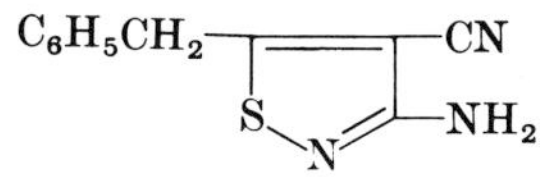	93	156
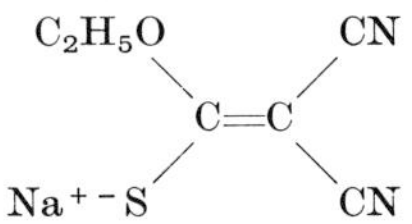	H_2NCl	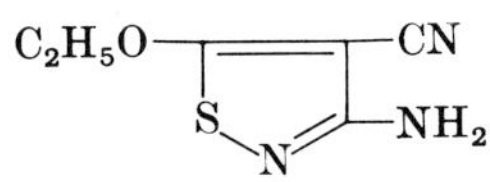	77	156
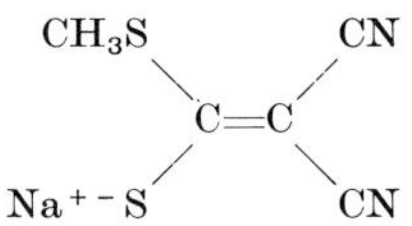	H_2NCl (or H_2NOSO_3H)	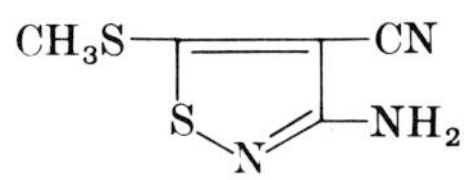	72	156

again the cyclization of the initially formed intermediate **2** involves an amine-to-nitrile addition (156).

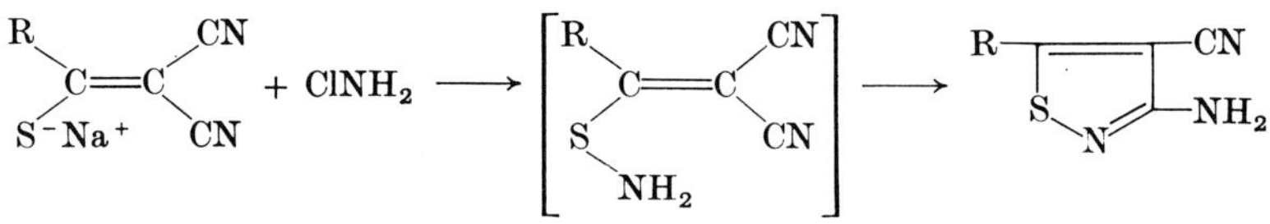

(**1**) (**2**) (**3**)

$R = CH_3, C_2H_5O, CH_3S, C_6H_5CH_2$

J. SYNTHESIS OF 1,2,3-TRIAZOLE *o*-AMINONITRILES (Table XIV)

Alkyl and aryl azides react with malononitrile in the presence of base (alkoxide ion or triethylamine) to give 1-substituted 4-cyano-5-amino-

1,2,3-triazoles (**1**) (343,344,529). The reaction probably occurs as follows:

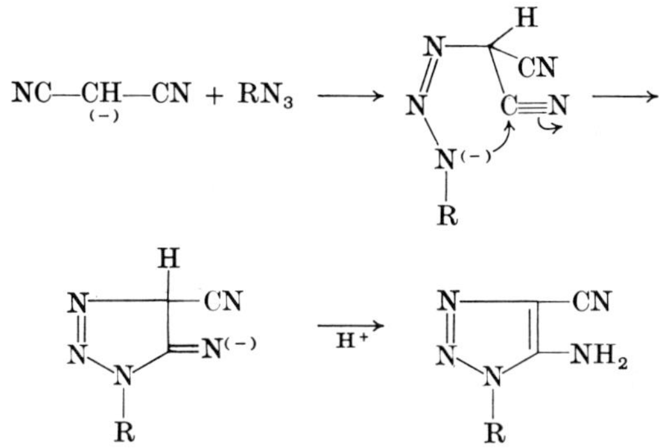

(**1**)

TABLE XIV

1,2,3-Triazole *o*-Aminonitriles

Reactants				
Malononitrile derivative	Other component	Product	Yield, %	Ref.
		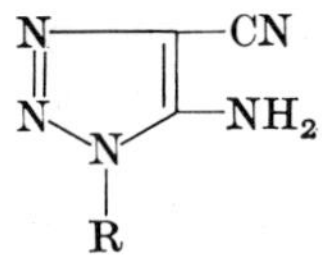 R		
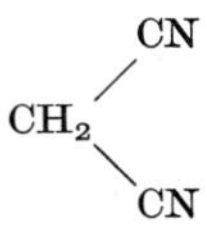	RN_3			
		C_6H_5	22	343
		p-ClC_6H_4	100	343
		p-$NO_2C_6H_4$	—	343
		$C_6H_5CH_2$	16	344
	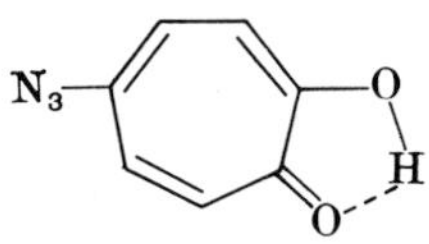	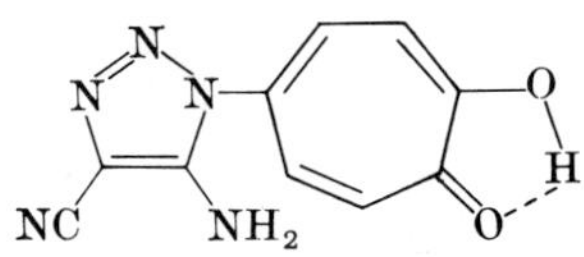	50–90	529

(*continued*)

TABLE XIV (*continued*)

Reactants				
Malononitrile derivative	Other component	Product	Yield, %	Ref.
$CH_2(CN)_2$ (*cont.*)	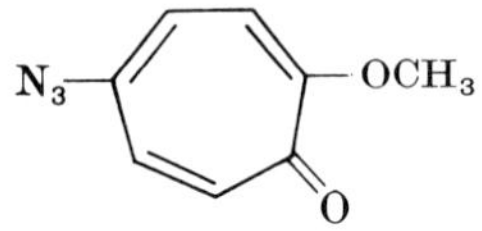	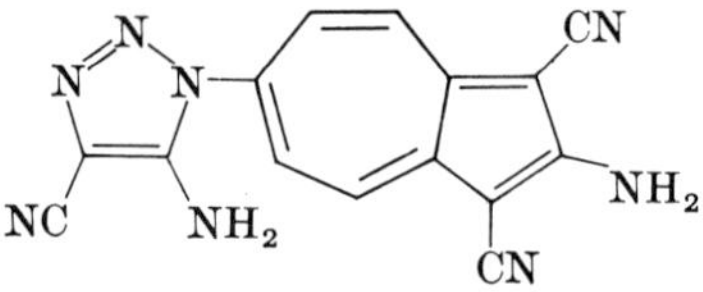	77	529
	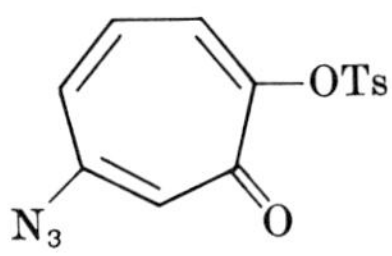	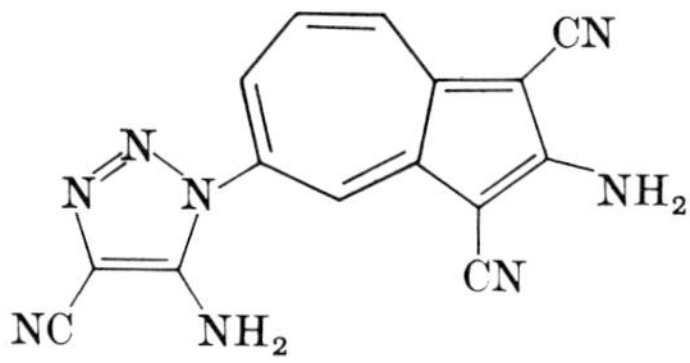	—	529

K. SYNTHESIS OF MISCELLANEOUS *S*-HETEROCYCLIC *o*-AMINONITRILES (Table XV)

Although carbon disulfide is known to undergo base-catalyzed condensations with active methylene compounds, the resulting dithio acids are unstable and must usually be isolated either in the form of their salts or as subsequent reaction products. In an ingenious application of this dithio acid formation to the synthesis of heterocyclic *o*-aminonitriles, it was found that the reaction of alkylidenemalononitriles (**1**) with carbon disulfide in the presence of triethylamine gave 5-cyano-6-amino-2*H*-thiopyranethiones (**2**). In the case of cyclohexanone, a "one-pot" synthesis of **2** (R = R′ = $(CH_2)_4$) resulted from reaction with carbon disulfide and malononitrile in the presence of triethylamine; this procedure could not, however, be extended to the use of other ketones. The only competing reaction in the above syntheses of **2** appeared to be dimerization of the alkylidenemalononitrile under the basic reaction conditions employed (see Chapter I, Section III-F), but this could be suppressed by the use of dimethylformamide as solvent (357).

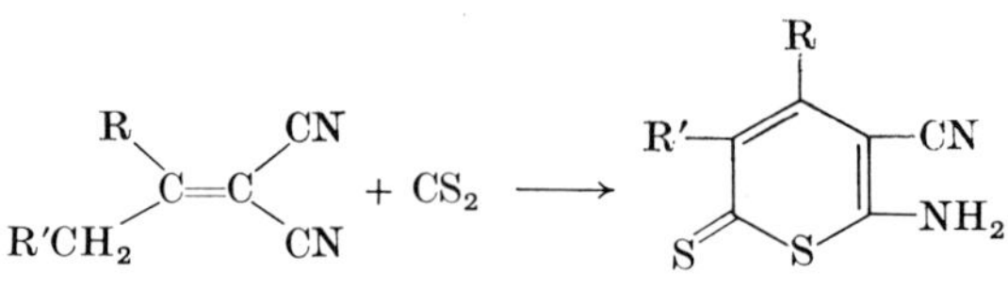

(**1**) (**2**)

4-Cyano-5-amino-1,2-dithiol-3-thione (**3**) has been prepared by (*a*) the condensation of malononitrile, carbon disulfide, and sulfur in dimethylformamide as solvent in the presence of di- or tri-ethylamine, (*b*) the condensation of cyanothioacetamide, carbon disulfide, and sulfur in dimethylformamide as solvent in the presence of triethylamine, (*c*) the condensation of cyanothioacetamide and carbon disulfide in dimethylformamide as solvent in the presence of triethylamine and iodine, and (*d*) treatment of the disodium salt of 1,1-dicyano-2,2-dimercaptoethylene (**4**) with sulfur, followed by acidification of the reaction mixture (216,358). It has been shown that this latter series of reactions yields an intermediate isothiazole (**5**) which undergoes a ring-opening, ring-reclosure sequence in acidic solution to generate the acid-stable dithiol-3-thione **3**. The latter compound is very sensitive to dilute alkali and readily loses sulfur to regenerate the disodium salt **4**. Thus the usual condensation reactions employing *o*-aminonitriles as intermediates for the annelation of heterocyclic rings, discussed in detail on pp. 226–307, are not applicable to **3** because of its sensitivity to base (Scheme 11).

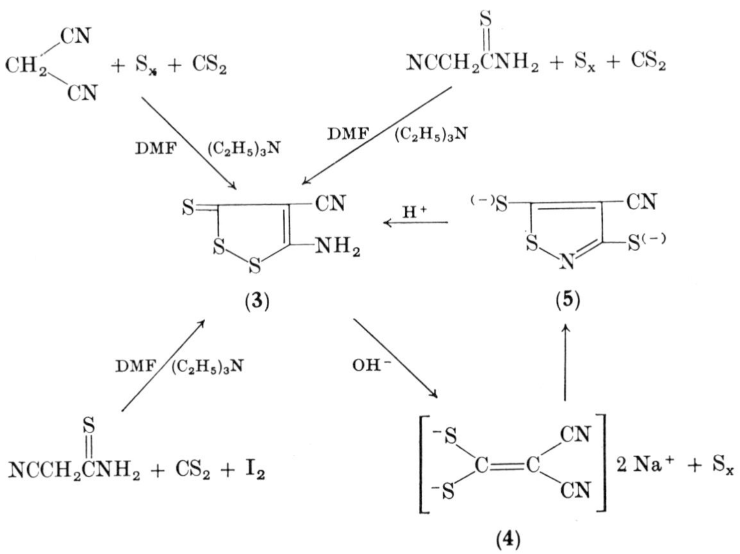

Scheme 11

A pseudo aromatic *o*-aminonitrile (without, however, the characteristic properties of truly aromatic *o*-aminonitriles) has been prepared in the thiothiophthene series. Condensation of 5-phenyl-1,2-dithiol-3-one (**6**)

TABLE XV

Miscellaneous *S*-Heterocyclic *o*-Aminonitriles

Reactants				
Malononitrile derivative (or other intermediate nitrile)	Other component	Product	Yield, %	Ref.
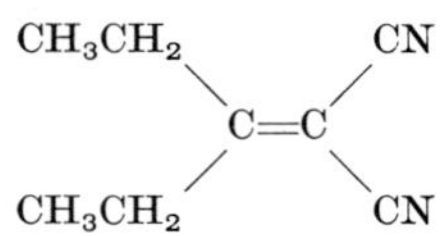	CS_2/DMF/Et_3N	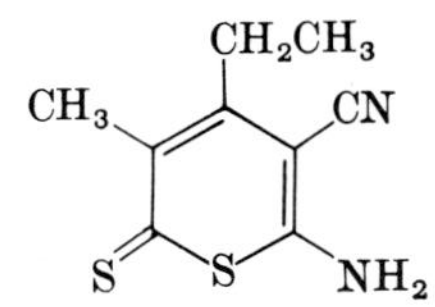	70	357
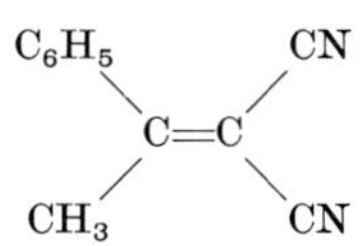	CS_2/DMF/Et_3N	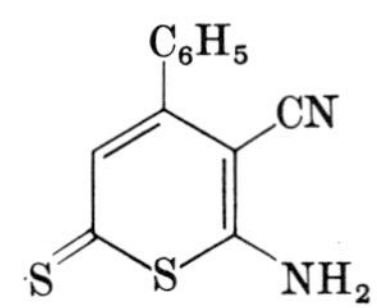	66	357
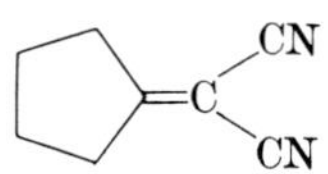	CS_2/DMF/Et_3N	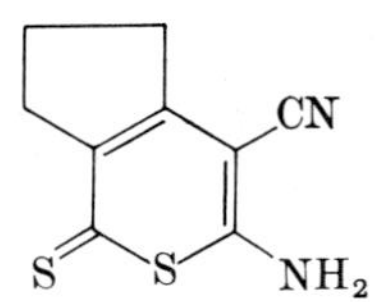	57	357

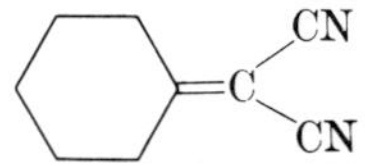	$CS_2/DMF/Et_3N$	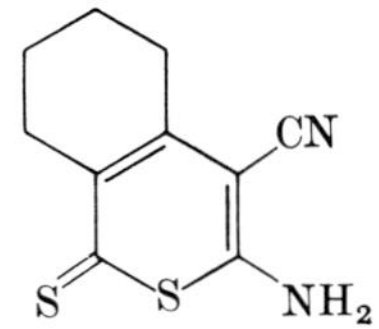	85	357
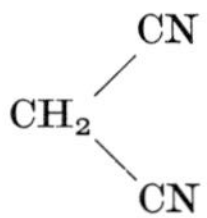	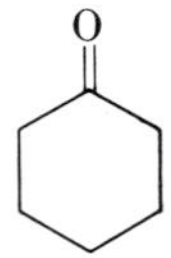 + CS_2/Et_3N	Same as above	61	357
	$CS_2/S/DMF/Et_3N$	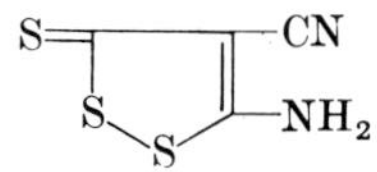	86	358
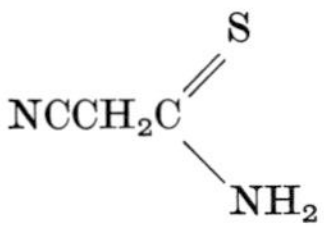	$CS_2/S/DMF/Et_3N$	Same as above	75	358
	$CS_2/I_2/DMF/Et_3N$	Same as above	92	358
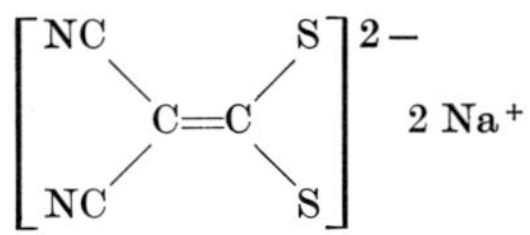	S/DMF	Same as above	56	358

(*continued*)

TABLE XV (*continued*)

Reactants				
Malononitrile derivative (or other intermediate nitrile)	Other component	Product	Yield, %	Ref.
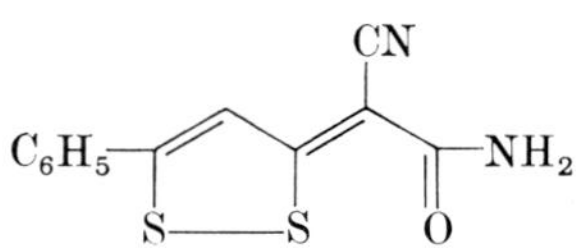	P_2S_5	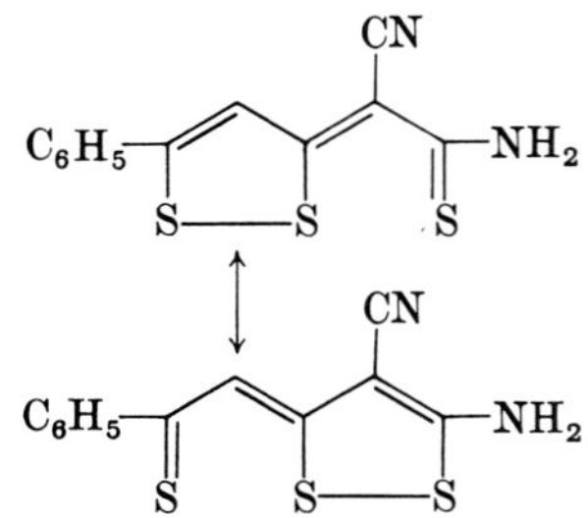	68	471

with malononitrile in the presence of phosphorus oxychloride gave the dinitrile **7**, which upon alkaline hydrolysis yielded the α-cyanoamide **8**. Reaction with phosphorus pentasulfide then gave a "thioamide" whose spectral properties suggested its formulation as the thiothiophthene **9**. The compound is acidic rather than basic; it was suggested that no-bond resonance contributes to electron delocalization via sulfur *d*-orbital expansion (471). The thiothiophthene **9** has alternately been prepared by condensation of 3-methylmercapto-5-phenyl-1,2-dithiolium iodide (**10**) with cyanothioacetamide (489) (Scheme 12).

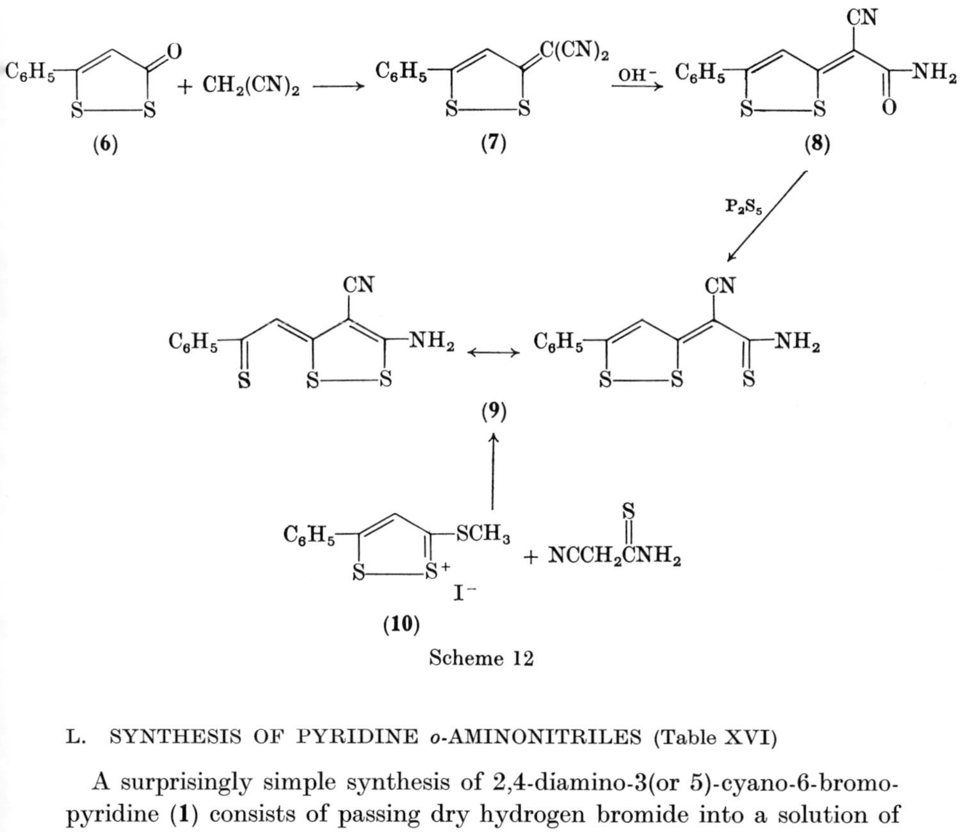

Scheme 12

L. SYNTHESIS OF PYRIDINE *o*-AMINONITRILES (Table XVI)

A surprisingly simple synthesis of 2,4-diamino-3(or 5)-cyano-6-bromopyridine (**1**) consists of passing dry hydrogen bromide into a solution of malononitrile in tetrahydrofuran (84). This remarkable transformation apparently proceeds by initial dimerization of malononitrile by the action of hydrogen bromide (malononitrile dimer is the sole product formed when hydrogen chloride rather than hydrogen bromide is used); a subsequent acid-catalyzed intramolecular cyclization of malononitrile dimer

TABLE XVI

Pyridine *o*-Aminonitriles

Reactants		Product			Yield, %	Ref.
Malononitrile derivative	Other component	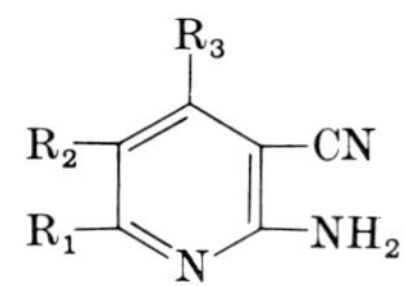				
		R_1	R_2	R_3		
$CH_2(CN)_2$	HBr	Br	H	NH_2	72	84
$NCCH_2C(NH_2)=C(CN)_2$	HBr	Br	H	NH_2[a]	72	84
$CH_2(CN)_2$	$CH_3COC(CH_3)=CHOH$ and NH_3	CH_3	CH_3	H	14–17	108
	$CH_3COCH_2COCH_3$ and NH_3	CH_3	H	CH_3	24	108

$C_2H_5O(H)C{=}C(CN)_2$	$Na^{+}\,{}^{-}HC(CN)_2$	C_2H_5O	CN	H	96	106
	KOH and CH_3OH	CH_3O	CN	H	29	246
	$CH_3SC({=}NH)NH_2$	CH_3S	CN	H	47	246
$H_2N(H)C{=}C(CN)_2$	H_2/Pd/C at 1000–1500 psi/77°	H	CN	H	54	251
$(NC)_2C{=}CH{-}\bar{C}(CN)_2$	KOH and CH_3OH	CH_3O	CN	H	85	246
	$CH_3SC({=}NH)NH_2$	CH_3S	CN	H	90	246
	HCl gas	Cl	CN	H	90	106
	HBr gas	Br	CN	H	93.5	106
	C_2H_5SH	C_2H_5S	CN	H	80	246
	$C_6H_5CH_2SC({=}NH)NH_2$	$C_6H_5CH_2S$	CN	H	45	246

(continued)

TABLE XVI (*continued*)

Reactants		Product				
Malononitrile derivative	Other component	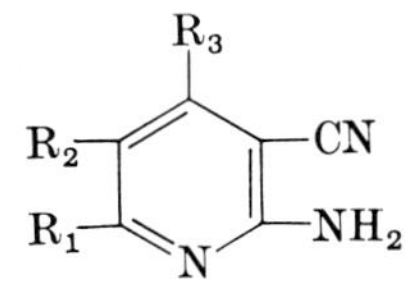			Yield, %	Ref.
		R_1	R_2	R_3		
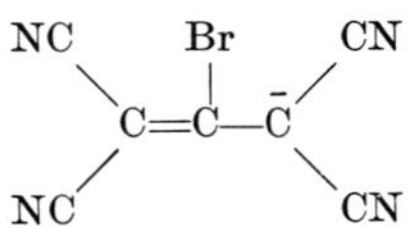	HBr gas	Br	CN	Br	93	106
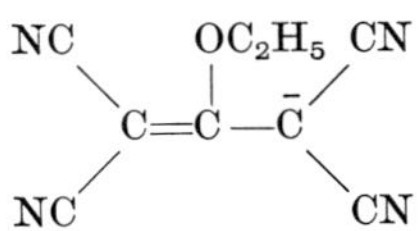	HCl gas	Cl	CN	C_2H_5O	77.7	106
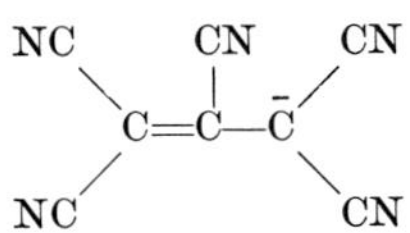	HCl gas	Cl	CN	CN	69.5	106
	HBr liquid	Br	CN	CN	90	106
	C_2H_5OH and H_2SO_4	C_2H_5O	CN	CN	75	106
	$(CH_3)_2CHOH$ and H_2SO_4	$(CH_3)_2CHO$	CN	CN	31	106

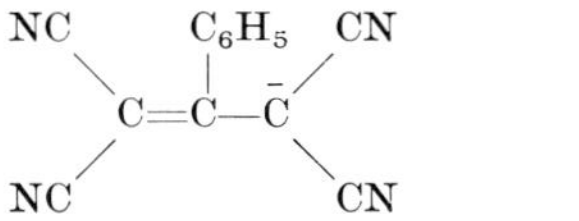	HCl gas	Cl	CN	C_6H_5	87.6	106
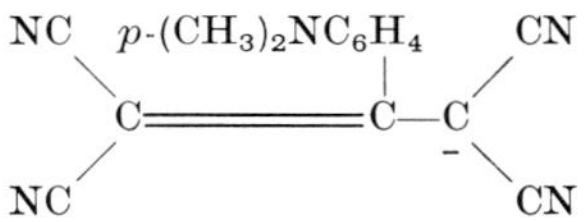	HCl gas	Cl	CN	p-$(CH_3)_2NC_6H_4$	86	106
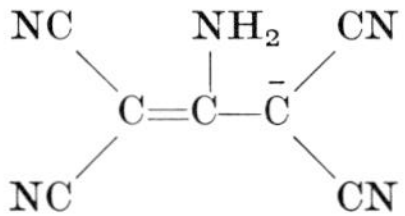	HCl gas	Cl	CN	NH_2	52	106
	HBr liquid	Br	CN	NH_2	82	106
	HI (aqueous)	I	CN	NH_2	72	106
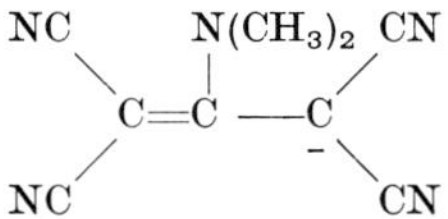	HCl gas	Cl	CN	$N(CH_3)_2$	10	106
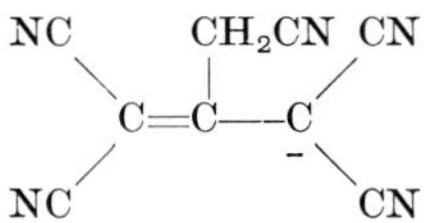	HCl conc.	Cl	CN	CH_2CN	57	555
	HBr	Br	CN	CH_2CN	93	555
	HI/acetone	I	CN	CN CH_3 OH $C{=}CCH_2C(CH_3)_2$	24	555

(continued)

TABLE XVI (*continued*)

Reactants		Product	Yield, %	Ref.
Malononitrile derivative	Other component			
$NCCH_2CONH_2$	$NaOC_2H_5$	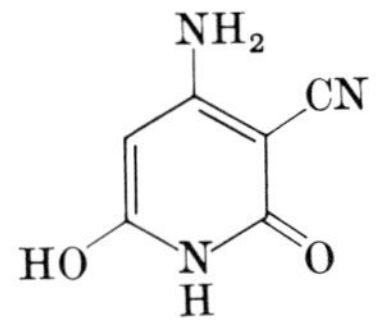	36	508
$H_2NC(=O)CH_2C(NH_2)=C(CN)-COOC_2H_5$	Na_2CO_3	Same as above	—	521

[a] CN group is either in 3 or 5 position.

then gives **1**. The fact that malononitrile dimer, independently prepared, could be substituted for malononitrile in the above synthesis of **1** substantiates this proposed reaction course.

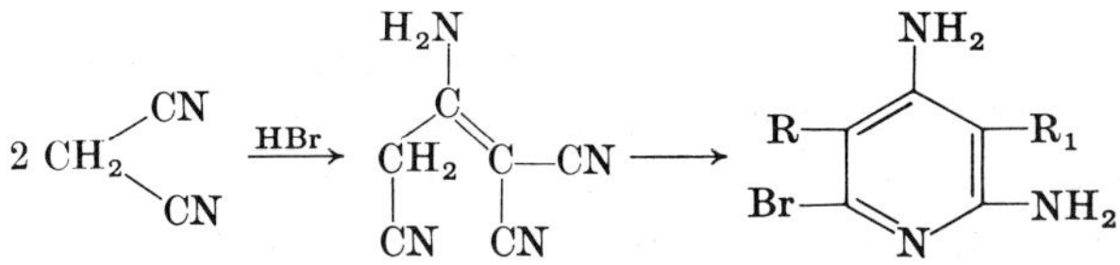

1 R = H, R_1 = CN or
R = CN, R_1 = H

It has been claimed (108) that 3-hydroxymethylene-2-butanone (**2**) upon treatment with methanolic ammonia, followed by addition of malononitrile, gives 2-amino-3-cyano-5,6-dimethylpyridine (**3**). This structural assignment was supported by diazotization to a compound (in extremely small yield and amount) whose mixed melting point gave no depression with an authentic sample of 3-cyano-5,6-dimethyl-2(1*H*)-pyridone. Despite this latter evidence, it seems probable that the product actually formed was 2-amino-3-cyano-4,5-dimethylpyridine (**5**). The reaction of α-hydroxymethylene ketones with ammonia to give

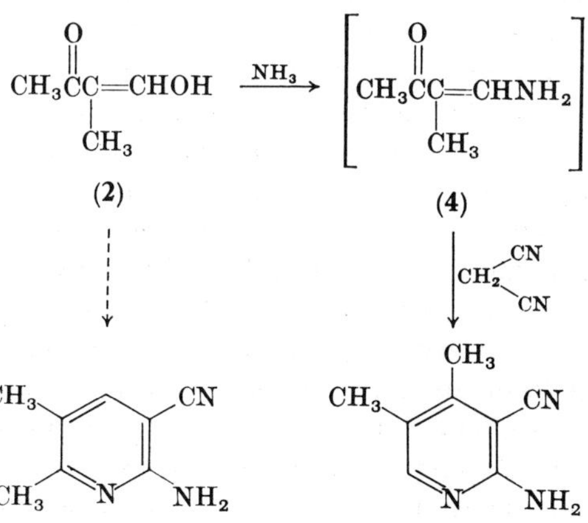

aminomethylene ketones is well established and in the above example must have given 3-aminomethylene-2-butanone (**4**) as an intermediate; subsequent condensation with malononitrile could only have given **5**.

An analogous reaction of acetylacetone first with dry ammonia and then with malononitrile gave 2-amino-3-cyano-4,6-dimethylpyridine (108), whose structure cannot be in doubt.

As mentioned on p. 105, the reaction of ethoxymethylenemalononitrile with *S*-methylthiourea gives 2-methylthio-4-amino-5-cyanopyrimidine (**6**) in a reaction characteristic of pyrimidine syntheses from the former versatile intermediate. A recent study of this classical reaction has led to the surprising conclusion that the nature of the product formed is dependent upon the solvent in which the condensation is run. Thus, although **6** was formed when the condensation was carried out in aqueous acetone, 2-amino-3,5-dicyano-6-methylthiopyridine (**11**, R = CH_3) was formed in moderate yield when the reaction mixture in aqueous acetone was brought slowly to neutrality with ammonium hydroxide. 2-Amino-3,5-dicyano-6-methoxypyridine (**12**, R = CH_3) was formed simply upon addition of aqueous potassium hydroxide to a solution of ethoxymethylene-malononitrile in methanol (151,246).

These apparently bewildering observations were unraveled as follows. When a solution of ethoxymethylenemalononitrile in methanol was added slowly to an equivalent of aqueous potassium hydroxide, the potassium salt of hydroxymethylenemalononitrile (**8**) was formed in excellent yield. On the other hand, if a solution containing one-half an equivalent of potassium hydroxide in methanol was added slowly to a solution of ethoxymethylenemalononitrile in methanol, ethyl formate was formed along with the potassium salt of 1,1,3,3-tetracyanopropene (**10**). These observations were explained by assuming that the initial reaction was hydration of the double bond of ethoxymethylenemalononitrile to give **7**, which can either lose ethanol to give **8**, or undergo a reverse aldol condensation to give ethyl formate and the potassium salt of malononitrile (**9**). It is well known that the latter compound reacts readily with ethoxymethylenemalononitrile to give **10** (Scheme 13).

In a remarkable series of reactions, 1,1,3,3-tetracyanopropene (**10**) can be converted to a variety of pyridine *o*-aminonitriles (106,246). For example, heating **10** (as its potassium salt) in a basic solution of aqueous methanol gives 2-amino-3,5-dicyano-6-methoxypyridine (**12**, R = CH_3) in 85% yield (246). In fact, the condensation of the sodium salt of malononitrile with ethoxymethylenemalononitrile gives 2-amino-3,5-dicyano-6-ethoxypyridine (**12**, R = C_2H_5) in 95% yield when the reaction is allowed to proceed without cooling; when the reaction is carried out at 0°, **10** is obtained in 90% yield. When this latter compound is treated with hydrogen chloride 2-amino-3,5-dicyano-6-chloropyridine (**13**) is formed (106). The corresponding 6-alkoxypyridines are readily formed upon

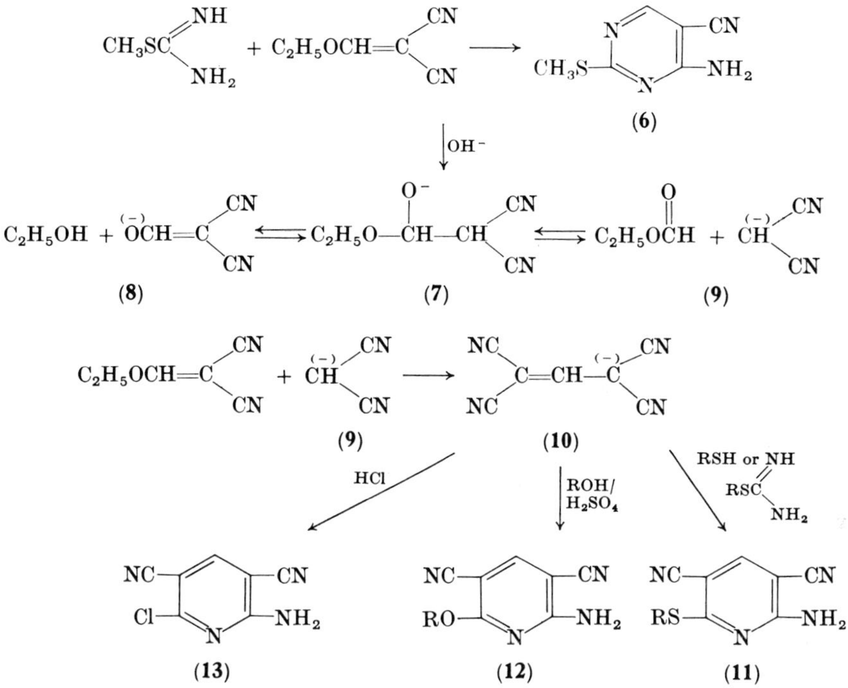

Scheme 13

treatment of **10** (as its potassium or ammonium salt) with alcohols in the presence of sulfuric acid.

All of these reactions appear to involve the initial formation from **10** of an imino ether or imino chloride, followed by intramolecular cyclization.

The only apparent conversion of an aliphatic enaminonitrile to an aromatic *o*-aminonitrile is the palladium-on-carbon dehydrogenation of *N*-nitroso or *N*-acetyl 3-cyano-4-amino-3-piperideine (prepared by nitrosation or acetylation, respectively, of the Thorpe-Ziegler cyclization product of bis-cyanoethylamine), which smoothly yields 3-cyano-4-aminopyridine (536).

M. SYNTHESIS OF QUINOLINE AND ISOQUINOLINE *o*-AMINONITRILES (Table XVII)

In an extension of the reaction described above in which 2-amino-3-cyanopyridines are formed by condensation of malononitrile with α-aminomethylene ketones, 2-amino-3-cyano-5,6,7,8-tetrahydroquinoline

TABLE XVII

Quinoline and Isoquinoline *o*-Aminonitriles

Reactants		Product	Yield, %	Ref.
Malononitrile derivative	Other component			
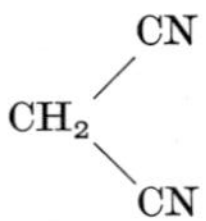	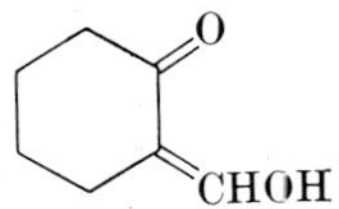 and NH_3	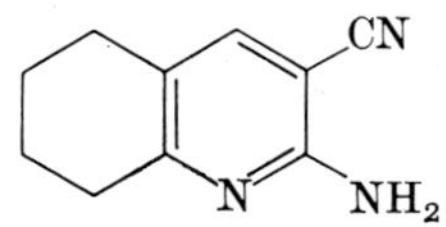	22–25	108
	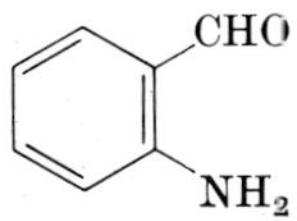 with $CH_3\bar{C}OO\overset{+}{N}H_4$ and CH_3CONH_2	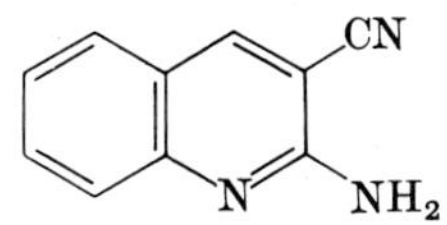	100	260
	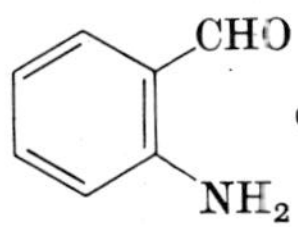 excess CH_3COOH, a "little" pyridine	Same as above	30[a]	261
	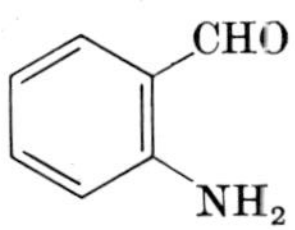 and piperidine	Same as above	85 —	162 14

	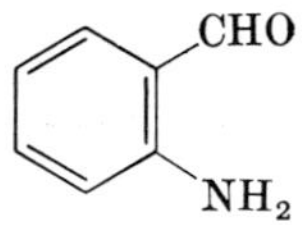 and pyridine	Same as above	76	31
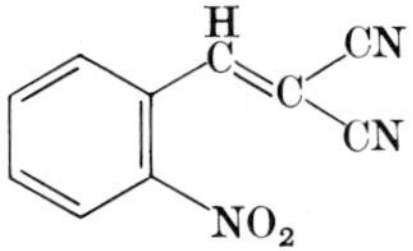	Fe/CH_3COOH	Same as above	70	172
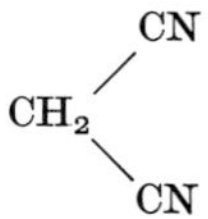	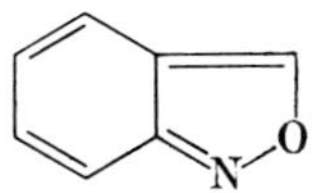 and piperidine	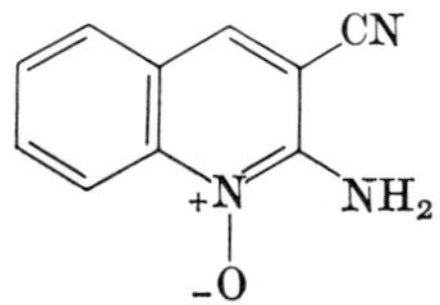	99	465
	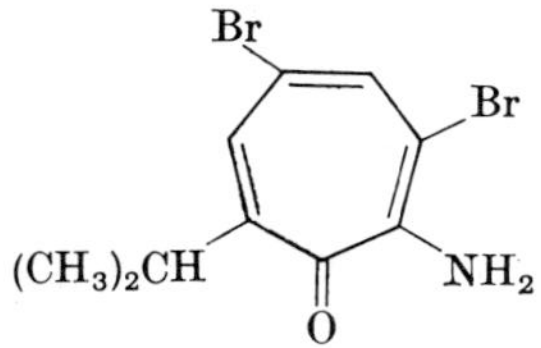	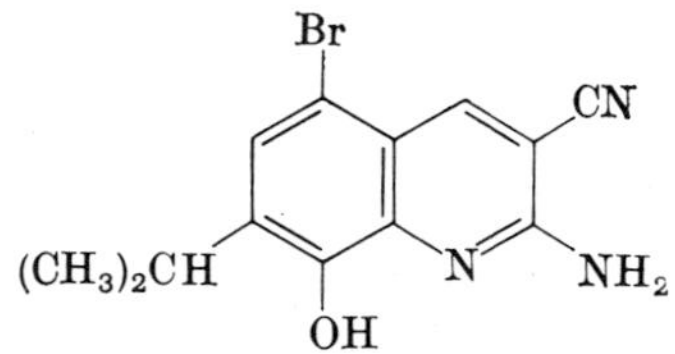	—	418
	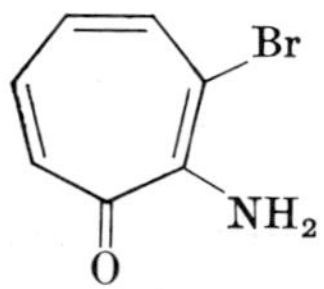	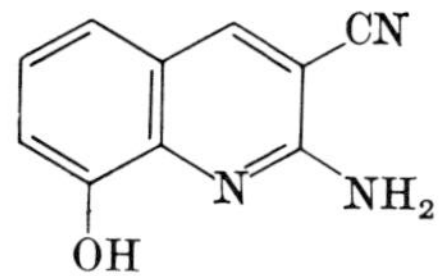	—	420

(continued)

TABLE XVII (*continued*)

Reactants				
Malononitrile derivative	Other component	Product	Yield, %	Ref.
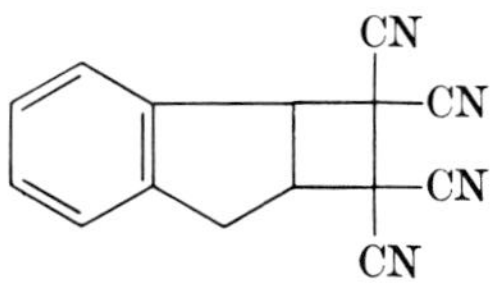	Pyrolyze	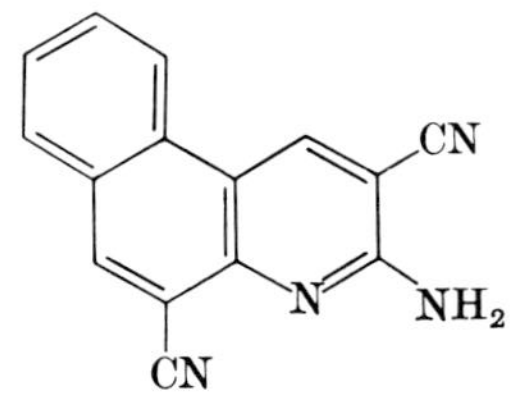	—	469
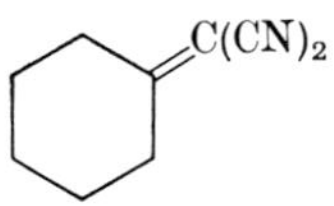	$HC(OC_2H_5)_3/(CH_3CO)_2O$, then NH_3	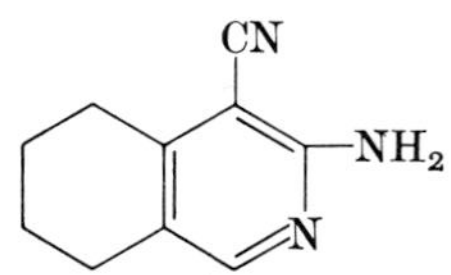	30–40	509
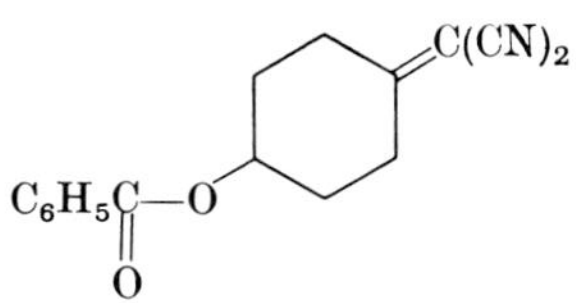	$HC(OC_2H_5)_3/(CH_3CO)_2O$, then NH_3	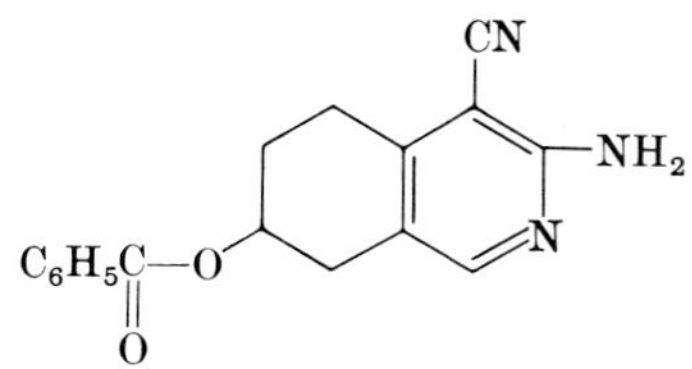	10	509

[a] About 30% of the intermediate ylidene product is also isolated.

TABLE XVIII

Pyrido(2,3-*b*)pyridine and Pyrido(2,3-*d*)pyrimidine *o*-Aminonitriles

Reactants: Malononitrile derivative	Reactants: Other component	Product	Yield, %	Ref.
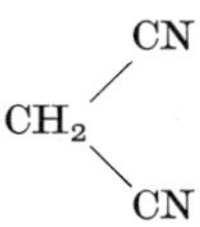	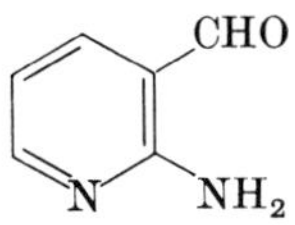	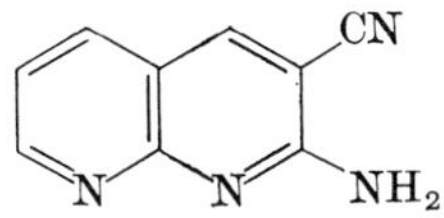	96	541
	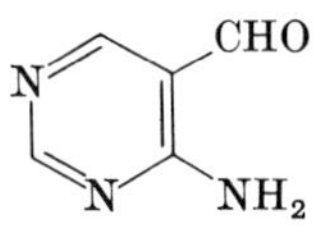	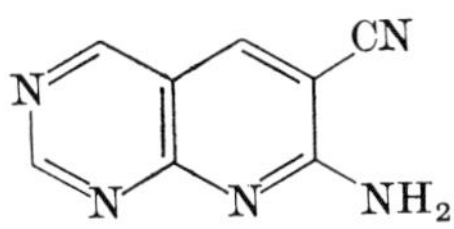	86	444
	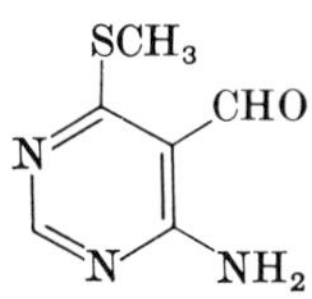	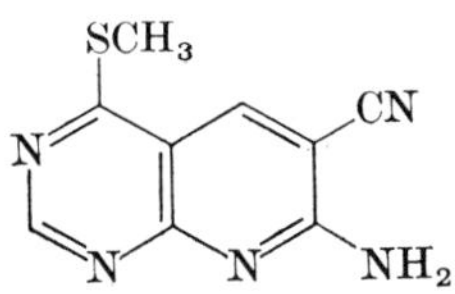	73	444
	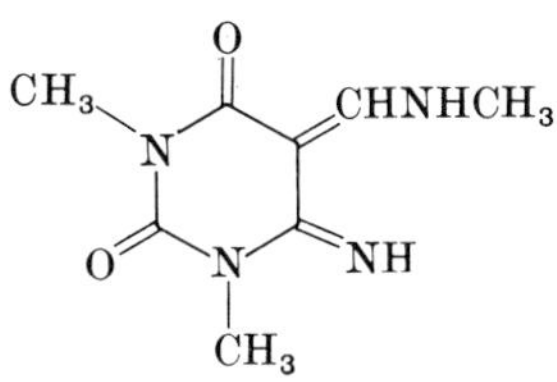	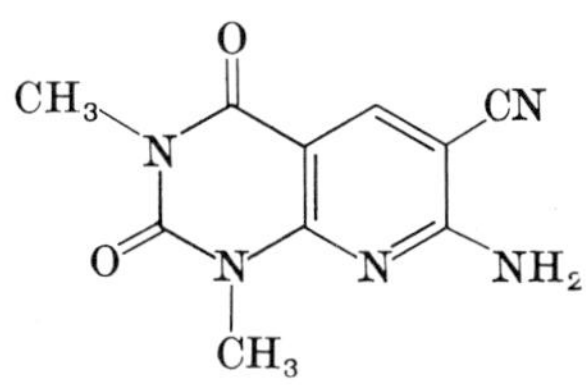	84	557

(**3**) is formed from α-hydroxymethylenecyclohexanone (**1**), ammonia and malononitrile (108). The initial condensation must be between malononitrile and the α-formyl grouping of **1** to give an intermediate alkylidenemalononitrile (**2**) which reacts subsequently with ammonia. Using the same reagents (cyclohexanone, malononitrile, ammonia, and ethyl formate or ethyl orthoformate), but reversing the order of the reactions, the isomeric 3-amino-4-cyano-5,6,7,8-tetrahydroisoquinoline (**6**) is formed.

Thus, the reaction of cyclohexylidenemalononitrile (**4**) with ethyl orthoformate in the presence of acetic anhydride gives **5**, which upon treatment with ammonia is cyclized to **6** (509).

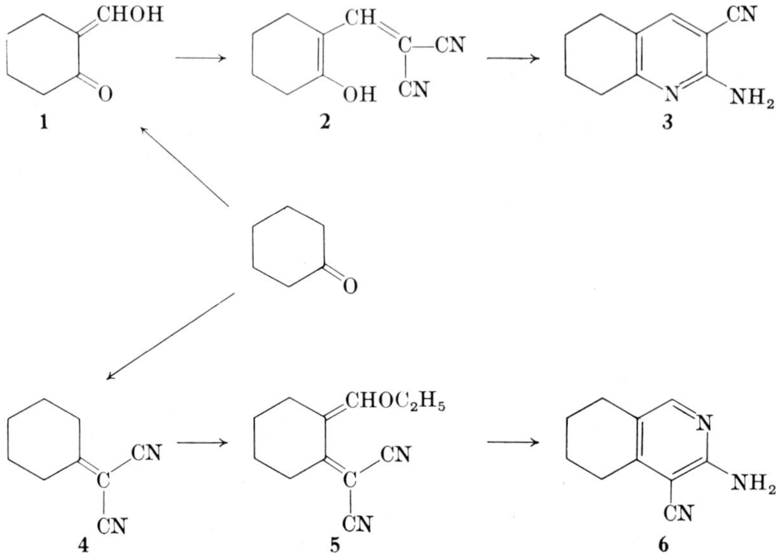

2-Amino-3-cyanoquinoline (**7**) itself arises by condensation of malononitrile with *o*-aminobenzaldehyde (**8**) in the presence of a basic catalyst (14,31,162,260,261), or alternately by reduction of *o*-nitrobenzylidenemalononitrile (**9**) (172).

It is well known that anthranil (**10**) reacts with alkoxides, strong alkali, barium hydroxide, etc., to give esters or salts of anthranilic acid (**12**), while certain other bases such as hydroxylamine and hydrazines give derivatives of *o*-hydroxylaminobenzaldehyde (**13**) (466). This curious difference in the oxidation states of the *o*-substituents in the products produced upon ring-opening of anthranil with different bases appears to find its rationale in the fate suffered by the respective adducts (**11**) formed by Michael-type addition of the base to the α,β-unsaturated anil

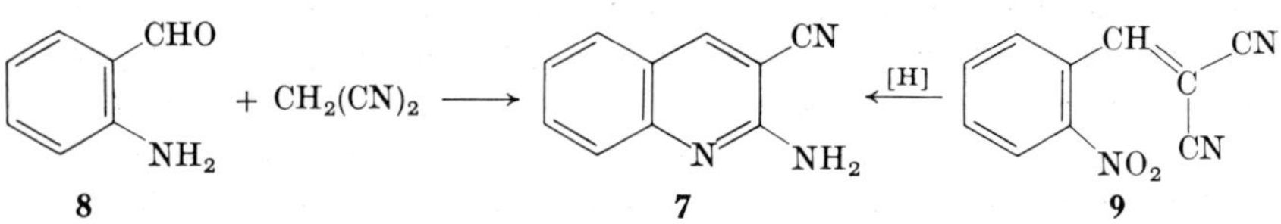

system present in **10**. Thus, loss of H_a leads to O—N cleavage and the formation of a derivative of anthranilic acid, while loss of H_b results in C—O bond cleavage and the formation of a derivative of *o*-hydroxylaminobenzaldehyde. The use of a nucleophile carrying both an acidic proton H_b and an electrophilic substituent capable of interacting with the hydroxylamino grouping produced upon ring-opening should lead to recyclization, with ring expansion. Among the nucleophiles utilized was malononitrile, which reacts exothermically with anthranil in the presence of a catalytic amount of piperidine to give 2-amino-3-cyanoquinoline 1-oxide (**14**) in quantitative yield (465) (Scheme 14).

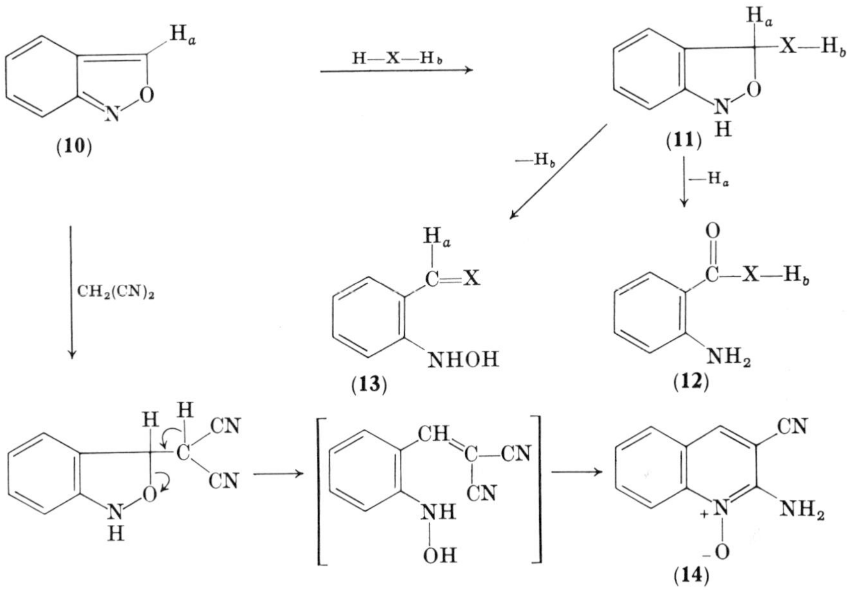

Scheme 14

The pyrolysis of the adduct between indene and tetracyanoethylene to give 3-amino-2,5-dicyanobenzo(*f*)quinoline (469) has already been discussed in Chapter I, Section III-F (The Thorpe-Ziegler Reaction).

N. SYNTHESIS OF PYRAZINE AND PTERIDINE *o*-AMINONITRILES (Tables XIX and XX)

Almost all naturally occurring pteridines are 6-substituted derivatives, and most are unsubstituted at position 7 (564,565). Their synthesis by the classical, most widely employed synthetic route to pteridines (condensation of a 4,5-diaminopyrimidine with an α,β-dicarbonyl compound

TABLE XIX

Pyrazine *o*-Aminonitriles

Reactants				
Malononitrile derivative	Other component	Product	Yield, %	Ref.
$H_2NCH(CN)_2 \cdot TsOH$	$\mathrm{HON{=}CH\overset{\overset{O}{\Vert}}{C}CH_3}$	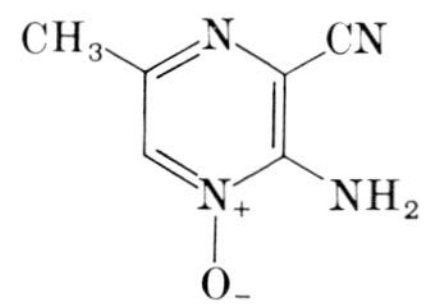	81	551
	$\mathrm{HON{=}CH\overset{\overset{O}{\Vert}}{C}CH{=}NOH}$	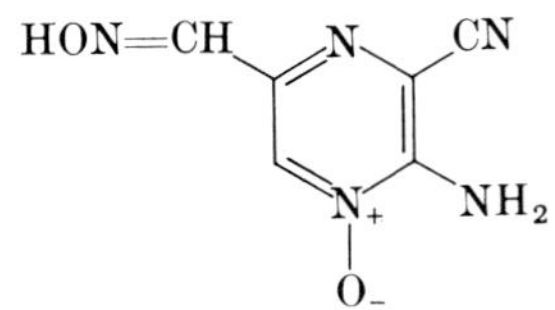	100	537
	$\mathrm{HON{=}CH\overset{\overset{O}{\Vert}}{C}C_3H_7}$	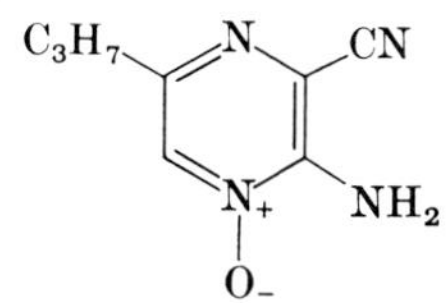	84	537

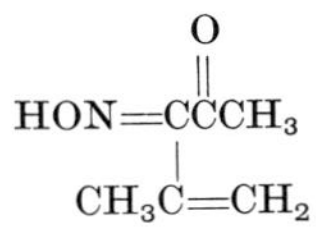	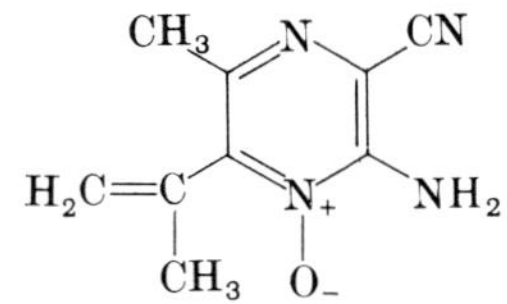	78	537
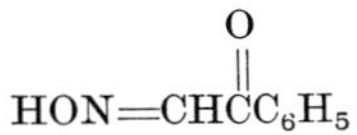	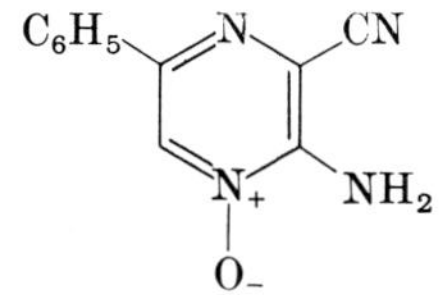	95	537
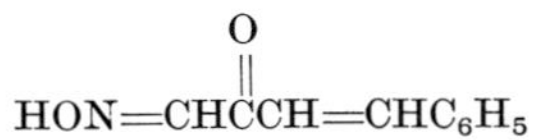	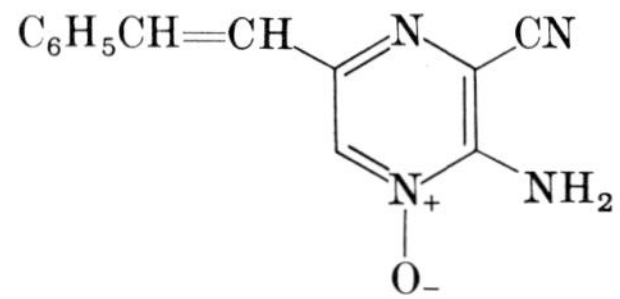	70	537
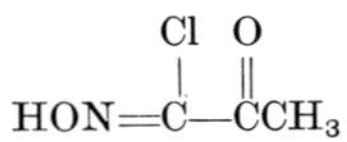	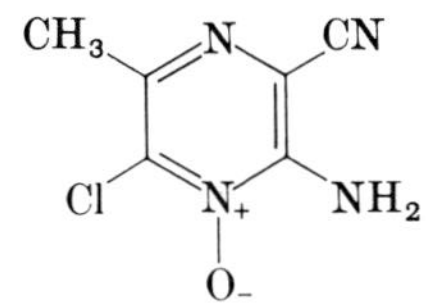	90	537

(continued)

TABLE XIX (*continued*)

Reactants				
Malononitrile derivative	Other component	Product	Yield, %	Ref.
$H_2NCH(CN)_2 \cdot TsOH$ (*cont.*)	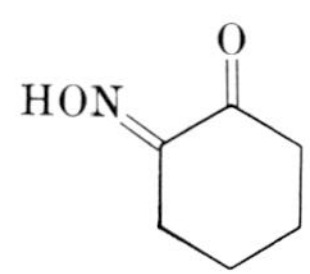	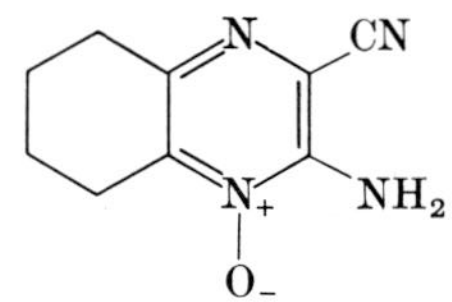	75	537
	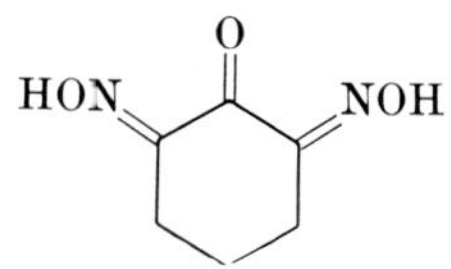	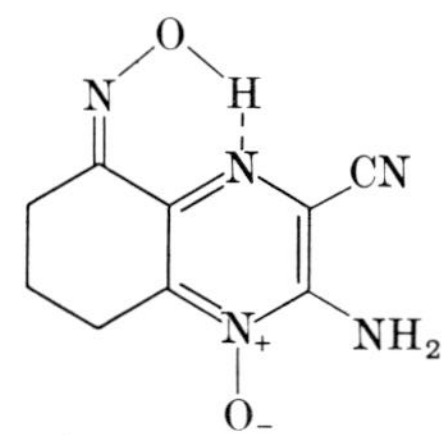	90	537

TABLE XX
Pteridine *o*-Aminonitriles

Reactants				
Malononitrile derivative	Other component	Product	Yield, %	Ref.
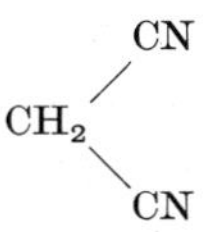	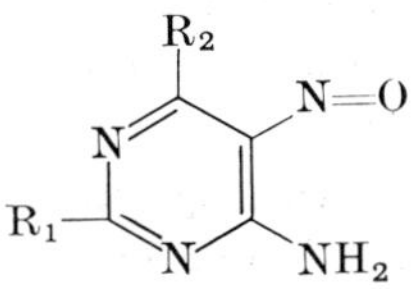	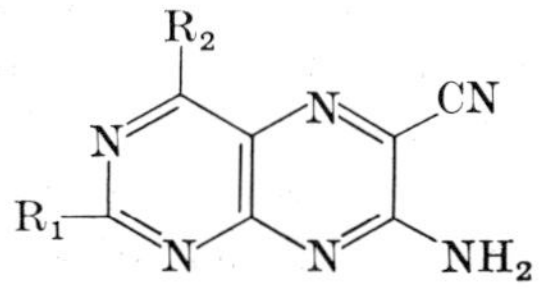		
	R_1	R_2		
	C_6H_5	NH_2	—	439
	p-ClC_6H_4	NH_2	—	439
	p-$CH_3C_6H_4$	NH_2	—	439
	o-ClC_6H_4	NH_2	—	439
	m-$CH_3C_6H_4$	NH_2	—	439
	o-$C_4H_9OC_6H_4$	NH_2	—	439
	m-BrC_6H_4	NH_2	—	439
	p-$C_2H_5C_6H_4$	NH_2	—	439
	m-Br-*p*-$CH_3OC_6H_3$	NH_2	—	439
	o-C_4H_9-*m*-$CH_3OC_6H_3$	NH_2	—	439
	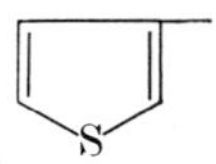	NH_2	—	439
	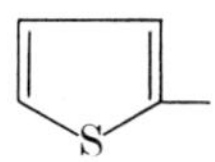	NH_2	—	439
	C_6H_5	$NHCH_3$	—	439
	C_6H_5	$N(CH_3)_2$	—	439
	C_6H_5	$N(C_4H_9)_2$	—	439
	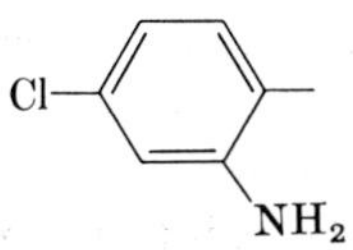	NH_2	—	439
	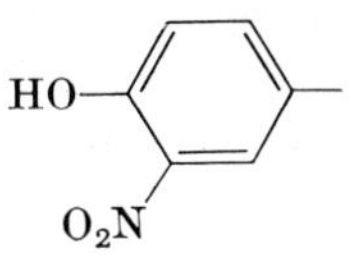	NH_2	—	439
	m-$NH_2C_6H_4$	NH_2	—	439

(502)), is unavoidably ambiguous when an unsymmetrical α,β-dicarbonyl compound is employed; a mixture of isomers results which is not only extremely difficult to separate but is often even undetectable by the usual chromatographic methods. In principle, this ambiguity may be avoided by the initial preparation of a pyrazine with a known substitution pattern and subsequent closure of the pyrimidine ring, but this approach has seen little application because of inaccessibility of the requisite pyrazine intermediates. A recent solution to this problem involves still another application of the malononitrile → *o*-aminonitrile condensation principle. Aminomalononitrile tosylate (566) (**1**) has been found to condense in high yield with α-oximinoketones to give 2-amino-3-cyanopyrazine 1-oxides (**2**). Subsequent cyclization with guanidine gives 2,4-diaminopteridine 8-oxides (**3**) (see pp. 272, 293), from which the natural pterins (2-amino-4(3*H*)-pteridinones) (**4**) may be prepared by deoxygenation and hydrolysis (537, 551).

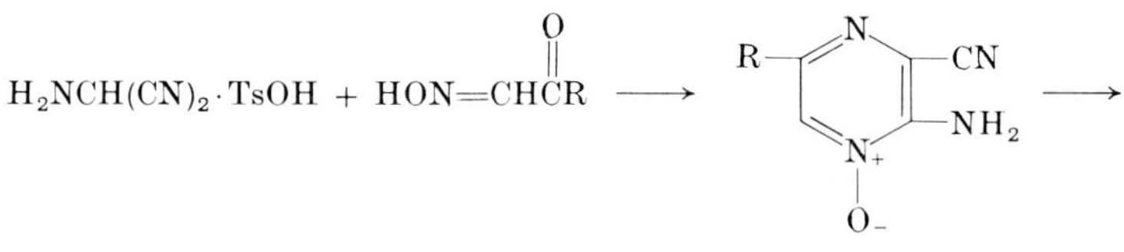

(**1**) (**2**)

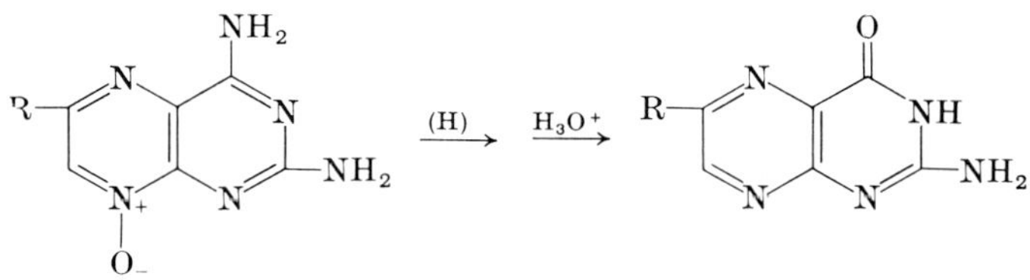

(**3**) (**4**)

In an earlier attempt to devise an unequivocal pteridine synthesis which avoided the ambiguity of the classical Isay procedure (501), Timmis first described (503) the condensation of 4-amino-5-nitrosopyrimidines (**5**) with active methylene compounds (504,505). In this reaction, the nitroso grouping functions as a carbonyl; the anil thus formed (**6**) suffers an intramolecular cyclization analogous in principle to the many terminal cyclization reactions discussed above. This concept has been applied to the synthesis of pteridine *o*-aminonitriles (**7**) by the use of malononitrile as the active methylene component (439). Although an analogous condensation would, in principle, lead to other condensed pyrazine *o*-aminonitriles, the reaction has seen little application outside of the pteridine field because of the peculiar availability of the requisite 4-amino-5-nitrosopyrimidines by direct nitrosation of 4-aminopyrimidines (215), or by the condensation of isonitrosomalononitrile with amidines (567).

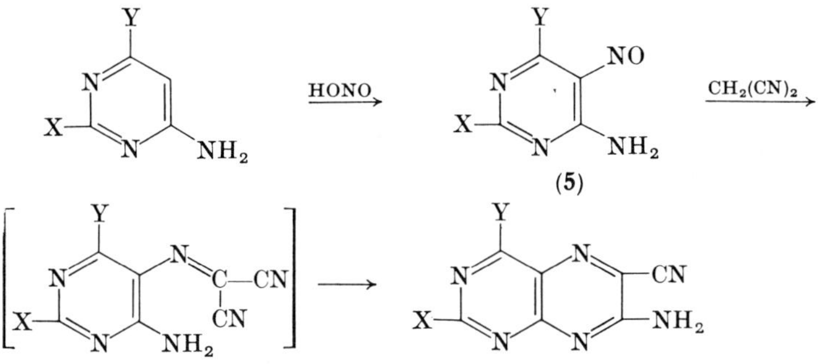

(6) (7)

O. ACID-CATALYZED CYCLIZATION OF ALKYLIDENEMALONONITRILES (Table XXI)

An effective synthesis of polycyclic aromatic *o*-aminonitriles involves acid-catalyzed cyclization of alkylidenemalononitriles. For example, 1,3-diphenylisopropylidenemalononitrile (**1**) has been cyclized to 1-amino-2-cyano-3-benzylnaphthalene (**2**) with a mixture of phosphoric acid, sulfuric acid, and phosphorus pentoxide (286). 2-(Cyclohexylidene)-cyclohexylidenemalononitrile (**3**) has been reported to cyclize in quantitative yield to **4** in concentrated sulfuric acid (434). A careful study of the conditions necessary to effect aminonitrile vs. aminoamide formation has been published (435), and the reaction has been briefly reviewed (436).

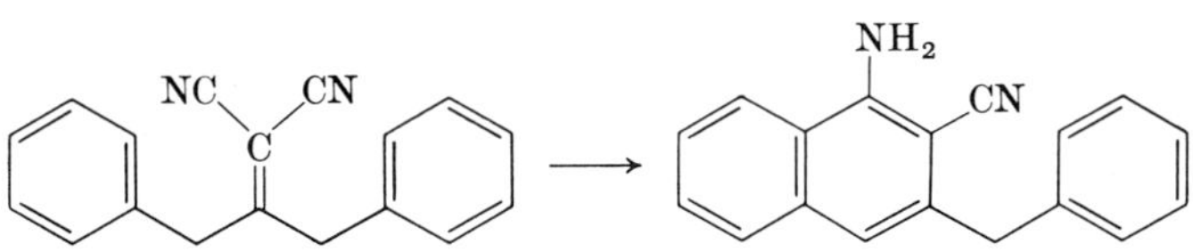

(1) (2)

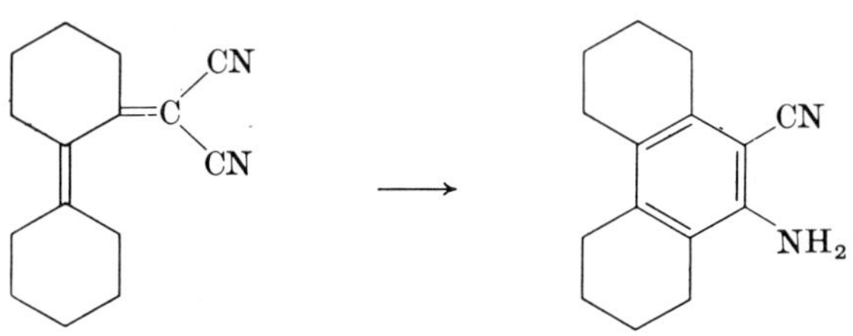

(3) (4)

TABLE XXI

o-Aminonitriles by Acid-Catalyzed Cyclization of Alkylidenemalononitriles

Starting material	Conditions	Product	Yield, %	Ref.
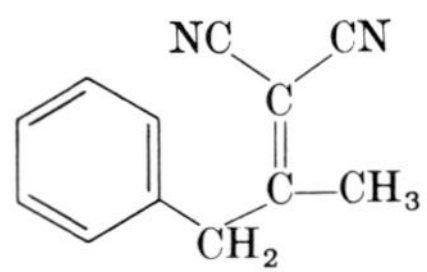	H_2SO_4/R.T./1 hr	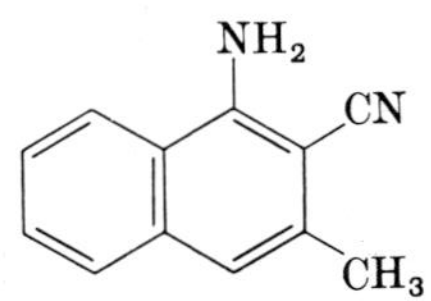	73	435
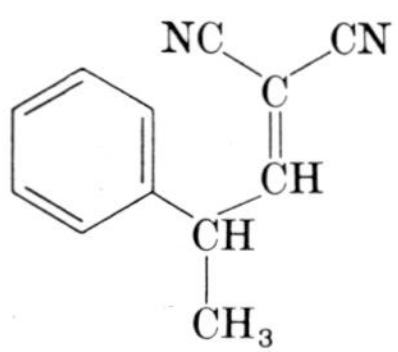	H_2SO_4/5–10°/1 hr	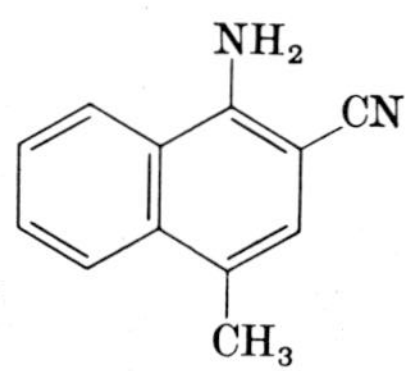	63	435
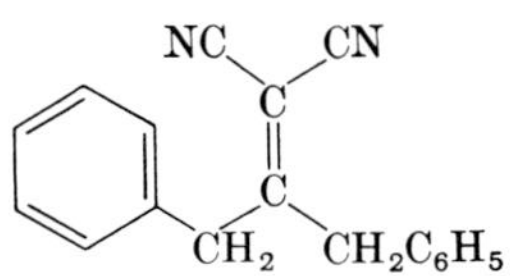	$H_3PO_4/H_2SO_4/P_2O_5$	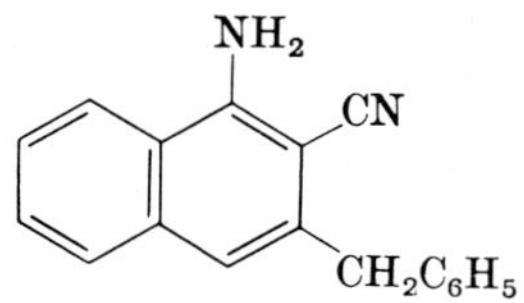	—	286

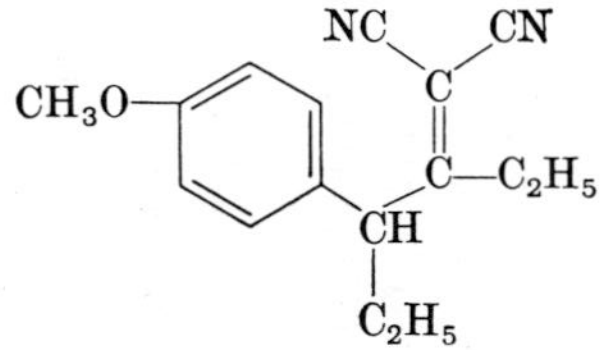	PPA/120°/20 min	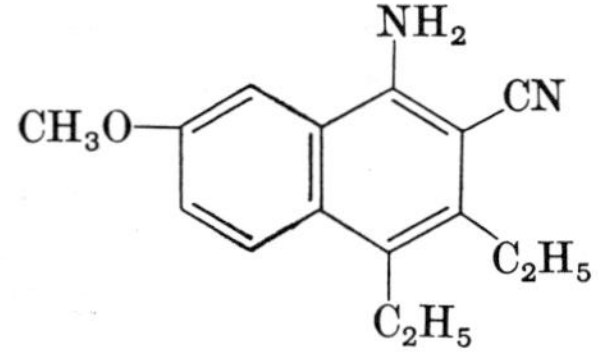	10	435
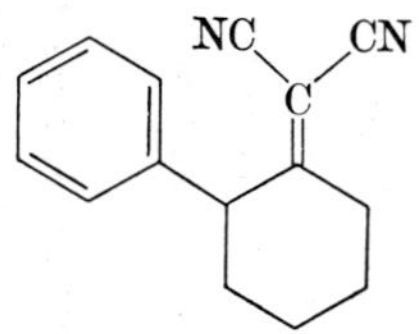	H_2SO_4/R.T./6 hr	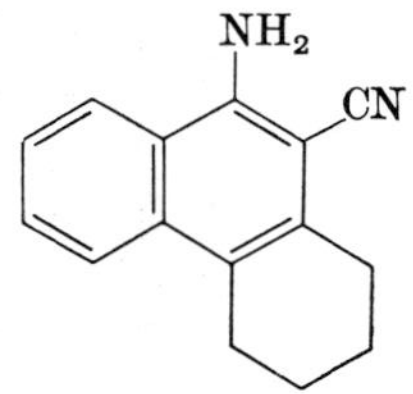	80	435
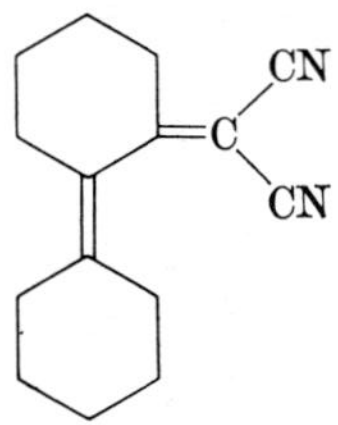	H_2SO_4	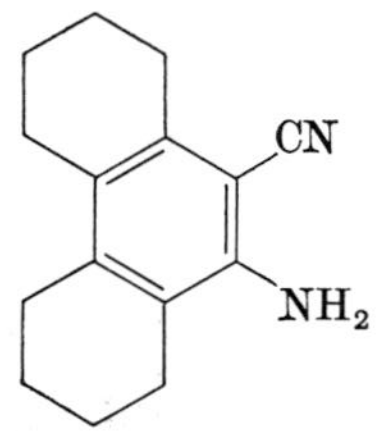	100	434

Condensation of cyclohexylidenemalononitrile (**5**) with the enamine **6** has been reported to give the octahydrobenzo(*h*)quinoline **7** (490).

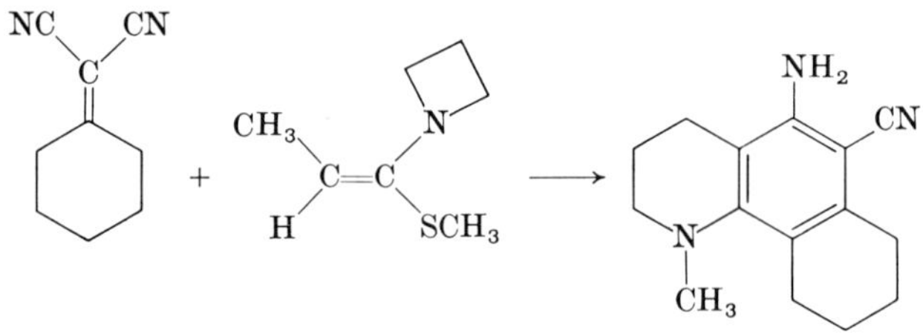

(**5**) (**6**) (**7**)

III. Synthesis of *o*-Aminonitriles by Dehydration of Amides and Related Transformations

Supplementary to the previously described synthetic routes to *o*-aminonitriles involving condensation and rearrangement reactions, a number of miscellaneous syntheses of aromatic and heterocyclic *o*-aminonitriles have been reported which involve the chemically prosaic processes of dehydration of oximes or amides, desulfurization of thioamides, and hydrolysis of urethanes and ureido intermediates. Table XXII lists a number of representative syntheses of this type. Included are several examples of 4,6-diamino-5-cyanopyrimidines, variously substituted in position 2, which were prepared directly from the corresponding 5-aldehyde by reaction with hydroxylamine in the presence of glacial acetic acid and sodium acetate; the aldehyde oxime formed *in situ* is smoothly dehydrated by the glacial acetic acid (458).

There appears to be but a single example of the application of the Hofmann degradation of amides to the preparation of an *o*-aminonitrile, presumably because of the unavailability of the requisite *o*-cyanocarboxamides. Thus, 1-phenyl-2-amino-3-cyanopyrrole (**3**) has been prepared in 55% yield from 1-phenyl-2-carbamoyl-3-cyanopyrrole (**2**) under normal Hofmann conditions; the amide was prepared from dihydromucononitrile (**1**) by the sequence of reactions shown in Scheme 15 (370) (see p. 187).

IV. Synthesis of *o*-Aminonitriles by Nucleophilic Displacement of Halogen, Oxygen, or Sulfur

Some use has also been made of standard aromatic nucleophilic displacement reactions in which an *o*-bromoamine (or protected amine) is converted with cuprous cyanide to the corresponding *o*-aminonitrile, or an *o*-halonitrile is converted with ammonia into the desired *o*-aminonitrile.

TABLE XXII

o-Aminonitriles by Dehydration of *o*-Aminoamides, and Related Transformations

Starting material	Conditions	Product	Yield, %	Ref.
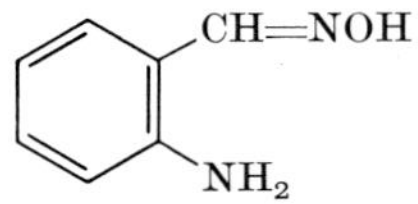	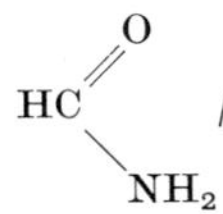 /heat	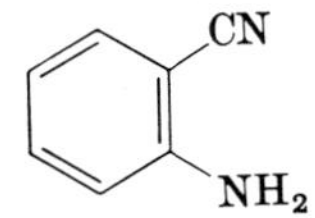	~50	128
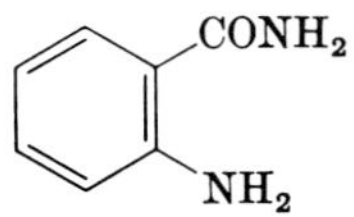	P_2O_5/sand/heat	Same as above	30–40	241
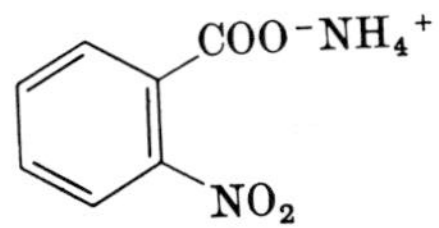	Heat	Same as above	—	263
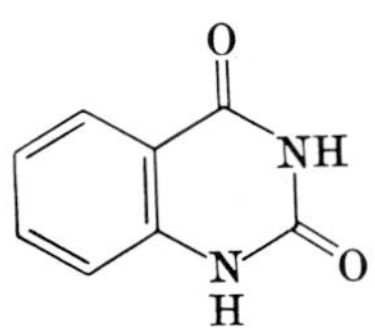	CaO/heat	Same as above	—	213

TABLE XXII (*continued*)

Starting material	Conditions	Product	Yield, %	Ref.
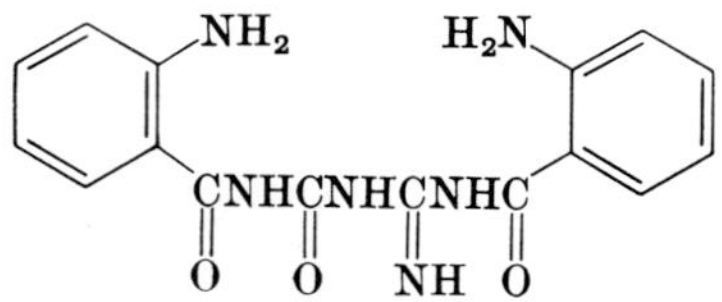	Zn dust, dry distillation	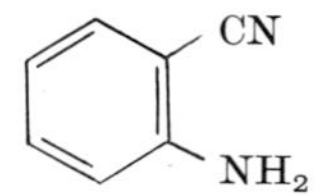	~50	331
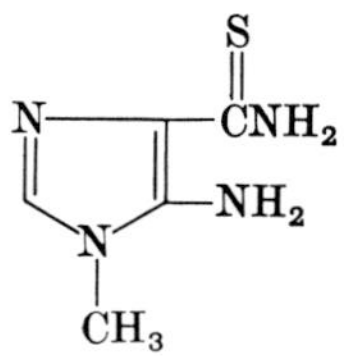	$HgCl_2$ and CH_3NH_2	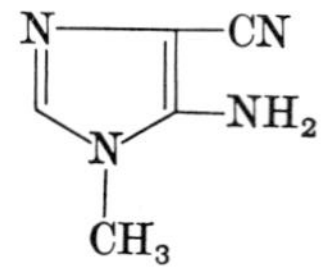	56	114
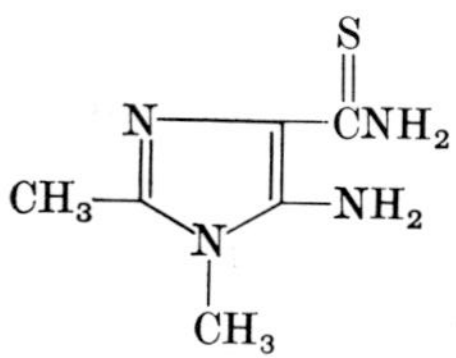	$HgCl_2$ and CH_3NH_2	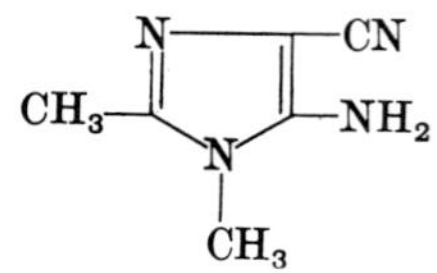	100	114
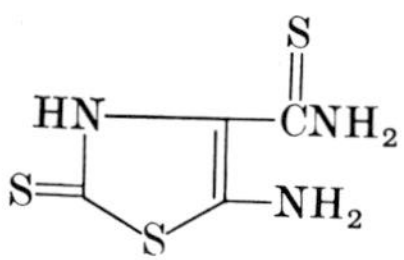	(1) ^{-}OH; (2) $(CH_3)_2SO_4$; (3) NaOH	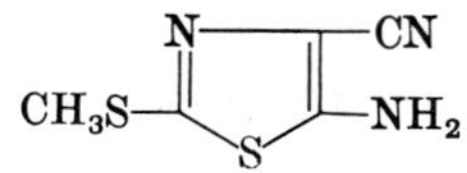	80	114

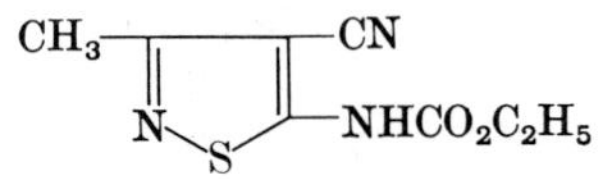	MgO/heat[a]	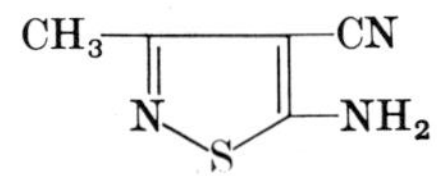	61	283
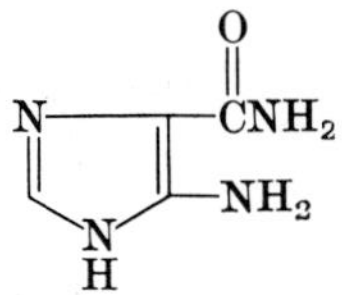	$SOCl_2$/pyridine	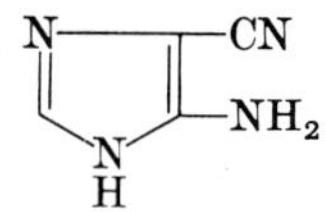	15 13	335 32
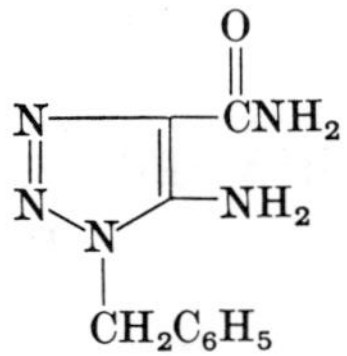	$ClCO_2C_2H_5$	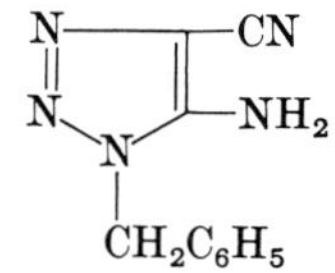	31	343
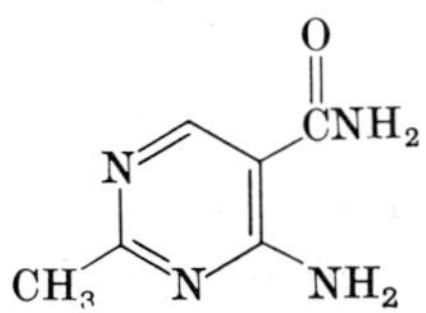	$POCl_3$	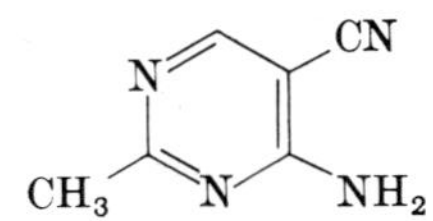	50	196

(continued)

TABLE XXII (*continued*)

Starting material	Conditions	Product	Yield, %	Ref.
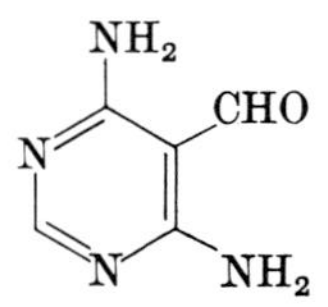	$H_2NOH/CH_3COOH/CH_3COONa$	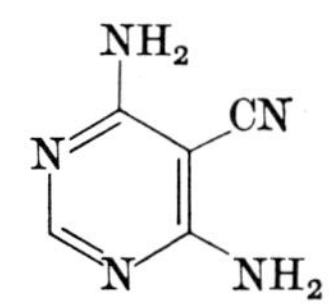	59	458
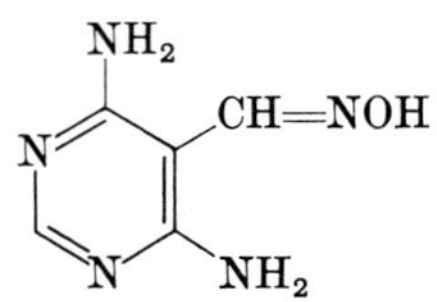	CH_3COOH	Same as above	70	458
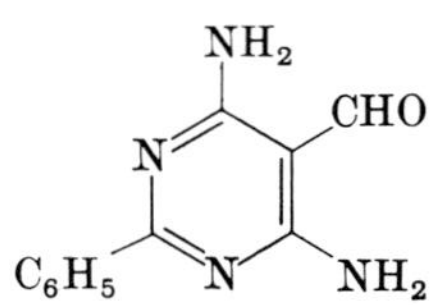	$H_2NOH/CH_3COOH/CH_3COONa$	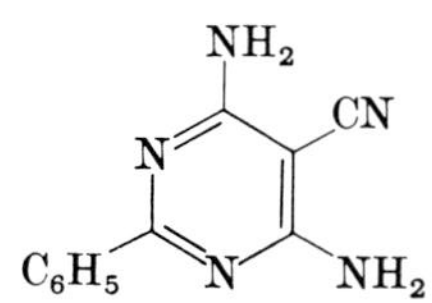	52	458
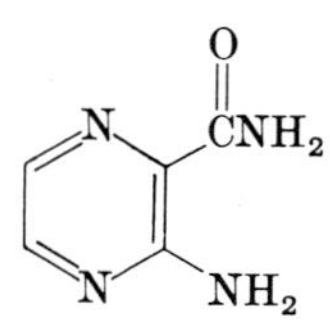	P_2O_5 in pyridine	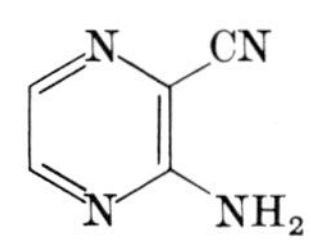	51.6	185

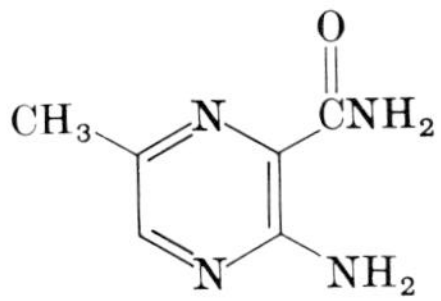	DMF/$POCl_3$	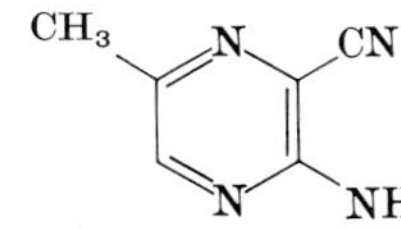 [b]	44	562
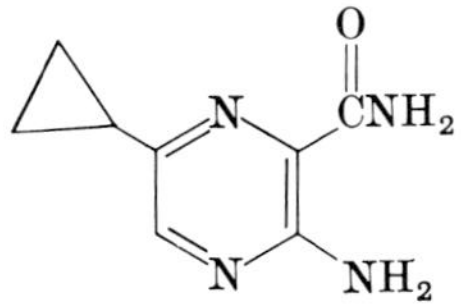	DMF/$POCl_3$	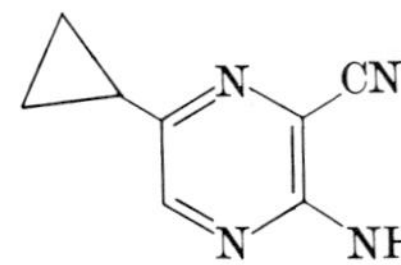 [b]	49	562
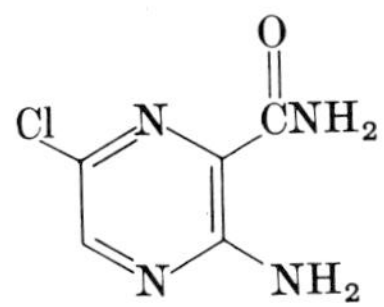	DMF/$POCl_3$	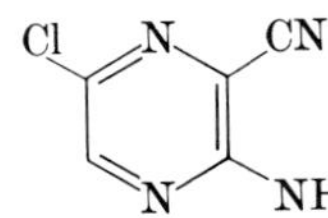 [b]	66	562
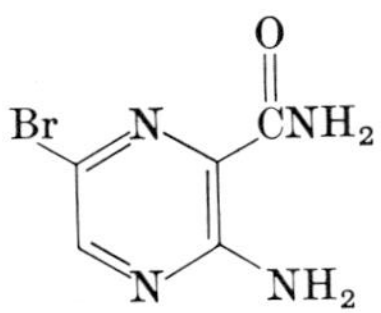	DMF/$POCl_3$	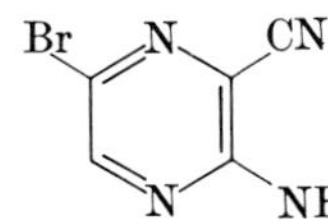 [b]	53	562

(*continued*)

TABLE XXII (*continued*)

Starting material	Conditions	Product	Yield, %	Ref.
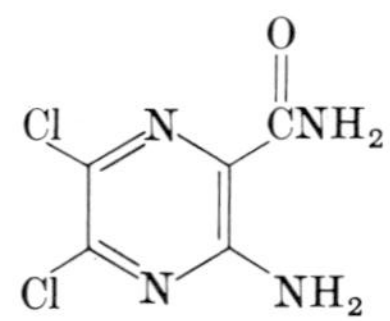	DMF/$POCl_3$	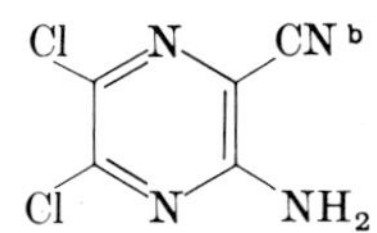 [b]	46	562

[a] This reaction is cited because of the unusual conditions employed for hydrolysis of the urethane grouping.

[b] Via the intermediate formation of the *o*-cyanodimethylaminomethyleneamino derivative, which was hydrolyzed with HCl to the *o*-aminonitrile.

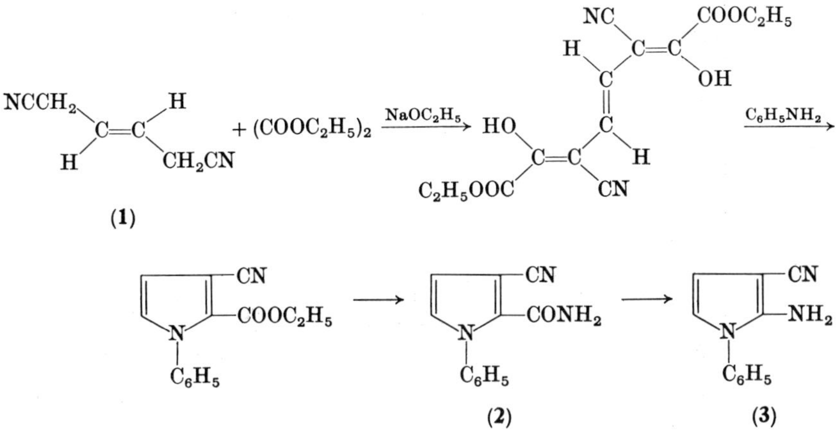

Scheme 15
(see p. 180)

These latter displacement reactions are particularly useful in the preparation of various pyrimidine and pyridine *o*-aminonitriles, where the intermediate *o*-halonitriles are readily accessible (see Table XXIII). It is possible to prepare cyclic enaminonitriles (e.g., 3-amino-4-cyano-3-pyrrolines (**6**)) by treatment of the corresponding α-cyanoketones with ammonium formate (76). Similar reaction conditions have been reported to yield 1-amino-2-cyanocyclohexene from α-cyanocyclohexanone (101), and an interesting series of 2-cyano-3-amino-Δ^2-steroids (**2**) has been prepared by the reaction of ammonium formate on the corresponding

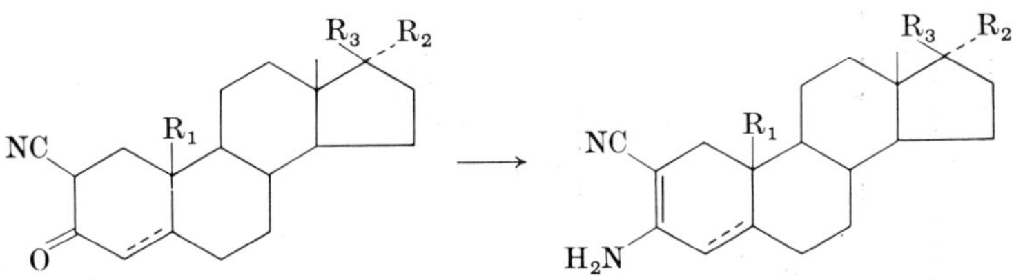

(**1**) (**2**)

2-cyano-3-keto precursors (**1**) (438). It should be pointed out, however, that the most convenient synthesis of the requisite α-cyanoketones is usually hydrolysis of the cyclic enaminonitriles (see Chapter I, Section IV), so that this circuitous sequence of interconversions has little synthetic merit except in those individual cases where the α-cyanoketone may be independently accessible. For example, the 4-cyano-3-oxopyrrolidines

TABLE XXIII

o-Aminonitriles by Replacement of –Hal or –O by $–NH_2$ or –CN

Starting material		Reagent(s)	Product		Yield, %	Ref.
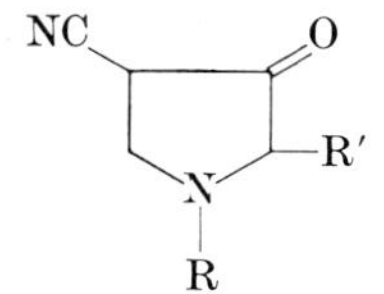			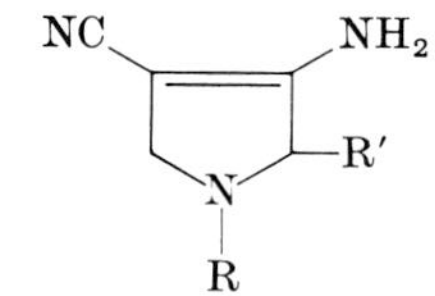			
R	R′		R	R′		
CH_3CO	H	$HCOO^-NH_4^+$	CH_3CO	H	71	76
CH_3CO	CH_3	$HCOO^-NH_4^+$	CH_3CO	CH_3	82	76
C_6H_5CO	CH_3	$HCOO^-NH_4^+$	C_6H_5CO	CH_3	85	76
CH_3CO	$C_6H_5CH_2$	$HCOO^-NH_4^+$	CH_3CO	$C_6H_5CH_2$	88	76
C_6H_5CO	$C_6H_5CH_2$	$HCOO^-NH_4^+$	C_6H_5CO	$C_6H_5CH_2$	85	76
CH_3	=O	Urea	CH_3	=O	—	539
C_4H_9	=O	Urea	C_4H_9	=O	50	539
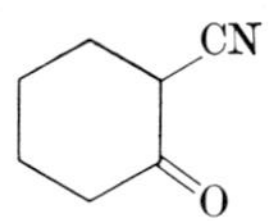		$(NH_4)_2CO_3$, then NH_3, then H_2O	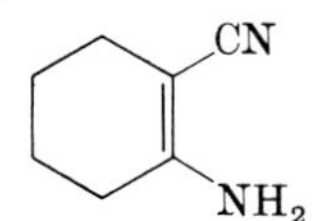		—	101
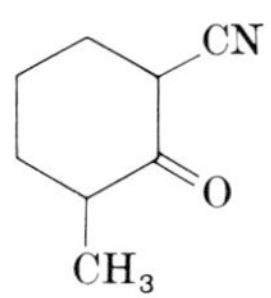		NH_3, then H_2O	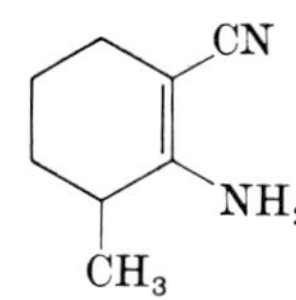		80	101

Reactant	Reagent	Product	R$_1$	R$_2$	R$_3$	Yield	Ref.
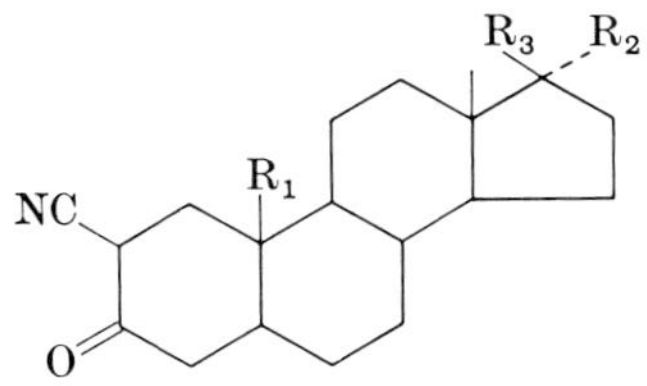	$HCOO^-NH_4^+$	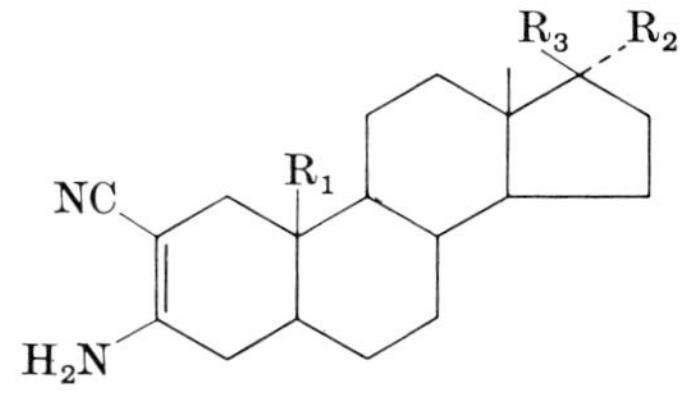					
			CH_3	H	OH	—	438
			CH_3	H	$OCOCH_3$	—	438
			H	H	OH	—	438
			H	H	$OCOCH_3$	—	438
			CH_3	CH_3	OH	—	438
			CH_3	H	C_8H_{17}	—	438
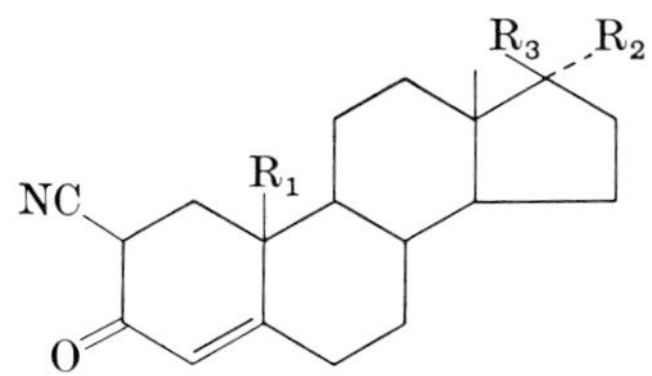		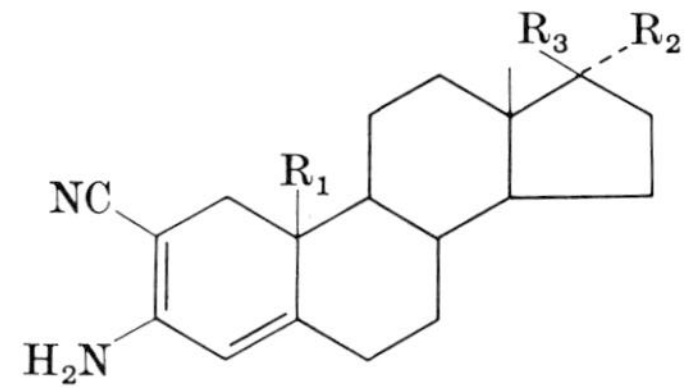					
			CH_3	H	OH	—	438
			H	H	OH	—	438
			H	H	$OCOCH_3$	—	438

(continued)

TABLE XXIII (*continued*)

Starting material				Reagent(s)	Product		Yield, %	Ref.
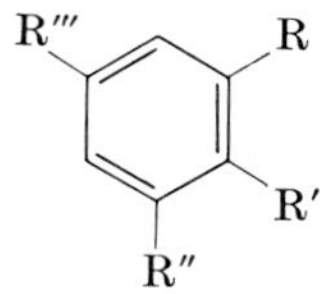					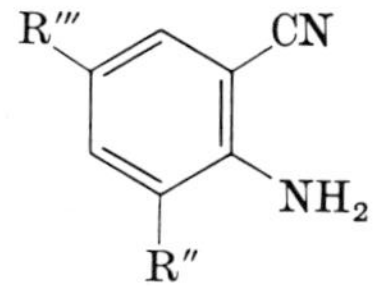			
R	R′	R″	R‴		R″	R‴		
Br	NH_2	H	H	CuCN in pyridine	H	H	83	311
I	NH_2	H	Cl	CuCN in pyridine	H	Cl	76	67
							72	563
CN	Cl	H	NO_2	NH_3/C_2H_5OH/pressure	H	NO_2	—	37
CN	Cl	H	NO_2	NH_3	H	NO_2	—	186
Cl	NH_2	H	Cl	CuCN in pyridine	H	CN	—	278
CN	OCH_3	NO_2	NO_2	NH_4OH	NO_2	NO_2	—	315
CN	Cl	H	NO_2	$NH_3/CH_3OH/170°$	H	NO_2	66	21
							—	37
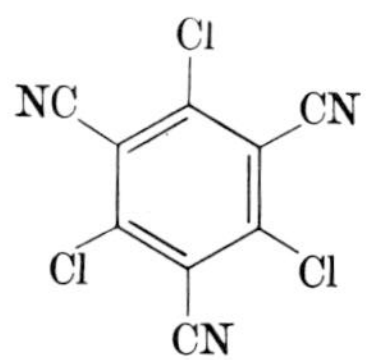				NH_3 in benzene	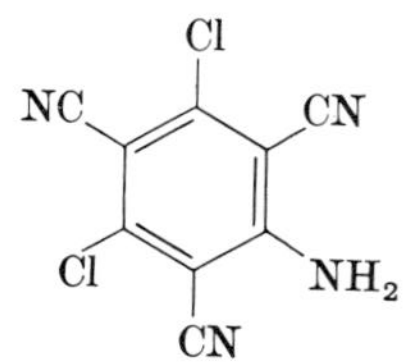		65	481
							—	482

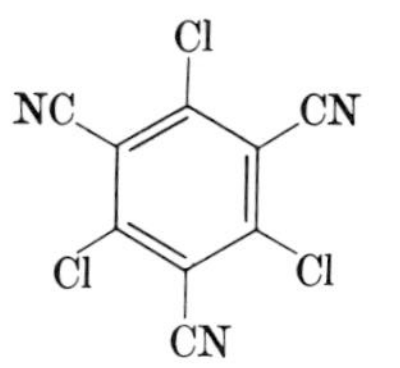	NH_3 in acetonitrile	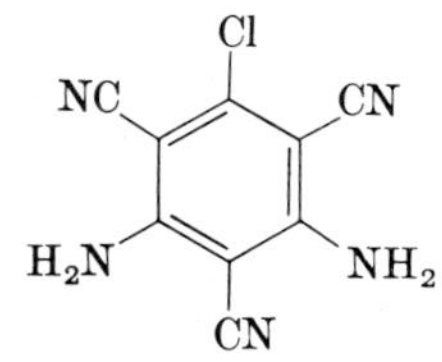	85 —	481 482
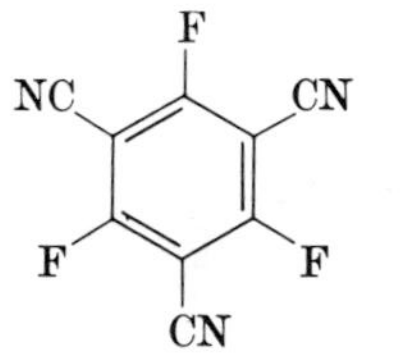	NH_3 in benzene	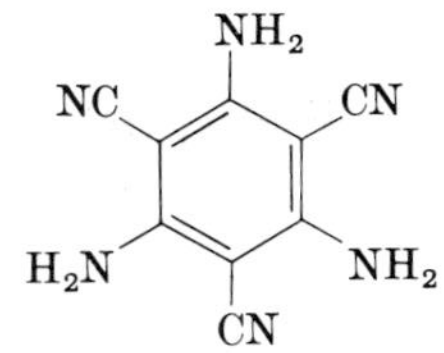	85 —	481 482
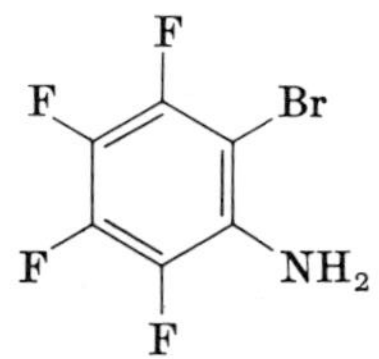	CuCN/DMF	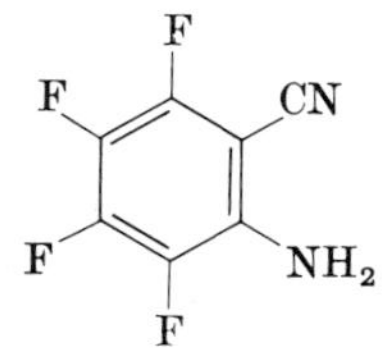	58	546
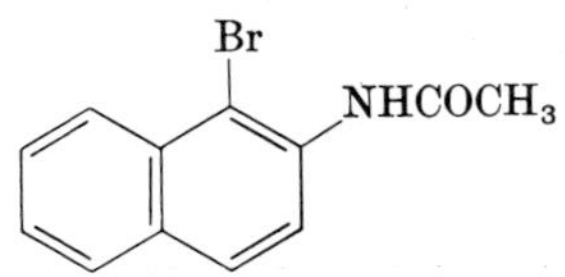	CuCN in pyridine	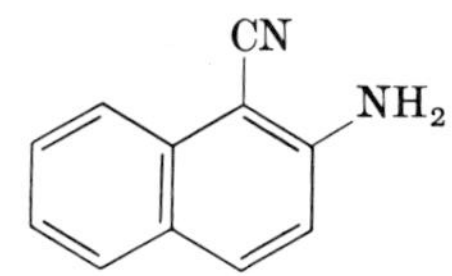	92 (overall)	267
	CuCN in pyridine		40 (replacement)	152
	CuCN in pyridine		—	212
	CuCN in pyridine	(after hydrolysis)	—	340
	CuCN in DMF		63 (overall)	289

(continued)

TABLE XXIII (*continued*)

Starting material	Reagent(s)	Product	Yield, %	Ref.
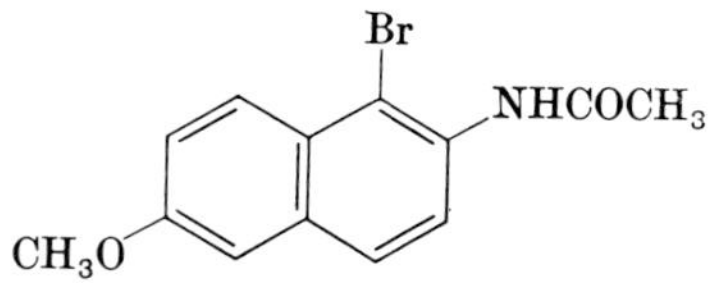	CuCN in pyridine	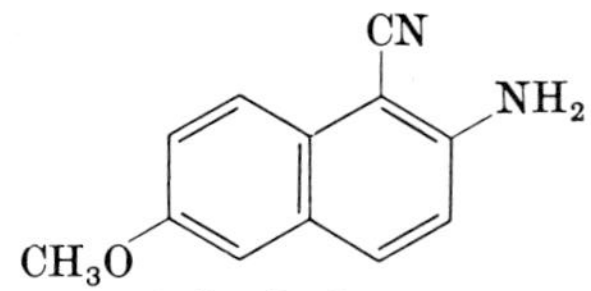 (after hydrolysis)	54 (overall)	340
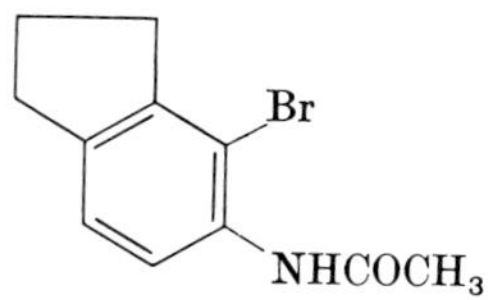	CuCN	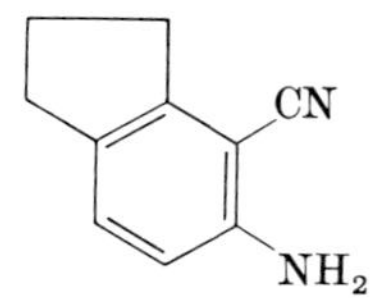 (after hydrolysis)	—	313
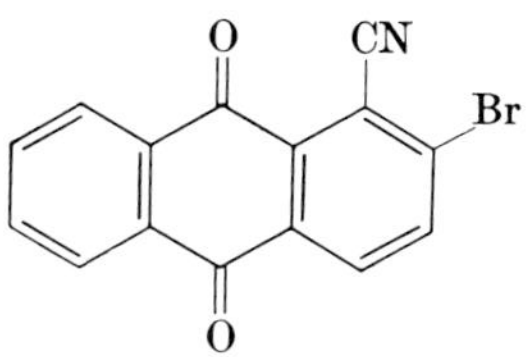	NH_3	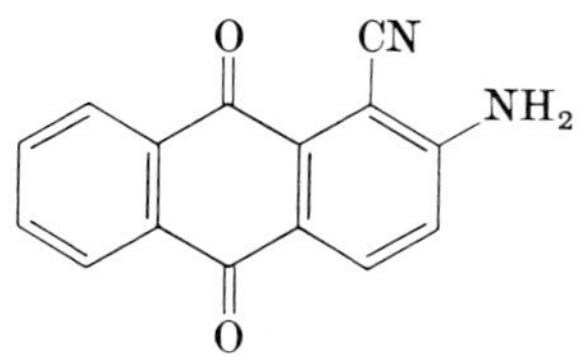	60	325

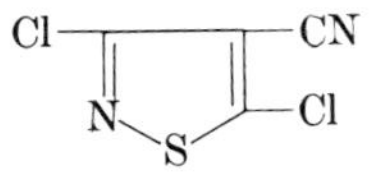

NH_3 and THF, Δ

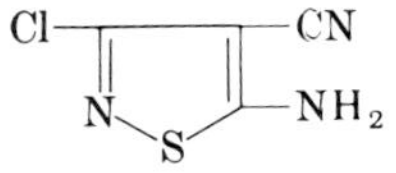

92 173

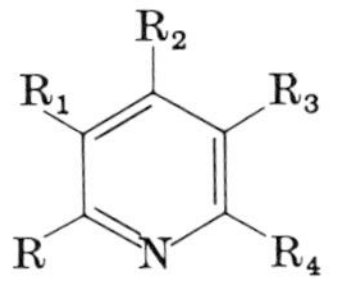

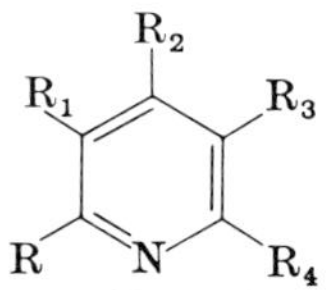

R	R_1	R_2	R_3	R_4	Reagent	R	R_1	R_2	R_3	R_4	Yield	Ref.
H	H	H	CN	Cl	NH_3 in C_2H_5OH	H	H	H	CN	NH_2	79	30
H	H	H	NH_2	Br	CuCN	H	H	H	NH_2	CN	32	78
H	Cl	H	NH_2	Br	CuCN	H	Cl	H	NH_2	CN	20	78
H	Br	H	NH_2	Br	CuCN	H	Br	H	NH_2	CN	25	78
CH_3	H	H	CN	Cl	NH_3/CH_3OH/pressure	CH_3	H	H	CN	NH_2	59	87
CH_3	H	CH_3	CN	Cl	NH_3/pressure	CH_3	H	CH_3	CN	NH_2	66	80
Cl	CN	H	CN	NH_2	NH_4OH	NH_2	CN	H	CN	NH_2	58	106
Cl	CN	CN	CN	NH_2	NH_3	NH_2	CN	CN	CN	NH_2	95	106
C_6H_5	H	C_6H_5	CN	Cl	NH_3	C_6H_5	H	C_6H_5	CN	NH_2	68	354
NH_2	CN	H	CN	SO_2CH_3	NH_4OH	NH_2	CN	H	CN	NH_2	57	151

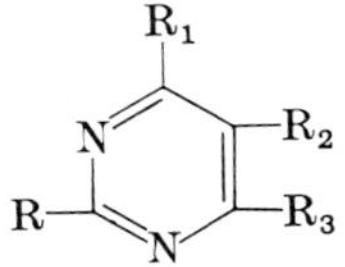

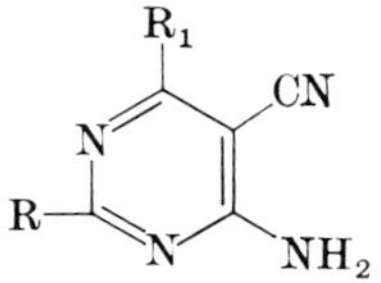

R	R_1	R_2	R_3	Reagent	R	R_1	Yield	Ref.
H	H	CN	Cl	NH_3	H	H	64	544
H	H	CN	Cl	$HC(=NH)NH_2 \cdot HOAc$	H	H	20	544

(*continued*)

TABLE XXIII (*continued*)

Starting material				Reagent(s)	Product		Yield, %	Ref.
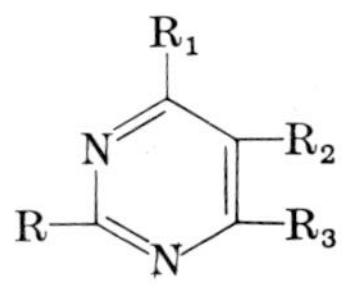 (*cont.*)					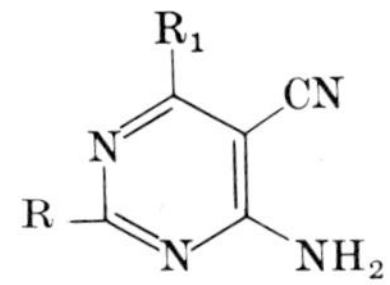			
R	R_1	R_2	R_3		R	R_1		
CH_3	H	CN	Cl	NH_3	CH_3	H	—	196
							—	375
CH_3	H	CN	CH_3O	Liq. NH_3	CH_3	H	—	63
CH_3	Cl	CN	NH_2	$Pd/CaCO_3/H_2$	CH_3	H	96	63
C_2H_5S	H	CN	Cl	NH_3	C_2H_5S	H	94	41
							—	363
CH_3	Cl	CN	Cl	NH_3 in C_2H_5OH	CH_3	Cl	84.5	63
CH_3	CH_3	CN	Cl	NH_3	CH_3	CH_3	—	48
CH_3	CH_3	Br	NH_2	CuCN	CH_3	CH_3	—	48
CH_3	C_2H_5O	CN	C_2H_5O	Liq. NH_3	CH_3	C_2H_5O	—	63
CH_3	Cl	CN	Cl	Liq. NH_3	CH_3	NH_2	74.7	63
							—	338
C_6H_5	Cl	CN	Cl	NH_3	C_6H_5	Cl	76	458
Cl	Cl	CN	Cl	NH_3	NH_2	Cl	49	458
NH_2	C_2H_5O	CN	Cl	NH_3	NH_2	C_2H_5O	47	561
$(CH_3)_2N$	H	CN	Cl	NH_3	$(CH_3)_2N$	H	—	416
$(C_2H_5)_2N$	H	CN	Cl	NH_3	$(C_2H_5)_2N$	H	—	416
n-C_4H_9NH	H	CN	Cl	NH_3	n-C_4H_9NH	H	—	416

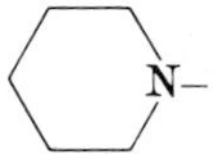	H	CN	Cl	NH_3		H	—	416
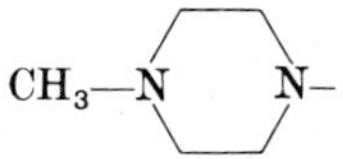	H	CN	Cl	NH_3	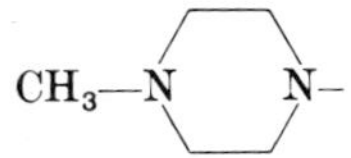	H	—	416
C_6H_5	C_6H_5	CN	Cl	NH_3	C_6H_5	C_6H_5	96	359
p-ClC_6H_4	p-ClC_6H_4	CN	Cl	NH_3	p-ClC_6H_4	p-ClC_6H_4	100	359
Cl	p-BrC_6H_4NH	CN	Cl	NH_3	NH_2	p-BrC_6H_4NH	70	348
Cl	p-IC_6H_4NH	CN	Cl	NH_3	NH_2	p-IC_6H_4NH	76	348
Cl	3,4-diClC_6H_3NH	CN	Cl	NH_3	NH_2	3,4-diClC_6H_3NH	63	348
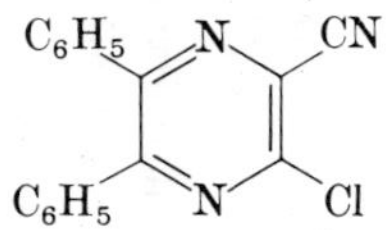				$CH_3COO^-NH_4^+$	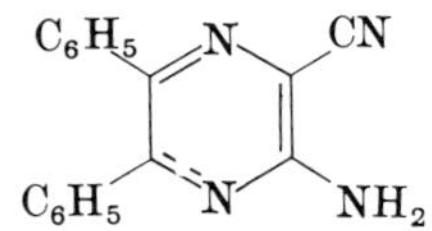		Low	8
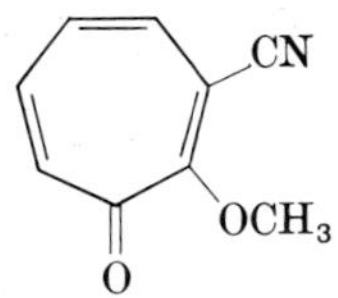				NH_3 (Liq.)	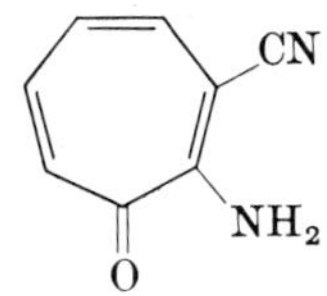		—	9

(**5**) requisite for the preparation of **6** were prepared by acylation of *N*-cyanoethyl derivatives of α-amino acid ethyl esters (**3**) to give **4**, followed by cyclization with sodium ethoxide in absolute ethanol (76).

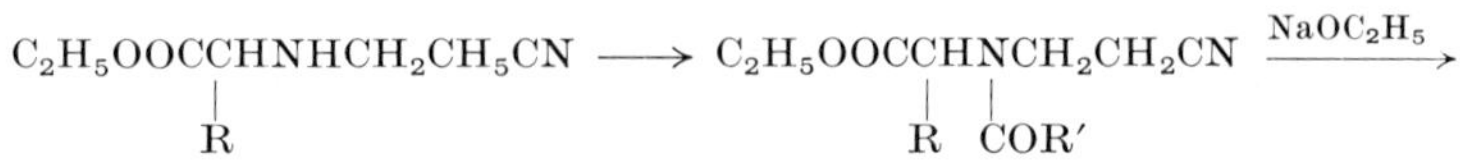

(**3**) (**4**)

(**5**) (**6**)

Surprisingly, there appear to be only two reported examples of the preparation of *o*-aminonitriles by application of the Sandmeyer reaction to an appropriately protected *o*-diamine. Thus, the *o*-phthalimido anilines **7**, on diazotization followed by a normal Sandmeyer sequence utilizing cuprous cyanide and finally cleavage of the phthalimido protecting grouping by hydrazine, gave the corresponding 2-aminobenzonitriles **8** in moderate yield (318).

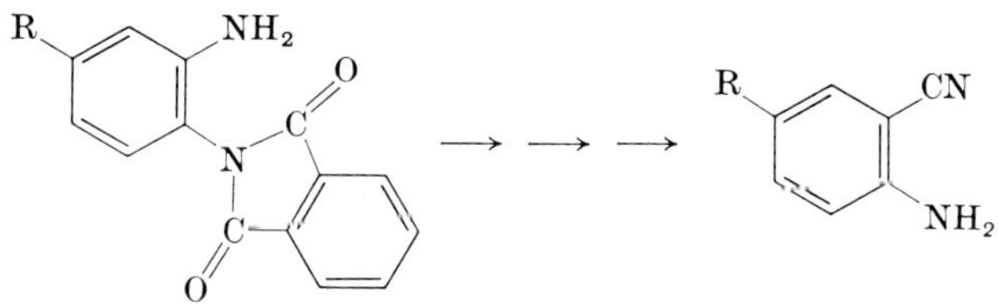

(**7**) R = H, OCH_3 (**8**)

V. Synthesis of *o*-Aminonitriles by Reduction of *o*-Nitronitriles

Since a variety of *o*-nitronitriles are readily accessible by reaction of *o*-chloronitro compounds with alkali cyanides, their reduction to *o*-aminonitriles would appear to be a simple and prosaic route to the latter derivatives (see Table XXIV). It has long been known, however, that the nature of the product actually formed depends critically upon the reduction conditions. Thus, although 2-nitrobenzonitrile (**1**) can be reduced to 2-aminobenzonitrile (**5**) with stannous chloride, metallic tin, or metallic zinc in concentrated hydrochloric acid, reduction with hydrazine and nickel or catalytic reduction using Raney nickel (183,321) or platinum

TABLE XXIV

o-Aminonitriles by Reduction of *o*-Nitronitriles

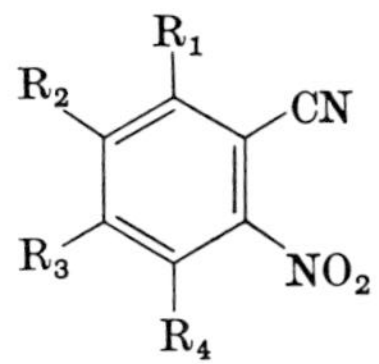

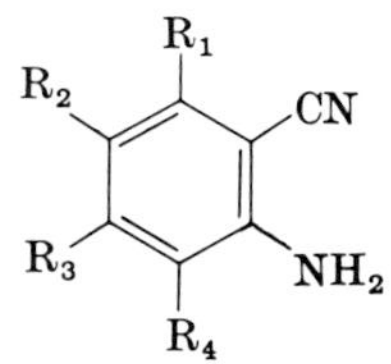

Starting material				Reduction conditions	Product				Yield, %	Ref.
R_1	R_2	R_3	R_4		R_1	R_2	R_3	R_4		
H	H	H	H	$SnCl_2/HCl$	H	H	H	H	85	123
				$SnCl_2/HCl$					82	67
				$SnCl_2/HCl$					72	137
				$SnCl_2/HCl$					—	569
				$SnCl_2/HCl$					—	333
				$SnCl_2/HCl$			(^{14}CN)		—	554
				Sn/HCl					—	264
				Sn/HCl gas/CH_3COOH					—	263
				Zn/HCl/C_2H_5OH					90	332
				Zn/HCl					50	40
				Zn/HCl					27	197
				Fe/12% HCl					—	305
				Fe/HCl or CH_3COOH					—	253

(continued)

TABLE XXIV (*continued*)

Starting material				Reduction conditions	Product				Yield, %	Ref.
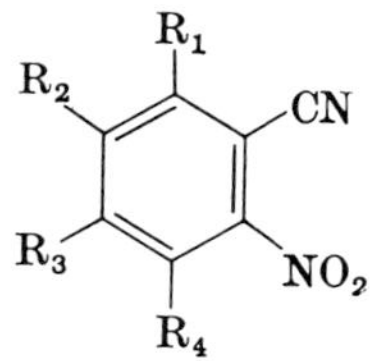					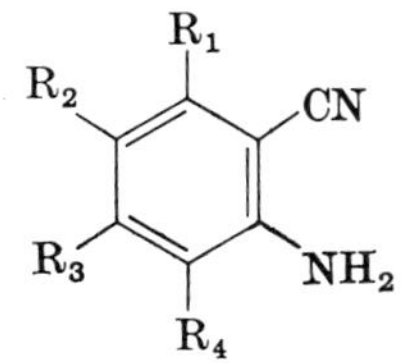					
R_1	R_2	R_3	R_4		R_1	R_2	R_3	R_4		
H	H	H	H		H	H	H	H		
				$PbCl_2/HCl$ (< 40°)					78	314
				Fe/H^+					—	183
				PtO_2/dioxan/H_2					20 (+amide)	256
				$Pd/BaSO_4/C_2H_5OH/H_2$					47	256
				$Pd/BaSO_4$/3:1 dioxan-C_2H_5OH/H_2					75	256
				Pd/cyclohexene/THF					59	317
				6:4-Ni:Cu/H_2/vapor phase reduction					20	308
CH_3	H	H	H	Red P/HI or $SnCl_2/HCl$ below 50°	CH_3	H	H	H	—	272
				P/HI					70	274 (275)
				$SnCl_2/HCl$					83	563
H	CH_3	H	H	Sn/HCl	H	CH_3	H	H	86	184

H	CH_2Cl	H	H	Zn/HCl	H	CH_3	H	H	75	174
H	H	CH_3	H	$SnCl_2/HCl$	H	H	CH_3	H	80	293
				$SnCl_2/HCl$					—	291
				$SnCl_2/HCl$					68	563
				$SnCl_2/HCl$					80	198, 199
				$SnCl_2/HCl/CH_3COOH$					90	294
H	H	H	CH_3	$SnCl_2/HCl$	H	H	H	CH_3	—	274
C_2H_5	H	H	H	$SnCl_2/HCl$	C_2H_5	H	H	H	60	316
H	H	H	$(CH_3)_3C$	$Pd/BaSO_4/H_2N\overset{S}{\overset{\Vert}{C}}NH_2/H_2$ in 3:1-dioxan-C_2H_5OH	H	H	H	$(CH_3)_3C$	23	257
				$Pd/BaSO_4$ in 3:1-dioxan-C_2H_5OH/H_2					23	256
H	CH_3	CH_3	H	$SnCl_2/HCl$	H	CH_3	CH_3	H	61	563
CH_3	H	H	CH_3	$Pd/C_2H_5OH/H_2$	CH_3	H	H	CH_3	73	323
CH_3O	H	H	H	Sn/HCl or $SnCl_2/HCl$	CH_3O	H	H	H	—	245
				$SnCl_2/HCl$					30	563
				$Pd/C/H_2$					69	121
H	CH_3O	H	H	$SnCl_2$	H	CH_3O	H	H	76	131
H	H	CH_3O	H	$SnCl_2/HCl$	H	H	CH_3O	H	50	563
									65	250
									67	131
H	H	C_2H_5O	H	$Zn/HCl/H_2O$	H	H	C_2H_5O	H	—	279
Cl	H	H	H	$SnCl_2/HCl$	Cl	H	H	H	70	312
				Fe/HCl					89	319
H	H	Cl	H	$SnCl_2$	H	H	Cl	H	73	132
									—	563
H	Cl	Cl	H	$SnCl_2/HCl$	H	Cl	Cl	H	63	563

(continued)

TABLE XXIV (*continued*)

Starting material				Reduction conditions	Product				Yield, %	Ref.
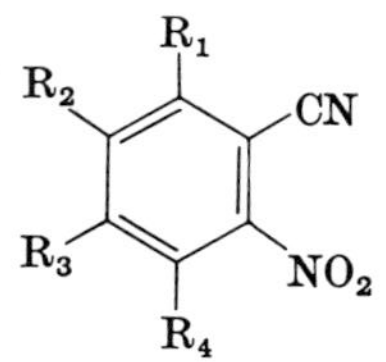					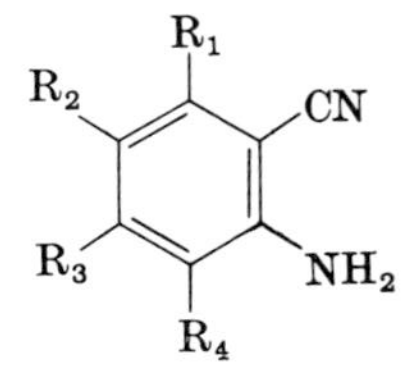					
R_1	R_2	R_3	R_4		R_1	R_2	R_3	R_4		
H	Br	H	H	Sn/HCl	H	Br	H	H	60	17
CH_3	H	Cl	H	Fe/CH_3COOH	CH_3	H	Cl	H	—	304
H	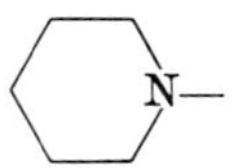	H	H	Sn/HCl	H	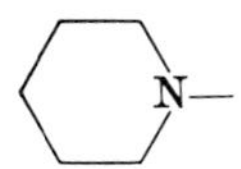	H	H	89	93, 94
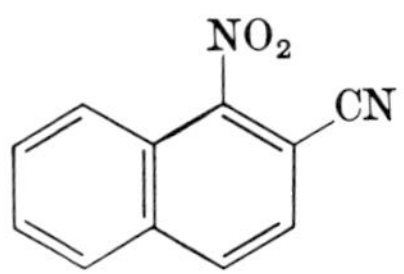				$Pd/BaSO_4/H_2$/dioxan Fe/CH_3COOH $Ni/H_2/CH_3COOC_2H_5/C_2H_5OH/H_2O$	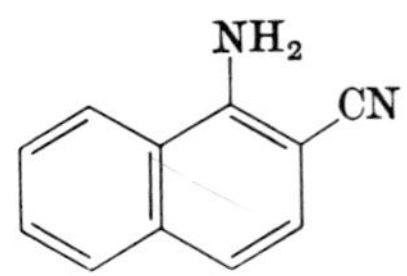				79 — —	280 70 68
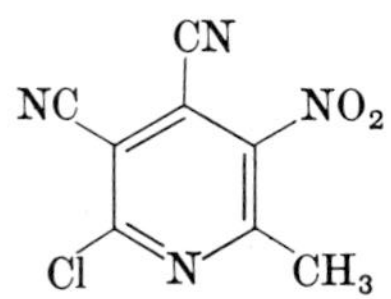				$PtO_2/(CH_3CO)_2O/H_2$ $PtO_2/C_2H_5OH/H_2$	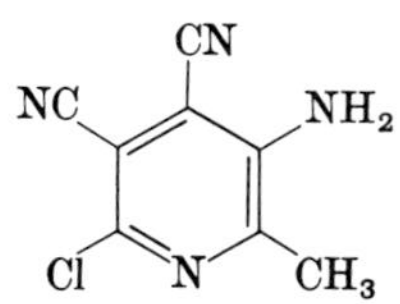				52 "poor"	77 86

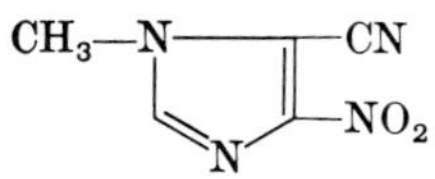	Sn/HCl Ni/H_2 (1) $PtO_2/C_2H_5OH/H_2$ → —NHOH (2) $Ni/C_2H_5OH/H_2$	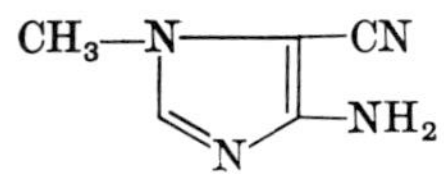	— 88 25 overall	98 136 24
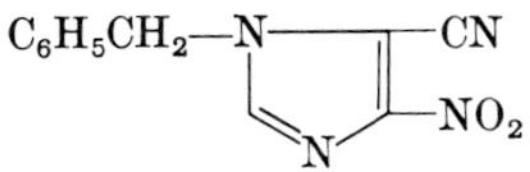	$Ni/C_2H_5OH/H_2$	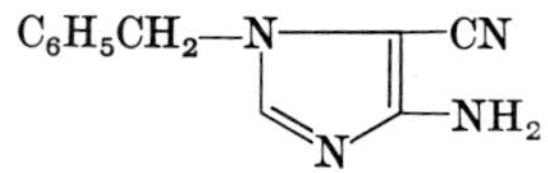	99	71
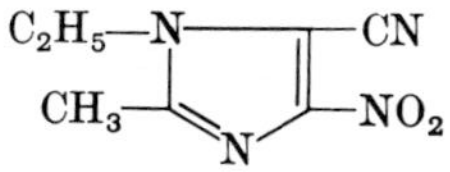	Sn/HCl	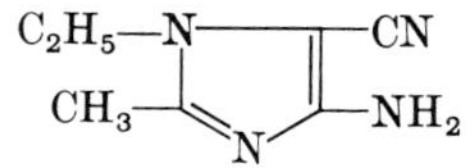	78	98
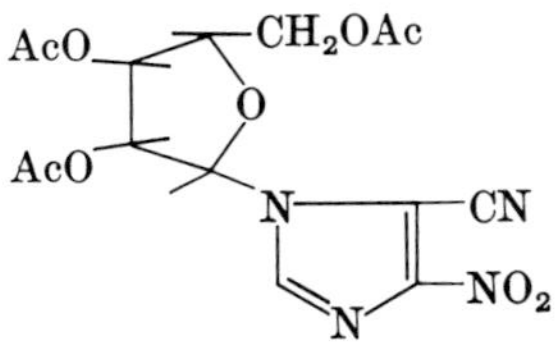	H_2/Ni(R) H_2/Pd/C	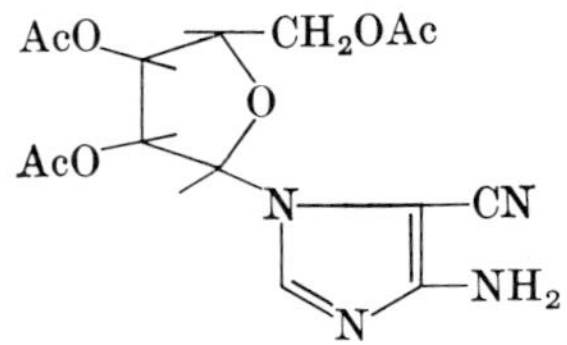	— 87	355 552

in alcohol gives only 2-aminobenzamide (**4**). Analogous results have been found with many other *o*-nitronitriles, which give *o*-aminonitriles under some conditions and *o*-aminoamides under other conditions. An examination of the course of reduction of 2-nitrobenzonitrile, using ^{18}O tracer studies (41,256), has shown that the amide oxygen atom originates from the *o*-situated nitro group and not from the solvent; the transfer involves initial reduction of the nitro group to a hydroxylamine intermediate (**2**) which is either reduced further to 2-aminobenzonitrile (**5**) (in solvents such as dioxane) or undergoes an intramolecular cyclization (in solvents such as ethanol) to 3-aminoanthranil (**3**). This compound, which was previously unknown, was actually isolated and characterized in the reduction of 2-nitrobenzonitrile either with hydrogen and platinum oxide in ethanol or with zinc dust and ammonium chloride (256). Further reduction of the latter compound (**3**) then gives 2-aminobenzamide (**4**).

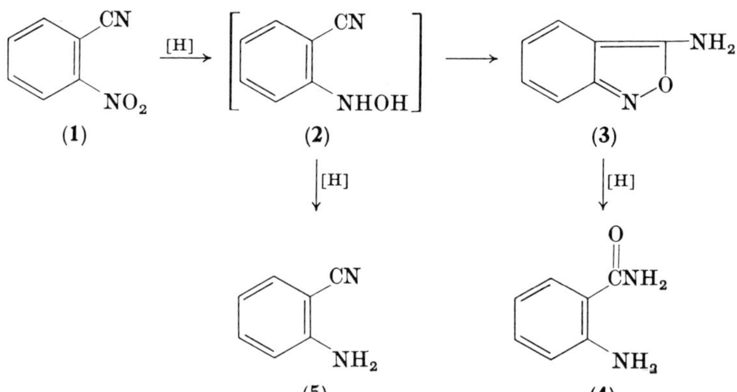

Reduction of the triazene **6** with stannous chloride and hydrochloric acid cleaves the diazoamino grouping to give 2-aminobenzonitrile and 3-aminoindazole (493), the latter presumably via 2-hydrazinobenzonitrile, which is known to cyclize spontaneously (120). Various other 2-cyanotriazenes and 2-cyanoazo compounds are readily reduced to 2-aminobenzonitriles (553), but the reactions are of no preparative value because of the relative inaccessibility of the requisite precursors.

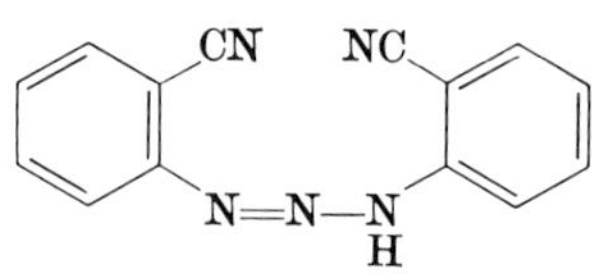

(**6**)

The interesting tris-aminonitrile 2,4,6-tricyano-1,3,5-triaminobenzene (**7**) has been prepared by dithionite reduction of 2,4,6-tricyano-1,3,5-triazidobenzene (as well as by hydrazinolysis of 2,4,6-tricyano-1,3,5-triphthalimidobenzene, and aminolysis of 2,4,6-tricyano-1,3,5-trifluorobenzene) (481,482).

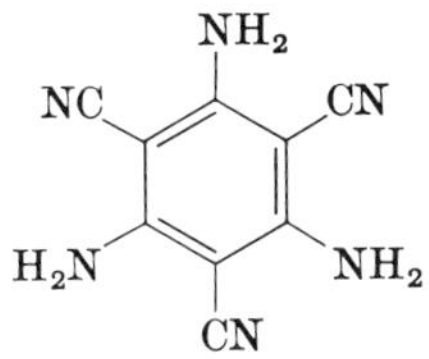

(**7**)

VI. Synthesis of *o*-Aminonitriles by the Bedford-Partridge Reaction

A convenient route to 2-aminobenzonitriles (**4**) is the Bedford-Partridge reaction (1), in which an isatin 3-oxime (**1**) is pyrolyzed under reduced pressure (see Table XXV). The resulting second-order Beckman rearrangement gives a 2-aminobenzonitrile directly, presumably via an intermediate isocyanate (**2**) which is hydrated by the water eliminated in the cleavage step. When the decomposition is carried out in the presence of phosphorus pentachloride (which reacts with the water evolved and thus prevents its subsequent participation), the intermediate isocyanate can actually be isolated (240); subsequent hydrolysis then gives the 2-aminobenzonitrile. Yields in this sometimes violent reaction are only modest (16–70%), but the ready availability of isatins and the simplicity of the decomposition step make the Bedford-Partridge reaction an attractive route to substituted 2-aminobenzonitriles. This degradation can also be applied to 2,2′-disubstituted pseudoindoxyls. For example,

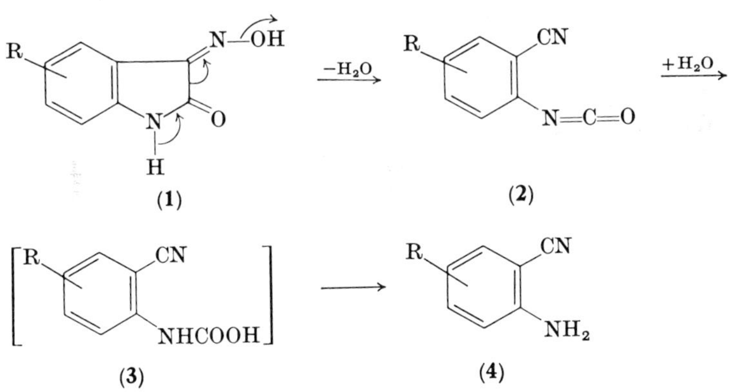

TABLE XXV

o-Aminobenzonitriles from the Bedford-Partridge Reaction

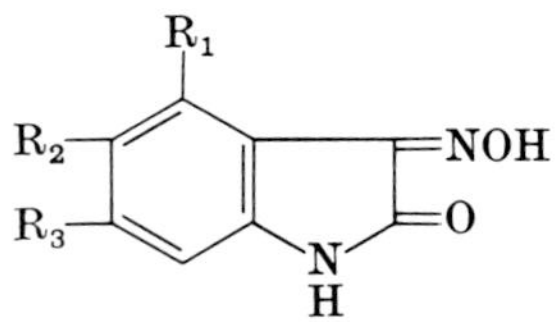

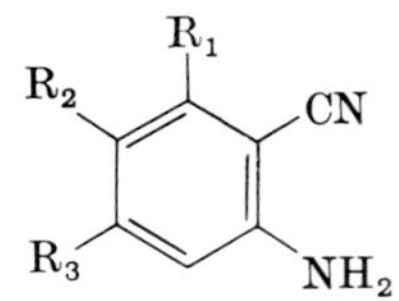

Isatin precursor R_1	R_2	R_3	Conditions	Product R_1	R_2	R_3	Yield, %	Ref.
H	H	H	Pyrolysis	H	H	H	65	1
			Pyrolysis				20	39
			Pyrolysis				—	15
			Pyrolysis				—	62
			Δ, nonaethylene glycol				—	320
			Δ, nonaethylene glycol				—	433
CH_3	H	H	Pyrolysis	CH_3	H	H	—	313
			Pyrolysis				—	320
H	CH_3	H	Pyrolysis	H	CH_3	H	49	1
			Pyrolysis				16	563
			Δ, phthalic acid diethylhexyl ester				75.4	433
			Pyrolysis				47	320
H	H	CH_3	Pyrolysis	H	H	CH_3	—	320

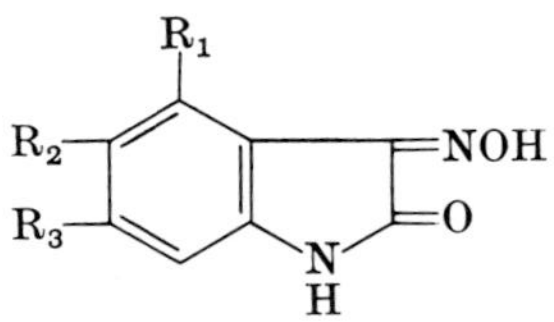				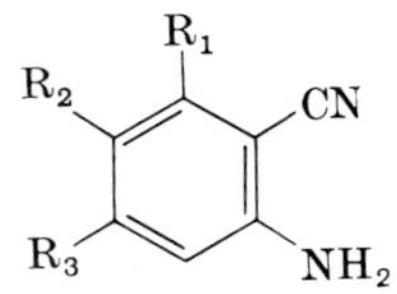				
R_1	R_2	R_3		R_1	R_2	R_3		
C_2H_5	H	H	Pyrolysis	C_2H_5	H	H	—	313
H	NO_2	H	Pyrolysis	H	NO_2	H	66	1
H	CH_3O	H	Pyrolysis from PCl_5	H	CH_3O	H	30	240
			Pyrolysis				51	468
H	Cl	H	Pyrolysis	H	Cl	H	52	1
			Δ, phthalic acid diethylhexyl ester				75	433
H	Br	H	Pyrolysis	H	Br	H	—	320
			Pyrolysis				43	468
H	F	H	Pyrolysis	H	F	H	—	320
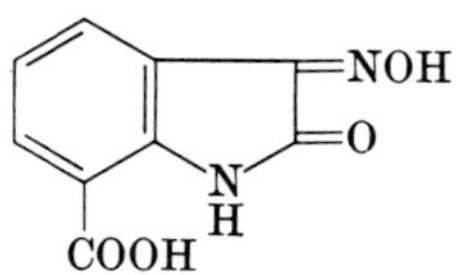			Pyrolysis	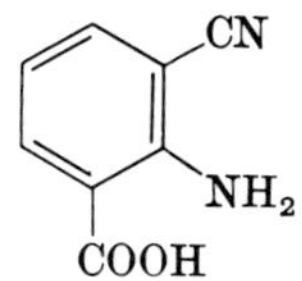			16	1
—$(CH_2)_4$—		H	Pyrolysis	—$(CH_2)_4$—		H	—	313
H	-benzo-		Pyrolysis	H	-benzo-		70	65

(continued)

TABLE XXV (*continued*)

Isatin precursor	Conditions	Product	Yield, %	Ref.
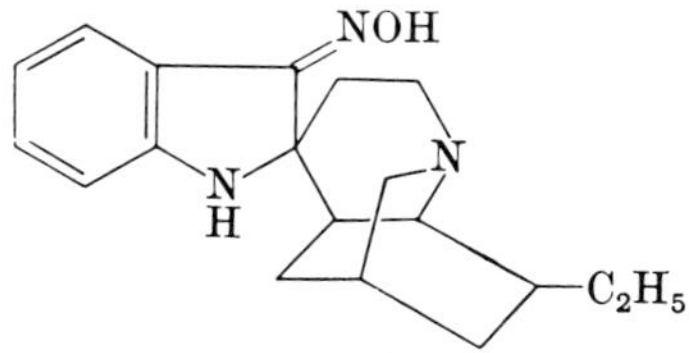	TsCl/pyridine	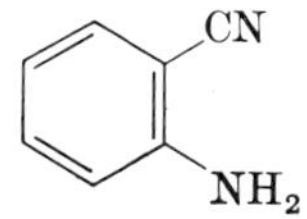 (isolated as N-acetyl derivative)	27	240
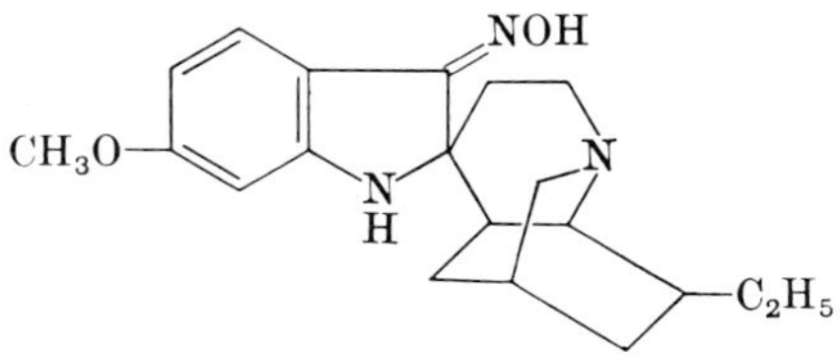	TsCl/pyridine	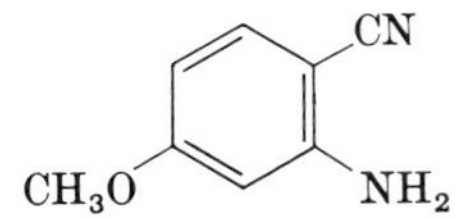 (isolated as N-acetyl derivative)	11	240
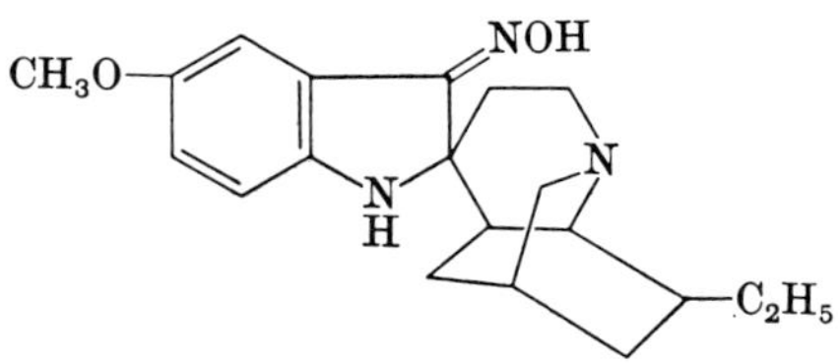	TsCl/pyridine	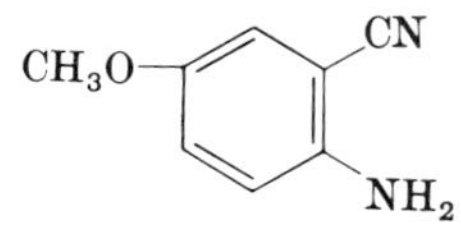 (isolated as N-acetyl derivative)	65	240
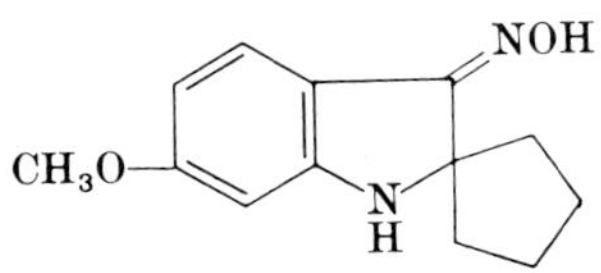	TsCl/pyridine	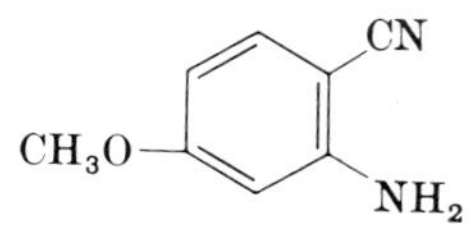 (isolated as N-acetyl derivative)	—	240

6-methoxy-2,2-tetramethylene pseudoindoxyl oxime (**5**) was converted into cyclopentanone and 4-methoxy-2-aminobenzonitrile by heating the oxime tosylate in aqueous pyridine. Through this rearrangement the alkaloids ibogaine, ibogamine, and tabernathine were then interrelated; conversion to their pseudoindoxyls, formation of their oximes and rearrangement furnished the corresponding 2-aminobenzonitriles and the same tricyclic ketone **6** (240).

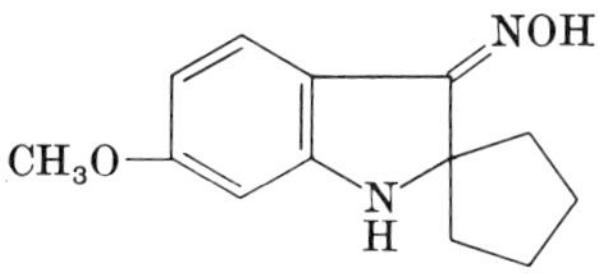

(5)

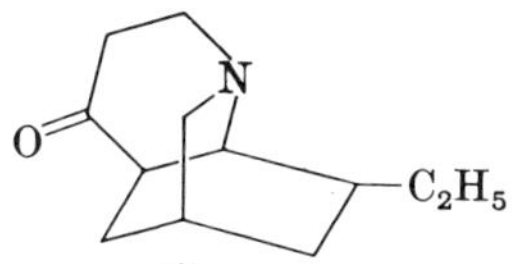

(6)

VII. Synthesis of *o*-Aminonitriles by Ring-Cleavage Reactions

A. OF THIOPHENES

A perhaps unique example of a ring isomerization leading to an *o*-aminonitrile (in which, however, the aminonitrile functionality is not produced in the isomerization, but simply survives the rearrangement unchanged) is the formation of 2-amino-3,4-dicyano-5-mercaptopyrrole (**3**) by treatment of 2,5-diamino-3,4-dicyanothiophene (**1**) with aqueous sodium hydroxide (102,104). This overall isomerization is the result of addition of base to the 2 (or 5) position of **1** followed by cleavage of the S—C bond to give the open chain intermediate **2a**,**b** which recyclizes irreversibly to give the stable anion of the mercaptopyrrole **3**. This isomerization is similar in principle to the base-catalyzed rearrangements of thiazoles to imidazoles and *m*-thiazines to mercaptopyrimidines which are discussed in detail on pp. 299–306.

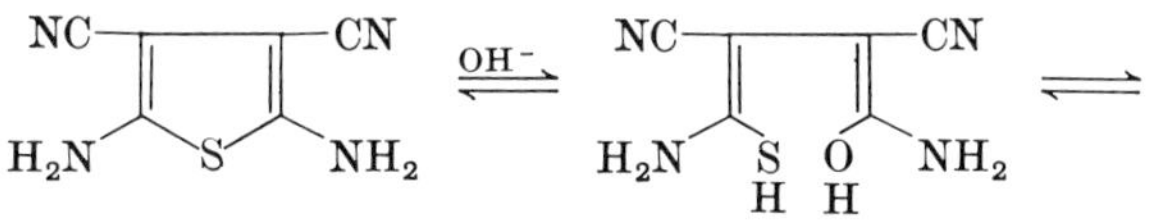

(1) **(2a)**

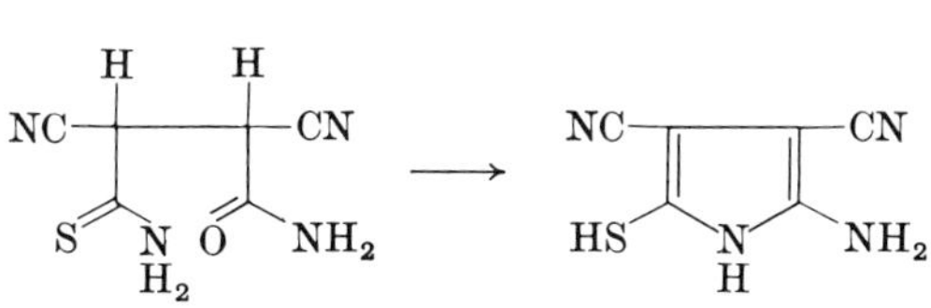

(2b) **(3)**

B. OF *s*-TRIAZINE

A convenient synthesis of a variety of 4,5-disubstituted pyrimidines from *s*-triazine (**1**) and a variety of active methylene compounds has been described by two groups (165,166,259). Thus, the condensation of *s*-triazine with malononitrile in ethanol, followed by recrystallization of the crude product, gave 4-amino-5-cyanopyrimidine (**4**) in 21.3% yield (259). In an independent investigation (166), it was found that these components in the presence of sodium ethoxide gave **4** in 80% yield, and that modifications of the reaction conditions gave rise to intermediates which aided in a clarification of this remarkable conversion. Thus, when *s*-triazine (**1**) and malononitrile were mixed in ethanol solution in the absence of base, an exothermic reaction occurred leading to *N*-(2,2-dicyanovinyl)formamidine (**2**), which was converted in high yield to **4** with a catalytic amount of base or even just upon boiling with water. Treatment of **1** with malononitrile in dimethylformamide gave aminomethylenemalononitrile (**3**), which itself upon further reaction with *s*-triazine in the presence of sodium ethoxide in ethanol likewise gave 4-amino-5-cyanopyrimidine (**4**) in good yield. These results strongly suggest that both **2** and **3** are intermediates in the base-catalyzed conversion of *s*-triazine (**1**) to **4**, and, in fact, both of these compounds were actually isolated from the early stages of a base-catalyzed condensation of **1** with malononitrile.

The probable course of the reaction of *s*-triazine (**1**) with malononitrile is depicted in Scheme 16.

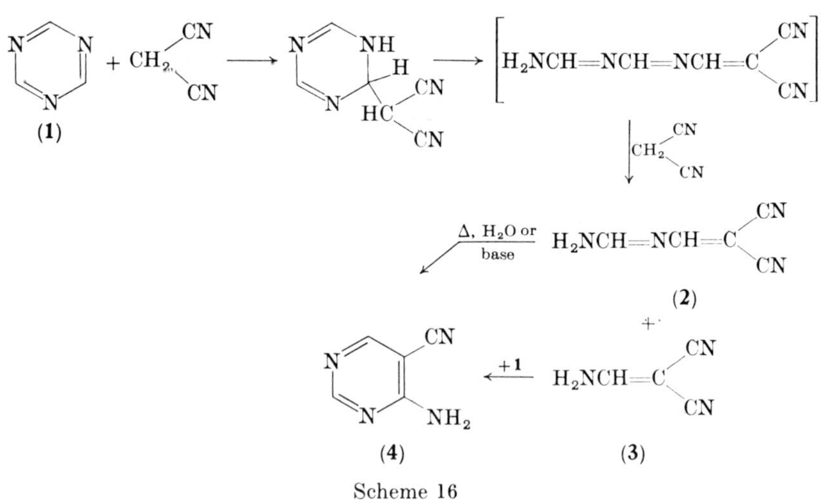

Scheme 16

In a closely related utilization of *s*-triazine (**1**) as an intermediate for pyrimidine syntheses, it was found (165) that condensation with ethyl cyanoacetimidate (**5**) likewise gave 4-amino-5-cyanopyrimidine (**4**) in 67% yield. It has been suggested that this conversion involves initial nucleophilic (enamine type) addition of **5** to one of the strongly polarized C=N bonds of **1** to give the adduct **6** which upon ring opening to **7** and subsequent cyclization with loss of ethanol and HCN gives 4-amino-5-cyanopyrimidine (**4**). The possibility that the reaction might alternately proceed through the intermediates **8** and **9** as depicted in Scheme 17 has not been rigorously excluded.

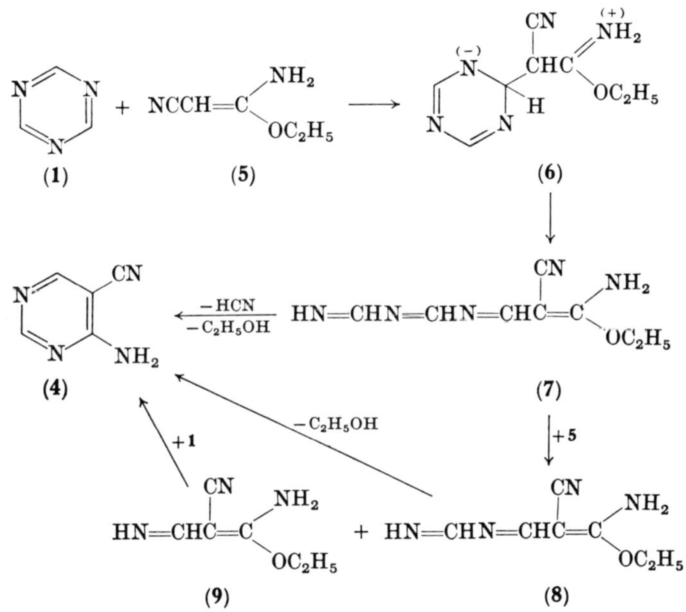

Scheme 17

C. OF PTERIDINES

A facile synthesis of *o*-aminonitriles in both aromatic and heterocyclic systems was discovered by Taylor (23) during an investigation of pteridine chemistry. For example, the reaction of 6,7-diphenyl-4(3*H*)-pteridinethione (**1**, R = C_6H_5) with chloroacetic acid and sodium bicarbonate gave 2-amino-3-cyano-5,6-diphenylpyrazine (**5**, R = C_6H_5) and thioglycolic acid. Under similar conditions, 4(3*H*)-pteridinethione (**1**, R = H) gave 2-amino-3-cyanopyrazine (**5**, R = H). The suggested reaction course for this unusual pyrimidine cleavage reaction involved the initial formation of a 4-carboxymethylthio derivative (**2**) which then, in a reversible step,

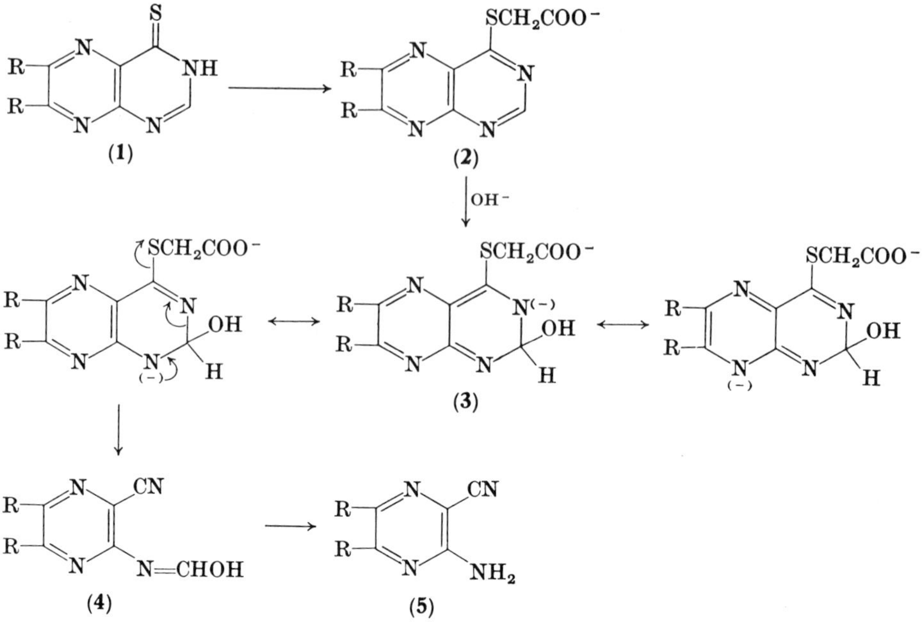

adds hydroxide ion at C_2 to give the resonance-stabilized anion **3**. Irreversible elimination of thioglycolic acid would then give the formylaminonitrile **4** which on hydrolysis in the sodium bicarbonate solution yields the isolated aminonitrile **5**.

Further experiments with other condensed pyrimidinethiones showed (23) that 4(3*H*)-pyrido(2,3-*d*)pyrimidinethione (**6**) gave 2-aminonicotinonitrile (**7**), and 4(3*H*)-pyrido(4,3-*d*)pyrimidinethione (**8**) gave 4-aminonicotinonitrile (**9**). Analogous results were obtained by using methyl

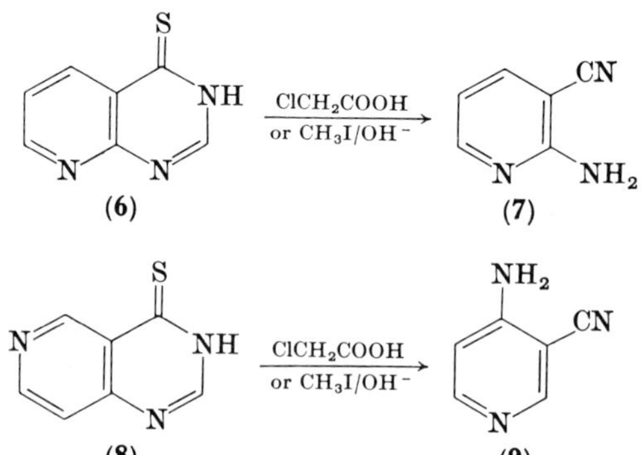

iodide rather than chloroacetic acid in the presence of alkali; methyl mercaptan, rather than thioglycolic acid, was then eliminated. These observations supported the hypothesis that cleavage of fused 4(3*H*)-pyrimidinethiones to *o*-aminonitriles could be accomplished provided that (*a*) the anion formed by attack of hydroxide ion at the C_2 position of the pyrimidine ring be capable of stabilization by appropriate structural features in the remainder of the molecule, and (*b*) the substituent at position 4 (i.e., the alkylthio substituent) be capable of ejection with its bonding electron pair in an irreversible cleavage step. These conclusions were substantiated not only by the observed stability under the above conditions of a number of isomeric fused 4(3*H*)-pyrimidinethiones which were incapable of satisfying condition (a) (for example, 9-methyl-6(1*H*)-purinethione and 4(3*H*)-pyrido(3,4-*d*)pyrimidinethione) but also by the

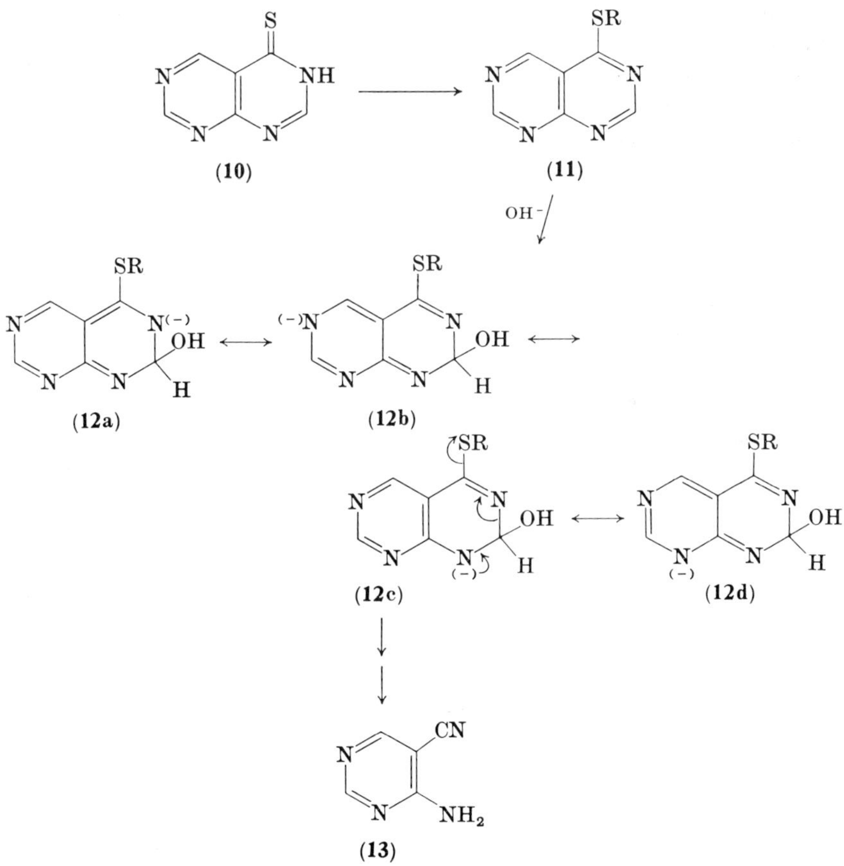

remarkable *lability* of 4(3*H*)-pyrimido(4,5-*d*)pyrimidinethione (**10**) which is capable of stabilizing the anionic intermediate **12a–d** (analogous to **3**) by participation of each of the four heterocyclic nitrogen atoms. In this latter case, treatment of **10** with methyl iodide and sodium carbonate for only a few minutes at room temperature gave 4-amino-5-cyanopyrimidine (**13**) as the sole product; the intermediate 4-methylthio derivative **11** ($R = CH_3$) was isolated by employing an equivalent amount of alkali and operating at 0°, but it was rapidly converted to **13** with aqueous sodium carbonate at 20°.

These considerations were also extended to a number of isomeric bz-nitro-4(3*H*)-quinazolinethiones (23). Thus, 6-nitro-4(3*H*)-quinazolinethione (**14**) and 8-nitro-4(3*H*)-quinazolinethione (**16**) were converted with chloroacetic acid and potassium carbonate to 5-nitro-2-aminobenzonitrile (**15**) and 3-nitro-2-aminobenzonitrile (**17**), respectively. By contrast, no nitrile was obtained with 7-nitro-4(3*H*)-quinazolinethione under similar conditions, and 4-methylthio-5-nitroquinazoline was recovered unchanged under conditions which successfully converted the isomeric 4-methylthio-6-nitroquinazoline to 5-nitro-2-aminobenzonitrile. Thus, provided that the appropriate structural features are built into

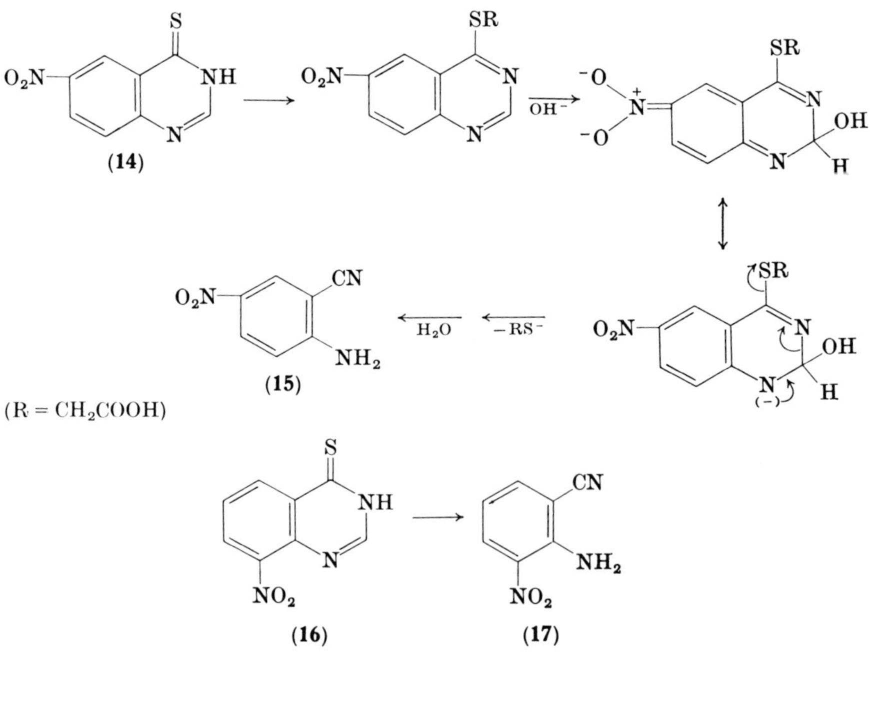

(**14**)

(**15**)

($R = CH_2COOH$)

(**16**) (**17**)

fused 4(3*H*)-pyrimidinethiones to satisfy condition (*a*) above, and provided in addition that a satisfactory synthetic route is available to the requisite condensed pyrimidine heterocycles, this cleavage reaction provides a facile, if somewhat circuitous, route to a variety of interesting *o*-aminonitriles.

In the context of ring cleavage reactions leading to *o*-aminonitriles, it should be mentioned that 4,6-diamino-5-formylpyrimidine (**19**), the immediate precursor to 4,6-diamino-5-cyanopyrimidine (**20**) by reaction with hydroxylamine, glacial acetic acid and sodium acetate (458) (see Chapter II, Section III), is itself made by a remarkably efficient, acid-catalyzed ring cleavage of 4-aminopyrimido(4,5-*d*)pyrimidine (**18**), which is made in turn by the reaction of formamidine acetate on 4-amino-5-cyanopyrimidine (**13**). This sequence of reactions effectively adds an amino group to position 6 of the latter compound, and is outlined in Scheme 18.

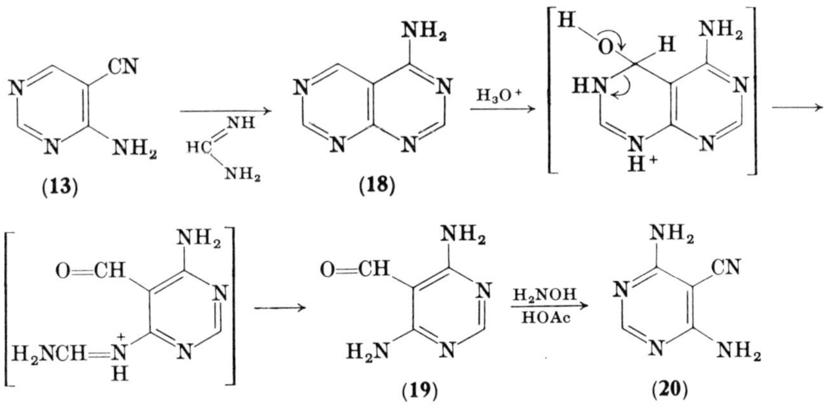

Scheme 18

D. OF DIELS-ALDER ADDUCTS

An ingenious synthesis of some substituted 2-aminobenzonitriles (**2**) involves the Diels-Alder addition of maleic anhydride to a series of 2-amino-3-cyanofurans (**1**) (see pp. 126–129 for preparation). The intermediate Diels-Alder adduct (*exo* ?) suffers ring cleavage and subsequent dehydration to give the observed product (258).

A Diels-Alder reaction between 2-aminofuran (**5**) and *cis*-4-hydroxy-crotononitrile (**4**) (both formed *in situ*) has been suggested as a key step in the remarkable formation of *trans*-3,4-dihydro-2-cyano-3-hydroxy-methyl-4-hydroxyaniline (**6**) from the reaction of epichlorohydrin (**3**)

TABLE XXVI

o-Aminonitriles by Rearrangement or Ring Cleavage Reactions

Starting material	Conditions	Product	Yield, %	Ref.
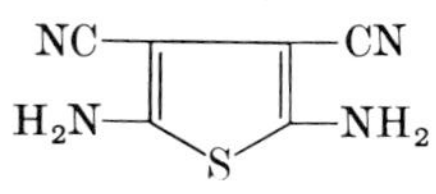	Aqueous NaOH	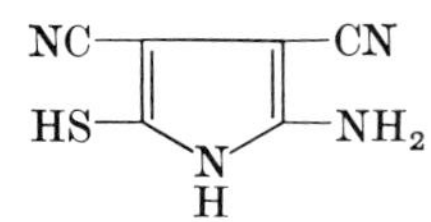	—	102
			61	104
			—	10
			—	394
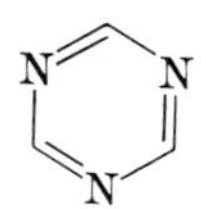	$CH_2(CN)_2/NaOCH_3/C_2H_5OH$	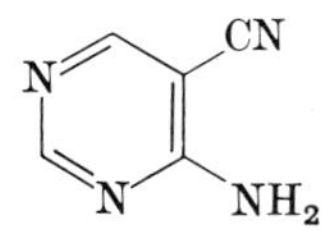	80	166
	$CH_2(CN)_2/C_2H_5OH$/heat		21.3	259
	$NCCH_2C(=NH)OC_2H_5$		67	165
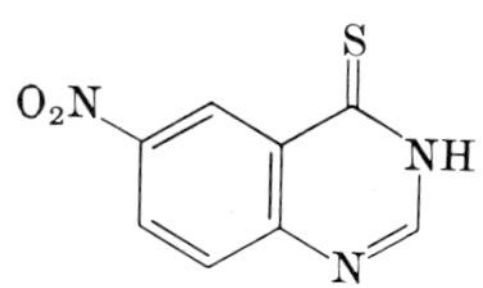	$K_2CO_3/ClCH_2COOH/H_2O$	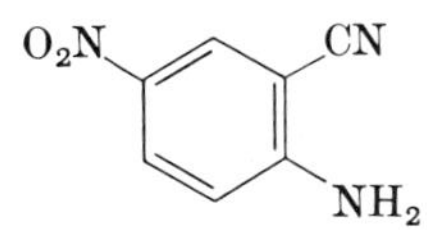	55	23
	KOH/H_2O/dioxan		9	23

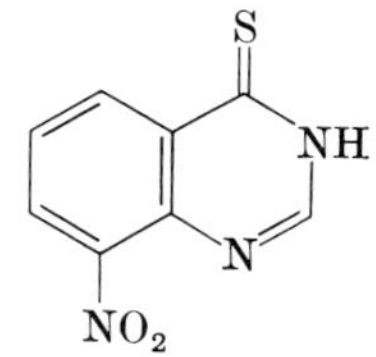	$K_2CO_3/ClCH_2COOH/H_2O$	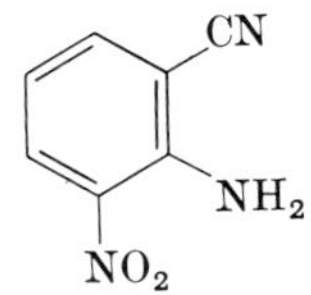	11	23
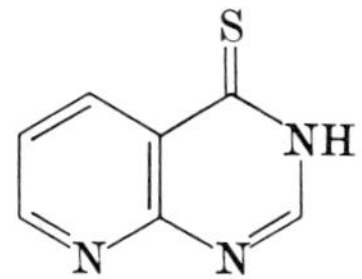	$Na_2CO_3/ClCH_2COOH/H_2O$	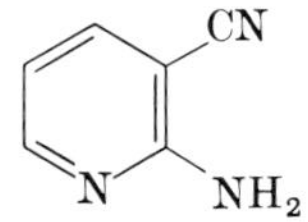	28	23
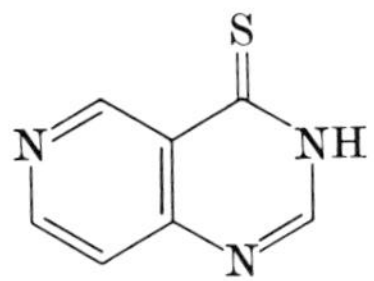	$Na_2CO_3/ClCH_2COOH/H_2O$	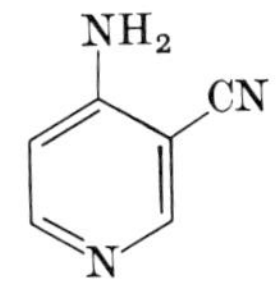	40	23
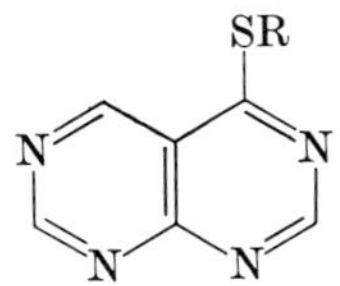	R = H: $NaOH/H_2O$ R = CH_3: $NaOH/H_2O$	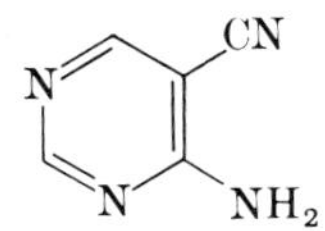	49 82	23 23

(continued)

TABLE XXVI (*continued*)

Starting material	Conditions	Product	Yield, %	Ref.
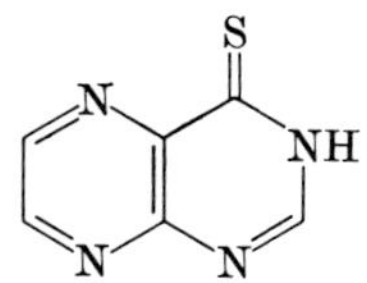	$NaHCO_3/ClCH_2COOH/H_2O$	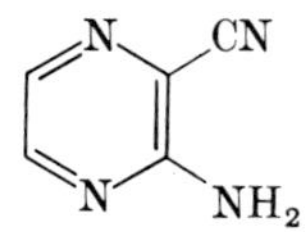	50	23
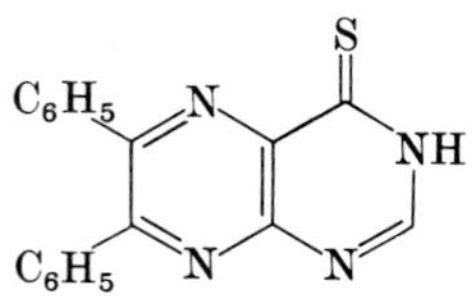	$NaHCO_3/ClCH_2COOH/H_2O$	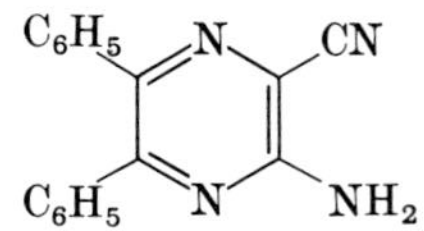	70	23
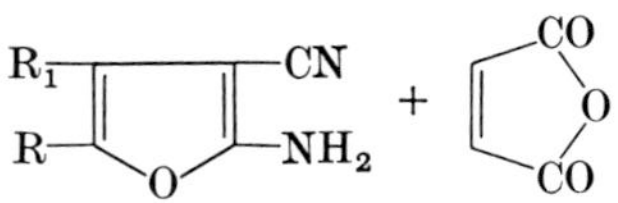	$(CH_3CO)_2O$/heat	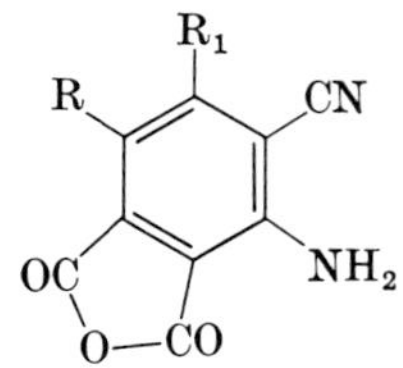		
R R_1		R R_1		
H CH_3		H CH_3	60	258
CH_3 CH_3		CH_3 CH_3	56	258
—$(CH_2)_4$—		—$(CH_2)_4$—	43	258
C_6H_5 C_6H_5		C_6H_5 C_6H_5	53	258

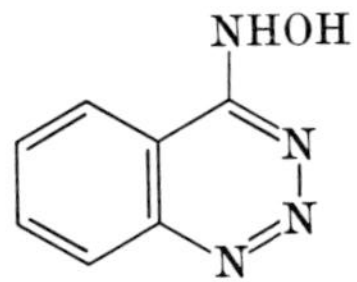	Ni(R)/H_2O	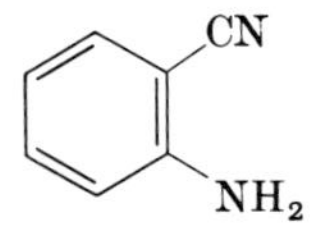	—	134
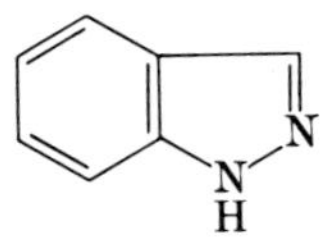	Irradiation $NaNH_2$ CuCl	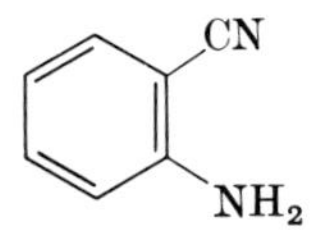	1 — 4.2	547 475,569 569
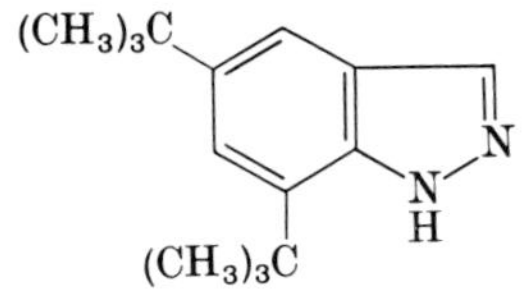	Irradiation	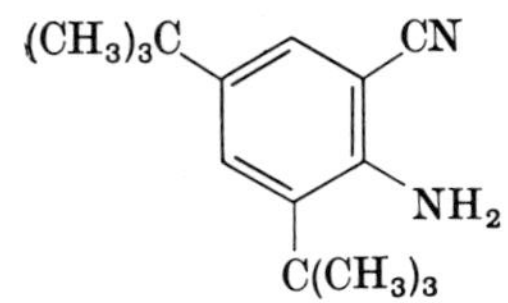	2–13	547

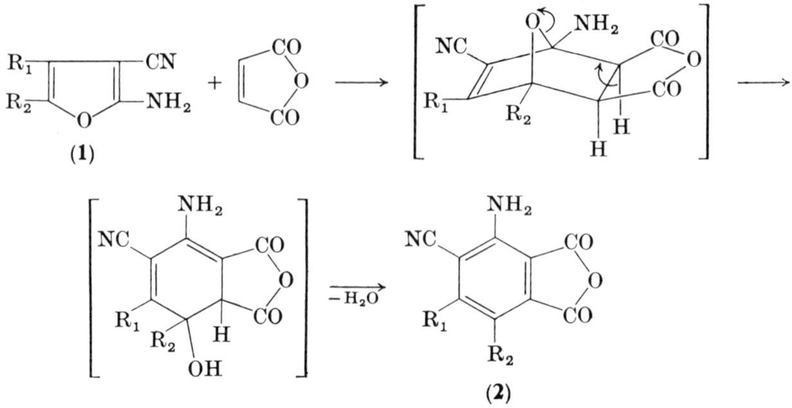

with cyanide ion. The reaction pathway shown in Scheme 19 has been advanced for this unexpectedly efficient conversion (470).

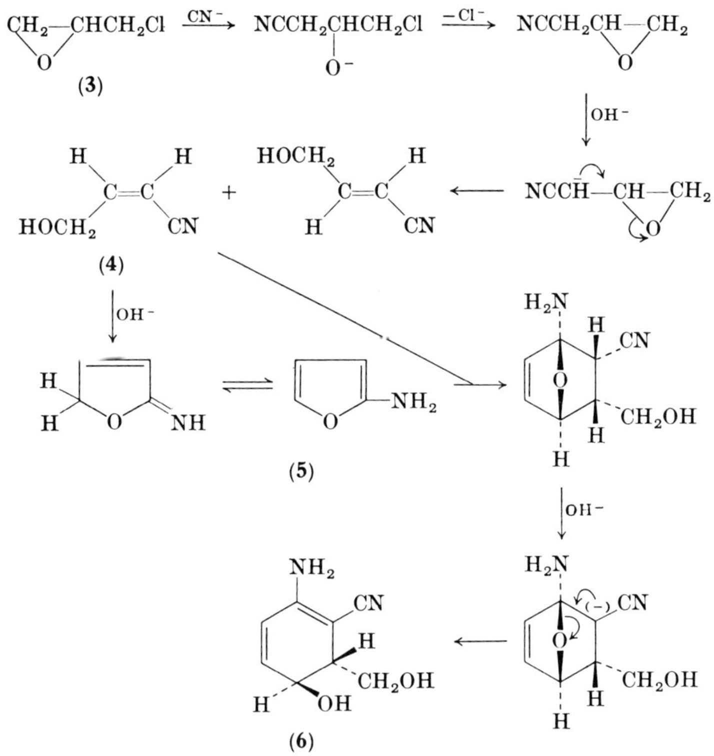

Scheme 19

E. MISCELLANEOUS

An interesting (if impractical) synthesis of 2-aminobenzonitrile follows the reaction sequence pictured in Scheme 20, starting with 2-aminobenzamide (134). The final ring cleavage, which is effected by boiling with Raney nickel in aqueous solution, probably involves preliminary fission of the very labile 2,3 *N*—*N* bond leading to 2-hydrazinobenzamidoxime, which could give 2-aminobenzonitrile either by subsequent reduction to the amidine followed by loss of ammonia (amidines of this type are unstable in aqueous alkaline solution and readily revert to nitrile and ammonia) or by reduction of the hydrazine grouping and loss of hydroxylamine.

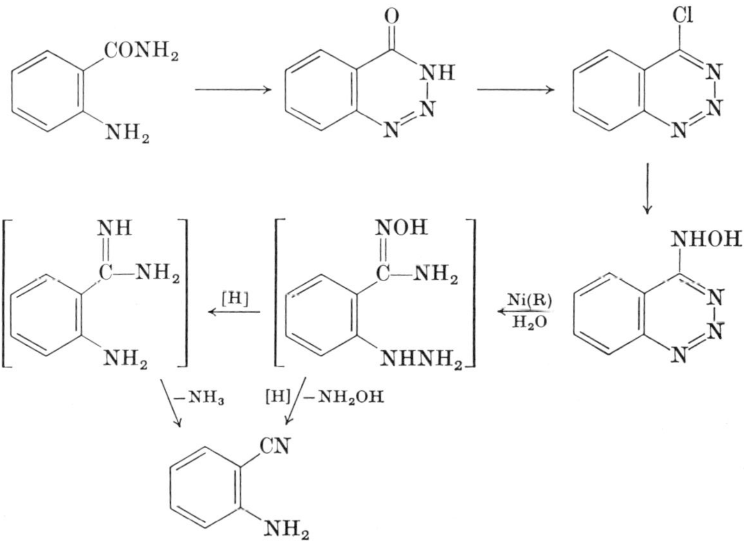

Scheme 20

VIII. Reactions of *o*-Aminonitriles

A. HYDROLYSIS AND RELATED REACTIONS

Although hydrolysis (usually with acid) of enaminonitriles leads first to α-cyanoketones by initial hydrolysis of the amino group, and then eventually to ketones by hydrolysis of the cyano group and subsequent decarboxylation (see Chapter I, Section IV), hydrolysis with either acid or base of aromatic *o*-aminonitriles leads first to *o*-aminocarboxamides and then to *o*-amino acids (32,78,104,122,177,185,368,484,507,541). The reaction has been used for the synthesis of anthranilic acid labeled with

^{14}C in the carboxylate grouping (314); numerous other examples can be found of direct conversion of *o*-aminonitriles to *o*-aminocarboxylic acids (70,160,174,274,318). Hydrolysis of *o*-aminonitriles to the corresponding acids followed by decarboxylation has been used not only as a synthesis of certain amines (e.g., 4-aminopyrimidine (**2**) from 4-amino-5-cyanopyrimidine (**1**)), but also as a means of structure proof for some condensa-

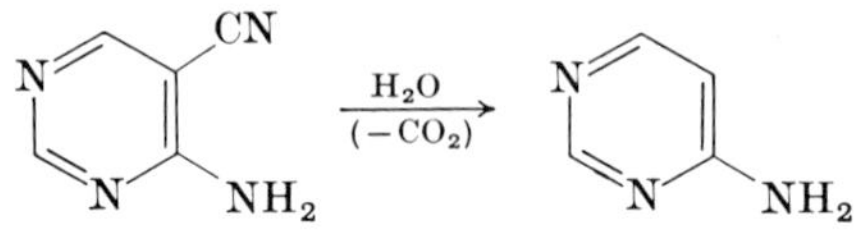

(**1**) (**2**)

tion products. Thus, the product formed upon high pressure catalytic reduction of aminomethylenemalononitrile (**3**) was shown to be 2-amino-3,5-dicyanopyridine (**4**) by hydrolysis and decarboxylation to 2-aminopyridine (**5**) (251). Pyrazolo(2,3-*a*)imidazolidine (**7**) has been prepared by

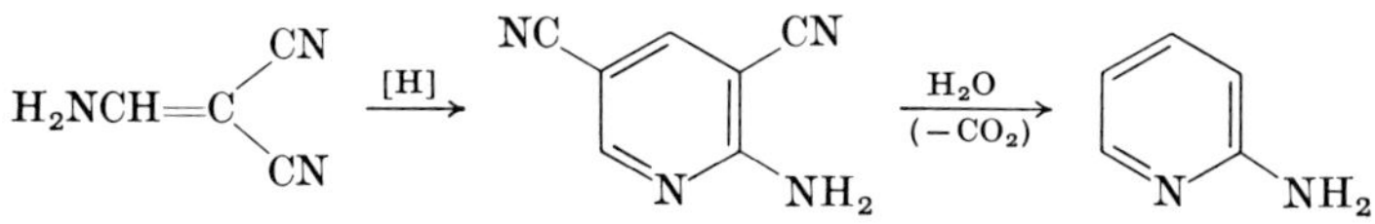

(**3**) (**4**) (**5**)

treatment of 1-(2-hydroxyethyl)-4-cyano-5-aminopyrazole (**6**) with sulfuric acid; since the latter compound is readily prepared from ethoxymethylenemalononitrile and 2-hydroxyethylhydrazine, this simple sequence provides a convenient route to the bicyclic heterocycle (150).

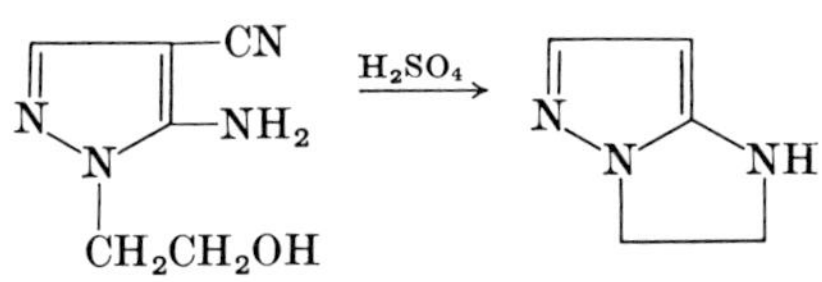

(**6**) (**7**)

Some interesting examples of selectivity in the hydrolysis of *o*-aminonitriles have been reported (125,133,151,353). For example, although 2-amino-3,5-dicyanopyridine (**4**) is hydrolyzed by $0.1N$ potassium hydroxide in aqueous solution to the dicarboxamide (**8**), hydrolysis in dimethylsulfoxide as solvent yields 2-amino-5-cyanonicotinamide (**10**) in almost quantitative yield. This selectivity has been attributed to a selective solvation of the less hindered 5-nitrile grouping by dimethylsulfoxide (**9**) thus protecting it against attack by alkali (133,151). Some importance has also been attributed to a possible stabilization of the intermediate to hydrolysis by hydrogen bonding with the 2-amino grouping. Treatment of **4** with 0.1*N* potassium hydroxide in acetone

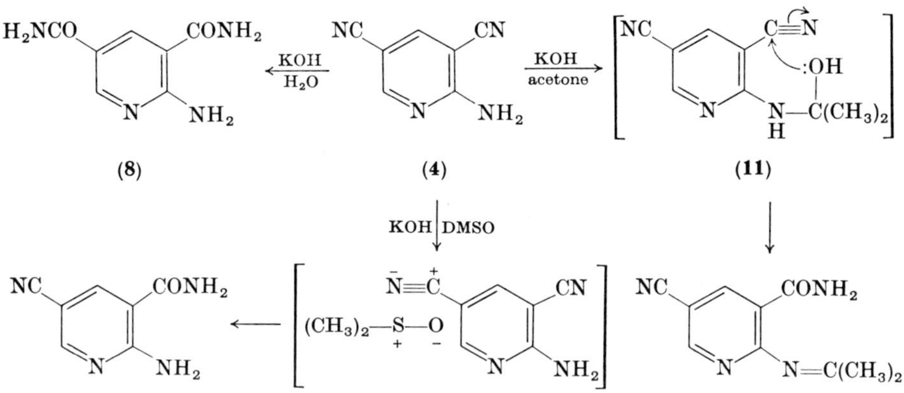

(8) (4) (11)

(10) (9) (12)

gives in almost quantitative yield the isopropylidene derivative of 2-amino-5-cyanonicotinamide (**12**). Since 2-amino-5-cyanonicotinamide itself does not react with acetone, it is clear that hydrolysis of the nitrile group cannot precede formation of the Schiff base, and an intramolecular hydrolysis via the hemiaminal **11** has been suggested (151).

In some cases attempted hydrolysis of *o*-aminonitriles leads to ring cleavage. For example, 4-cyano-5-aminoisoxazole (**14**) undergoes a ring cleavage reaction characteristic of all isoxazoles unsubstituted at position 3 (525) giving, with ammonia, the ammonium salt **15**, and with aniline *sym*-diphenylurea (**13**) (176). 2-Amino-3-cyano-4-acetyl-5-methylfuran

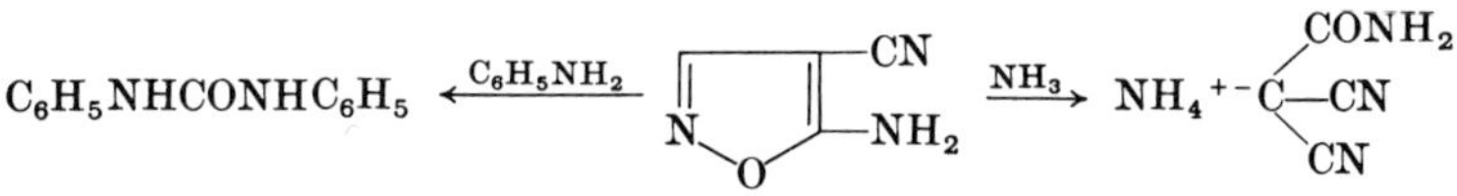

(13) (14) (15)

(**16**) undergoes ring opening with alkali to **17** (accompanied, however, by concomitant hydrolysis of the nitrile grouping), which can be recyclized to the furan upon acidification (276).

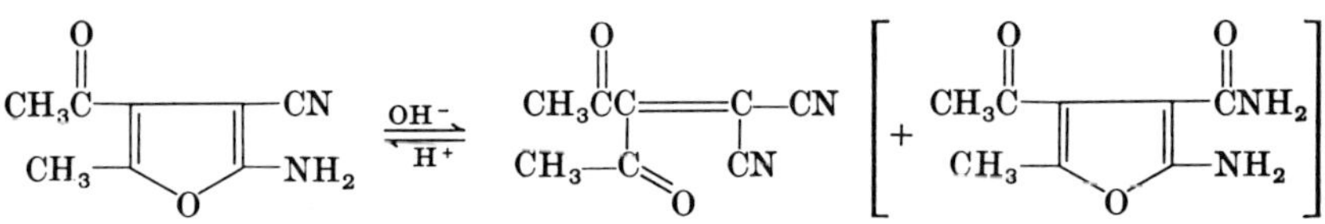

(16) (17)

In exceptional cases the amino group of *o*-aminonitriles undergoes hydrolysis preferentially (analogous to the hydrolysis of cyclic enaminonitriles). For example, the thiopyranthione **18** is hydrolyzed by methanol and acid to **19**, in which not only has the amino grouping been hydrolyzed but the thione grouping methylated. Treatment of **18** with alkali results in ring cleavage followed by subsequent recyclization to give the pyridone **20** (357).

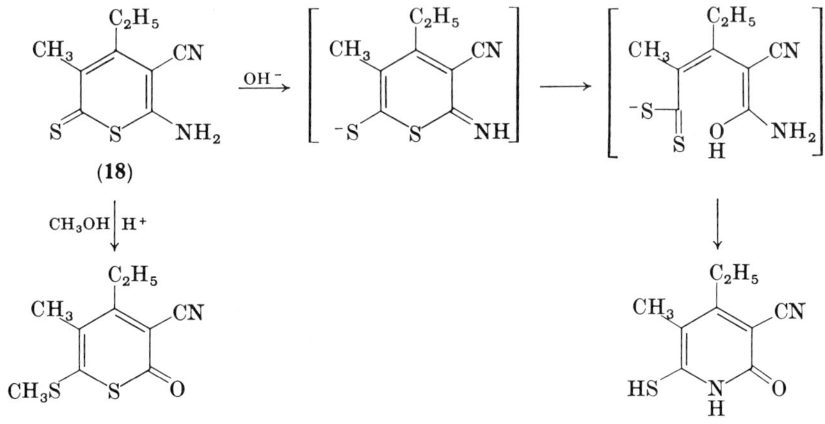

(18) (19) (20)

The utilization of *o*-aminonitriles for the preparation of condensed heterocyclic systems can involve either direct interaction with a second component with no isolation of acyclic intermediates, leading directly to the final product (see Chapter II, Section VIII-G to I), or an initial reaction of the *o*-aminonitrile with a reagent at either the amino or cyano group, followed by a second but intramolecular reaction leading to cyclization. Since the nitrile group in an *o*-aminonitrile is deactivated toward nucleophilic additions relative to a simple aromatic nitrile, attention has been focused primarily on initial reactions at the amino grouping followed by addition to the nitrile in the terminal cyclization step. Nevertheless, some synthetic utilization of *o*-aminonitriles for the preparation of interesting heterocyclic systems has followed the first route. For example, 2-aminobenzonitrile reacts with hydroxylamine to give 2-aminobenzamidoxime (**21**) (40,252) and subsequent diazotization of this compound apparently leads to a benzo-1,2,3-triazine, whose structure has been alternately proposed as the 4-hydroxylamino derivative **22** (40) and the 4-amino-3-oxide structure **23** (134) (Scheme 21). The latter appears to be correct, since the reaction of 4(3*H*)-benzo-1,2,3-triazinethione with hydroxylamine gave a product different from the product of diazotization of **21** (134). However, the curious reactions which **23** (or **22**) undergoes have not been satisfactorily rationalized in terms of either structure, and a reinvestigation would appear to be in order.

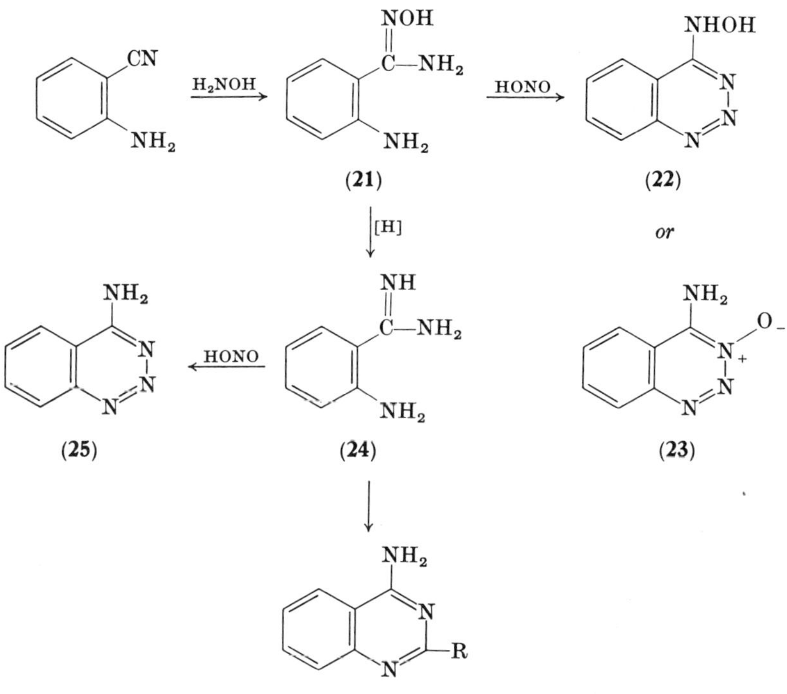

Scheme 21

Although it has been stated that 2-aminobenzonitrile cannot be converted to 2-aminobenzamidine (**24**) (129), this conversion can indeed be carried out by high pressure reduction of 2-aminobenzamidoxime (**21**) (252). Diazotization of **24** gives 4-aminobenzo-1,2,3-triazine (**25**), and one-carbon reagents convert it to 4-aminoquinazolines (**26**) (252). Similarly, hydrogen sulfide in the presence of pyridine and triethylamine has been shown to add to 2-amino-4-chlorobenzonitrile to give the thioamide in good yield (129); subsequent diazotization of the thioamide gave the corresponding 4(3*H*)-benzo-1,2,3-triazinethione. Addition of hydrogen sulfide in the presence of pyridine to 2-aminobenzonitrile gives 2-aminobenzthioamide, which on treatment with triethylorthoformate gives 4(3*H*)-quinazolinethione (95,96). An analogous sequence has been used to convert 4-amino-5-cyanopyrimidine to 4(3*H*)-pyrimido(4,5-*d*)pyrimidinethione (130,444).

Although 4-amino-5-cyanopyrimidine reacts with hydrogen sulfide satisfactorily in the presence of amines, 4-amino-5-cyano-6-methylpyrimidine is recovered unchanged (444). This failure appears to be due to steric hindrance at the cyano grouping, since 2-methyl-4-amino-5-cyanopyrimidine is also readily converted to the thioamide with hydrogen sulfide either in the presence of ammonia (462) or triethanolamine (444). 2-Amino-3-cyano-5-chloropyrazine readily adds methyl mercaptan in the presence of alkali to give the methyl thiocarboximidate, and hydrazine to give the corresponding amidrazone (562). *o*-Aminonitriles also appear to form imino ethers normally (562).

The reaction of 2-aminobenzonitrile with methylmagnesium iodide and phenylmagnesium bromide has been reported (442) to give the corresponding *o*-aminoketones by hydrolysis of the intermediate imines. Yields are better than 85%, and this appears to be a reasonable route to *o*-aminoketones.

B. REDUCTION

The nitrile grouping in *o*-aminonitriles may be reduced either to an aldimine or to an aminomethyl derivative without complications due to the *o*-amino grouping. This reaction would be without interest were it not for the intermediacy of 2-methyl-4-amino-5-aminomethylpyrimidine in the preparation of thiamin (Vitamin B_1). As a consequence, much study has been given to the reduction of 2-methyl-4-amino-5-cyanopyrimidine (42,46,49,55,57,87,196,203,255,362,379,381–387,392,393,397), and to the preparation of various thiamin analogs derived from other 4-amino-5-cyanopyrimidines by reduction (2,41,48,51,53,63,254,338,346,347,367,388,393). The reader is referred to numerous reviews on the chemistry of thiamin or to the papers cited above for details of these reductions.

4-Amino-5-cyanopyrimidine (readily available from malononitrile and formamidine) has been utilized in a convenient synthesis of the versatile intermediate 4-amino-5-formylpyrimidine by controlled reduction in the presence of complexing agents such as nickel chloride. The intermediate aldimine precipitates from the hydrogenation mixture as a nickel complex; the aldehyde is generated from the complex by treatment with the disodium salt of ethylenediaminetetraacetic acid (444). The same principle has been used for the synthesis of 2-methyl- and 2-ethyl-4-amino-5-formylpyrimidine from the corresponding aminonitriles by reduction in the presence of nickel chloride and ammonia, followed by decomposition of the insoluble aldimine nickel complexes with aqueous acid (444). Reduction of 4-amino-5-cyanopyrimidine with Raney nickel and hydrogen in the presence of semicarbazide gives a good yield of the semicarbazone of the aldehyde (444); reduction in 2*N* hydrochloric acid with hydrogen and palladium-on-carbon catalyst gives the free aldehyde in 49% yield (544).

C. ALKYLATION

Alkylation of *o*-aminonitriles has apparently received scant attention. 2-Aminobenzonitrile itself has been reported to yield both mono- and dialkyl derivatives upon treatment with ethyl iodide (264), and analogous alkylations would be expected with other aromatic *o*-aminonitriles. Alkylation is obviously complicated with heterocyclic compounds because of competing reaction with the ring nitrogens, and has been little studied.

D. DIAZOTIZATION

Aromatic *o*-aminonitriles can be diazotized normally, and the resulting diazonium salts coupled to give azo dyes (295–303,306,307,309,310), or subjected to the normal Sandmeyer conditions resulting in the replacement of the diazonium grouping by chlorine (232,272), hydrogen (232), or a cyano grouping. The latter replacement has been used as a synthesis of phthalonitriles (40,184,291,294) and as a means of establishing the orientation of substituents in the nitration of *p*-methylacetanilide (291). Reduction of the diazonium salt formed by diazotization of 2-aminobenzonitrile was formerly claimed to give 2-cyanophenylhydrazine (**1**) (128) but the product was later shown to be 3-aminoindazole (**2**) (119,120). In fact, diazotization of 2-aminobenzonitrile followed by reduction of the diazonium salt provides a convenient route to 3-aminoindazole and related derivatives (244).

It is well known that amino groups *o*- or *p*- to a ring nitrogen atom in heterocyclic compounds (e.g., 2- and 4-aminopyridines) normally undergo

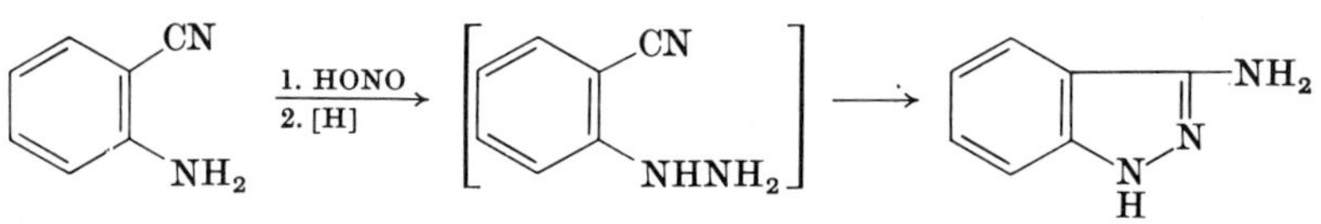

(**1**) (**2**)

hydrolysis of the amino grouping upon treatment in aqueous solution with nitrous acid. The same reaction has been observed for several heterocyclic *o*-aminonitriles; the result is selective hydrolysis of the amino grouping. For example, treatment of 2-amino-3-cyano-7-phenylpyrido(2,3-*b*)pyridine (**3**) with nitrous acid gave **4** (162). An analogous hydrolysis upon diazotization was observed with 2,4-diamino-3-cyanopyrido(2,3-*b*)quinoline (**5**) to give **6**; it is interesting to note that a selective hydrolysis of the 2-amino grouping took place. An analogous selectivity was observed in the treatment of **5** with sulfuric acid to give **7** (172). Similarly, 2-amino-3-cyanoquinoline is converted to 3-cyanocarbostyril (172), and 2-phenyl-4,6-diamino-5-cyanopyrimidine is selectively hydrolyzed to 2-phenyl-4-amino-5-cyano-6(1*H*)-pyrimidinone with nitrous acid (122).

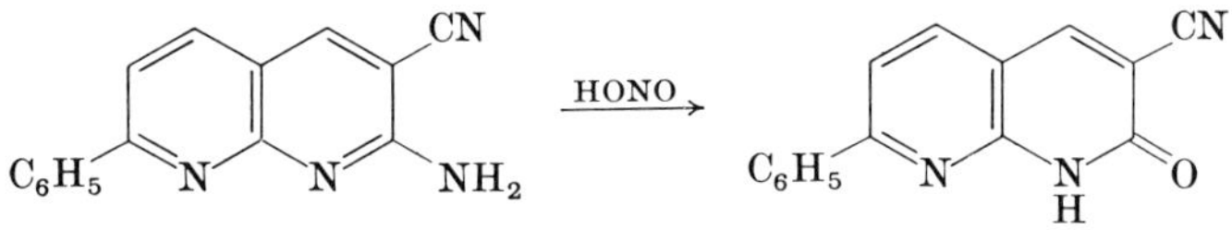

(**3**) (**4**)

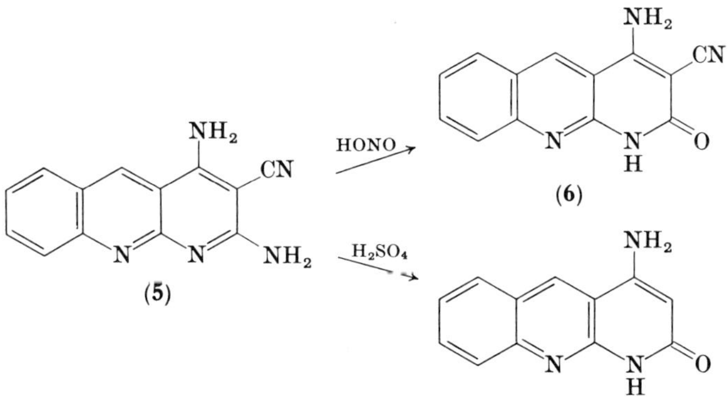

(**5**) (**7**)

E. SYNTHESIS OF FUSED PYRIMIDINES: ACYLATION OF *o*-AMINONITRILES

Both aromatic and heterocyclic *o*-aminonitriles react with formic acid and with the usual acylating agents to give *o*-formylamino and *o*-acylamino nitriles (137,240,247,250,278). This reaction is in itself trivial, but the *o*-acylaminonitriles so formed are of considerable interest because of their conversion by acid or base to condensed pyrimidines. These cyclization reactions were initially explored by Bogert and co-workers

(123) for the preparation of 4(3*H*)-quinazolones. Thus, 2-aminobenzonitrile is formylated with 98% formic acid to 2-formylaminobenzonitrile (**1**, R = H) which is converted in reasonable yield to 4(3*H*)-quinazolone

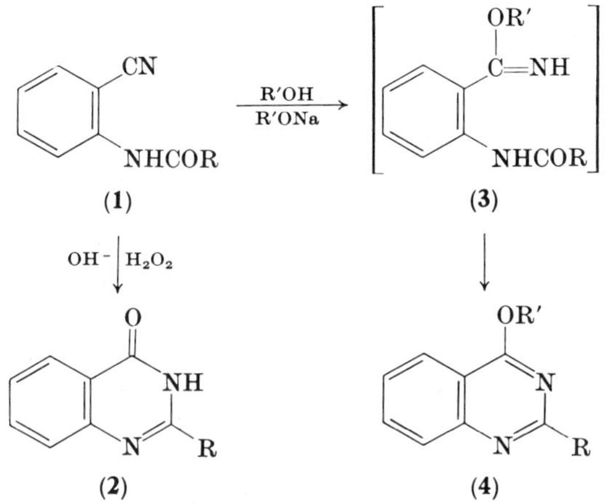

(**2**, R = H) by treatment with alkaline hydrogen peroxide. The reaction proceeds by initial hydration of the nitrile grouping to the carboxamide, which then undergoes cyclization in the alkaline medium. The reaction is equally applicable to 2-acylaminobenzonitriles and gives 2-substituted derivatives of 4(3*H*)-quinazolone (**2**). These latter compounds can alternately be prepared by direct heating of 2-aminobenzonitrile with acid anhydrides, but the conditions required are severe (sealed tube at 275–280°) and the conversion proceeds in lower yield. The reaction of *o*-acylaminonitriles with alkaline hydrogen peroxide has become a standard and widely employed method for the synthesis of quinazolones and condensed pyrimidones; examples of its application may be found by reference to the literature (79,126,131,132,140,168,293,532). In some cases the *o*-acylaminonitrile is simply heated with aqueous sodium hydroxide, and the yields are sometimes surprisingly good, as in the conversion of 1-cyano-2-acetylaminonaphthalene to 2-methyl-4(3*H*)-benzo(*f*)quinazolone in 80% yield (152). Aqueous potassium hydroxide has been used for the conversion of *o*-cyanophenylurea to 2,4(1*H*,3*H*)-quinazolinedione (332). 4(3*H*)-Quinazolinethiones are available analogously by treatment of *o*-acylaminonitriles with hydrogen sulfide; this reaction sequence is an alternative to initial conversion of the *o*-aminonitrile to the corresponding *o*-aminothioamide followed by heating with an acid anhydride (45,249).

A further variant is treatment of 2-aminobenzonitrile with an acid anhydride and sodium sulfide (45). The anhydride first acylates the *o*-aminonitrile, the acid thus formed liberates hydrogen sulfide from the sodium sulfide, and the hydrogen sulfide then adds to the nitrile group. Spontaneous cyclization leads to 4(3*H*)-quinazolinethione. Similarly, heating of 2-aminobenzonitrile with thioacids results in initial acylation with liberation of hydrogen sulfide which then adds to the nitrile grouping, with the same overall result. However, both reactions are best carried out in sealed tubes to prevent the escape of the hydrogen sulfide from the reaction mixture, and these conditions restrict the attractiveness of the procedures. Quinazoline derivatives substituted at position 4 with selenium are prepared in analogous fashion by treatment of *o*-acylaminonitriles with hydrogen selenide (127).

A closely related quinazoline synthesis from *o*-acylaminonitriles was discovered in the course of a routine attempt to effect deacylation with sodium methoxide in absolute methanol. Thus, 2-formylaminobenzonitrile (**1**, R = H) is converted in 40% yield under the above conditions to 4-methoxyquinazoline (**4**, R = H, R′ = OCH_3); the product is accompanied by 42% of deacylation product (2-aminobenzonitrile). Substantially higher yields of 4-alkoxyquinazolines result when *o*-acylaminonitriles are treated with alkoxide ion. Thus, 2-methyl-4-methoxyquinazoline is formed in 85% yield from 2-acetylaminobenzonitrile and sodium methoxide; the corresponding 4-ethoxy (82%), 4-benzyloxy (70%), and 4-phenoxy (89%) derivatives (**4**) were prepared in analogous fashion. Heating 2-cyano(methoxycarbonyl)aminobenzene (**5**) with sodium methoxide in absolute methanol gave 4-methoxy-2(1*H*)-quinazolone (**6**) in 95% yield; the reaction was shown to involve the alcohol used as solvent by the observation that 2-cyano(*ethoxy*carbonyl)aminobenzene was converted to **6** by sodium methoxide in absolute methanol. It was suggested that these reactions involved the intermediate addition of methoxide ion to the nitrile grouping to give an imino ether (analogous to **3**) which underwent intramolecular dehydration. This novel route to 4-alkoxyquinazolines was claimed to be substantially more convenient than the more usual routes involving 4-chloro or 4-methylthio intermediates (61,67).

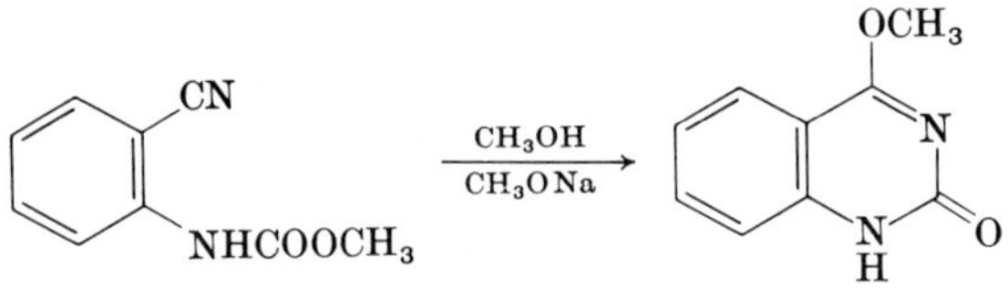

(**5**) (**6**)

The use of sodium methoxide or other alkoxides rather than sodium hydroxide does not always lead to 4-alkoxyquinazolines, however; treatment of *N*-phenyl-*N'*-(*o*-cyanophenyl)urea (**7**) with sodium methoxide leads to the quinazolone **8**, and analogous treatment of *o*-cyanophenylurea (**9**) with sodium methoxide gives 4-amino-2(1*H*)-quinazolone (**10**) (61). This route to quinazolines is discussed in detail on pp. 294–299.

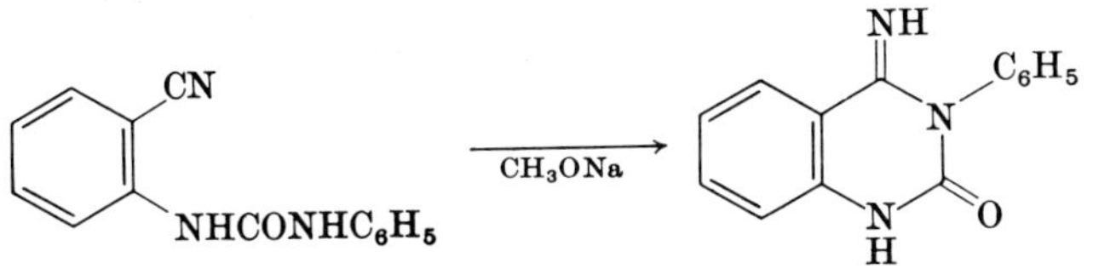

(**7**) (**8**)

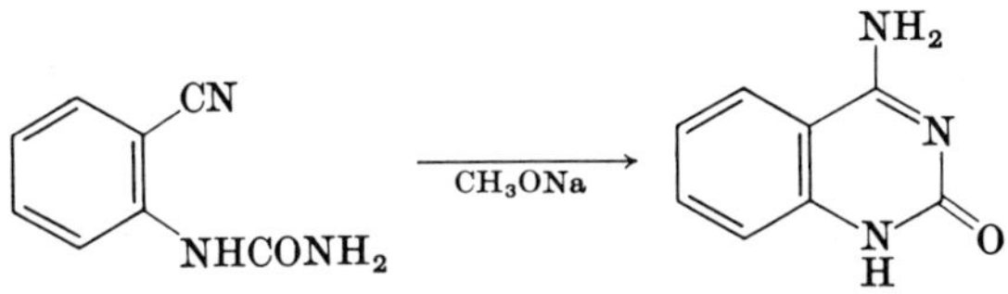

(**9**) (**10**)

The method of choice for the conversion of *o*-acylaminonitriles to 4(3*H*)-quinazolones is treatment with dry hydrogen chloride gas in absolute ethanol. For example, 1-cyano-2-acetylaminonaphthalene (**11**) is converted to 3-methyl-1(2*H*)-benzo(*h*)quinazolone (**12**) in quantitative yield (167). This reaction was the key step in the recent total synthesis of the azasteroids **14a**,**b** from the indicated precursors (**13a**,**b**) (340).

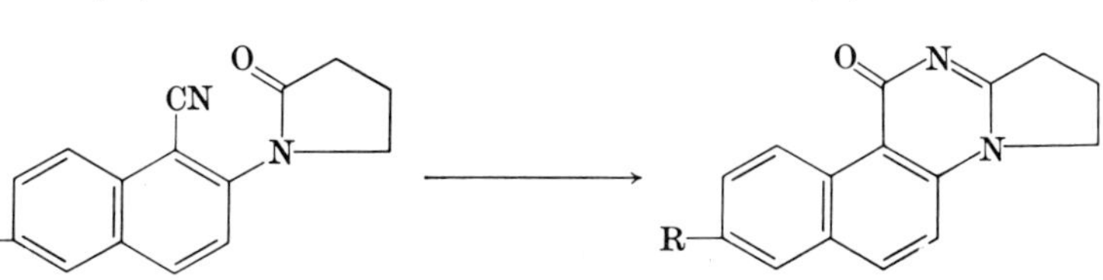

(**11**) (**12**)

(**13a**) R = H
(**13b**) R = OCH_3

(**14a**) R = H
(**14b**) R = OCH_3

An interesting variation on the acid-catalyzed cyclization of *o*-acylaminonitriles to 4(3*H*)-quinazolones is the treatment of *o*-aminonitriles with dimethylformamide saturated with dry hydrogen chloride (457). The products isolated after evaporation of the reaction mixture to dryness and addition of water are fused 4(3*H*)-pyrimidinones. Thus, 2-aminobenzonitrile gives 4(3*H*)-quinazolone, and 1-methyl-4-cyano-5-aminopyrazole (**15**) gives 1-methyl-4(5*H*)-pyrazolo(3,4-*d*)pyrimidinone (**16**, R = H). Substitution of dimethylacetamide for dimethylformamide with the latter aminonitrile gives 1,6-dimethyl-4(5*H*)-pyrazolo(3,4-*d*)-pyrimidinone (**16**, R = CH_3).

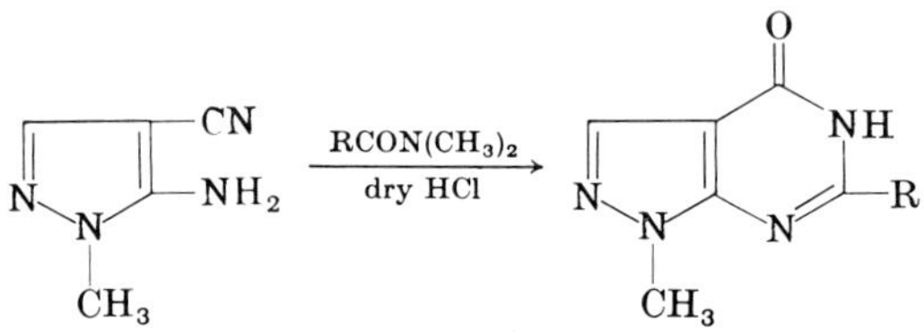

(**15**) (**16**)

Among the various mechanisms which have been considered for this conversion, that depicted in Scheme 22 appears to be the most reasonable (457), and involves an initial formylation of the amino group with the dimethylformamide–hydrogen chloride complex (Vilsmeier-Haack formylation) (or acetylation in the case of dimethylacetamide) to give an *o*-acylaminonitrile, followed by an acid-catalyzed intramolecular cyclization to a 4-imino-*m*-oxazine (**17**), which is the intramolecular equivalent of the imino ether formed in the presence of ethanol (e.g., **3**). Treatment

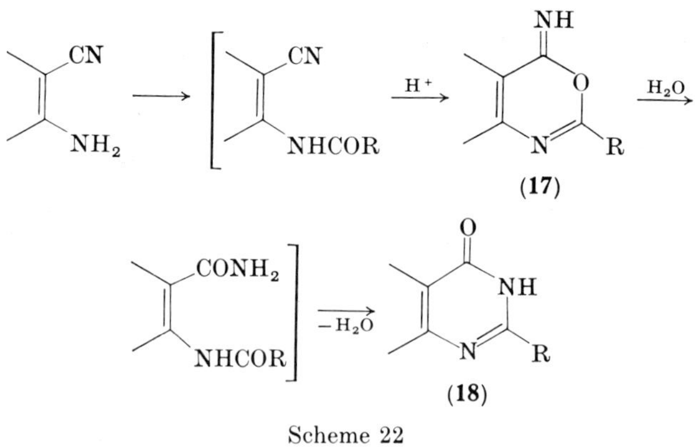

(**17**)

(**18**)

Scheme 22

of the *m*-oxazine with water during the workup results in hydrolysis to an *o*-acylaminocarboxamide (**18**) which is known to undergo ready intramolecular dehydration to a fused 4(3*H*)-pyrimidinone.

Acidic conditions can also be used for the cyclization of *o*-cyanoureides (e.g., **19** to **20**) (98), but these conditions offer no advantage over simple aqueous alkaline cyclization.

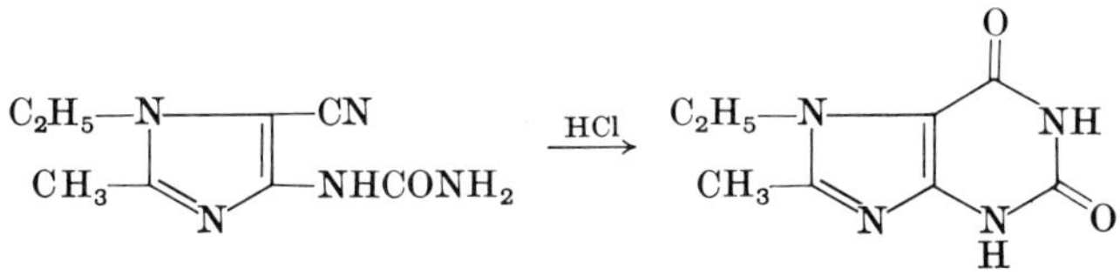

(**19**) (**20**)

An obvious and widely employed alternative synthesis of condensed 4(3*H*)-pyrimidinones or thiones is initial conversion of the *o*-aminonitrile to an *o*-aminoamide or thioamide, followed by treatment with an appropriate reagent to insert the missing carbon atom. Since the intrinsic interest in this reaction sequence deals only peripherally with the chemistry of *o*-aminonitriles, it will not be treated here in detail; the following examples are merely illustrative of this approach to condensed pyrimidines from *o*-aminonitriles. For example, a large number of pyrazolo(3,4-*d*)-pyrimidines have been prepared from 3-amino-4-cyanopyrazoles by initial conversion to the carboxamide followed by cyclization with formamide (3), urea (148,405), and various other reagents (115,124,144,145,372). Pyrimido(4,5-*d*)pyrimidines have likewise been made from 4-amino-5-cyanopyrimidines by initial hydration to the *o*-aminocarboxamide followed by cyclization with ethyl orthoformate and acetic anhydride or with diethyl carbonate (130,161). Condensed 4(3*H*)-pyrimidinethiones are available in analogous fashion by initial conversion of the *o*-aminonitrile to the thioamide followed by cyclization. Among the many examples reported are syntheses of pteridines (178), pyrimido(4,5-*d*)-pyrimidines (130), and quinazolines (45). Condensed triazinones may be prepared from *o*-aminocarboxamides by treatment with nitrous acid, and this reaction has been used to convert pyrazole aminonitriles to pyrazolo-(3,4-*d*)-1,2,3-triazinones (85,157).

F. SYNTHESIS OF FUSED PYRIDINES: ACYLATION OF *o*-AMINONITRILES

Although treatment of *o*-acylaminonitriles with dilute aqueous alkali results in the formation of 4(3*H*)-quinazolones by initial hydration of the nitrile to a carboxamide followed by intramolecular dehydration as discussed above, treatment of the same *o*-acylaminonitriles with *anhydrous* base (sodamide in liquid ammonia) gives excellent yields of condensed 4-amino-2(1*H*)-pyridones (59). For example, 2-acetylaminobenzonitrile (**1**) is readily cyclized in 90% yield under these conditions to 4-amino-carbostyril (**2**). This cyclization appears to be applicable as well to cyclic enaminonitriles; 2-acetylamino-1-cyanocyclopentene (**3**) is converted with sodamide in liquid ammonia in 96% yield to 4-amino-6,7-dihydro-2(1*H*)-pyrindinone (**4**). This cyclization must represent one of the early examples of the intermediacy of dianions in intramolecular reactions, since the formation of a 4-aminopyridine must involve the intermediate **5**, which undergoes intramolecular nucleophilic addition to the nitrile grouping. The claim (59) that 2-acetylamino-1-cyanocyclohexene undergoes an analogous cyclization to the isomeric 4-amino-5,6,7,8-tetrahydro-2(1*H*)-quinolinone appears to be spurious, for a reinvestigation of this reaction has shown (456) that the properties of the alleged quinolinone correspond with those of authentic 2-acetylamino-1-cyanocyclohexene, which was not obtained as a crystalline intermediate by the original authors. No obvious explanation for the failure of the cyclization in this latter case is apparent. It is interesting to note that although 1-amino-2-

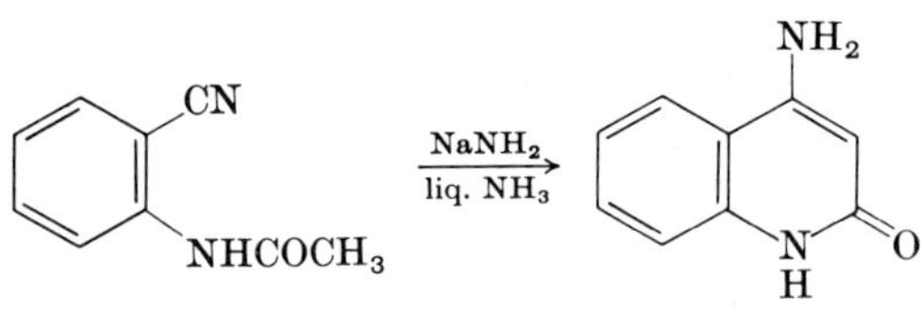

(**1**) (**2**)

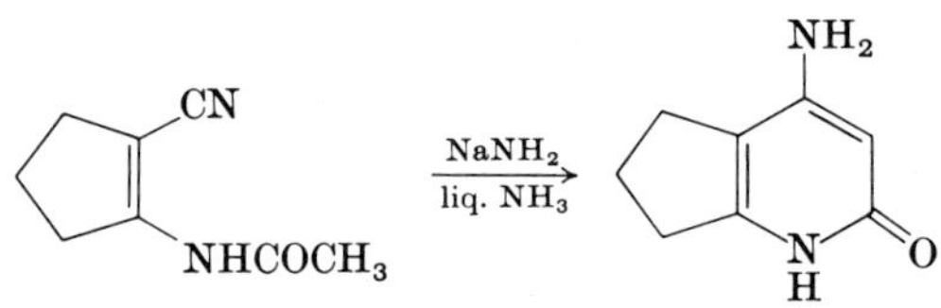

(**3**) (**4**)

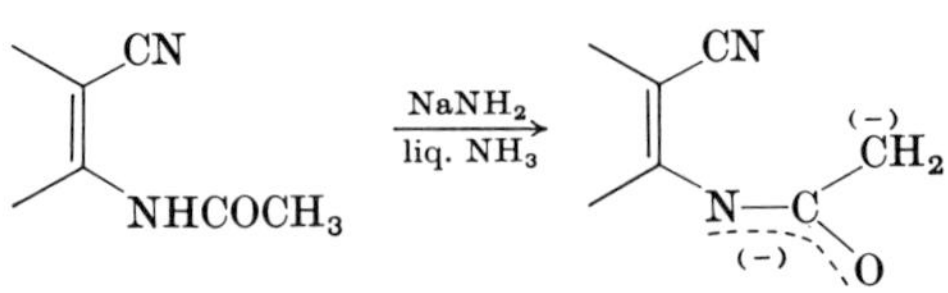

(**5**)

cyanocyclohexene can be acylated without complications with acetic anhydride, the use of acetyl chloride and pyridine leads to an exothermic reaction and the formation of 1-acetylamino-2-carbamoylcyclohexene (456).

G. SYNTHESIS OF FUSED 4-AMINOPYRIMIDINES

1. *By Dimerization of o-Aminonitriles. Reaction with Nitriles*

During an extensive investigation of the chemistry of amidines, Cooper and Partridge (118) discovered that heating the *p*-toluenesulfonic acid salt of 2-aminobenzonitrile (**1**) gave a mixture of tricycloquinazoline (**5**) and the mono *p*-toluenesulfonic acid salt of a dimer of 2-aminobenzonitrile. Structure **3b** (6,12-diaminophenhomazine) was proposed for this dimer, since it could be prepared independently by treatment of 6,12-dichlorophenhomazine (**3a**) with ammonia at 120–150°. In a reinvestigation of this reaction, Huisgen and Bast (569) found that the dimer was, in fact, 2-(*o*-aminophenyl)-4-aminoquinazoline (**4**); this same compound could be prepared by heating 2-aminobenzonitrile (**1**) or indazole with sodamide in high boiling, inert solvents [ring opening of indazole to 2-aminobenzonitrile with sodamide was demonstrated independently (570)]. Both 2-aminobenzonitrile (**1**) and the dimer **4** were converted to tricycloquinazoline (**5**) by heating at 300° with sodamide; the dimer **4** therefore appears to be an intermediate in the formation of **5** from **1**.

These reactions are summarized in Scheme 23. The dimer **4** undoubtedly arises by an initial amine → nitrile addition to give the intermediate amidine **2**, which then undergoes a second but intramolecular amine → nitrile addition which results in closure of the quinazoline ring. The formation of tricycloquinazoline (**5**) directly from **4** on heating with sodamide must involve an initial fragmentation to give some 2-aminobenzonitrile, which then reacts with **4** to give **5** (probably via another intermediate amidine such as **6a**). An interesting variant of these tricycloquinazoline syntheses is the reaction of 2-aminobenzonitrile with the *p*-toluenesulfonic acid salt of methyl anthranilate (320); the quinazolone **7**, which arises by initial amidine formation from the two precursors, followed by intramolecular loss of methanol, then reacts with a second mole of 2-aminobenzonitrile to give **6b**, which cyclizes to **5**.

The conversion of **3a** to **4** by treatment with ammonia deserves separate comment. We suggest that 6,12-diaminophenhomazine (**3b**) (proposed by Cooper and Partridge as the structure of the dimer (118)) is probably formed as an intermediate, but that this nonplanar, nonaromatic diazacyclooctatetraene opens to the amidine **2** by cleavage of one of the amidine

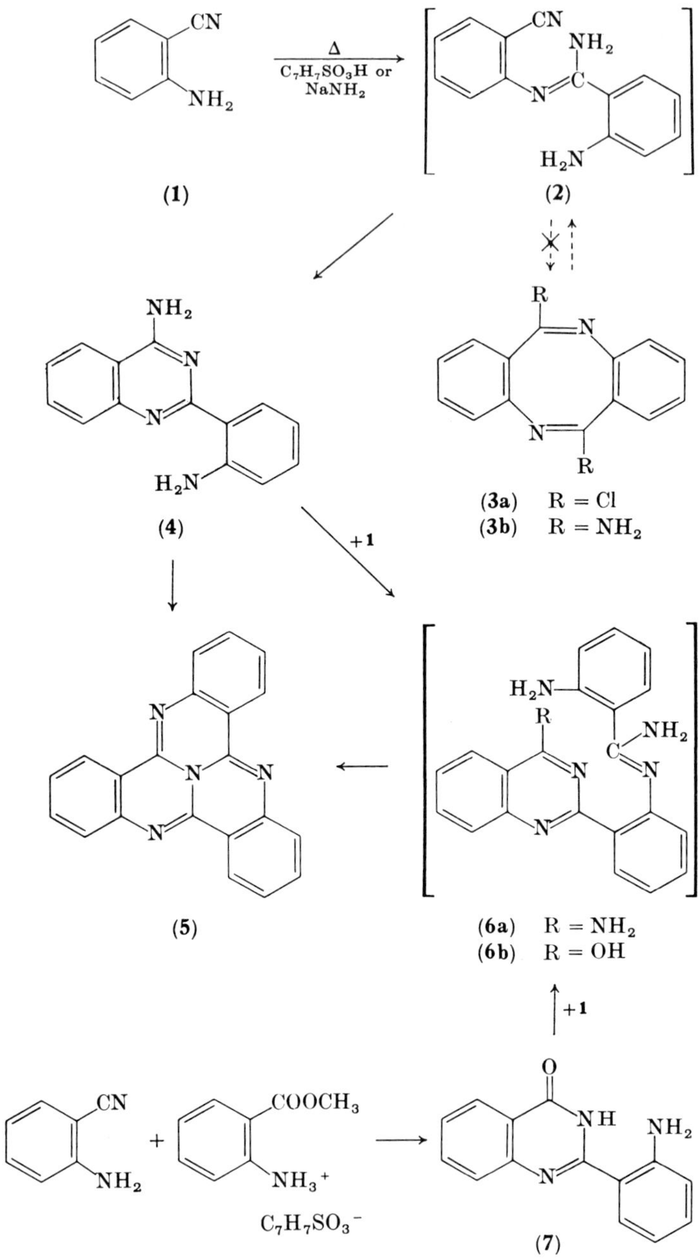
CN
NH2
Δ
C7H7SO3H or
NaNH2
CN
NH2
N
C
H2N
(1)
(2)
NH2
N
N
H2N
(4)
R
N
N
R
(3a) R = Cl
(3b) R = NH2
+1
N
N
N
N
(5)
H2N
R
NH2
C
N
N
N
(6a) R = NH2
(6b) R = OH
+1
CN
NH2
+
COOCH3
NH3+
C7H7SO3−
O
NH
NH2
N
(7)

Scheme 23

bonds (reversal of the amine → nitrile addition reaction which normally results in amidine formation). Recyclization of **2** then gives the aromatic quinazoline **4**.

The dimerization of *o*-aminonitriles to fused 4-aminopyrimidines appears to be a general reaction, and was first observed and recognized as such in 1958 (27) upon treatment of 2-chloronicotinonitrile with alcoholic ammonia. In addition to 2-aminonicotinonitrile (**8**), a high melting, insoluble fluorescent solid was obtained, the amount of which increased as the severity of the reaction conditions increased until at sufficiently high temperatures it became the major product. Furthermore, 2-aminonicotinonitrile (**8**) was shown to be an intermediate in the formation of the high-melting compound and, indeed, **8** could be converted to the yellow fluorescent solid not only by heating with ammonia but also by refluxing with sodium ethoxide in ethanol, simple pyrolysis, or heating its *p*-toluenesulfonic acid salt. Analysis and molecular weight determinations indicated that the yellow solid was a dimer of **8**, and degradation studies established its structure as 2-(3-(2-aminopyridyl))-4-aminopyrido(2,3-*d*)pyrimidine (**9**). This dimerization probably proceeds by initial condensation of the amino group of one molecule of the *o*-aminonitrile **8** with the nitrile group of a second molecule to give the intermediate amidine **10**, which then undergoes a second, but intramolecular, amine → nitrile condensation to give the observed product. Although amidine formation from amines and nitriles generally is not favored under basic conditions, it would appear

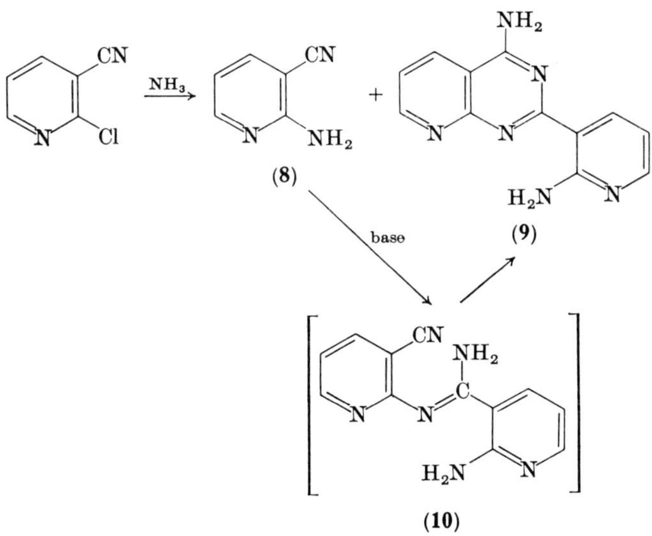

that cyclization to a low-energy, highly aromatic fused 4-aminopyrimidine derivative in the terminal step provides sufficient driving force to displace an otherwise unfavorable initial equilibrium.

An analogous dimerization was observed during the preparation of 2-amino-5-nitrobenzonitrile (**11**) by treatment of 2-chloro-5-nitrobenzonitrile with alcoholic ammonia (21). Simple replacement of the activated chloro group by an amino group takes place at 150°, but when the reaction mixture is allowed to reach 170°, there was isolated an extremely insoluble, high-melting yellow by-product which did not contain a nitrile grouping, but which was isomeric with 2-amino-5-nitrobenzonitrile. As was the case with 2-aminonicotinonitrile discussed above, the same product could be obtained in good yield simply by treatment of 2-amino-5-nitrobenzonitrile with alcoholic ammonia at 180°. This product was shown by degradation and independent synthesis to be 2-(2-amino-5-nitrophenyl)-4-amino-6-nitroquinazoline (**12**) (21).

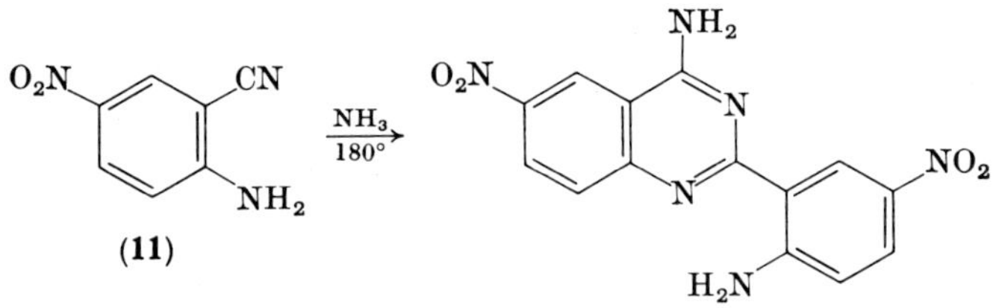

(**11**) (**12**)

By contrast, however, neither 2-aminobenzonitrile nor 2-amino-5-bromobenzonitrile underwent analogous dimerizations under mild alkaline conditions, and it thus appeared that the dominant factor determining dimerization was the ability of the nitrile grouping to undergo nucleophilic attack rather than the basicity of the attacking amino group. It was thus predicted that 2-aminobenzonitrile, precisely because it did *not* dimerize under mild alkaline conditions, should readily undergo *mixed* condensations with more reactive nitriles. Thus, 2-aminobenzonitrile

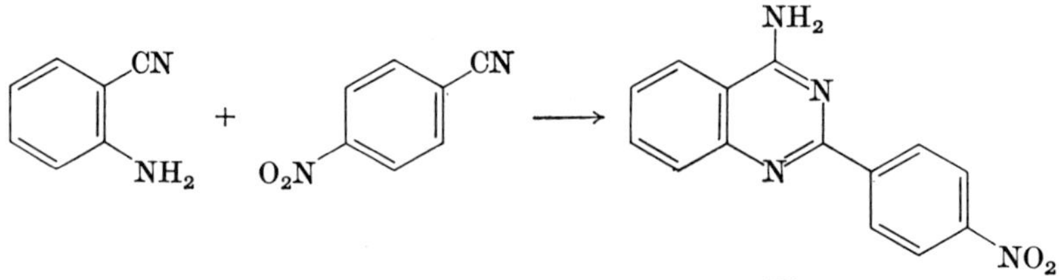

(**13**)

reacted readily with 4-nitrobenzonitrile to give in good yield 2-(4-nitrophenyl)-4-aminoquinazoline (**13**). Moreover, the dependence of the course of the condensation reaction on the proclivity of the participating nitrile group to undergo nucleophilic attack was clearly demonstrated by the observation that a mixture of 2-amino-5-nitrobenzonitrile (**11**) and 4-nitrobenzonitrile yielded exclusively the product of mixed condensation, 2-(4-nitrophenyl)-4-amino-6-nitroquinazoline (**14**); none of the dimer (**12**) was formed.

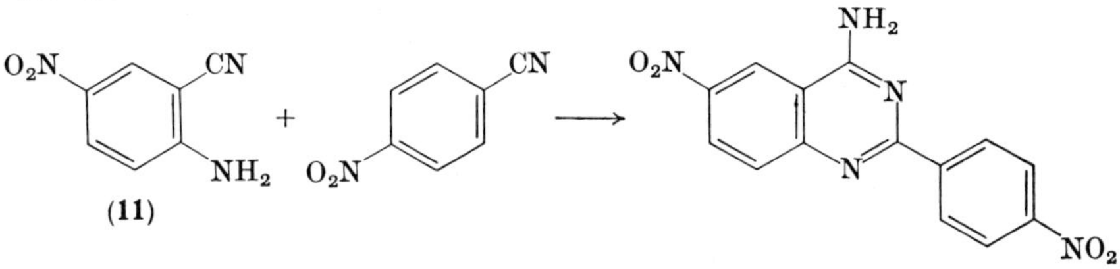

(**11**)

(**14**)
(no **12**)

These reactions have been extended to a useful and general synthesis of fused 4-aminopyrimidines by condensation of *o*-aminonitriles with aliphatic and aromatic nitriles (17). In general, those *o*-aminonitriles which undergo dimerization readily react in mixed condensations only with very reactive nitriles, while *o*-aminonitriles undergoing dimerization with difficulty are correspondingly less selective in undergoing mixed condensations with other nitriles. With heterocyclic *o*-aminonitriles the reaction provides a simple and versatile method for the preparation of purines (**15**), pyrazolo(3,4-*d*)pyrimidines (**16**), and other condensed 4-aminopyrimidine systems (17,21,27,467,532).

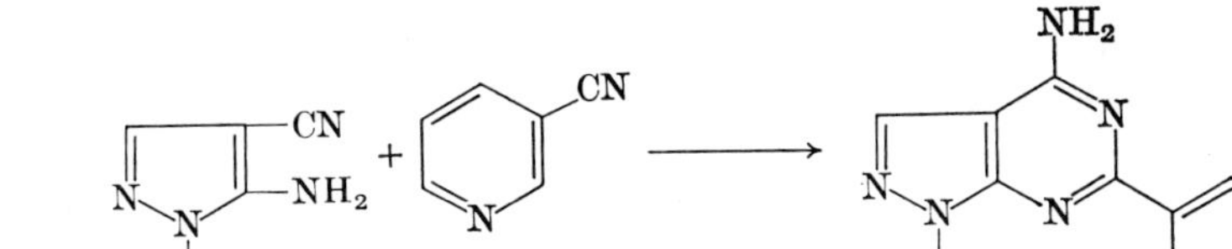

(**15**)

(**16**)

2. *By Reaction of o-Aminonitriles with Orthoformate Esters and Amines* (see Table XXVII)

All of the above reactions involving the interaction of an *o*-aminonitrile either with itself to give a dimer or with a trapping nitrile to give a condensed 2-substituted-4-aminopyrimidine presumably involve the intermediate formation of an amidine by an amine → nitrile addition. The conditions under which the above reactions are carried out, however, preclude the isolation of such an intermediate, for it would either revert by dissociation to the starting materials (thus reversing what is undoubtedly an unfavorable equilibrium in any case), or cyclize by intramolecular amine → nitrile addition. As a general synthetic route to fused 4-aminopyrimidines, the above reaction possesses the disadvantage that the substituent introduced into the 2-position is itself dependent upon the electrophilicity of the participating R—CN reactant, and by the fact that dimerization either accompanies or supersedes a mixed condensation in those cases where the nitrile grouping is relatively unreactive. It is clear that an alternative synthesis of *o*-cyanoamidines which did not involve the direct use of R—CN might offer both greater versatility and the possibility of isolation of the intermediate, provided that mild conditions for its formation could be found. These considerations suggest the possible utilization of imino ethers, since their reaction with amines to give amidines is one of the few general syntheses of amidines (519) that occurs even at 0°. The intermediacy of an imino ether for the synthesis of an *o*-cyanoamidine from an *o*-aminonitrile requires either the reaction of the latter with an externally formed imino ether or the conversion of the latter to an imino ether (by reaction of the amino group with an ortho ester) followed by reaction with an amine. The former route appears not to have received much attention, but the latter has been developed into a useful and versatile synthesis of condensed pyrimidines (20) (see Table XXVII).

The reaction sequence leading from *o*-aminonitriles to condensed pyrimidines via the intermediacy of imino ethers can be illustrated with 1-methyl-4-cyano-5-aminopyrazole (**1**). Condensation of this material with a mixture of ethyl orthoformate and acetic anhydride yields the crystalline ethoxymethyleneamino intermediate **2** which upon treatment in benzene solution with alcoholic methylamine results in the immediate separation of the *o*-cyanoamidine **3**. It will be noted that this compound is structurally analogous to the presumed intermediate in the *o*-aminonitrile dimerizations and nitrile condensations discussed above. When the amidine **3** is heated in dry pyridine, it is rapidly converted into an isomer (**4**) in which the desired intramolecular amine → nitrile addition has taken place. This same condensed pyrimidine can alternately be prepared directly and rapidly from the imino ether **2** by treatment with ethanolic methylamine at room temperature. However, extended treatment of **4** with ethanolic methylamine or exposure to the atmosphere for 24 hr at room temperature results in its further isomerization to 1-methyl-4-methylaminopyrazolo(3,4-*d*)pyrimidine (**5**). The rearrangement of **4** to **5** can readily be effected in water or dilute sodium hydroxide; no reaction occurs upon prolonged refluxing in aqueous acid. The initial ring closure of **3** to **4** also appears to be base-catalyzed, for the conversion is readily carried out in dry pyridine or in ethanol containing excess methylamine, although **3** is completely stable in refluxing dry ethanol or even upon vacuum sublimation (20).

The conversion of *o*-aminonitriles to condensed 4-substituted-aminopyrimidines by initial conversion to the ethoxymethyleneamino derivative followed by treatment with an aryl or alkyl amine has seen wide application since its original introduction in 1960 (20). It is not only unnecessary

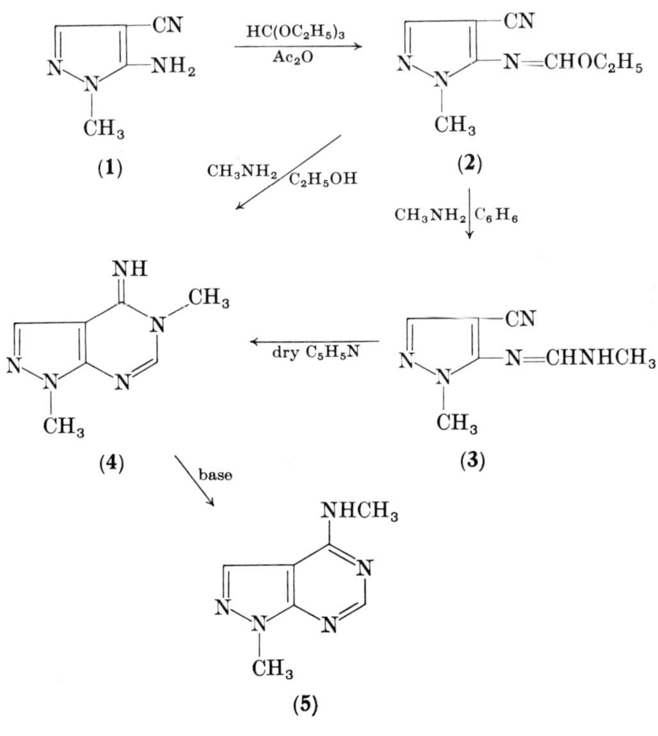

but often impossible to isolate the intermediate *o*-cyanoamidine corresponding to **3**. In fact, the unrearranged, initially formed intramolecular cyclization product (the 3-substituted-4-iminopyrimidine corresponding to **4**) is also seldom isolated, since it may rearrange to the 4-substituted-amino pyrimidine (corresponding to **5**) under the conditions of its formation unless particular precautions are taken. The conversion of an *o*-aminonitrile to a fused 4-amino (or substituted amino) pyrimidine can in practice be carried out without the isolation of any intermediate, and this conversion has seen particularly wide application with the use of ammonia as the reacting amine. Representative examples of the wide scope and application of this method of pyrimidine synthesis from *o*-aminonitriles are given in Table XXVII.

A key feature of the above pyrimidine synthesis is the rearrangement of a 3-substituted-4-iminopyrimidine to a 4-substituted-amino pyrimidine. This isomerization is commonly referred to as the Dimroth rearrangement, and its mechanism has been the subject of numerous investigations (520). It appears that, under the above conditions, this rearrangement involves the initial nucleophilic addition of a base (ammonia, alkyl amine, alkali) to position 2 of the condensed 3-substituted-4-iminopyrimidine **6**, followed by fission of the $C_2—N_3$ bond. The resulting amidine **7a** suffers free rotation around the C—C bond joining the amidine grouping to the aromatic or heterocyclic ring, and subsequent recyclization, with participation of the unsubstituted nitrogen of the amidine grouping (**7b**) and concomitant expulsion of the original attacking nucleophile, results in

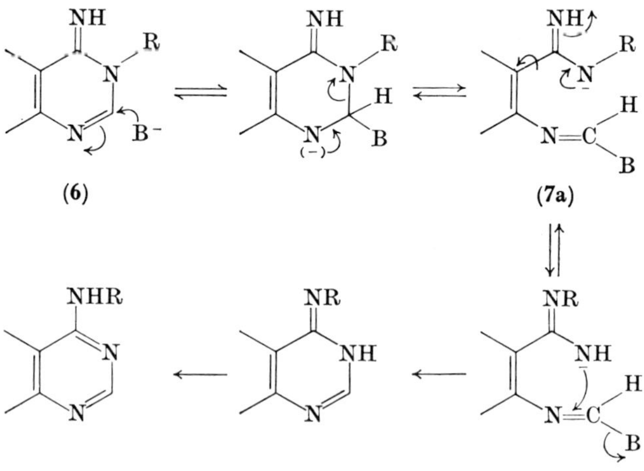

(6) (7a) (8) (7b)

isomerization with the formation of the fully aromatic 4-substituted-amino pyrimidine (**8**). Since analogous isomerizations have been observed under thermal conditions, and the rate of some isomerizations appears to be independent of the concentration of base, the above interpretation may not be relevant to all known Dimroth rearrangements.

When ethyl orthoformate is used for the formation of the intermediate imino ether (e.g., **2**), the final fused 4-aminopyrimidine obviously carries no substituent in position 2. Although the use of other ortho esters would allow the introduction of substituents at position 2, there appear to be few examples (552) of such reactions, perhaps because orthoformate esters are considerably more reactive toward amines than alkyl or aryl ortho esters (526).

Illustrative of the utility of the conversion of *o*-aminonitriles to fused 4-aminopyrimidines, employing ammonia as the amine, are the syntheses of the aglycone of toyocamycin (3-cyano-4-aminopyrrolo-(2,3-*d*)pyrimidine (**14**)) (10), and of 4-amino-5-substituted pyrimidines from 2-amino-3-cyanothiophenes (91,341). In the first (Scheme 24), tetracyanoethylene (**9**) was converted by the action of hydrogen sulfide to 2,5-diamino-

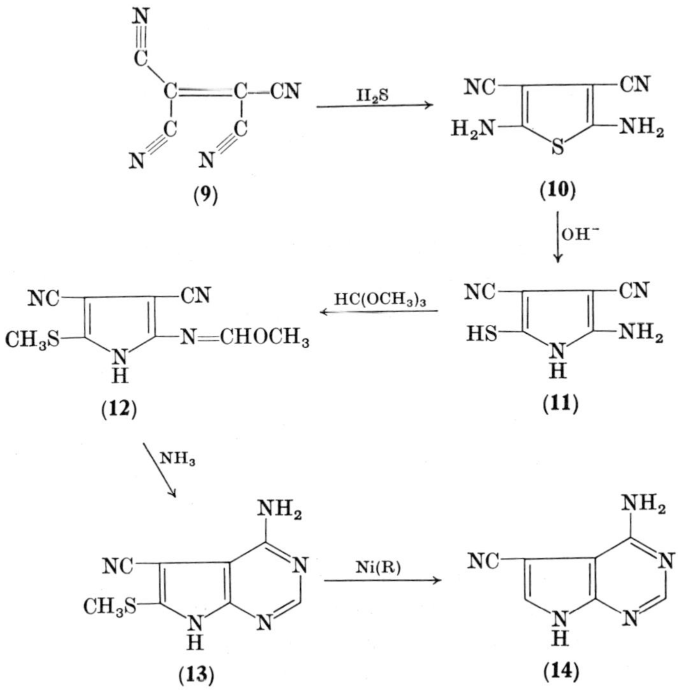

Scheme 24

3,4-dicyanothiophene (**10**) which was rearranged with alkali to 2-amino-3,4-dicyano-5-mercaptopyrrole (**11**) (102,104). Treatment of the latter compound with methyl orthoformate followed by alcoholic ammonia gave 4-amino-5-cyano-6-methylmercaptopyrrolo(2,3-*d*)pyrimidine (**13**) which gave the desired toyocamycin aglycone **14** upon desulfurization with Raney nickel (10). In the second, 2-amino-3-cyanothiophenes (**15**), prepared by the reaction of ketones with malononitrile and sulfur (see pp. 137–142), were condensed with ethyl orthoformate to give the intermediate 2-ethoxymethyleneamino derivatives (**16**) which were subsequently converted with ammonia to the isolated *o*-formamidinonitriles **17**. These were cyclized with sodium methoxide in dimethylformamide to give 4-aminothieno(2,3-*d*)pyrimidines (**18**) which on desulfurization with W-7 Raney nickel gave 4-amino-5-substituted-pyrimidines (**19**) (91,341).

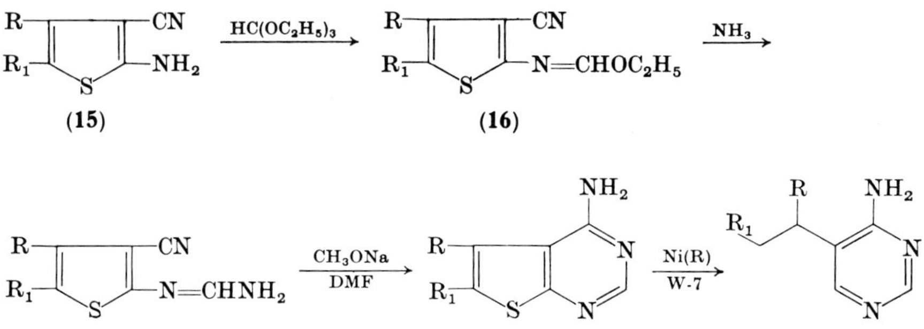

(15) (16) (17) (18) (19)

The first isothiazolo(3,4-*d*)pyrimidine (**21**) was prepared from 3-amino-4-cyano-5-methylisothiazole (**20**) by application of the ethyl orthoformate/acetic anhydride/ammonia procedure (156).

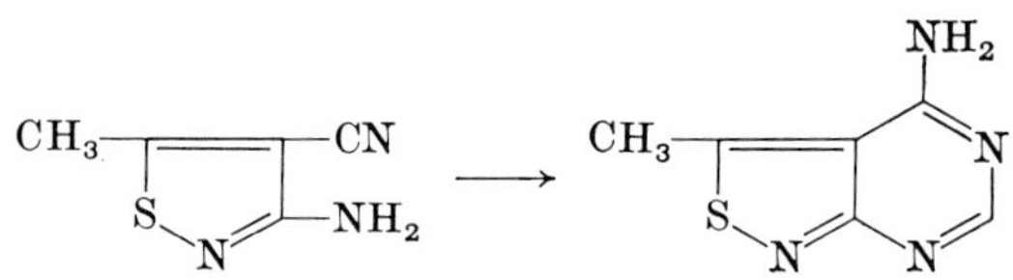

(20) (21)

The reaction of *o*-cyano ethoxymethyleneamino derivatives with amines other than ammonia or primary alkyl or aryl amines (or hydrazine) has been virtually neglected. The only example to our knowledge is the reaction of 1-phenyl-3-cyanomethyl-4-cyano-5-ethoxymethyleneamino-pyrazole (**22**) with guanidine, which surprisingly led to the 4-amino derivative **24**, perhaps by elimination of cyanamide from the initial cyclization product **23** (25).

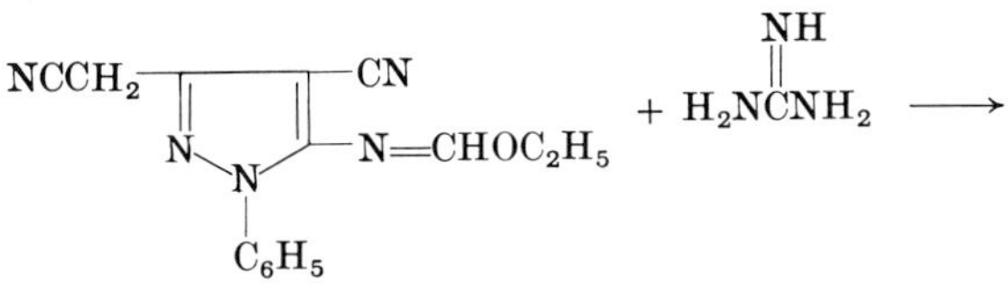

(22)

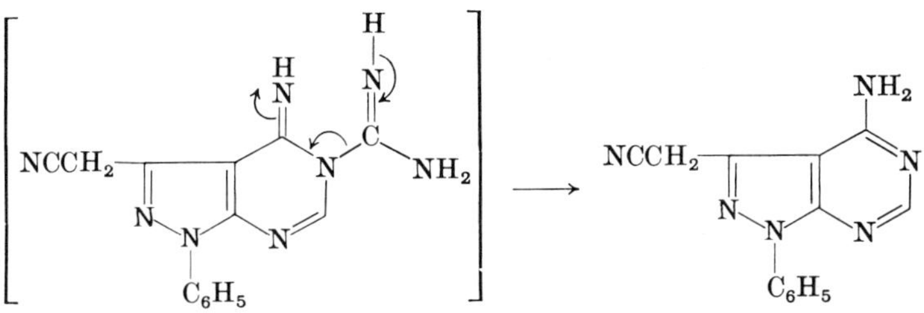

(23) **(24)**

3. *By Reaction of o-Aminonitriles with Amidines*

A further, alternative synthesis of *o*-cyanoamidines from *o*-aminonitriles is by direct interaction with an amidine with loss of ammonia or an amine (i.e., amidine exchange). The interaction of an *o*-aminonitrile with formamidine itself, for example, should give an intermediate formamidine (**1**), identical with the product resulting from the treatment of the corresponding *o*-ethoxymethyleneaminonitrile with ammonia. Conditions for amidine exchange are considerably more strenuous, however, than conditions for amidine formation from amines and imino ethers, and it would thus not be anticipated that the intermediate *N*-substituted formamidine **1** would be isolated under the reaction conditions. These predictions have not only been vindicated by actual experiment, but the interaction of *o*-aminonitriles with formamidine (usually as its acetate

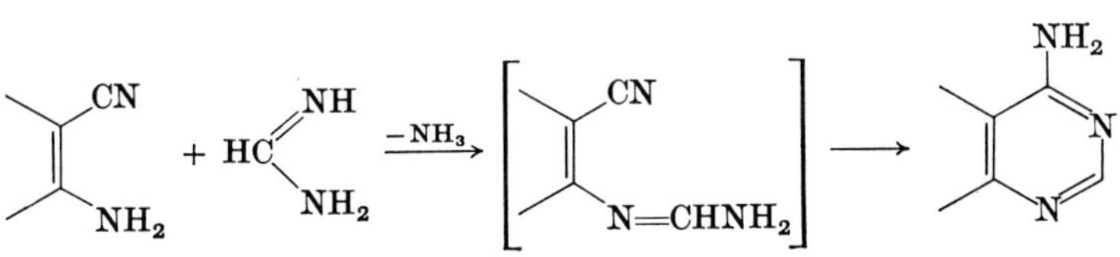

(1)

TABLE XXVII

Condensed 4-Aminopyrimidines from o-Aminonitriles and $HC(OC_2H_5)_3/NH_3$ (or RNH_2)

o-Aminonitrile	Procedure[a]	Product	Yield, %	Ref.
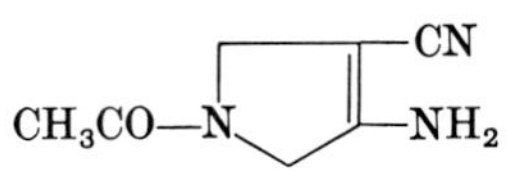	B	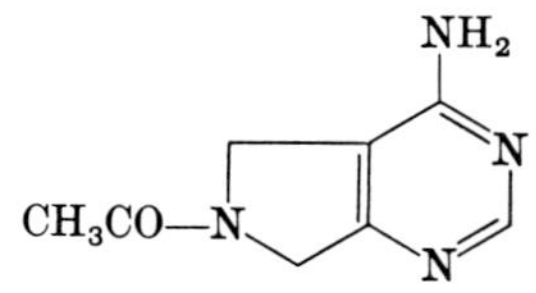	52	76
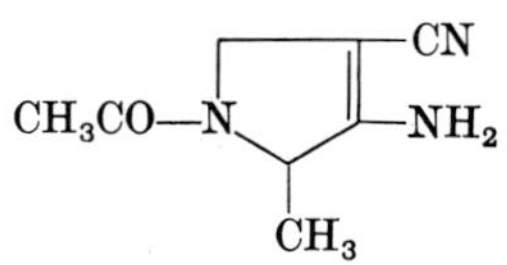	B	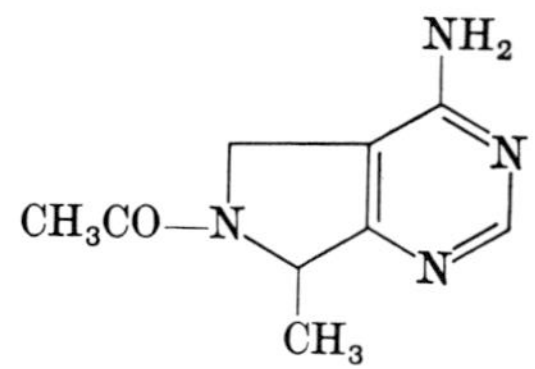	69	76
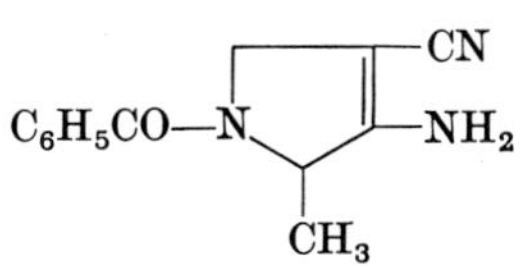	B	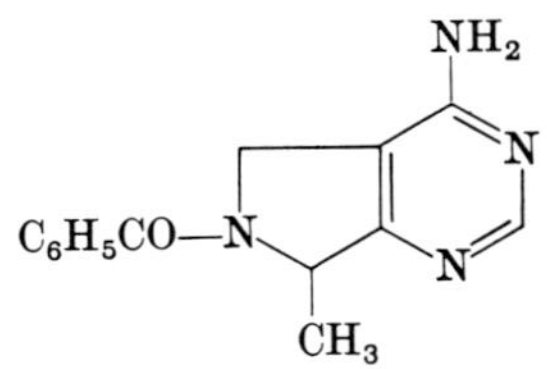	84	76

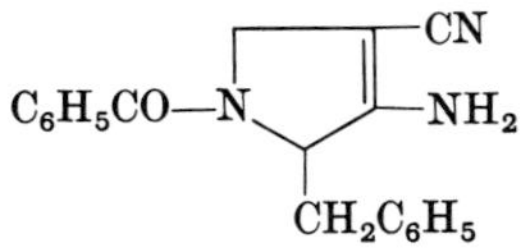	B	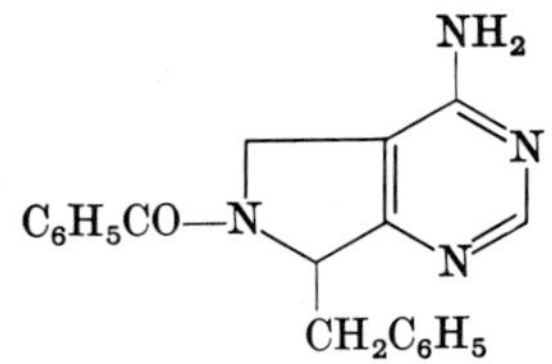	90	76
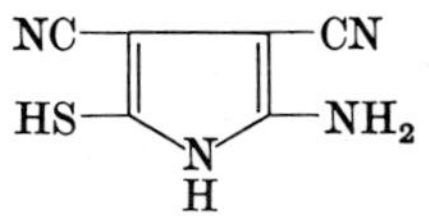	B (using $HC(OCH_3)_3$)	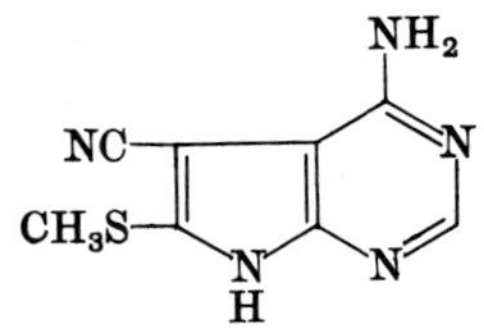	34	10
	B (using $HC(OCH_3)_3$ and CH_3NH_2)	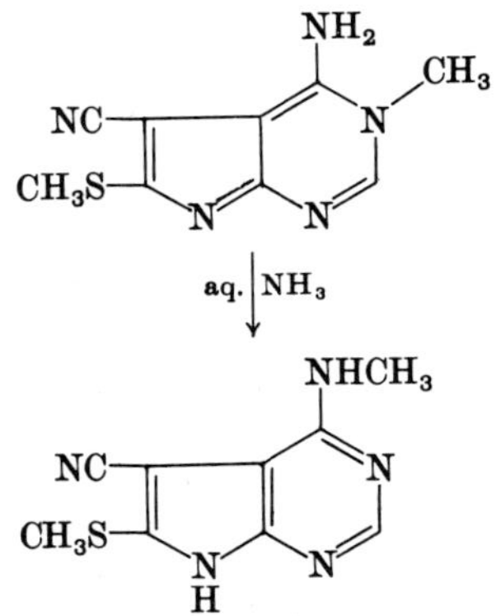	69	494
			—	

(continued)

TABLE XXVII (*continued*)

o-Aminonitrile	Procedure[a]	Product	Yield, %	Ref.
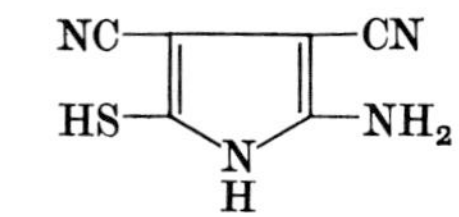	B	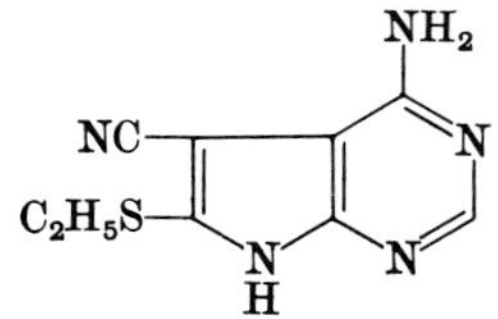	78	10
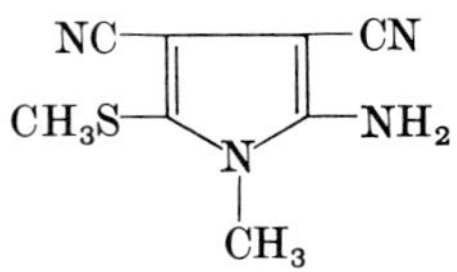	B (using $HC(OCH_3)_3$)	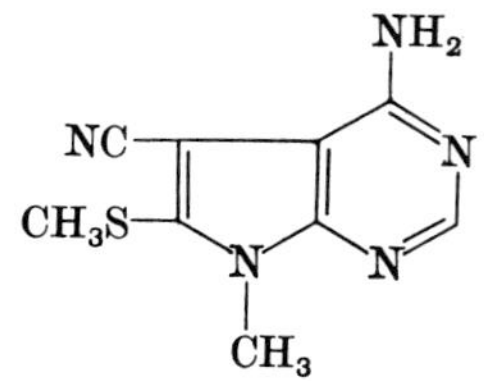	58	494
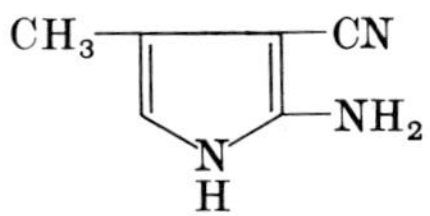	B	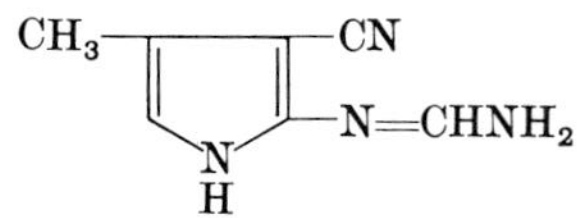	61	10
		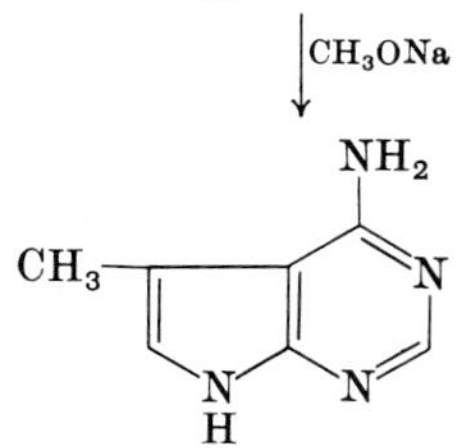	91	10

(CH_3NH_2)	B	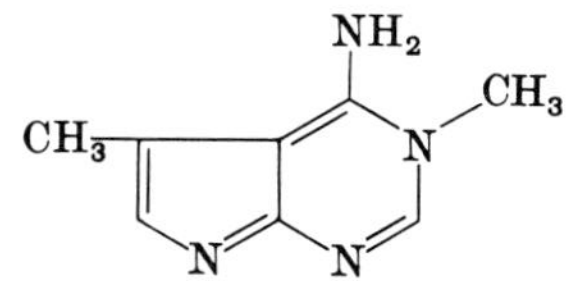	39	10
		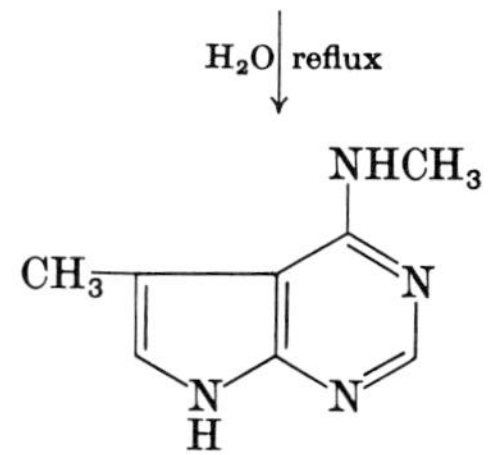	80	10
$((CH_3)_2NCH_2CH_2CH_2NH_2)$	B	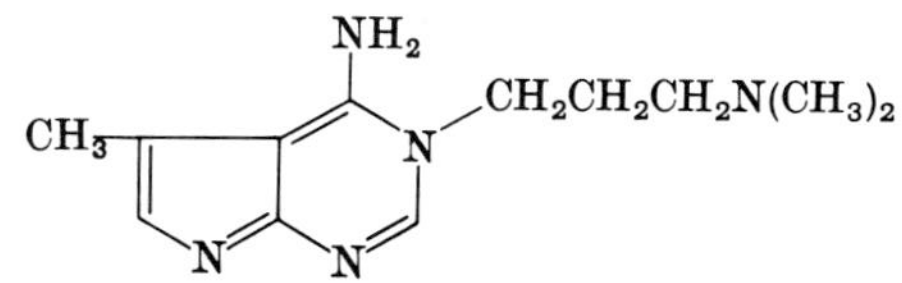	22	10
		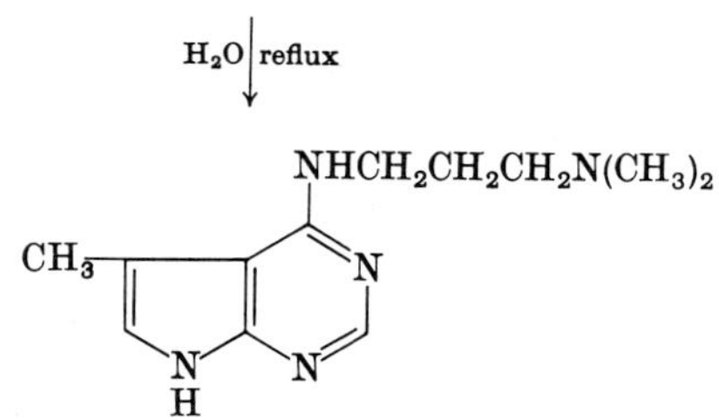	80	10

(continued)

TABLE XXVII (*continued*)

o-Aminonitrile	Procedure[a]	Product	Yield, %	Ref.
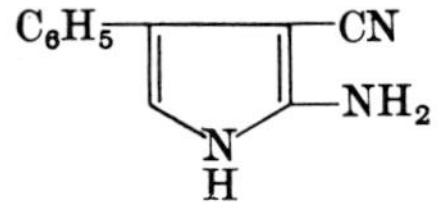	B	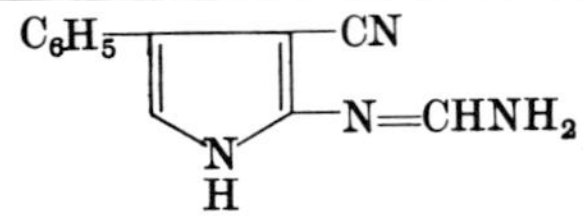	38	10
		↓ CH_3ONa 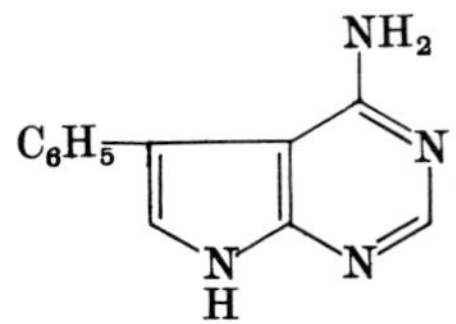	93	10
	B (CH_3NH_2)	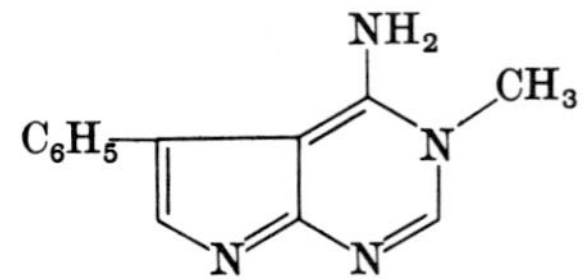	38	10
		↓ aq. C_2H_5OH reflux 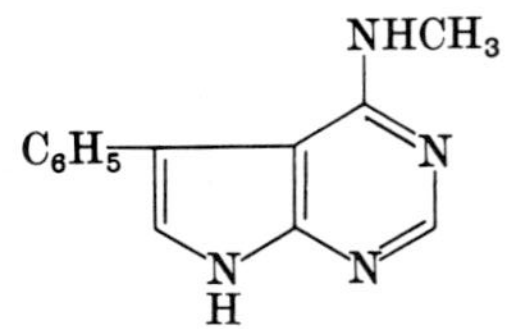	50	10

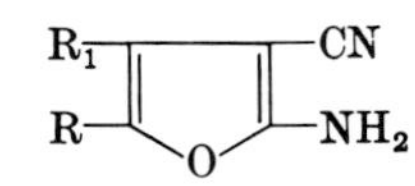

A

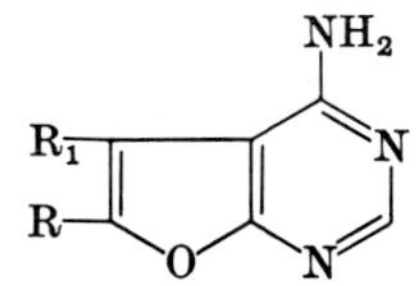

R	R_1		
C_6H_5	C_6H_5	70	492
$CH_3C_6H_4$	C_6H_5	55	492
C_6H_5	$CH_3C_6H_4$	53	492
$CH_3OC_6H_4$	C_6H_5	46	492
C_6H_5	$CH_3OC_6H_4$	42	492
$C_2H_5OC_6H_4$	C_6H_5	60	492
$C_6H_5C_6H_4$	C_6H_5	61	492
C_6H_5	$C_6H_5C_6H_4$	67	492
C_6H_5	ClC_6H_4	65	492
C_6H_5	BrC_6H_4	68	492

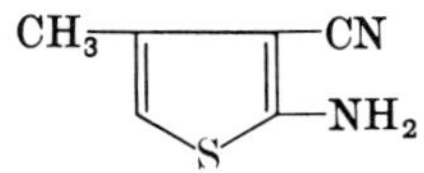

B

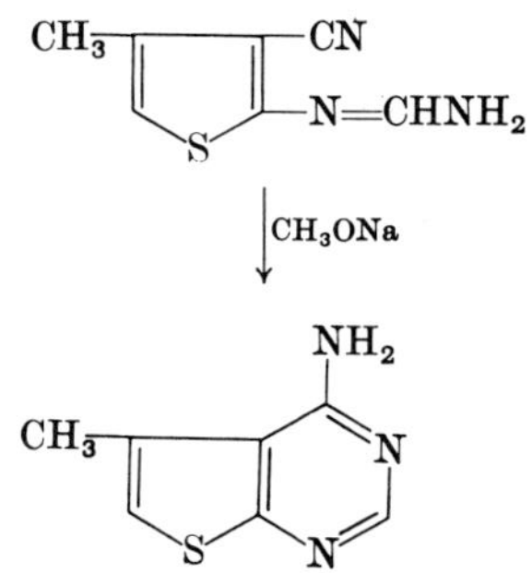

—	
82	91, 341

(*continued*)

TABLE XXVII (*continued*)

o-Aminonitrile	Procedure[a]	Product	Yield, %	Ref.
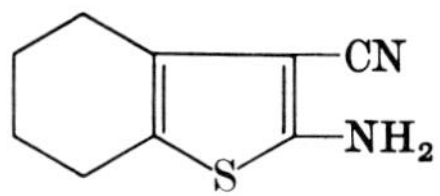	B	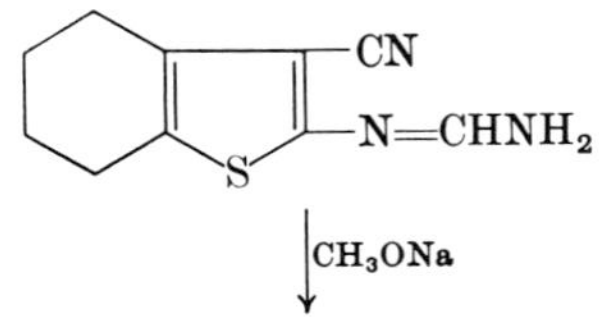	—	
		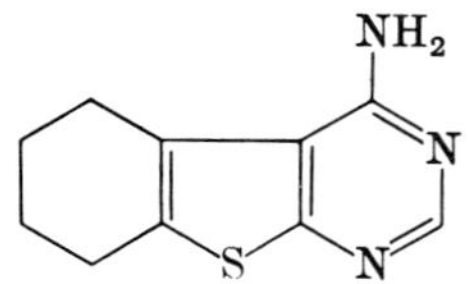	73	91, 341
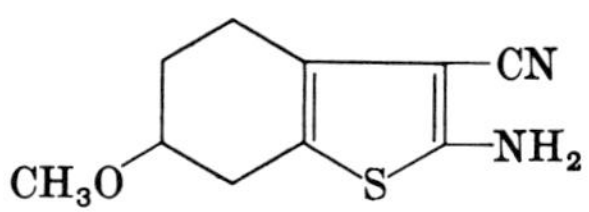	B	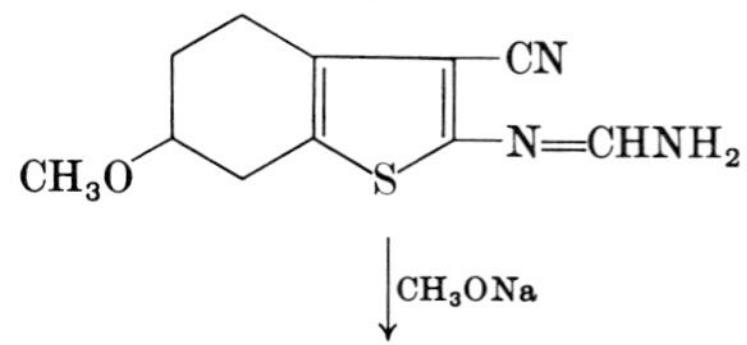	—	
		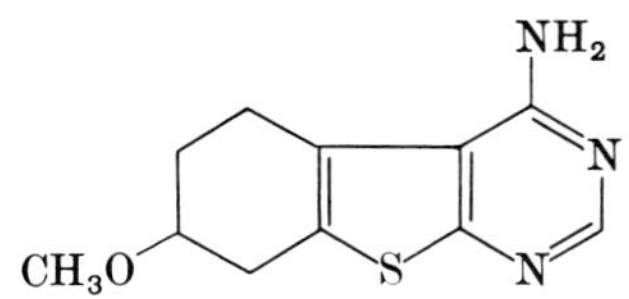	68	91, 341

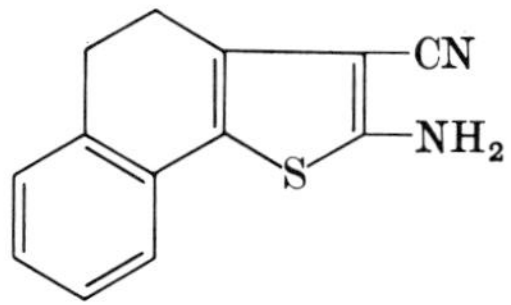	B	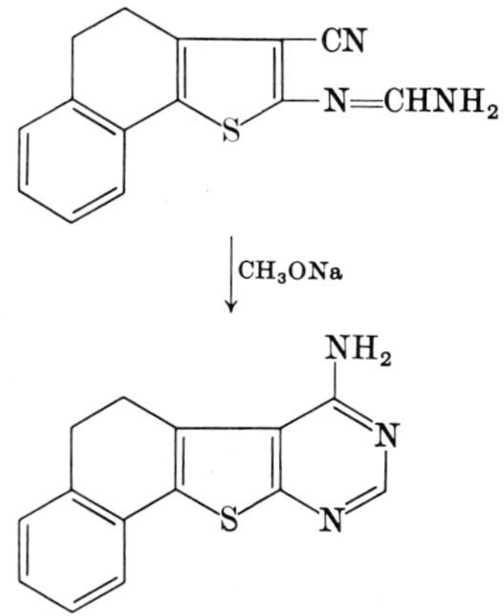	— 74	 91, 341
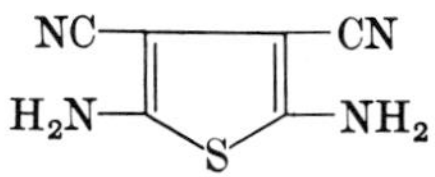	B	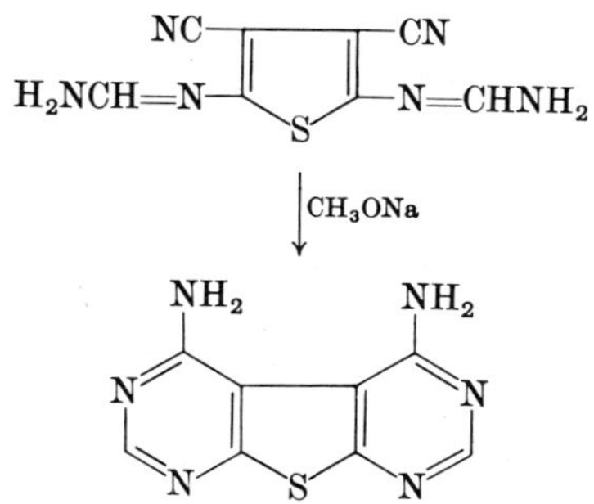	— 30	 91, 341

(continued)

TABLE XXVII (*continued*)

o-Aminonitrile	Procedure[a]	Product	Yield, %	Ref.
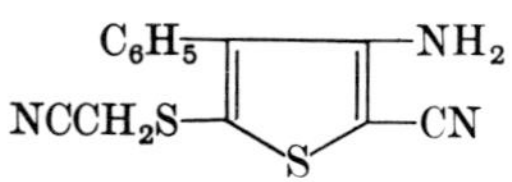	A	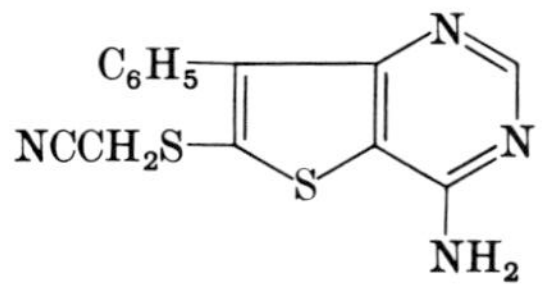	68	171
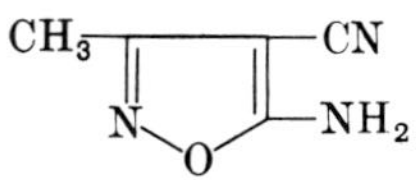	B	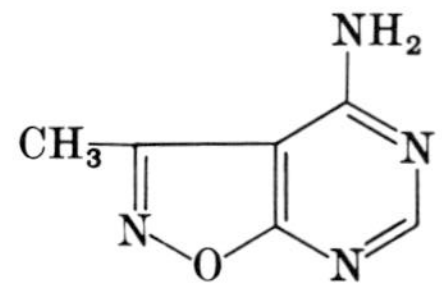	77	11
	A	Same as above	70	11
	A (CH_3NH_2)	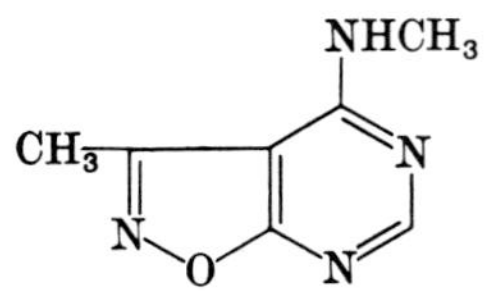	70	11
	A ($(CH_3)_2NCH_2CH_2CH_2NH_2$)	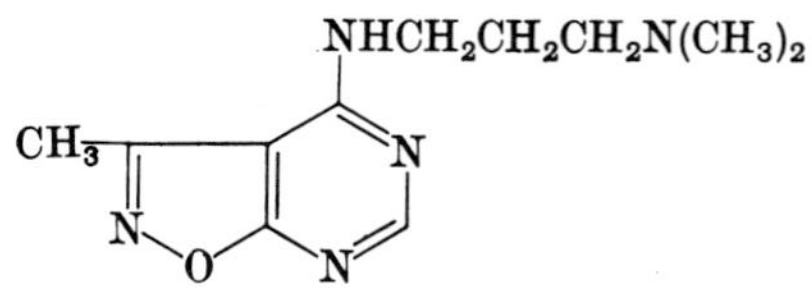	66	11

	A $((C_2H_5)_2NCH_2CH_2CH_2NH_2)$	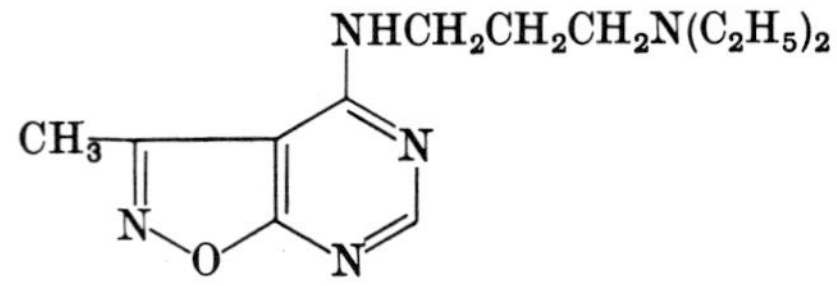	74	11
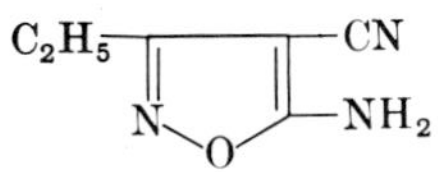	B	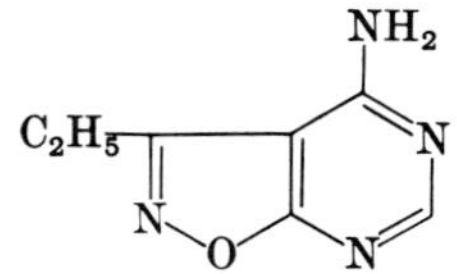	75	11
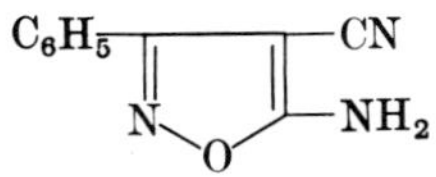	A	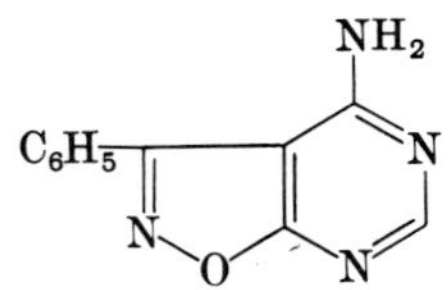	77	11
	A (CH_3NH_2)	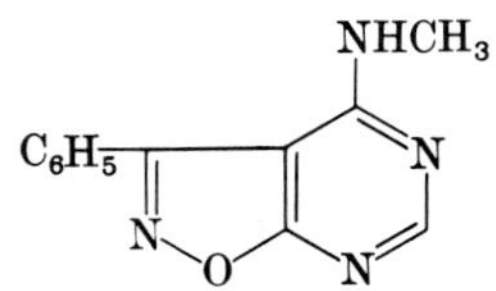	46	11
	A $((CH_3)_2NCH_2CH_2CH_2NH_2)$	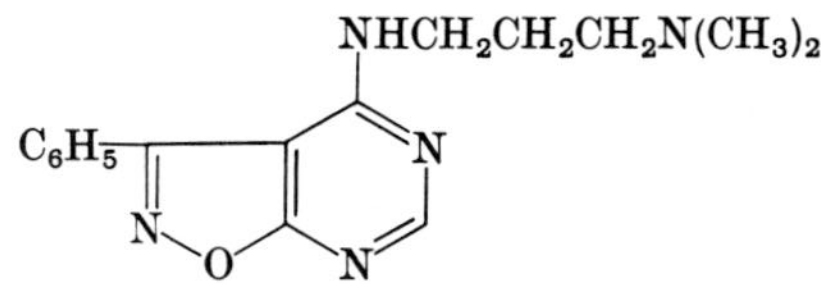	68	11

(*continued*)

TABLE XXVII (*continued*)

o-Aminonitrile	Procedure[a]	Product	Yield, %	Ref.
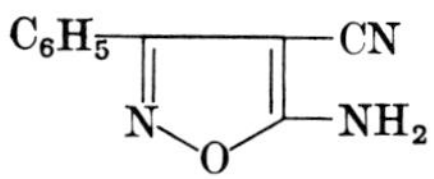	A ($((C_2H_5)_2NCH_2CH_2CH_2NH_2)$)	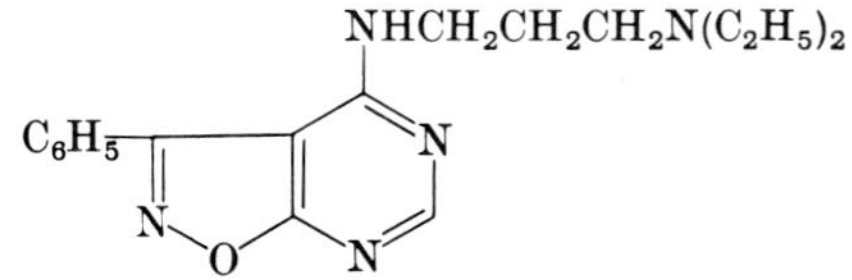	74	11
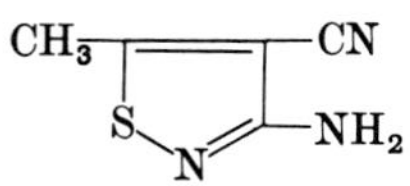	A	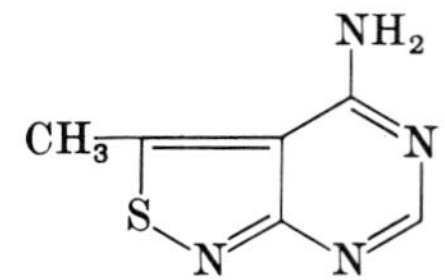	76	156
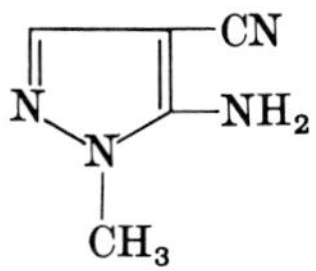	A	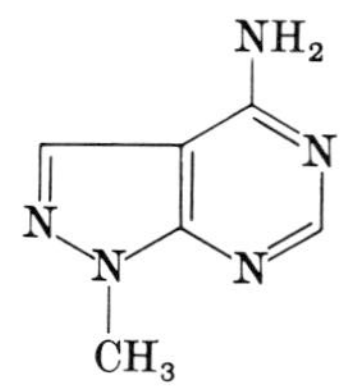	90	20
	A ((CH_3NH_2))	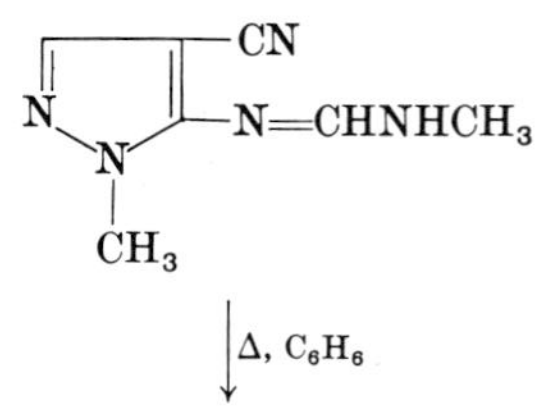	53	20

		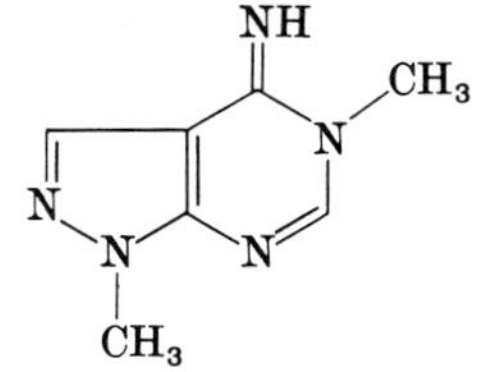	93	20
		↓ Δ, H_2O		
		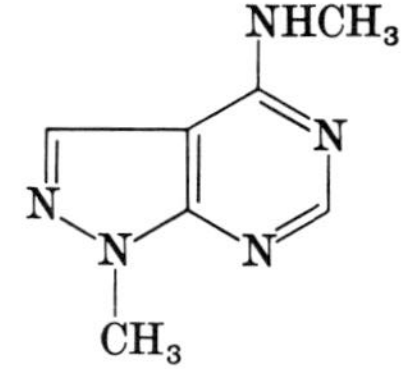	100	20
(n-$C_4H_9NH_2$)	A	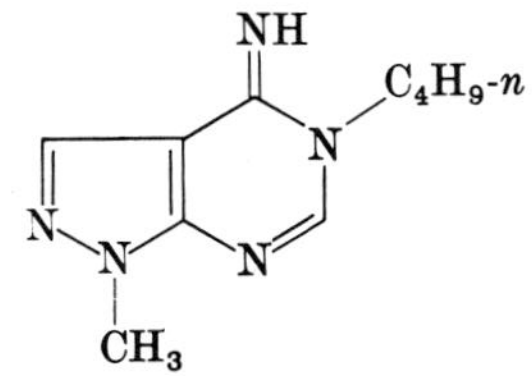	64	20
		↓ Δ, H_2O		
		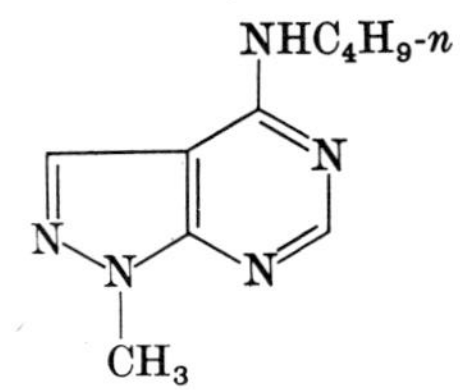	—	20

(continued)

TABLE XXVII (*continued*)

o-Aminonitrile	Procedure[a]	Product	Yield, %	Ref.
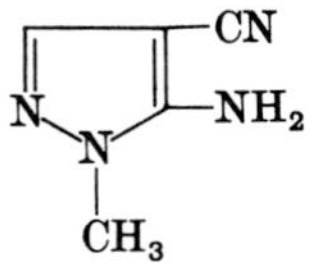	A (*d*-glucamine)	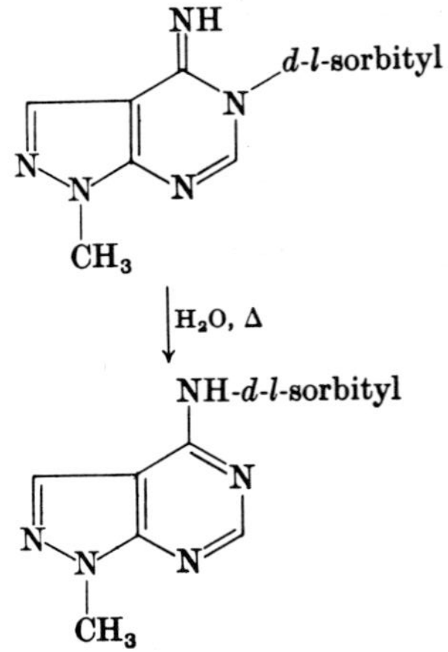	38	20
			—	20
	A (*d*-ribamine)	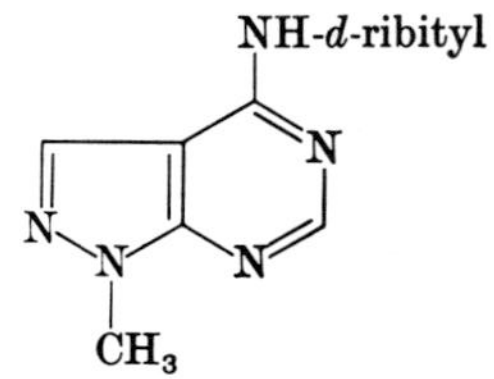	21	20
	A (H_2NNH_2)	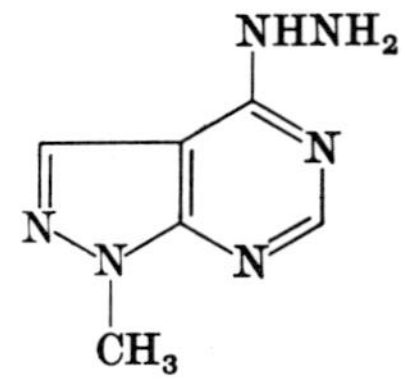	—	20

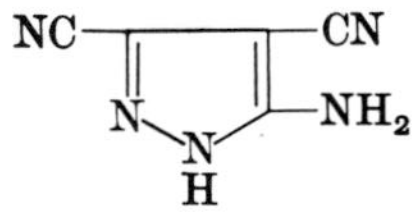		B	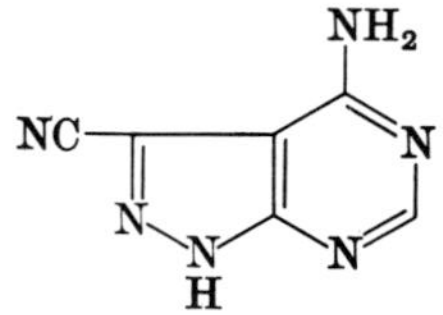	83	92
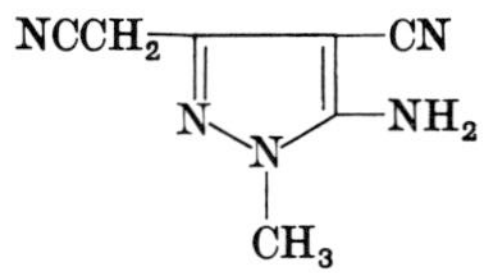		A	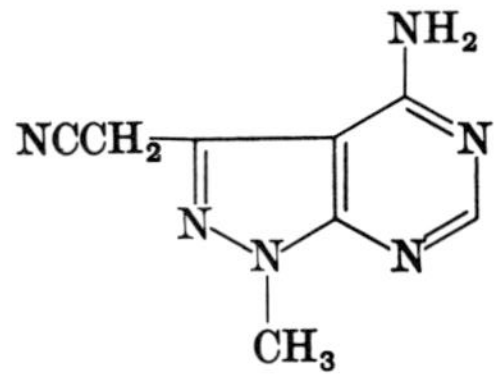	79	25
	(n-$C_4H_9NH_2$)	A	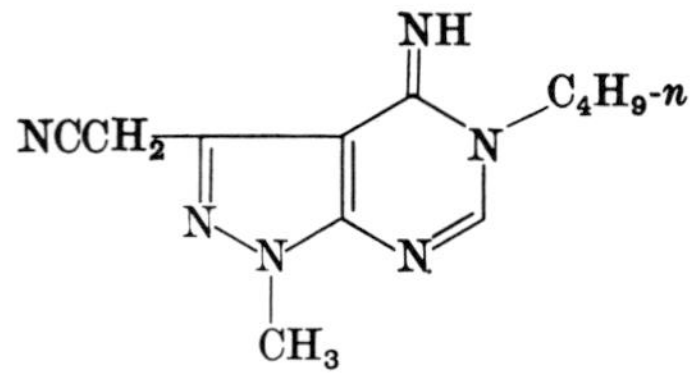	74	25
	(H_2NNH_2)	A	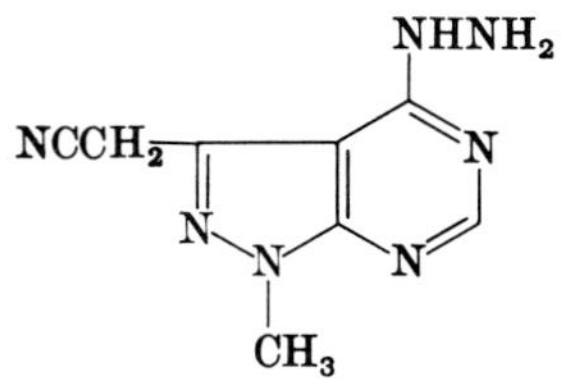 [b]	80	25

(continued)

TABLE XXVII (*continued*)

o-Aminonitrile	Procedure[a]	Product	Yield, %	Ref.
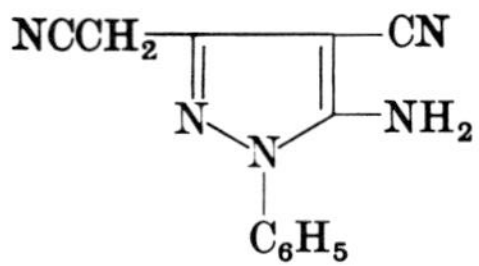	A	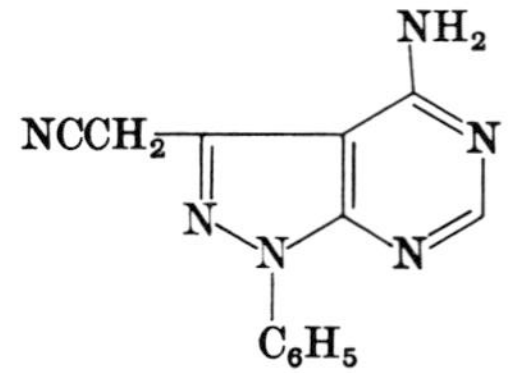	87	25
	A (n-$C_4H_9NH_2$)	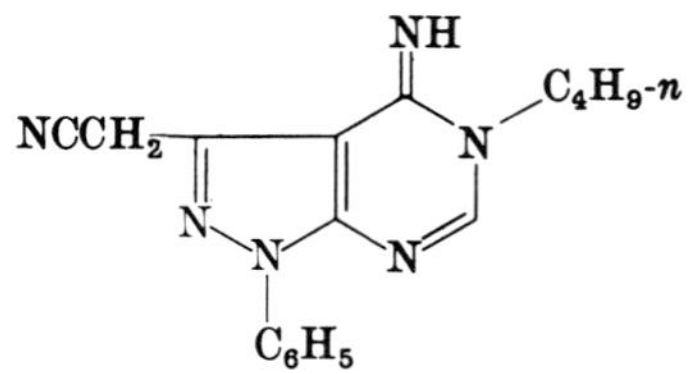	80	25
	A (H_2NNH_2)	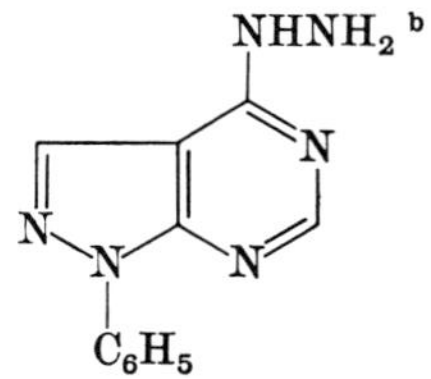 [b]	84	25
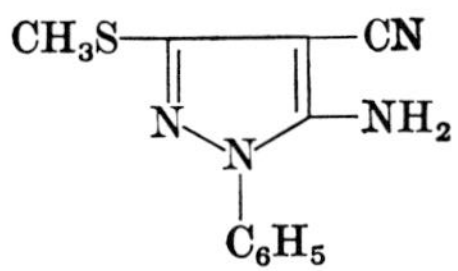	A	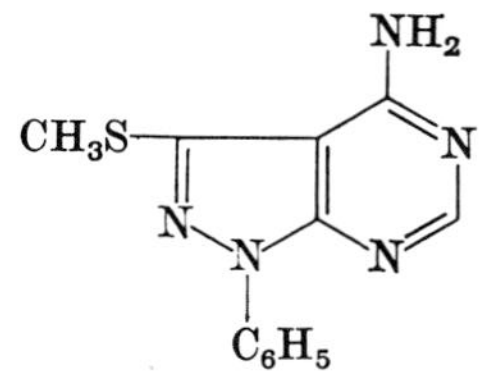	79	266

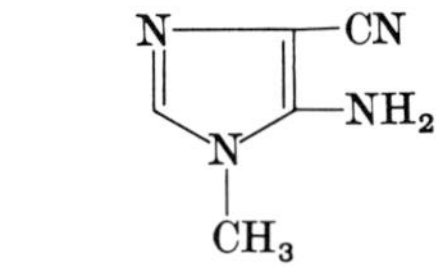		A	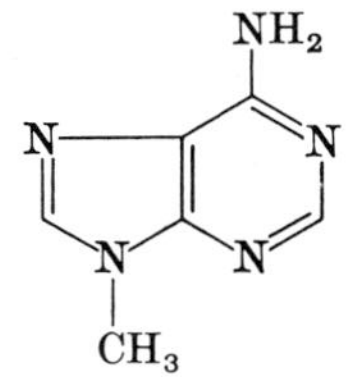	31	114
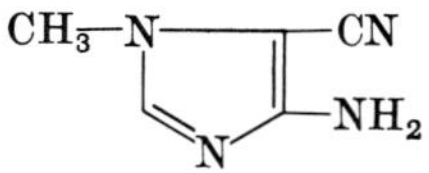		A	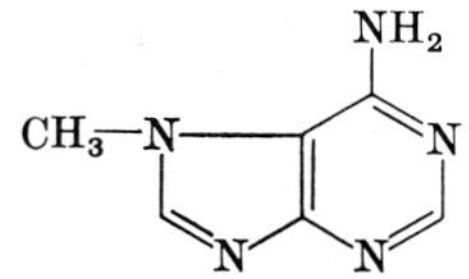	67	20
	(CH_3NH_2)	A	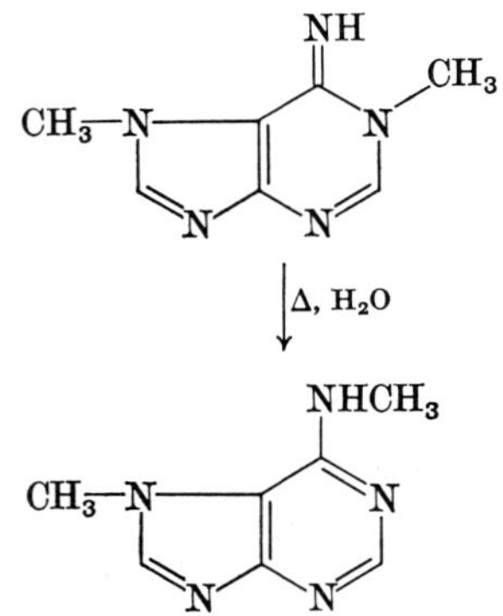	72 ~100	20 20

(continued)

TABLE XXVII (*continued*)

o-Aminonitrile	Procedure[a]	Product	Yield, %	Ref.
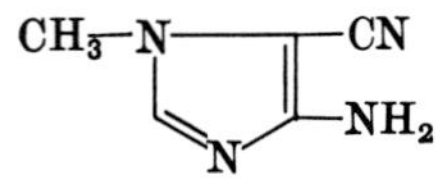	A (n-$C_4H_9NH_2$)	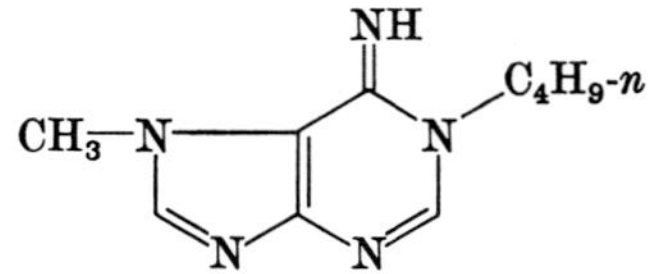	28	20
		↓ Δ, H_2O		
		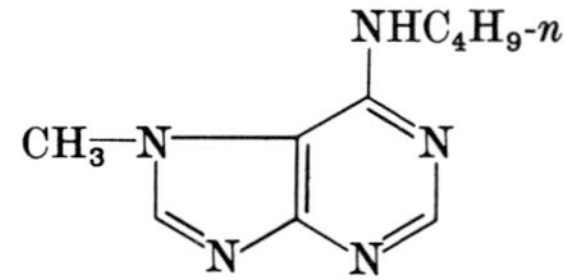	~100	20
	A (d-glucamine)	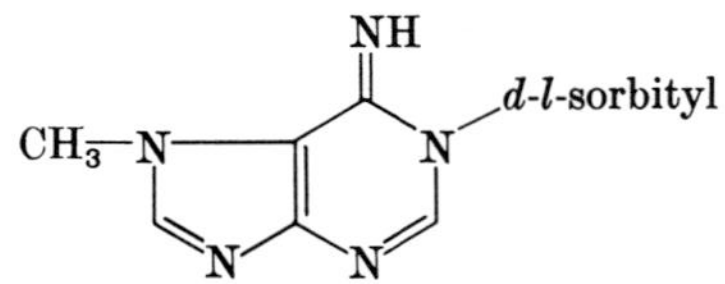	68	20
		↓ Δ, H_2O		
		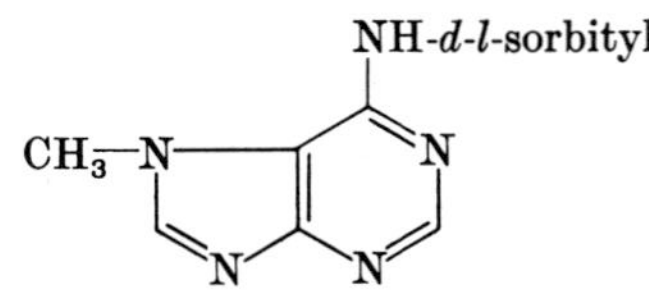	~100	20

	(H_2NNH_2)	A	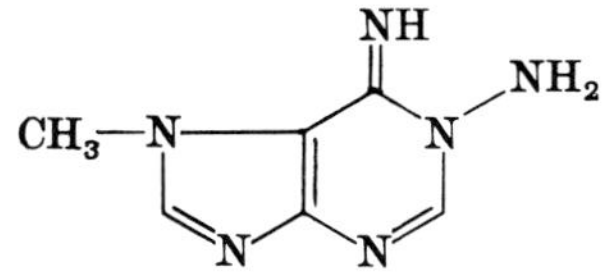	74	20
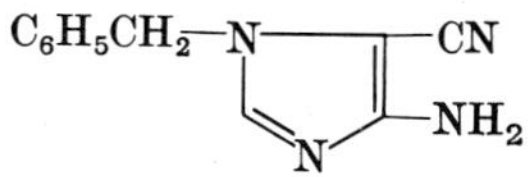		A	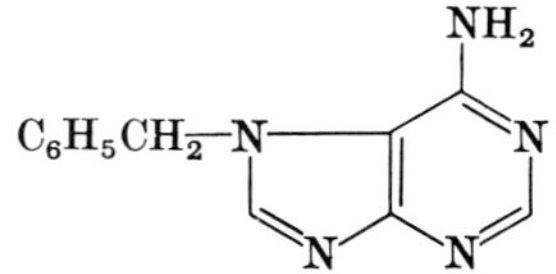	55	71
	$(C_6H_5CH_2NH_2)$	A	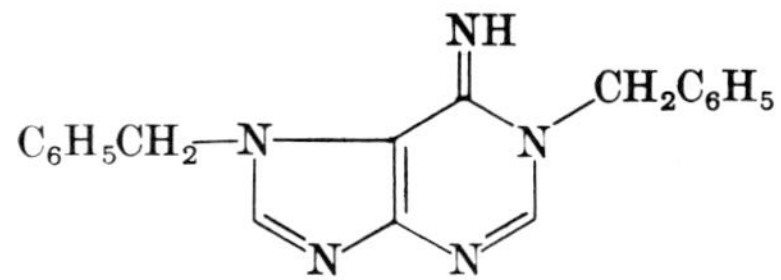	49	71
			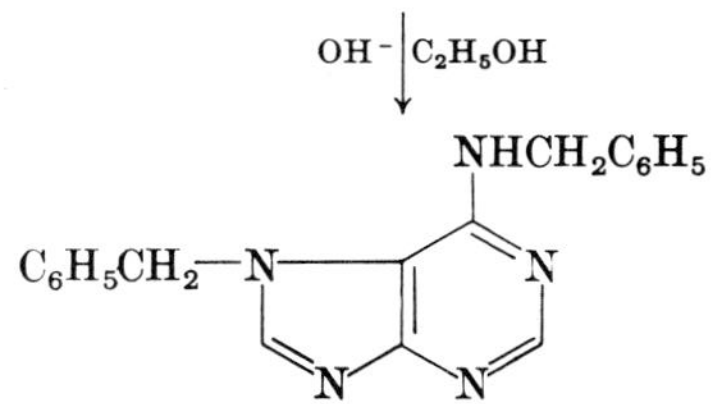	56	71

(continued)

TABLE XXVII (*continued*)

o-Aminonitrile	Procedure[a]	Product	Yield, %	Ref.
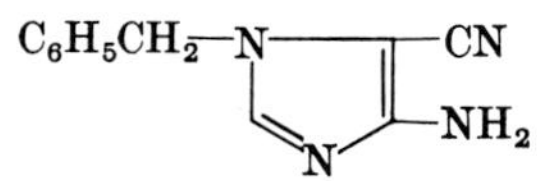	A 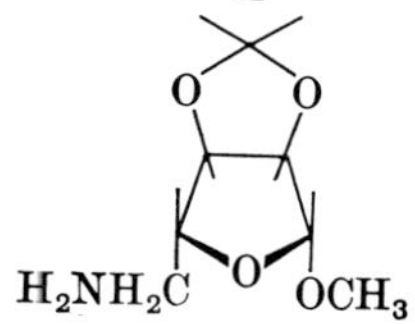	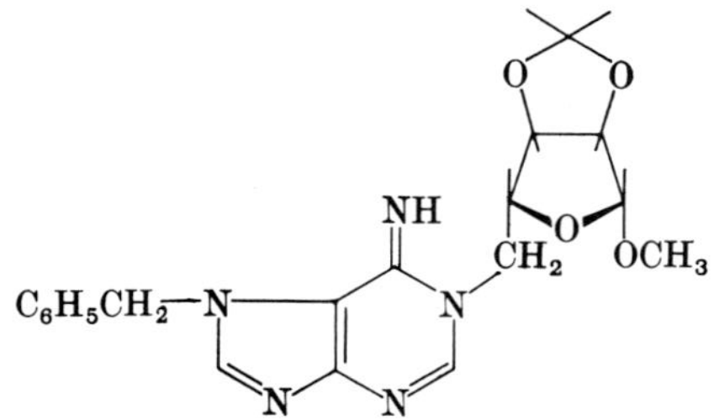	50	472
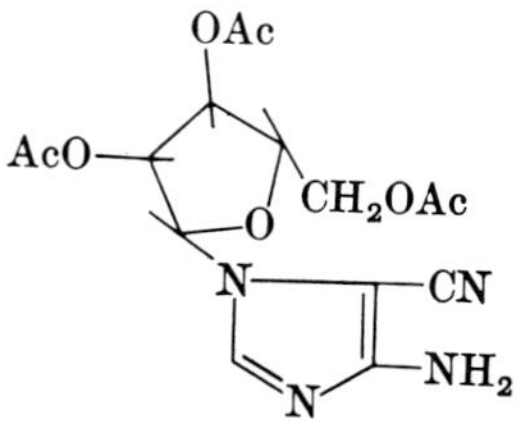	B A	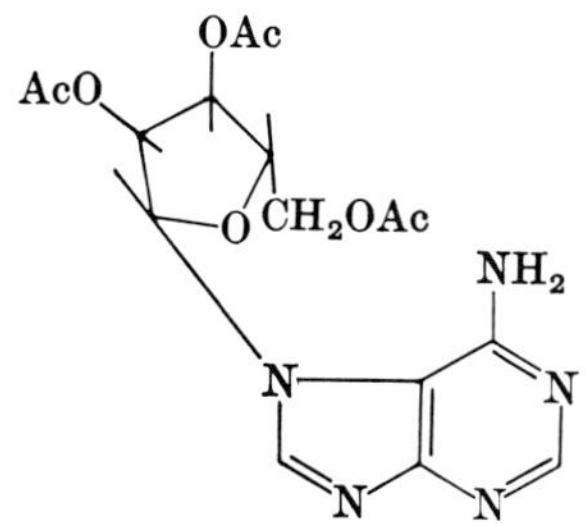	73 93	355 552
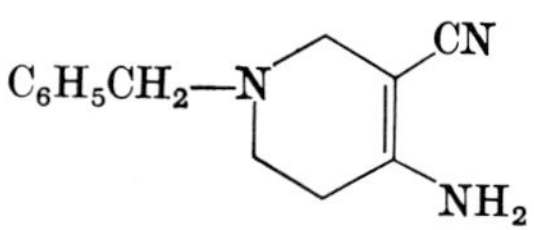	A	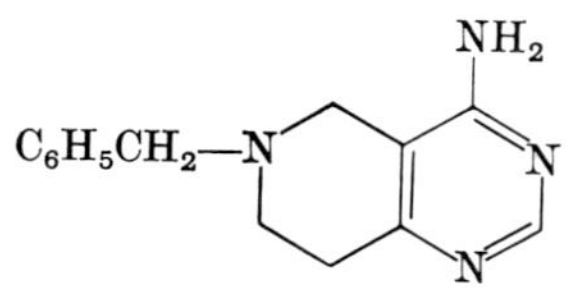	45	459

	B	Same as above	59	459
(CH_3NH_2)	A	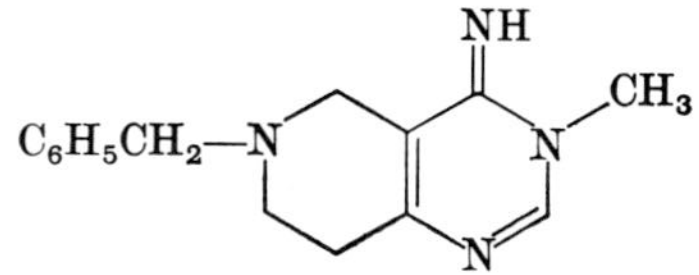	60	459
($C_6H_5CH_2NH_2$)	A	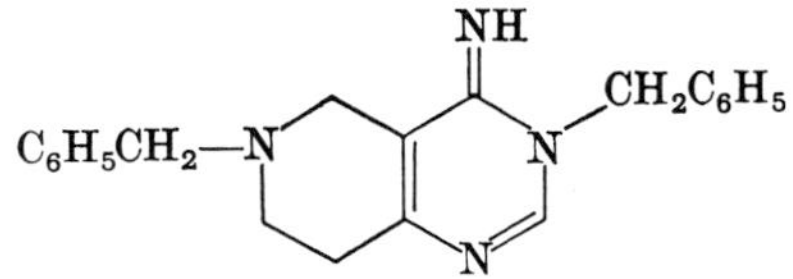	49	459
($C_6H_5CH_2CH_2NH_2$)	A	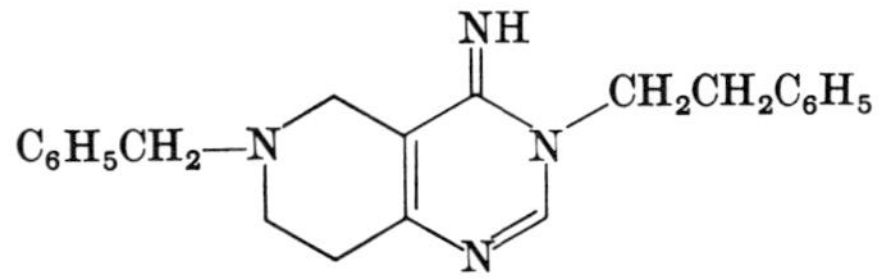	52	459
($C_6H_5NH_2$)	A	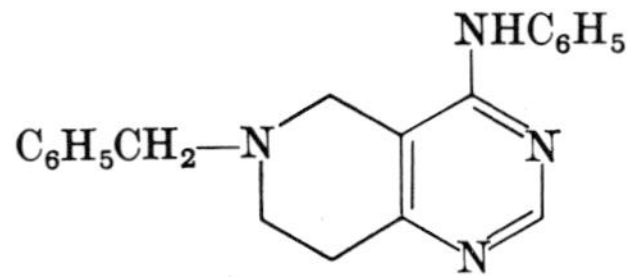	22	459
(p-$ClC_6H_4NH_2$)	A	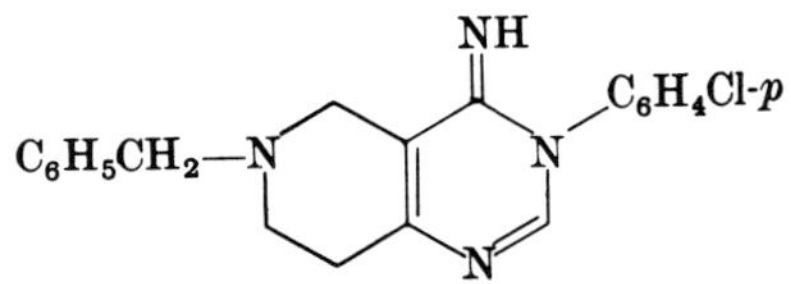	23	459

(continued)

TABLE XXVII (*continued*)

o-Aminonitrile	Procedure[a]	Product	Yield, %	Ref.
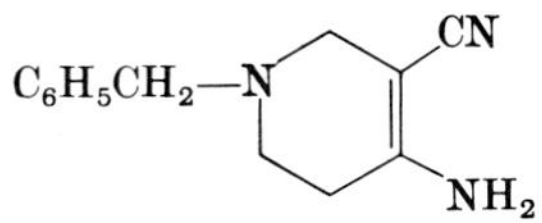 (H_2NNH_2)	A	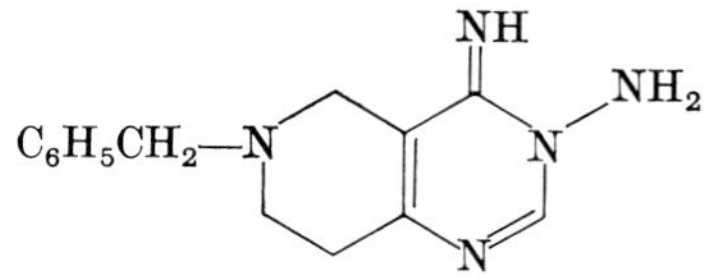	51	459
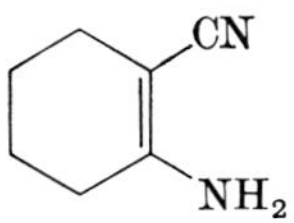	A	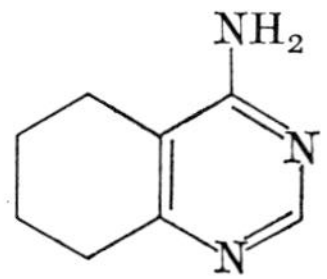	14	459
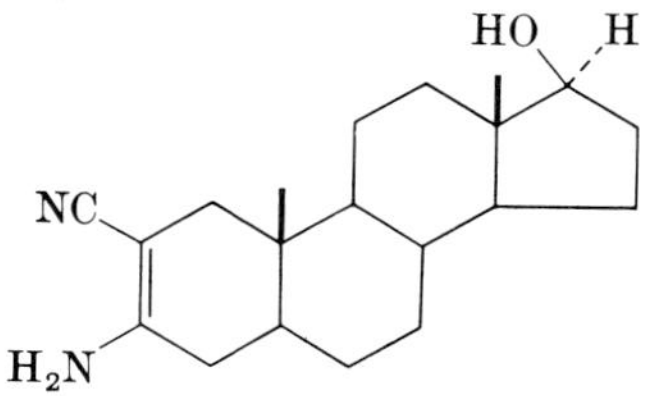	A	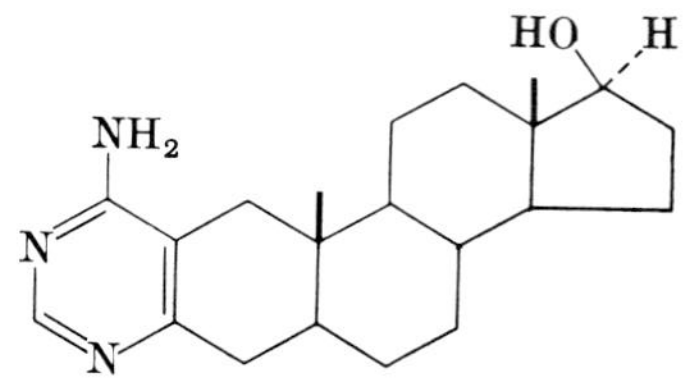	85	438
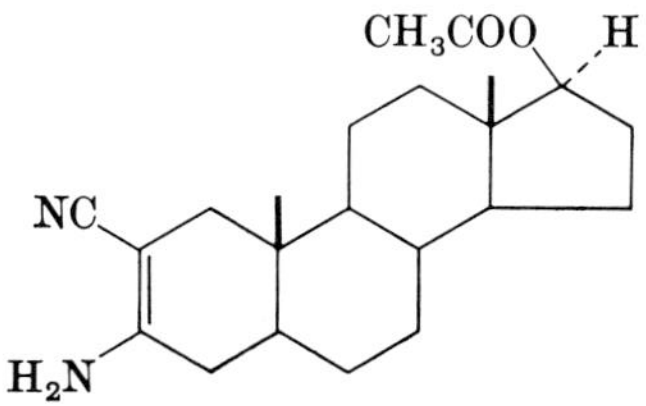	A	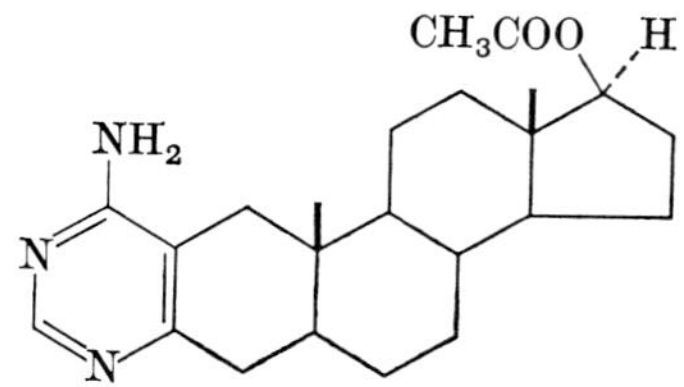	75	438

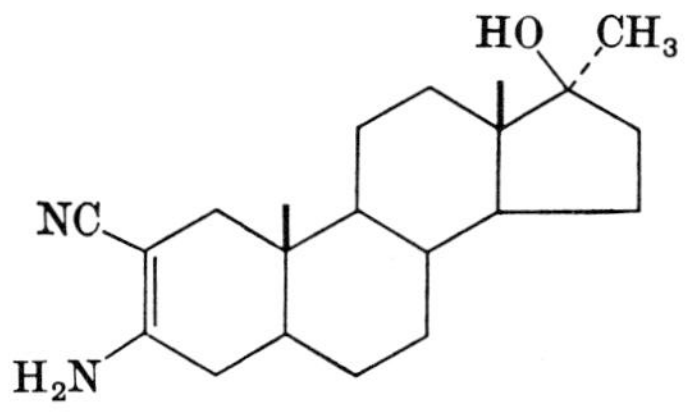	A	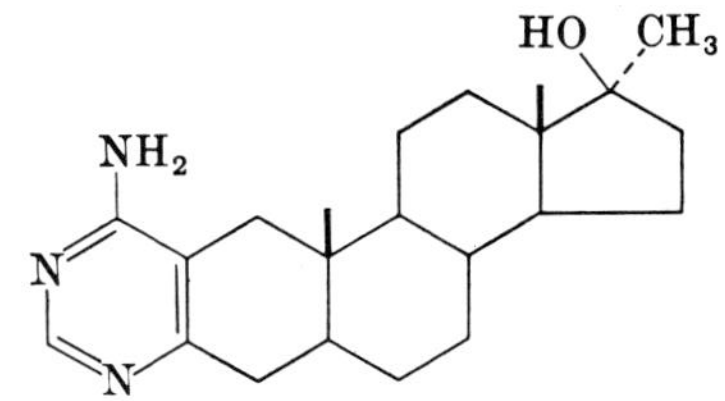	88	438
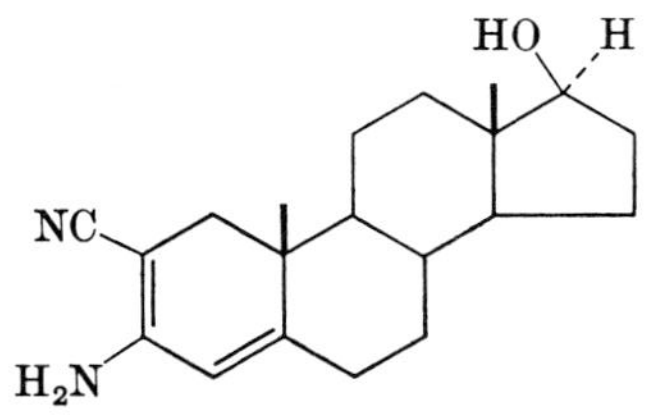	A	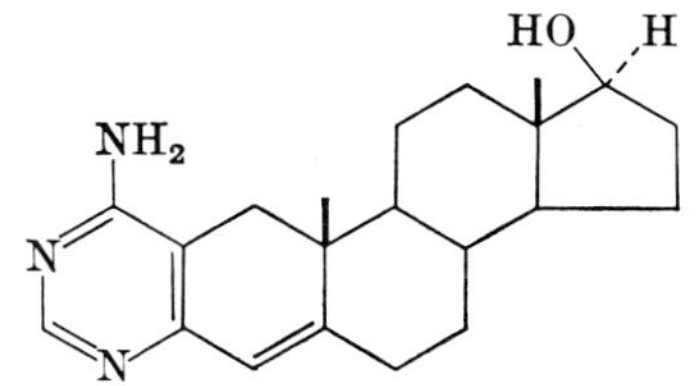	25	438
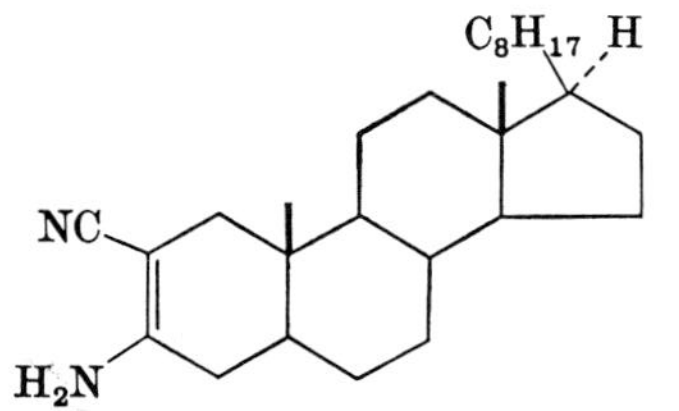	A	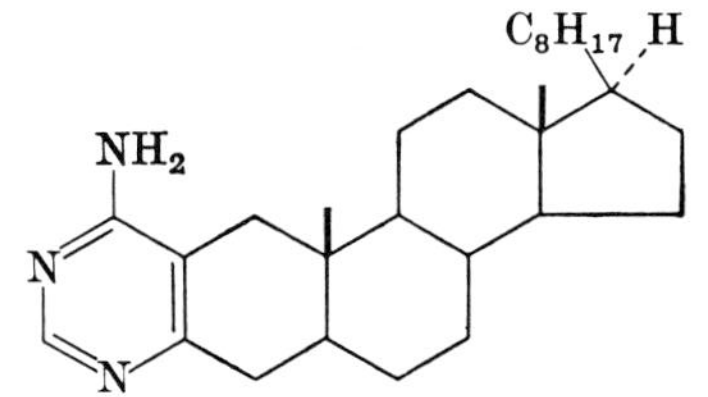	85	438

(continued)

TABLE XXVII (*continued*)

o-Aminonitrile	Procedure[a]	Product	Yield, %	Ref.
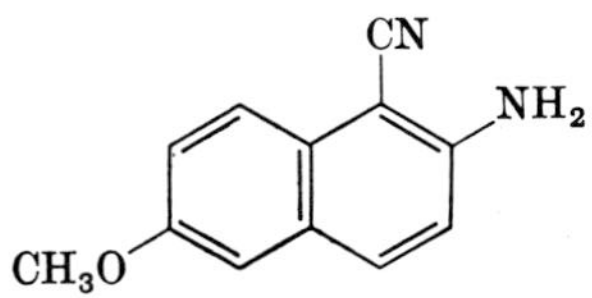	A	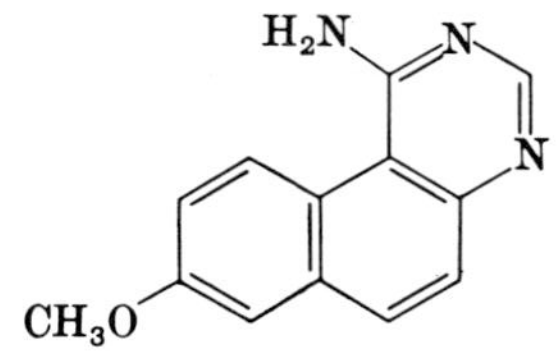	72	290
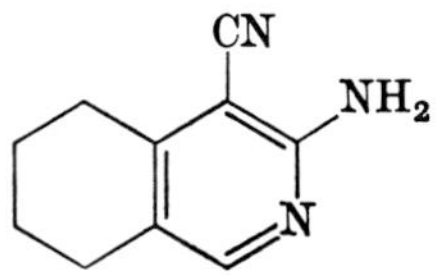	A	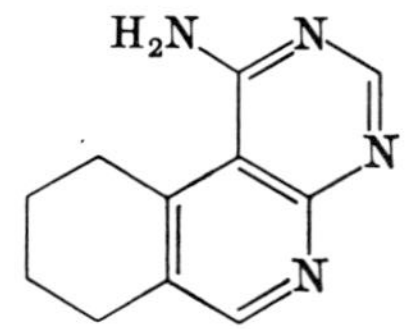	71	509
	$(H_2N(CH_2)_3N(C_2H_5)_2)$ A	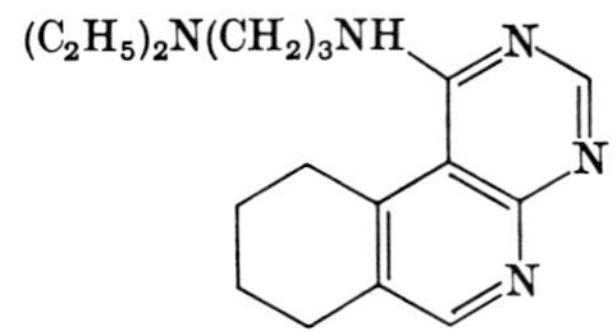	23	509
	A $(H_2NCH(CH_3)(CH_2)_3N(C_2H_5)_2)$	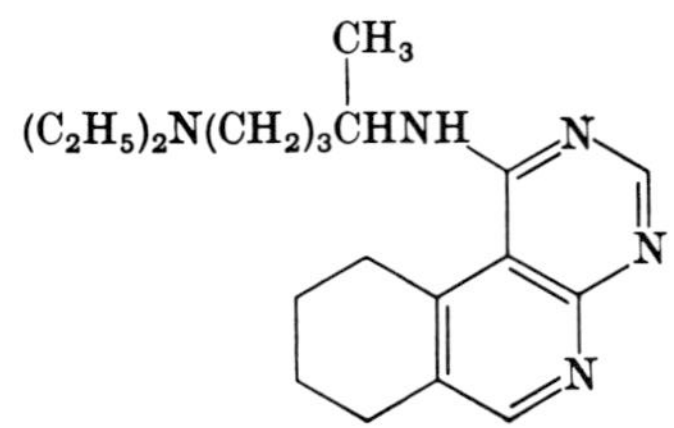	45	509

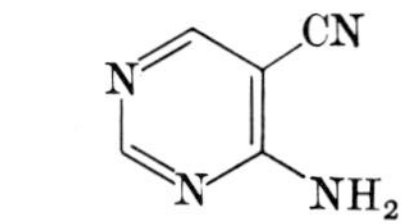		B	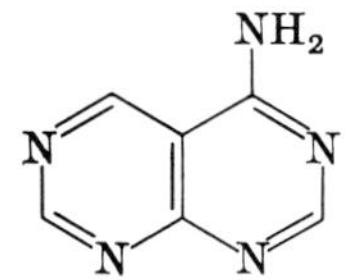	57	19
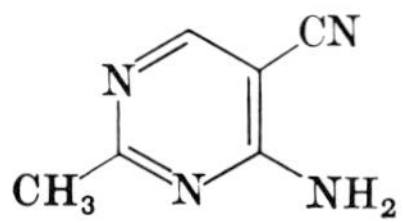		B	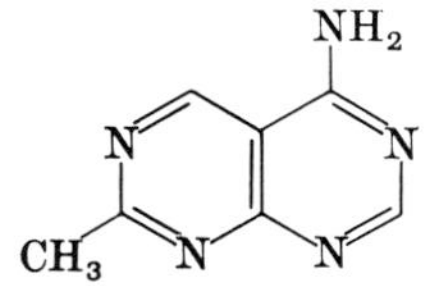	—	444
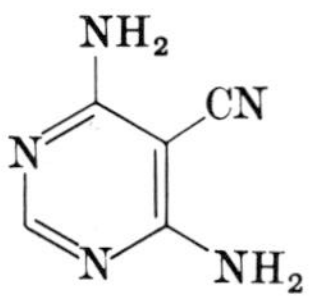		A	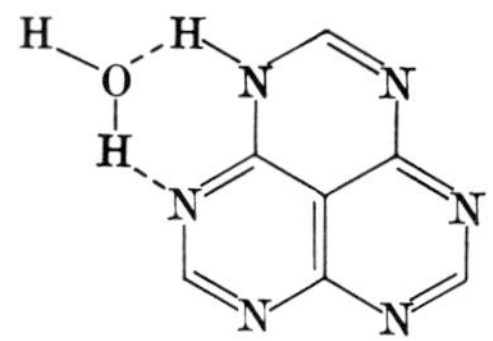	55	458
	(CH_3NH_2)	B	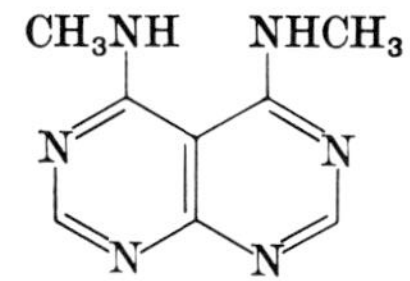	14	458

[a] (A) With isolation of the intermediate *o*-ethoxymethyleneamino derivative.
(B) Without isolation of the intermediate *o*-ethoxymethyleneamino derivative.

[b] This compound was originally believed to be (see ref. 25) the unrearranged 5-amino-4-imino derivative.

salt) (19) has seen wide application for the direct conversion of *o*-aminonitriles to fused 4-aminopyrimidines (see Table XXIX). Thus, 4-amino-5-cyanoimidazole (**2**) has been converted to adenine (**3**) upon treatment with formamidine acetate under "physiological conditions" (336), and (in better yield) by boiling with formamidine acetate in methyl cellosolve (32).

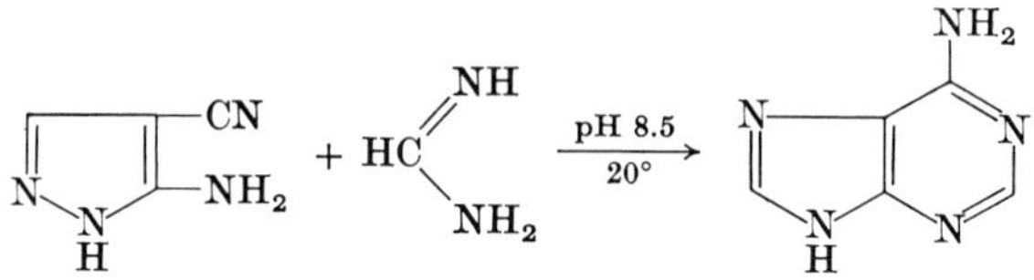

(**2**) (**3**)

4-Amino-5-cyanopyrimidine (**4**) is readily converted to 4-aminopyrimido(4,5-*d*)pyrimidine (**5**), also by formamidine acetate. Furthermore, since **4** itself is prepared by the reaction of 1 mole of malononitrile with 2 moles of formamidine (see p. 104), a one-step synthesis of the bicyclic product **5** was achieved directly from malononitrile and 3 moles of formamidine acetate (19).

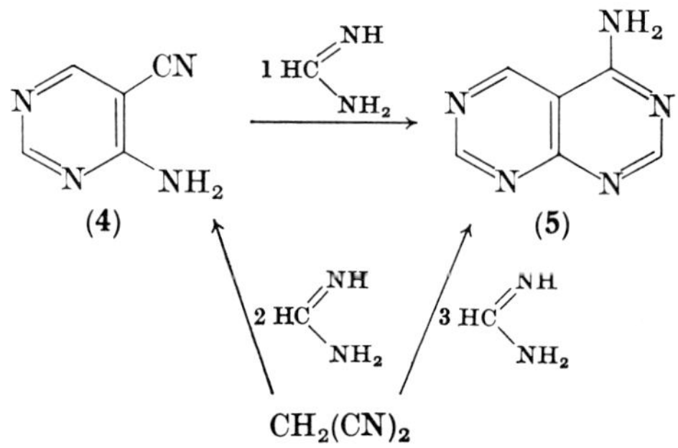

(**4**) (**5**)

The reaction of an *o*-aminonitrile with an *N*,*N*′-disubstituted formamidine should give, by amidine exchange, an *o*-cyanoamidine **6**, identical with the intermediate formed by the reaction of the corresponding ethoxymethyleneamino derivative with a substituted amine. As mentioned above, however, conditions of the amidine exchange reaction are sufficiently strenuous to bring about immediate cyclization of the intermediate to the 3-substituted-4-iminopyrimidine isomer **7**. However, the subsequent Dimroth rearrangement of this nonaromatic iminopyrimidine to a fused 4-substituted-aminopyrimidine (**8**) is itself base-catalyzed.

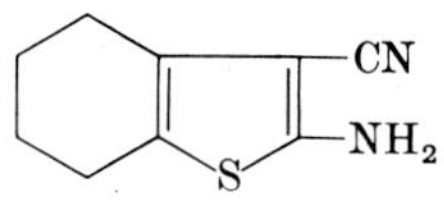	180–190°, 2 hr	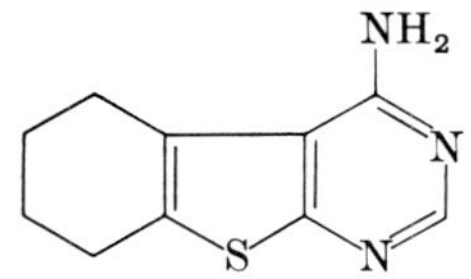	81	334
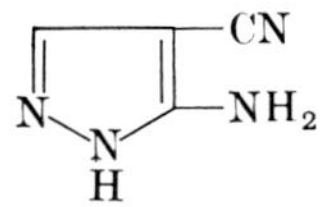	Reflux, 30 min	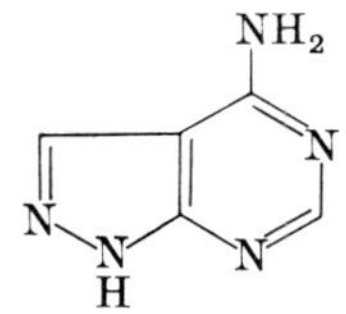	56	4
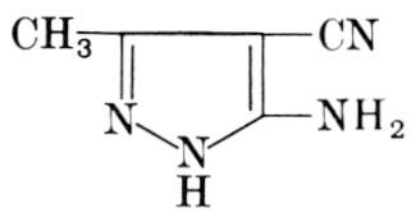	Reflux, 45 min	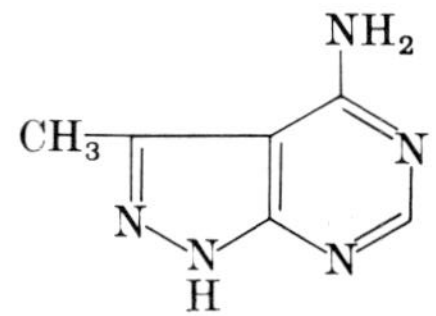	43	3
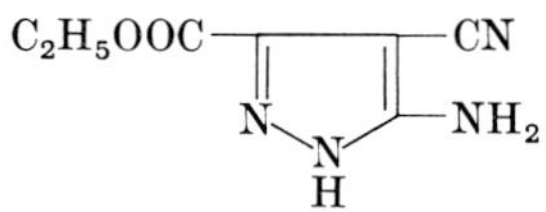	Reflux, 30 min, then 5% KOH, then sublimation	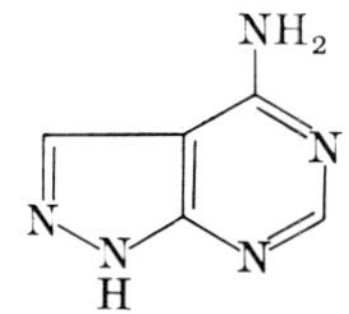	—	457

(continued)

TABLE XXVIII (*continued*)

o-Aminonitrile	Conditions	Product	Yield, %	Ref.
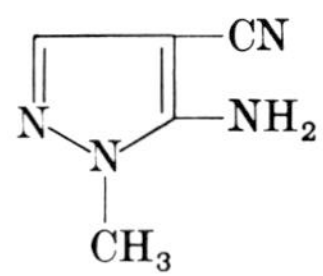	Reflux, 1 hr	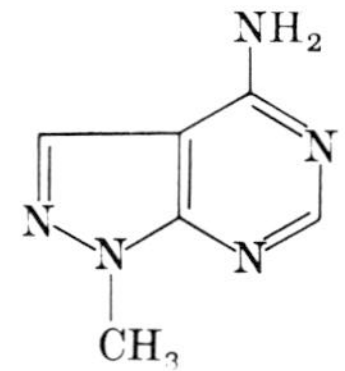	49	3
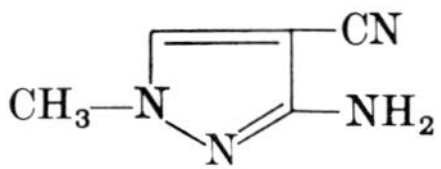	200°, 10 hr	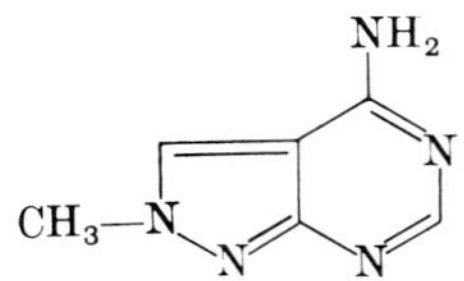	38	97
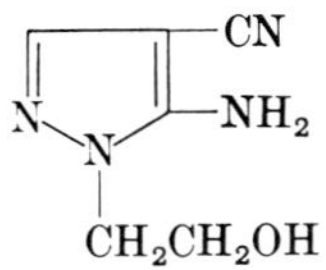	Reflux, 1½ hr	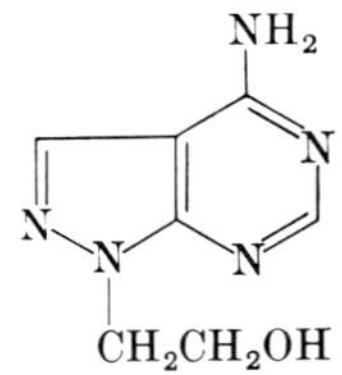	42.5	3
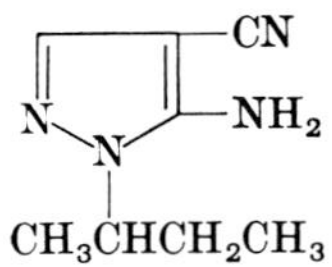	—	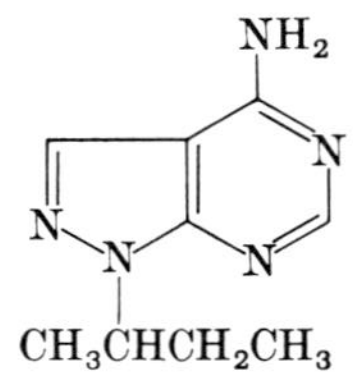	—	155

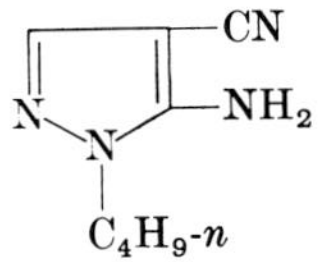	—	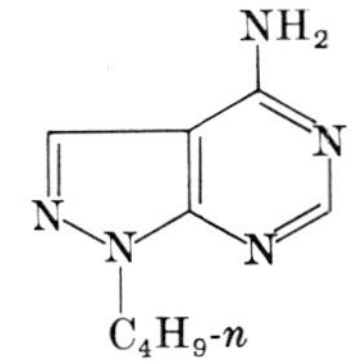	47	170
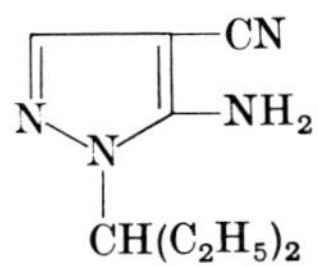	—	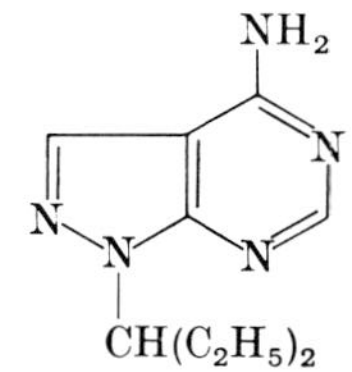	— —	155 159
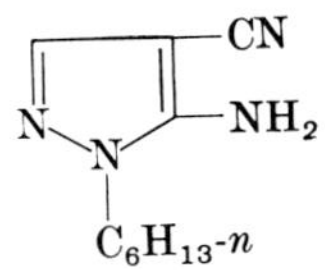	—	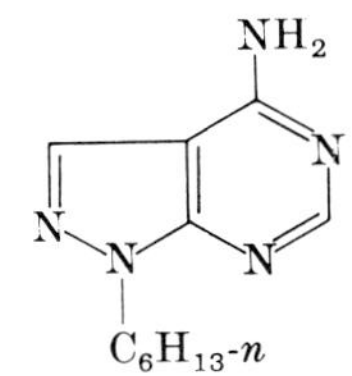	47	170
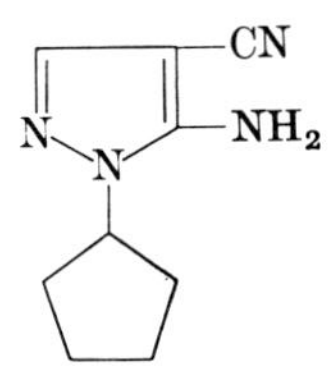	—	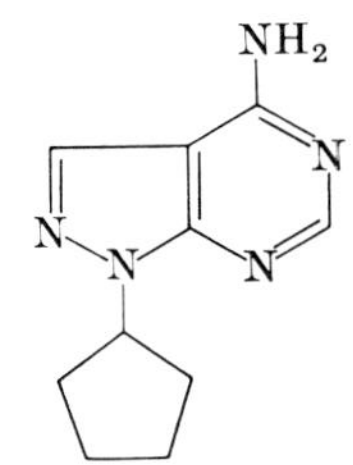	— —	155 159

(continued)

TABLE XXVIII (*continued*)

o-Aminonitrile	Conditions	Product	Yield, %	Ref.
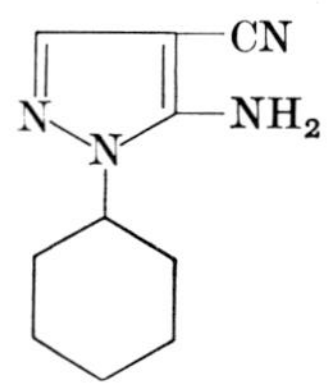	—	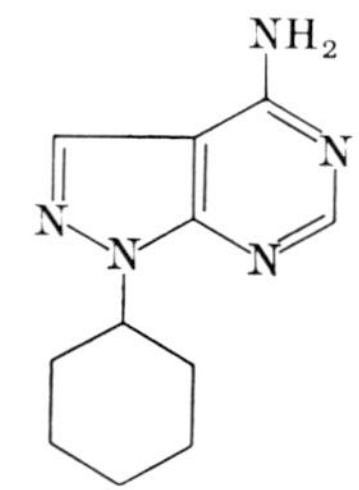	— 63	155 170
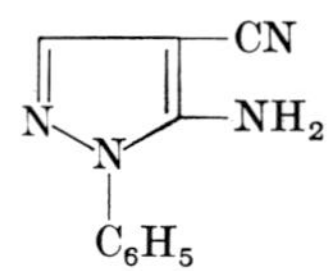	Reflux, 1 hr	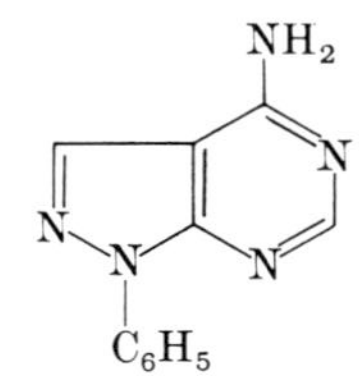	96	3
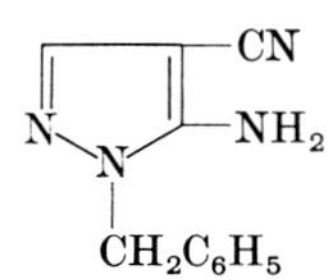	—	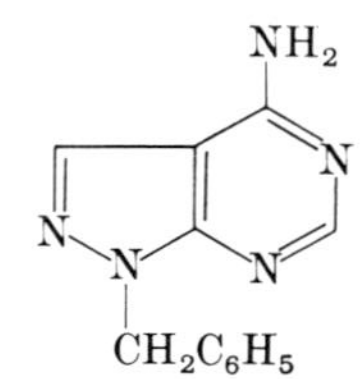	59	170

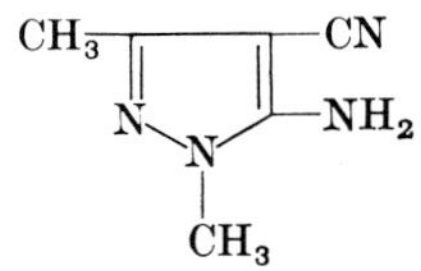	Reflux, 45 min	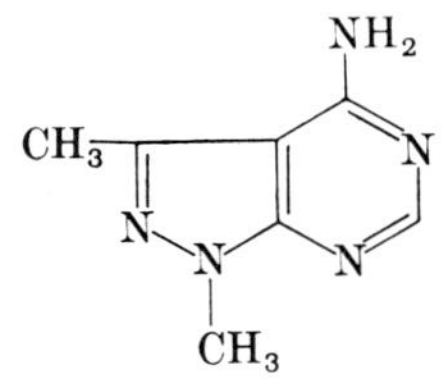	53	3
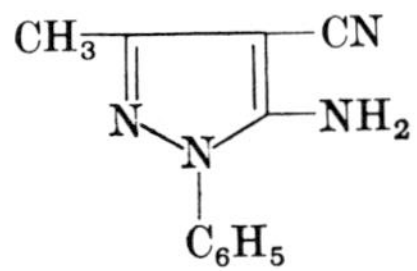	Reflux, 1 hr	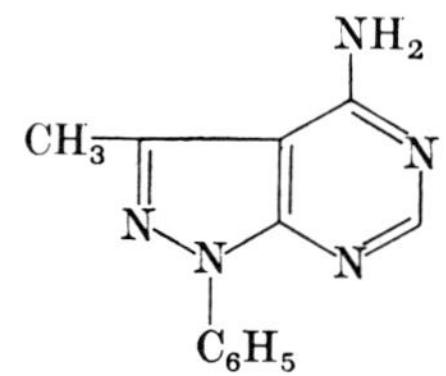	68	25
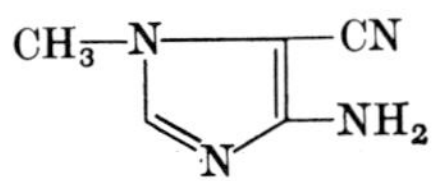	"Hot plate," 50 min	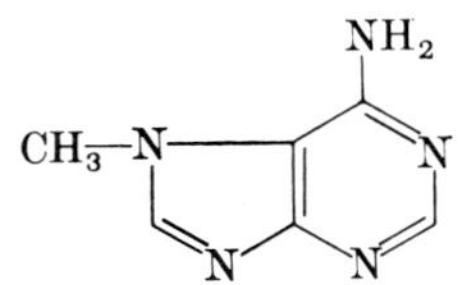	—	136
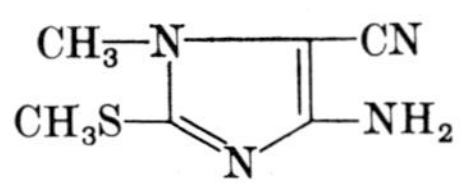	—	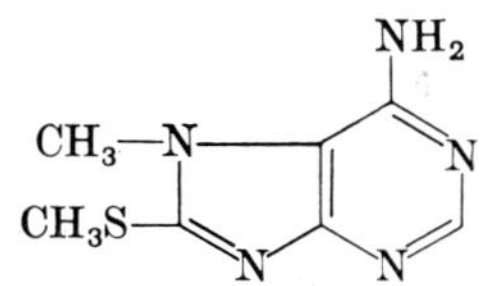	—	339

(continued)

TABLE XXVIII (*continued*)

o-Aminonitrile	Conditions	Product	Yield, %	Ref.
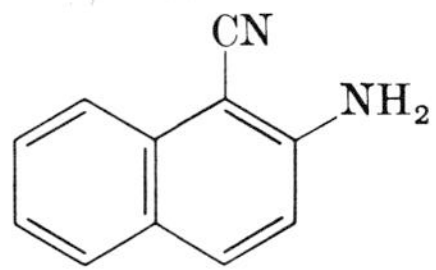	Reflux, 1 hr	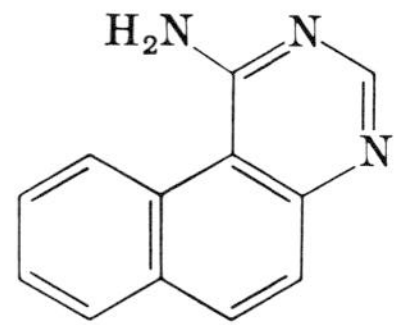	67	289
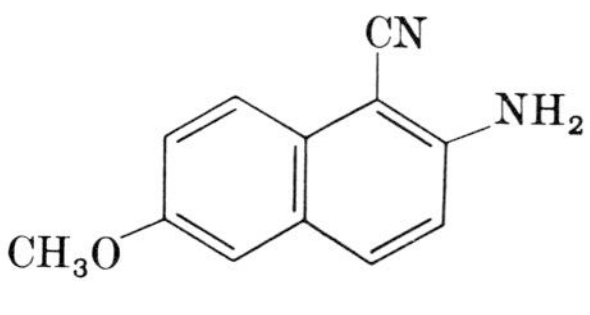	Reflux, 30 min	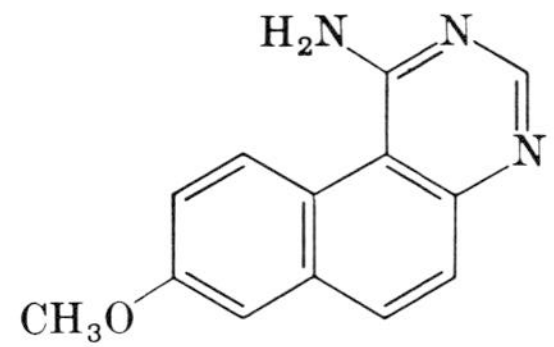	58	290
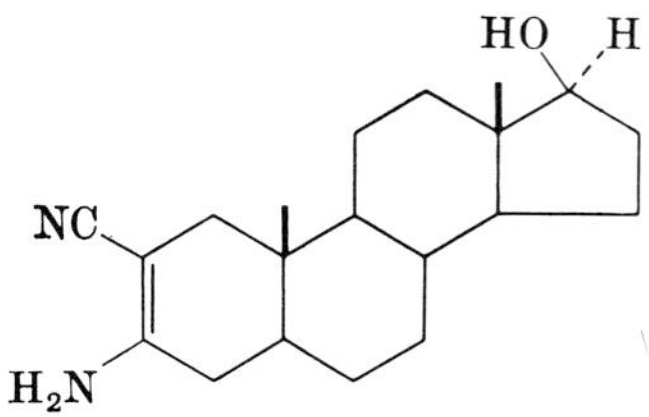	+ $HC(NHCHO)_3$ + TsOH	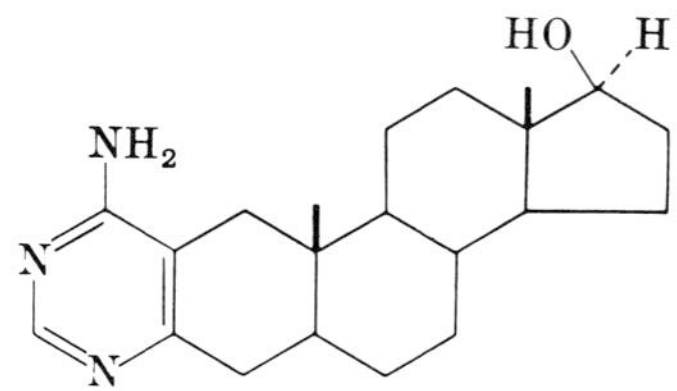	—	438

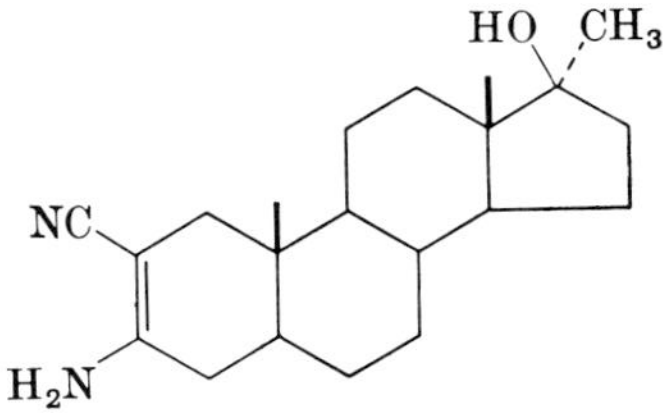	+ HC(NHCHO)$_3$ + TsOH	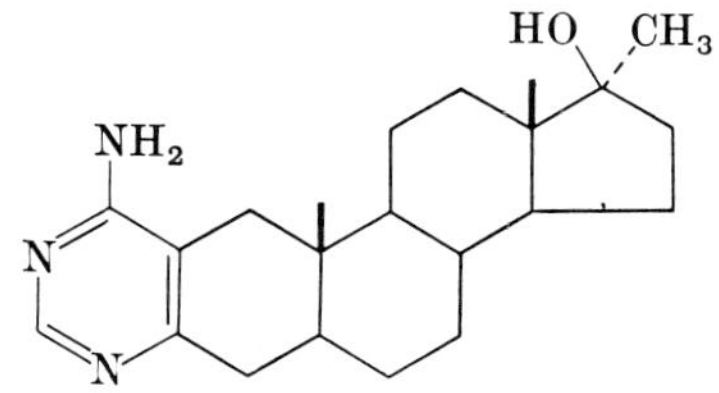	63	438
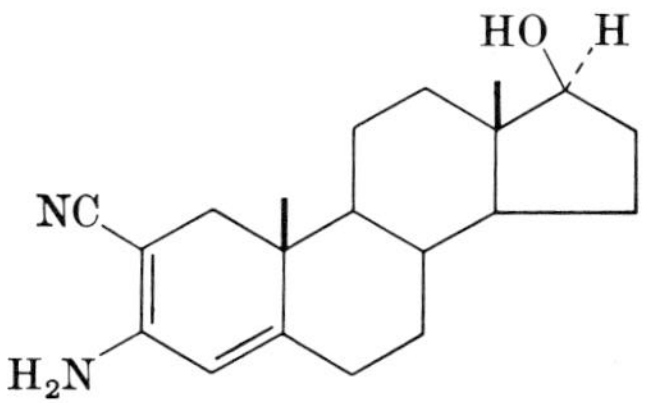	+ HC(NHCHO)$_3$ + TsOH	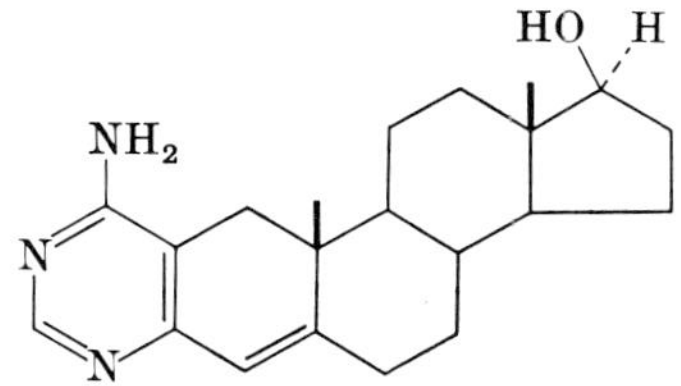	25	438
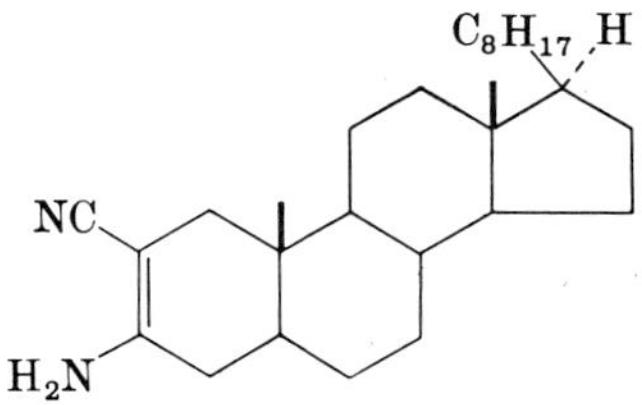	+ HC(NHCHO)$_3$ + TsOH	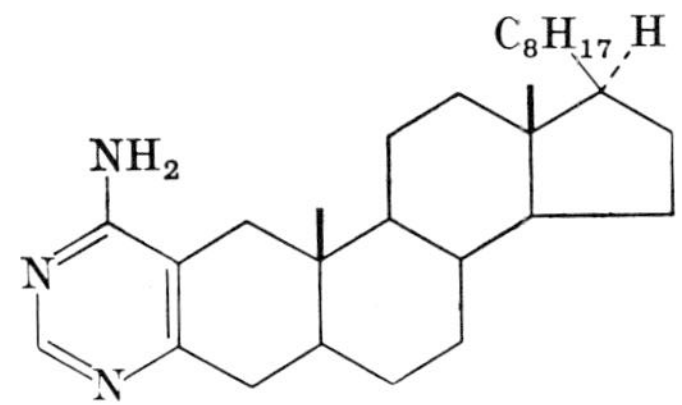	—	438
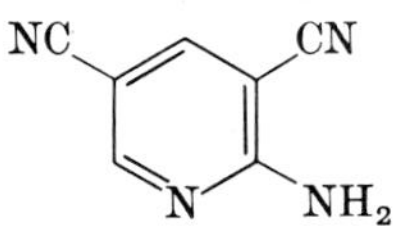	100°, 30 min, N_2	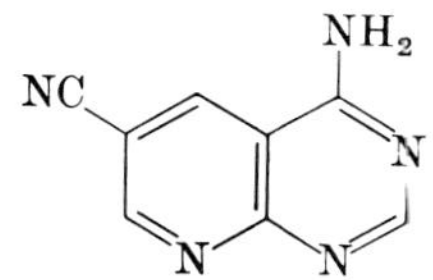	97	133

(continued)

TABLE XXVIII (*continued*)

o-Aminonitrile		Conditions	Product	Yield, %	Ref.
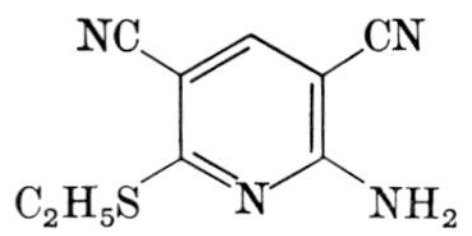		180°, 40 min	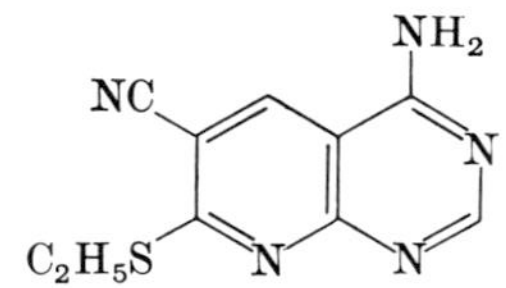	80	133
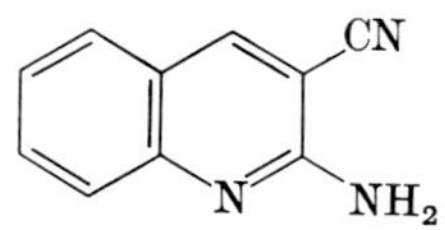		Reflux, 30 min	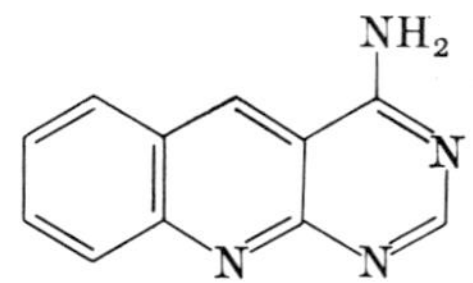	36	14
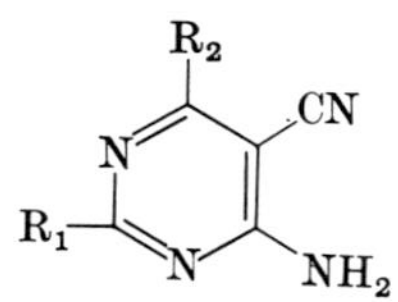			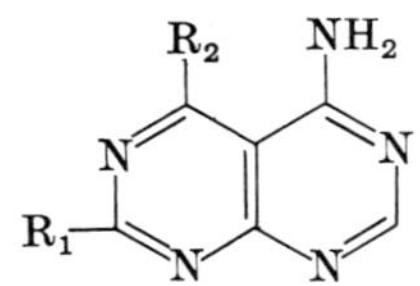		
R_1	R_2				
H	H	"Hot plate," 40 min		63	163
		Reflux, 30 min		16	22
CH_3	H	"Hot plate," 40 min		51	163
NH_2	H	Reflux, 30 min		42	22
		"Hot plate," 40 min		46	163

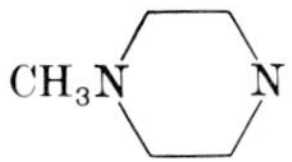	H	Reflux, 30 min		70	22
$(CH_3)_2N$	H	Reflux, 30 min		58	22
		—		—	139, 147
	H	Reflux, 30 min		29	22
$C_6H_5CH_2NH$	H	185°, 2 hr		7	22
$CH_3(CH_2)_5NH$	H	185°, 2 hr		23	22
HO	H	"Hot plate," 40 min		83	163
CH_3S	H	Reflux, 30 min		19	22
C_6H_5	H	Reflux, 30 min		25	22
C_6H_5	C_6H_5	185°, 6 hr		83	359
p-ClC_6H_4	p-ClC_6H_4	185°, 16 hr		79	359
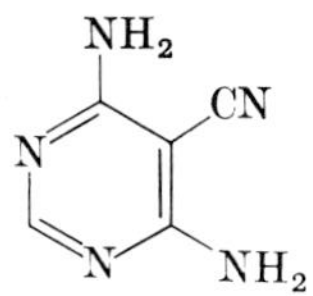		Reflux, 40 min, then CH_3COOH	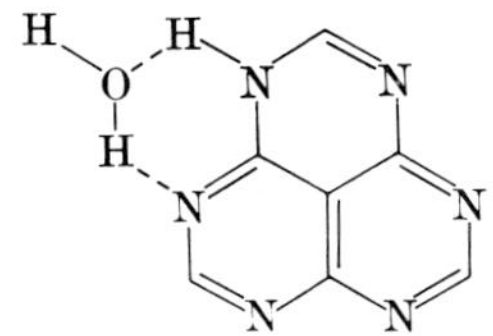	80	458
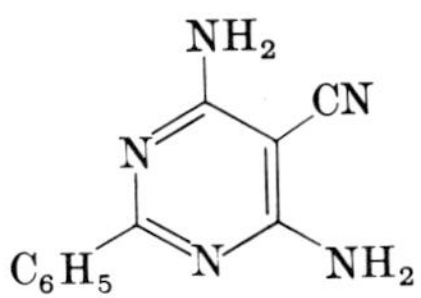		Reflux, 20 min	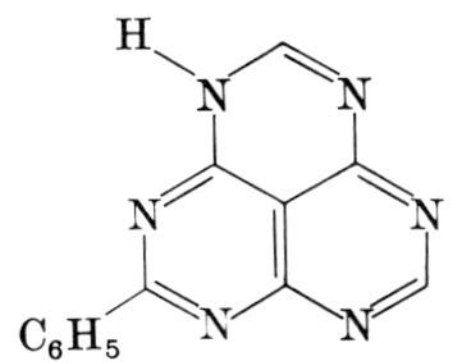	69	458

(*continued*)

TABLE XXVIII (*continued*)

o-Aminonitrile	Conditions	Product	Yield, %	Ref.
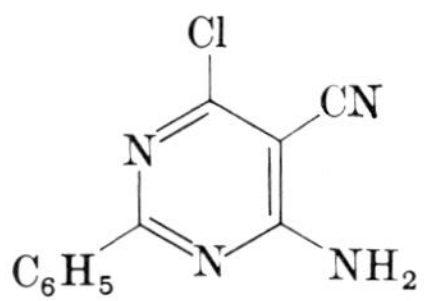	Reflux, 15 min	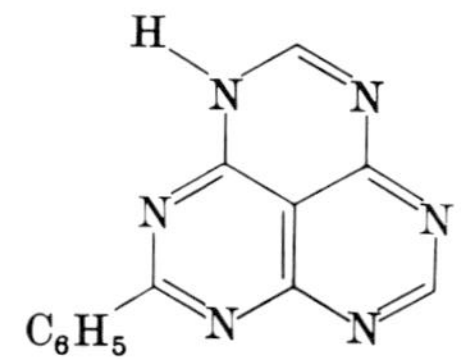	69	458
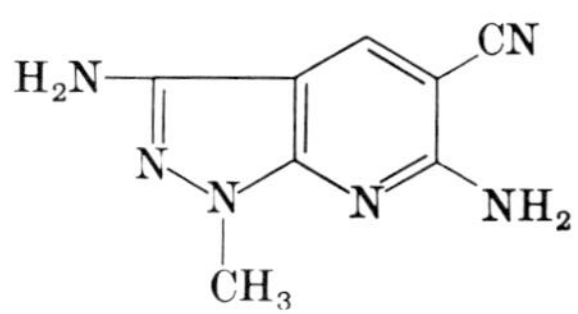	175°, 50 min	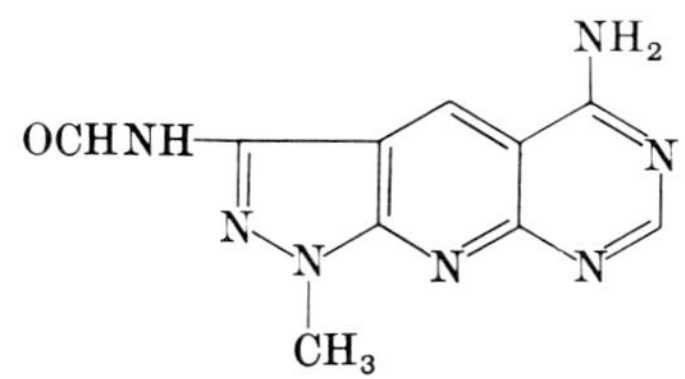	5	125
		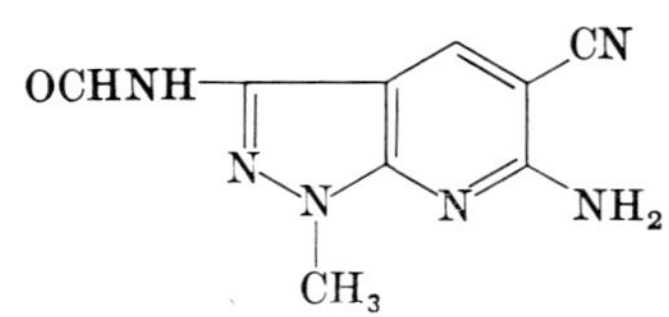	33	125

Some surprising failures with the guanidine cyclization have been reported. 2-Phenyl-4,6-diamino-5-cyanopyrimidine (**2**), for example, fails to give 2-phenyl-4,5,7-triaminopyrimido(4,5-*d*)pyrimidine (**3**) "in acceptable yield" (122); the latter compound had to be prepared alternately by reaction of guanidine with 2-phenyl-4-amino-5-cyano-6-methylmercaptopyrimidine (**4**). The reaction of 2,6-dimethyl-4-amino-5-cyanopyrimidine (**5**) with guanidine unexpectedly gave 2,5,7-triamino-4-methylpyrimido(4,5-*d*)pyrimidine (**6**), in which the 2-methyl group had been replaced by an amino group. It was assumed that either the starting material (**5**) or the initial ring-closed product (2,4-dimethyl-5,7-diaminopyrimido(4,5-*d*)pyrimidine) reacted with guanidine in a ring-opening reaction with the elimination of acetamidine. The intermediates **7** and **8** have been suggested for this cleavage step; both are stabilized by delocalization of charge on the pyrimidine ring nitrogen atoms (122).

An alternate and apparently more effective method for the direct conversion of an *o*-aminonitrile to a fused 2,4-diaminopyrimidine is fusion with cyanamide (in the presence of pyridine hydrochloride) or dicyandiamide (with the hydrochloride of the *o*-aminonitrile). The former

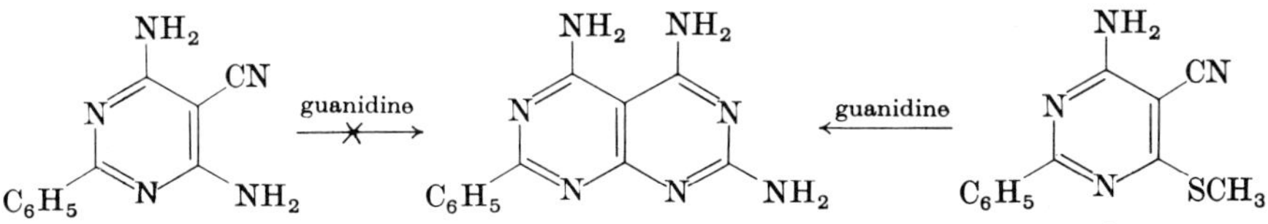

(2) **(3)** **(4)**

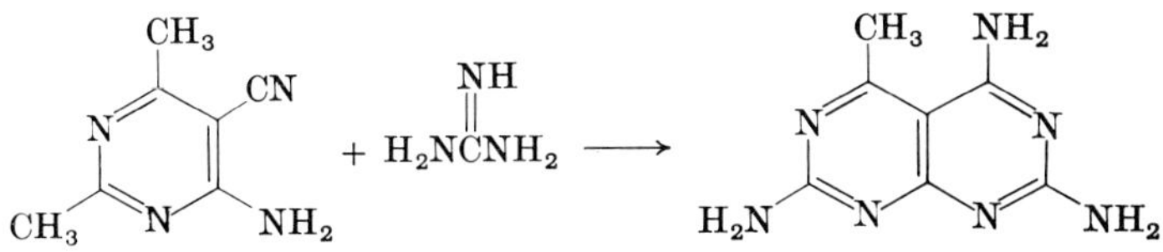

(5) **(6)**

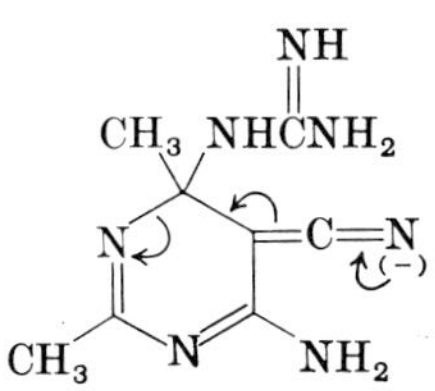

(7)

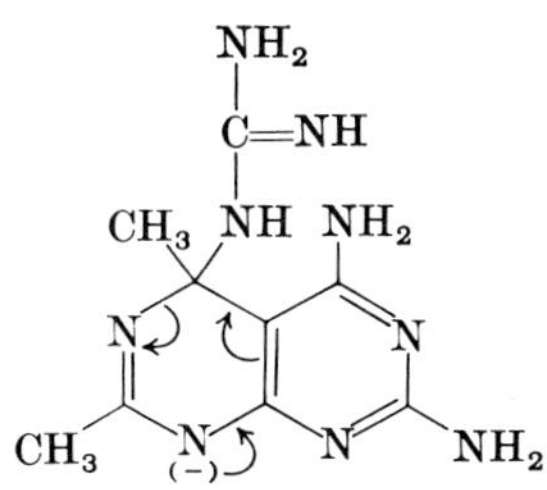

(8)

TABLE XXIX

Condensed 4-Aminopyrimidines from *o*-Aminonitriles and Amidines (Including Guanidine, Cyanamide, and Dicyandiamide)

o-Aminonitrile		Product		Amidine	Yield, %	Ref.
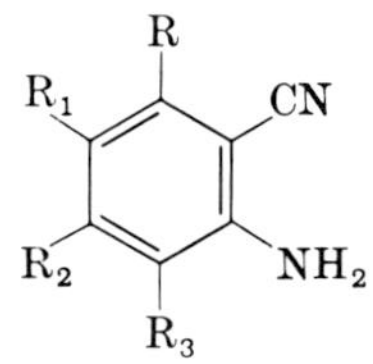		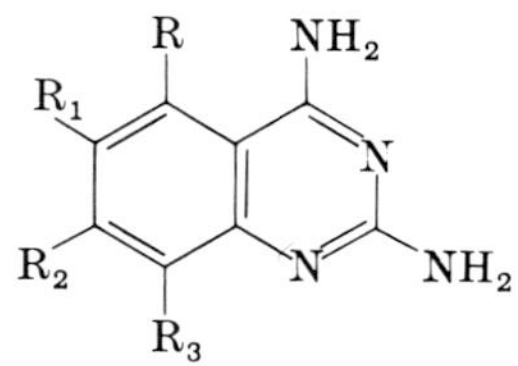				
R	R_1	R_2	R_3			
H	H	H	H	Cyanamide/pyridine·HCl	25	563
				Cyanamide (mineral acid)	—	142
				Dicyandiamide (mineral acid)	—	142
CH_3	H	H	H	Cyanamide/pyridine·HCl	33	563
				Dicyandiamide	30	313
				Guanidine	None	563
H	CH_3	H	H	Cyanamide/pyridine·HCl	31	563
				Guanidine	Trace	563
H	H	CH_3	H	Cyanamide/pyridine·HCl	36	563
				Dicyandiamide	24	563
H	CH_3	CH_3	H	Cyanamide/pyridine·HCl	45	563
C_2H_5	H	H	H	Dicyandiamide	25	313
—$(CH_2)_3$—		H	H	Dicyandiamide	25	313
—$(CH_2)_4$—		H	H	Dicyandiamide	30	313

H	Cl	H	H	Cyanamide/pyridine·HCl	37	563
				Dicyandiamide	33	563
				Guanidine	ca. 10	563
H	H	Cl	H	Cyanamide/pyridine·HCl	14	563
				Dicyandiamide	33	563
				Guanidine	ca. 10	563
H	Cl	Cl	H	Cyanamide/pyridine·HCl	9	563
				Dicyandiamide	ca. 10	563
				Guanidine	ca. 10	563
CH_3O	H	H	H	Cyanamide/pyridine·HCl	16	563
H	H	CH_3O	H	Cyanamide/pyridine·HCl	32	563
				Dicyandiamide	31	563
				Guanidine	Trace	563
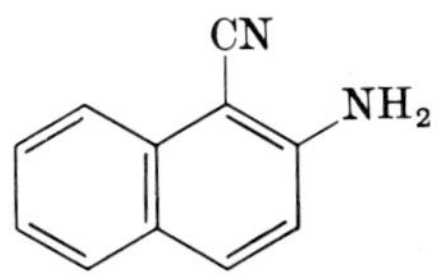		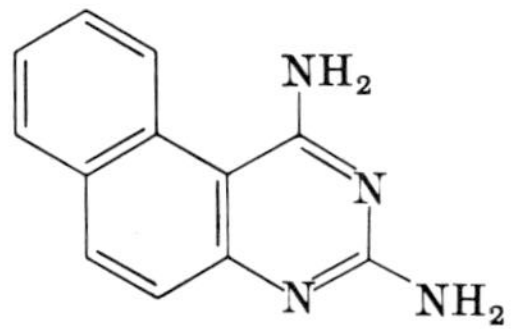		Guanidine	38	289
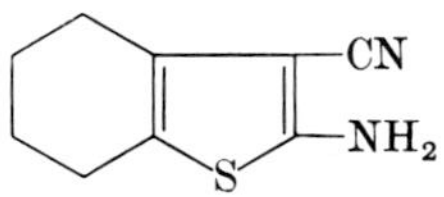		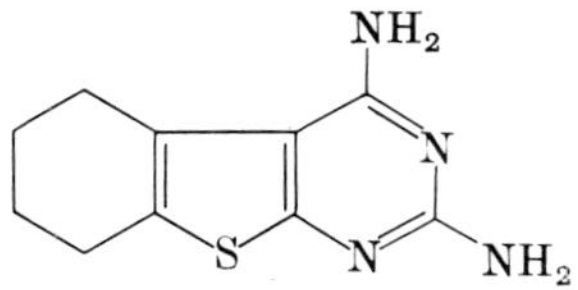		Cyanamide/pyridine·HCl	—	334

(continued)

TABLE XXIX (*continued*)

o-Aminonitrile		Product		Amidine	Yield, %	Ref.
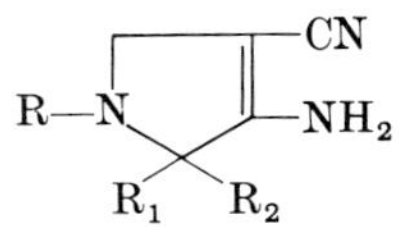		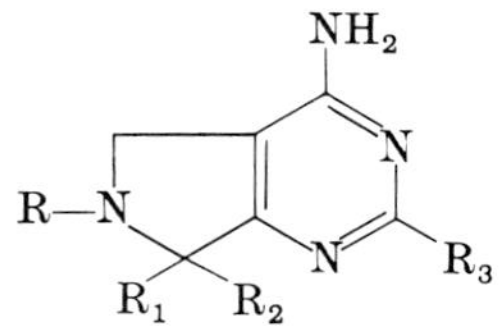				
R	R_1	R_2	R_3			
CH_3CO	H	H	H	Formamidine	17	76
CH_3CO	CH_3	H	NH_2	Guanidine	71	76
C_6H_5CO	CH_3	H	H	Formamidine	6	76
C_6H_5CO	CH_3	H	NH_2	Guanidine	71	76
C_6H_5CO	$C_6H_5CH_2$	H	H	Formamidine	34	76
C_6H_5CO	$C_6H_5CH_2$	H	NH_2	Guanidine	66	76
CH_3	CH_3	H	NH_2	Guanidine	—	116
					68	13
CH_3	C_6H_5	H	NH_2	Guanidine	63	13
$C_6H_5CH_2$	CH_3	H	NH_2	Guanidine	52	13
CH_3	CH_3	CH_3	NH_2	Guanidine	—	116
					67	13
$C_6H_5CH_2$	CH_3	CH_3	NH_2	Guanidine	60	13
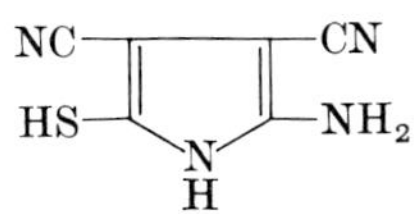		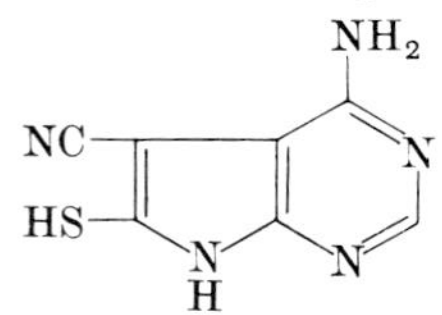		Formamidine	82	10

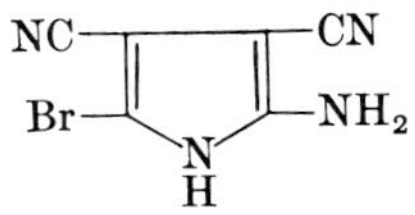	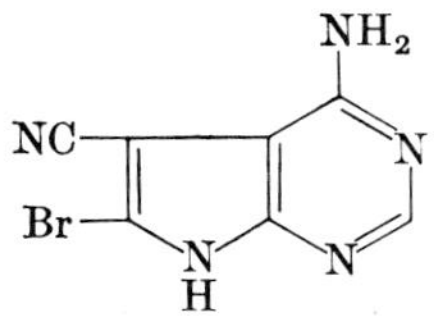	Formamidine	65	550
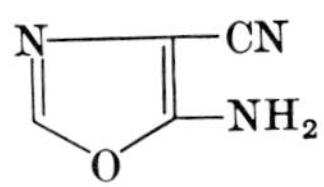	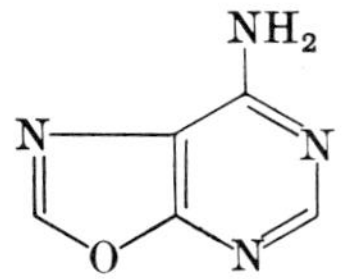	Formamidine	26	32
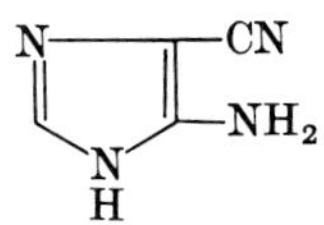	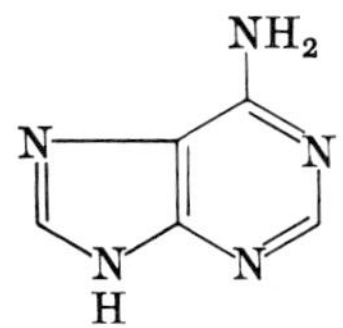	Formamidine	3 68	336 32, 335
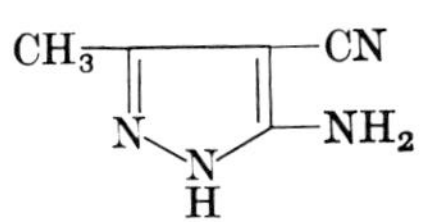	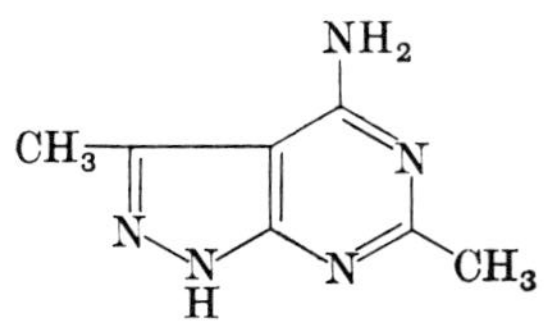	Acetamidine	—	467

(*continued*)

TABLE XXIX (*continued*)

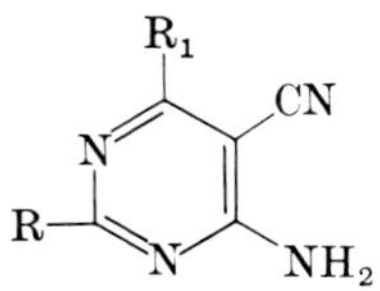

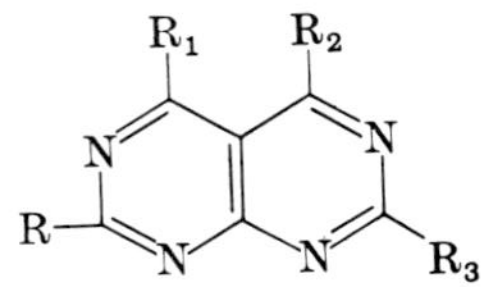

o-Aminonitrile R	R$_1$	Product R$_2$	R$_3$	Amidine	Yield, %	Ref.
H	H	NH_2	H	Formamidine	57	19
H	H	NHC_3H_7-*n*	H	*N*,*N*′-Di(*n*-propyl)formamidine	61	444
H	H	NHC_4H_9-*n*	H	*N*,*N*′-Di(*n*-butyl)formamidine	40	444
H	H	NHC_6H_{11}	H	*N*,*N*′-Dicyclohexylformamidine	34	444
H	H	NHC_6H_5	H	*N*,*N*′-Diphenylformamidine	58	444
H	H	NH_2	NH_2	Guanidine	78	22
CH_3	H	NH_2	H	Formamidine	42	444
CH_3	H	NH_2	NH_2	Guanidine	74	22
H	CH_3	NHC_4H_9-*n*	H	*N*,*N*′-Di(*n*-butyl)formamidine	50	444
C_2H_5S	H	NH_2	CH_3	Acetamidine	25	22
C_2H_5S	H	NH_2	C_6H_5	Benzamidine	15	22
C_2H_5S	H	NH_2	NH_2	Guanidine	40	22
H	NH_2 ($NHCH_3$ in product)	$NHCH_3$	H	*N*,*N*′-Dimethylformamidine	10	458
NH_2	H	NH_2	NH_2	Guanidine	68	22
NH_2	H	NH_2	C_6H_5	Benzamidine	4.2	122
n-C_3H_7NH	H	NH_2	NH_2	Guanidine	92	22

n-C_4H_9NH	H	NH_2	NH_2	Guanidine	72	22
n-$C_6H_{13}NH$	H	NH_2	NH_2	Guanidine	92	22
$HOCH_2CH_2NH$	H	NH_2	NH_2	Guanidine	33	22
$(CH_3)_2N$	H	NH_2	NH_2	Guanidine	61	22
	H	NH_2	NH_2	Guanidine	50	22
$C_6H_5CH_2NH$	H	NH_2	NH_2	Guanidine	73	22
C_6H_5NH	H	NH_2	NH_2	Guanidine	78	22
C_6H_5	H	NH_2	NH_2	Guanidine	94	22
H	C_6H_5	NH_2	NH_2	Guanidine	64	122
C_6H_5	NH_2	NH_2	NH_2	Guanidine	—	141
					Trace	122
C_6H_5	NH_2	NH_2	$N(CH_3)_2$	*N*,*N*-Dimethylguanidine	—	141
NH_2	C_6H_5	NH_2	NH_2	Guanidine	—	141
					70	122
NH_2	$(CH_3)_2N$	NH_2	C_6H_5	Benzamidine	—	141
$(CH_3)_2N$	C_6H_5	NH_2	NH_2	Guanidine	—	141
CH_3S	C_6H_5	NH_2	NH_2	Guanidine	—	141
C_6H_5	CH_3S (NH_2 in product)	NH_2	NH_2	Guanidine	25	122
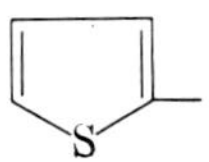	NH_2	NH_2	NH_2	Guanidine	—	141
NH_2	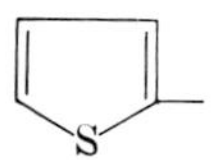	NH_2	NH_2	Guanidine	—	141

(*continued*)

TABLE XXIX (*continued*)

o-Aminonitrile		Product		Amidine	Yield, %	Ref.
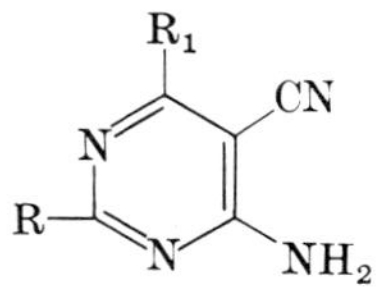		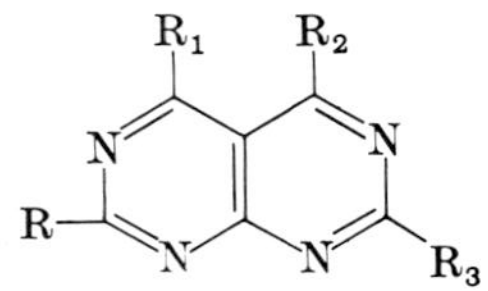				
R	R_1	R_2	R_3			
CH_3 (NH_2 in product)	CH_3	NH_2	NH_2	Guanidine	14	122
NH_2	C_2H_5	NH_2	NH_2	Guanidine	17	122
C_6H_5	C_6H_5	NH_2	NH_2	Guanidine	37	122
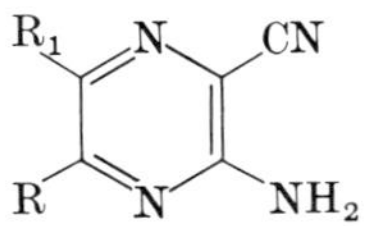		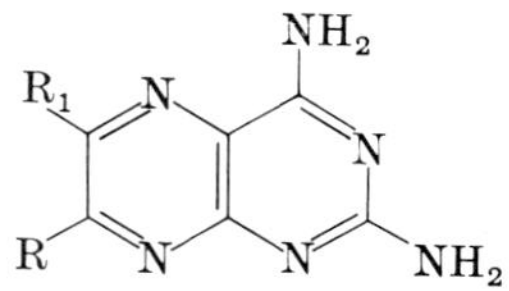				
R		R_1				
H		CH_3		Guanidine	39	562
H		Cyclopropyl		Guanidine	85	562
H		Cl		Guanidine	75	562
H		Br		Guanidine	50	562
NH_2		Cl		Guanidine	92	562
C_2H_5NH		Cl		Guanidine	86	562

$(CH_3)_2CHNH$	Cl	Guanidine	79	562
$CH_2{=}CHCH_2NH$	Cl	Guanidine	83	562
$(CH_3)_2N$	Cl	Guanidine	66	562
$(C_2H_5)_2N$	Cl	Guanidine	84	562
$(CH_3)_2CHO$	Cl	Guanidine	52	562

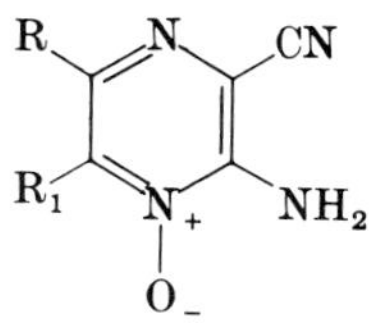

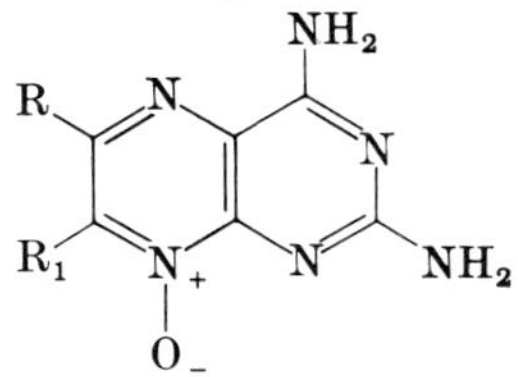

R	R_1			
CH_3	H	Guanidine	84	551
$CH{=}NOH$	H	Guanidine	79	537
$CH_3CH_2CH_2$	H	Guanidine	86	537
CH_3	$CH_2{=}C(CH_3)$	Guanidine	63	537
C_6H_5	H	Guanidine	100	537
$C_6H_5CH{=}CH$	H	Guanidine	97	537

conditions were effective with a range of substituted 2-aminobenzonitriles; the dicyandiamide reaction appeared to be comparable both in facility and in yield (563). A comparative study of the same substituted 2-aminobenzonitriles with guanidine, however, showed that halogenated derivatives were somewhat more satisfactory, while the reaction proceeded poorly or not at all with methyl and methoxy-substituted 2-aminobenzonitriles (563). Since the guanidine cyclization has been shown to be entirely satisfactory with pyrimidine (22,122,141,444,458) and pyrazine (537,551,562) *o*-aminonitriles, it would appear that the propensity of the nitrile grouping to undergo nucleophilic addition (e.g., that the nitrile grouping be attached to an electron-withdrawing nucleus) is the determining factor. In fact, intermediate *N*-amidinocarboxamidines have been isolated in the reaction of guanidine with a number of pyrazine *o*-aminonitriles (562,564). These observations are consistent with the failure of cyclic enaminonitriles such as 1-cyano-2-amino-3,4-dihydronaphthalene and 1-cyano-2-aminoindene to react even with cyanamide (563). Although 1-cyano-2-aminonaphthalene fails to react with dicyandiamide, even under drastic conditions, it does react with cyanamide in the presence of pyridine hydrochloride to give 1,3-diaminobenzo(*f*)quinazoline in fair yield (289).

6. *By Reaction of o-Aminonitriles with Urea and Thiourea*

Fused 4-amino-2(1*H*)-pyrimidinones (**1**) are readily prepared from *o*-aminonitriles by fusion with urea or, less commonly, by reaction with urethane (85,98). Urea fusion has been used with cyclic enaminonitriles (79), aromatic *o*-aminonitriles (289), heterocyclic enaminonitriles such as

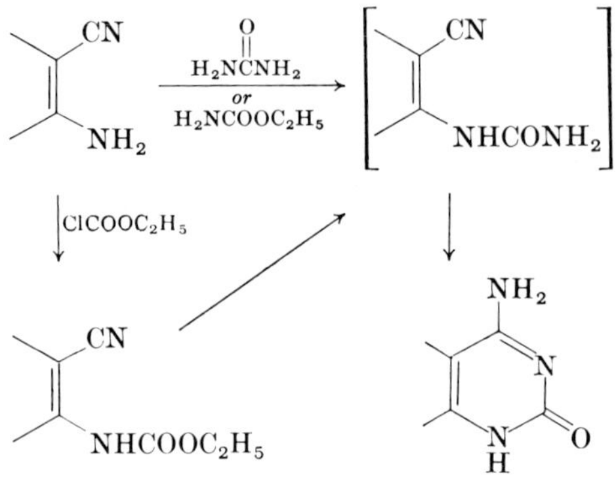

(**2**) (**1**)

3-amino-4-cyano-3-pyrrolines (116), pyrazole aminonitriles (4,124,169), 2-amino-3-cyanoquinoline (14), and condensed thiophene *o*-aminonitriles (334). A simple modification of this procedure which employs milder reaction conditions is condensation of the *o*-aminonitrile with ethyl chloroformate to give the urethane **2** followed by treatment with ammonia (438). A further extension of this general principle is the reaction of an *o*-aminonitrile with diethyl oxalate. The resulting amide (**3**) can be converted to a condensed 2-carbethoxy-4-aminopyrimidine (**5**) by initial methylation to give an imino ether (**4**) followed by treatment with ammonia (25).

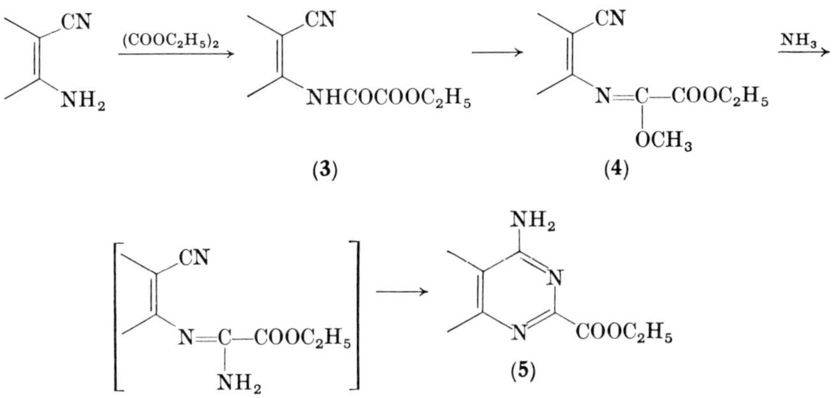

Analogously, condensation of *o*-aminonitriles with thiourea has been used as a means of synthesis of condensed 4-amino-2(1*H*)-pyrimidinethiones (**6**) (4,13,116,124,179,289).

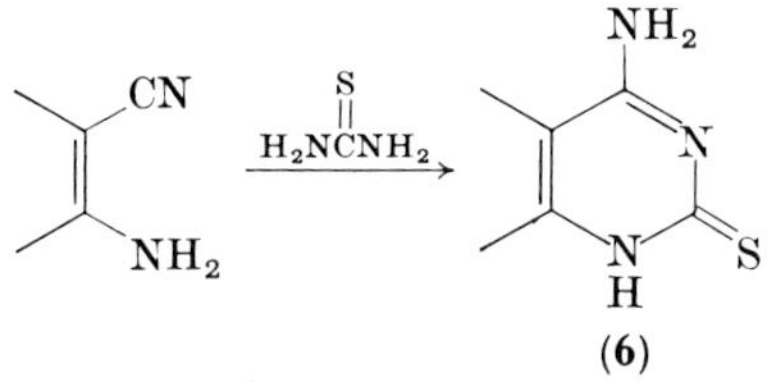

7. *By Reaction of o-Aminonitriles with Isothiocyanates and Isocyanates*

Intermediates in the above reactions of *o*-aminonitriles with urea and thiourea are the corresponding monosubstituted ureas and thioureas, which undergo a subsequent intramolecular addition to the nitrile group to give a condensed 4-aminopyrimidine derivative. Although the use of potassium cyanate or potassium thiocyanate should lead to the same intermediates, their use has apparently not been studied even though this represents a general synthesis of monosubstituted ureas and thioureas. However, the reaction of *o*-aminonitriles with aryl and alkyl isothiocyanates and isocyanates has been studied in some detail and represents

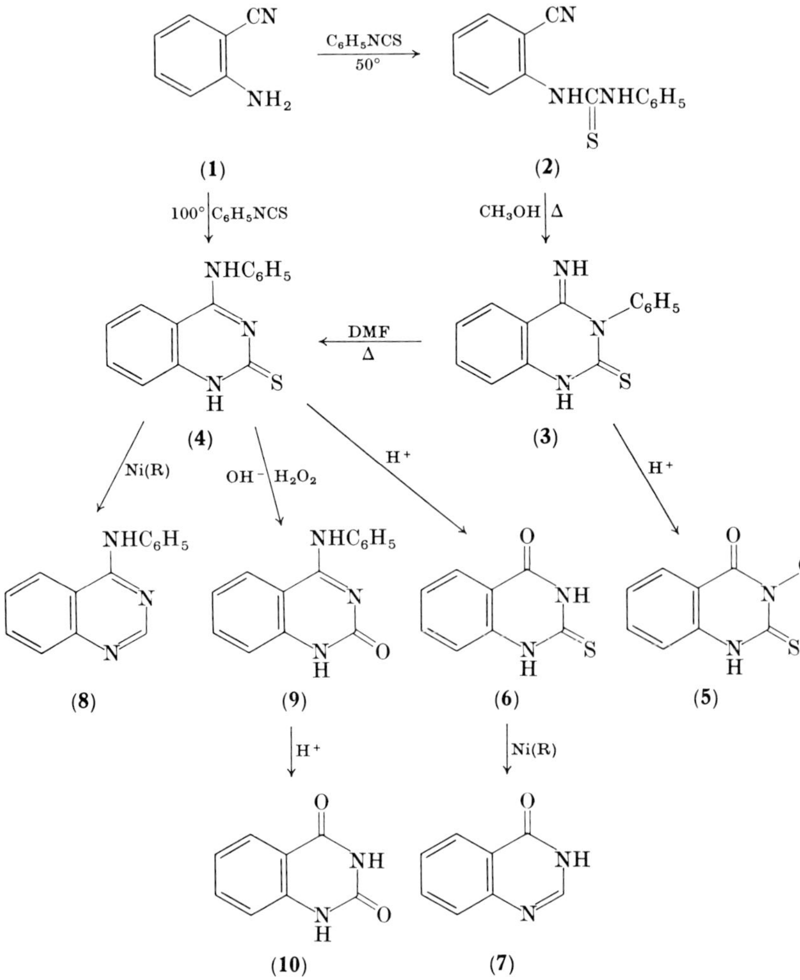

Scheme 26

a versatile route to condensed pyrimidines (15,290,459). Thus, 2-aminobenzonitrile (**1**) reacts with phenyl isothiocyanate at 50° in the absence of solvent to give *N*-phenyl-*N'*-(*o*-cyanophenyl)thiourea (**2**) in high yield. This compound upon short boiling in methanol is converted in quantitative yield to 3-phenyl-4(3*H*)-imino-2(1*H*)-quinazolinethione (**3**) which in turn, upon refluxing with aqueous dimethylformamide, rearranges to 4-anilino-2(1*H*)-quinazolinethione (**4**). Furthermore, the reaction of **1** with phenyl isothiocyanate at elevated temperatures rather than at 50° results in a vigorous and exothermic reaction with the formation of **4** in quantitative yield. The sequential formation from **1** of **2**, then **3**, and finally **4** is dramatically shown by determination of the melting point of **2**; it melts, then resolidifies, melts again at the melting point of **3**, resolidifies, and finally melts a third time at the melting point of **4** (Scheme 26). Acid hydrolysis of 3-phenyl-4(3*H*)-imino-2(1*H*)-quinazolinethione (**3**) gives 3-phenyl-4(3*H*)-oxo-2(1*H*)-quinazolinethione (**5**), while acid hydrolysis of the 4-anilino derivative (**4**) gives 4(3*H*)-oxo-2(1*H*)-quinazolinethione (**6**). Desulfurization of the latter compound gives 4(3*H*)-quinazolone (**7**) while desulfurization of 4 gives 4-anilinoquinazoline (**8**) (15). It will be recalled that the latter compound is alternatively prepared from 2-aminobenzonitrile (**1**) either by heating with N,N'-diphenylformamidine or by treatment with ethyl orthoformate followed by reaction with aniline and subsequent rearrangement (see Chapter II, Section VIII-G-2). Finally, hydrolytic desulfurization of 4-anilino-2(1*H*)-quinazolinethione (**4**) gives 4-anilino-2(1*H*)-quinazolone (**9**), which in turn is readily hydrolyzed to **10**.

The hydrolysis of the above 4-anilino and 4-imino derivatives to the corresponding 4(3*H*)-quinazolones can be carried out under mild acidic conditions and thus offers an attractive alternative to the synthesis of

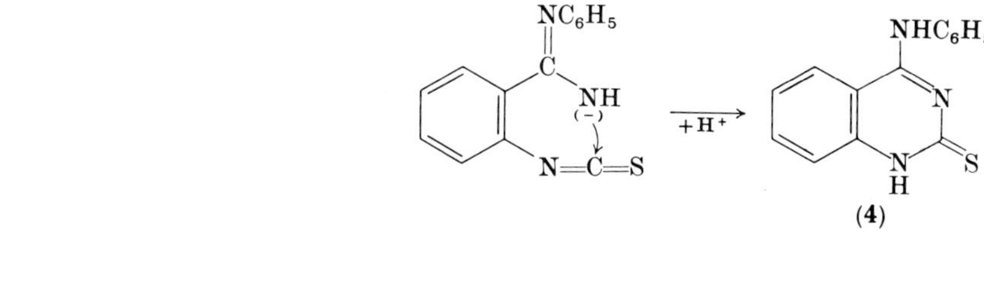

these compounds from the *o*-aminonitrile involving initial conversion to an amide (or substituted amide) followed by an appropriate cyclization (15,290).

The conversion of **3** to **4** is analogous in principle to the above-discussed rearrangement of 3-substituted-4(3*H*)-iminopyrimidines, the original products of the reaction of an *o*-cyano ethoxymethyleneamino derivative with a primary amine (i.e., the Dimroth rearrangement). This rearrangement also appears to be base-catalyzed and proceeds via a ring-opening, ring-closure sequence which is depicted on p. 297. The basic catalyst in the direct conversion of 2-aminobenzonitrile to **4** may well be the intermediate imine **3**.

Treatment of 2-aminobenzonitrile (**1**) with allyl isothiocyanate leads directly to the nonrearranged quinazolinethione **11**, which upon treatment with hydrochloric acid in acetic acid cyclizes to the thiopegan derivative **12**; in this case neither hydrolysis of the imino grouping nor rearrangement to a 4-substituted aminoquinazoline was apparently observed (158).

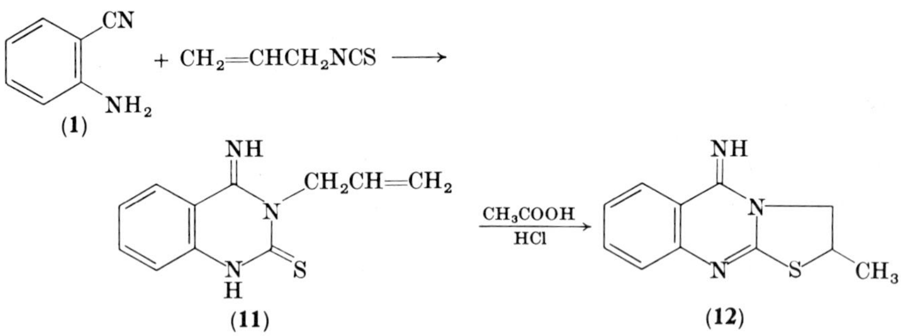

2-Aminobenzonitrile (**1**) reacts analogously with phenyl isocyanate (15,290). The conditions can be controlled to give either the noncyclized urea **13** (61) or 3-phenyl-4(3*H*)-imino-2(1*H*)-quinazolone (**14**) (15). The former intermediate can be independently cyclized to **14** with sodium ethoxide, but all attempts to rearrange the latter to 4-anilino-2(1*H*)-quinazolone (**15**) were unsuccessful. However, direct interaction of **1** with one equivalent of phenyl isocyanate at 135° gave the cyclized, rearranged quinazoline derivative **15** in 52% yield (15). Although it seems clear that the conversion of 2-aminobenzonitrile (**1**) to **15** must proceed through **14**, heating the latter compound with phenyl isocyanate failed to bring about rearrangement, apparently because of decomposition under the reaction conditions.

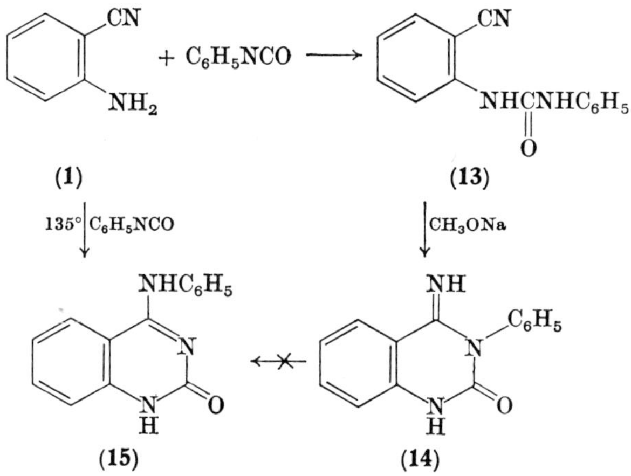

H. SYNTHESIS OF FUSED 4(3*H*)-PYRIMIDINETHIONES (4-MERCAPTO-PYRIMIDINES)

1. By Reaction of o-Aminonitriles with Orthoformate Esters and Hydrogen Sulfide

The rearrangement of condensed 3-substituted-4-iminopyrimidines to the 4-substituted amino isomers discussed above (the Dimroth rearrangement) has a close analogy in the known rearrangement of *m*-thiazines to pyrimidines (460–462), of 2-aminothiophenes to 2-mercaptopyrroles (102, 104) and of 5-aminothiazoles to 2-mercaptoimidazoles (463). All of these isomerizations involve a ring-cleavage, ring-closure sequence which is probably base-catalyzed. A useful, general synthesis of condensed 4(3*H*)-pyrimidinethiones based upon this rearrangement principle has recently been developed (95,96), starting with an *o*-aminonitrile (**1**). These latter compounds are converted under the usual conditions to ethoxymethyleneamino derivatives (**2**) which, without isolation, are treated with an ethanolic solution of sodium hydrosulfide. Workup with water followed by acidification gives the fused pyrimidinethione **3** in a high state of purity and in yields generally in excess of 90%. The reaction may proceed by addition of hydrosulfide anion to the nitrile grouping to give an intermediate *o*-ethoxymethyleneamino thioamide (**4**) which subsequently cyclizes with loss of ethanol. Alternately, initial displacement of ethanol from the ethoxymethyleneamino derivative **2** would give an intermediate *o*-cyanothioformamide (**5**) which could cyclize under the basic reaction conditions to the *m*-thiazine **6**. This should rearrange under the alkaline reaction conditions to give the fused 4(3*H*)-pyrimidinethione (**3**).

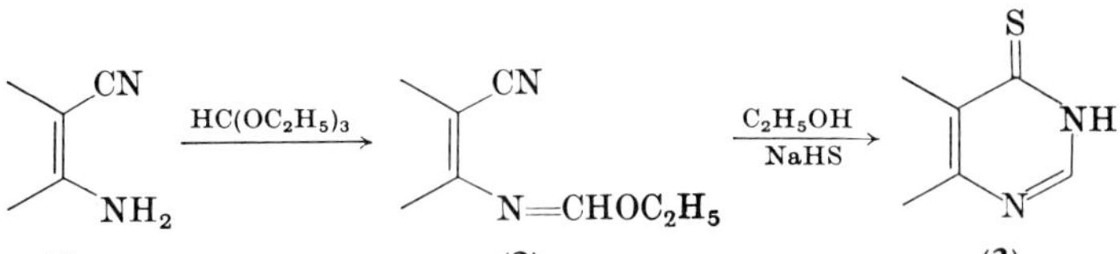

(1) (2) (3)

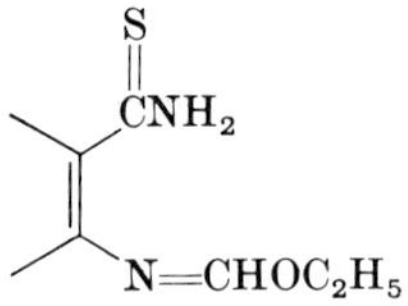

(4)

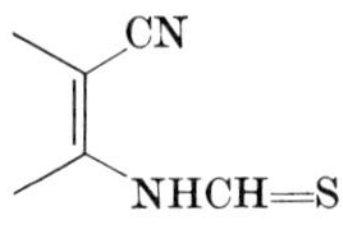

(5)

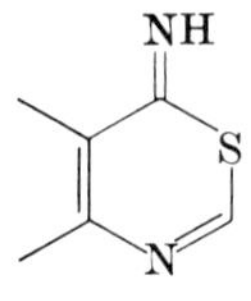

(6)

On the other hand, under different reaction conditions the conversion of an *o*-aminonitrile to a fused 4(3*H*)-pyrimidinethione may involve other intermediates. Treatment of **7** (R = H) with acetic anhydride and ethyl orthoformate, followed by dissolution of the crude ethoxymethyleneamino derivative in pyridine and addition of hydrogen sulfide (conditions originally described for this 4(3*H*)-pyrimidinethione synthesis (459,480)) resulted in the separation of what appeared to be the thioformylamino thioamide **8** (R = H). This upon subsequent treatment with base gave the fused 4(3*H*)-pyrimidinethione **9** (R = H). However, the same

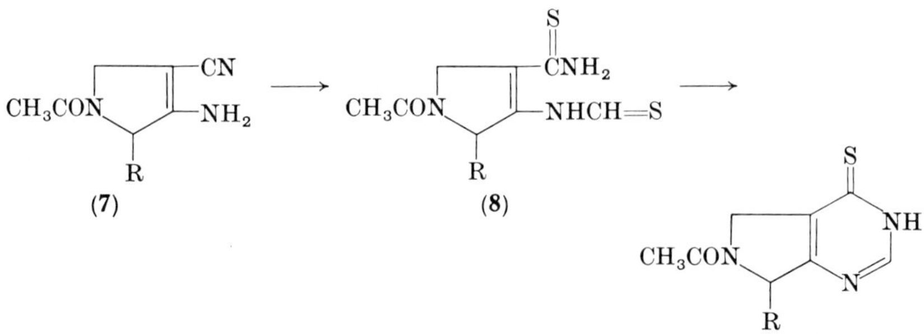

(7) (8) (9)

reaction conditions employed with **7** (R = CH_3) gave the pyrimidinethione **9** (R = CH_3) directly (479). It is not known whether **8**, which is isolated in this instance because of its fortuitous insolubility in pyridine, arose from hydrogen sulfide cleavage of an initially formed *m*-thiazine, or whether it is the product of the direct reaction of the ethoxymethyleneamino derivative with 2 moles of hydrogen sulfide.

The effectiveness of the ethyl orthoformate/sodium hydrosulfide sequence may be illustrated by the preparation of 7(6*H*)-cyclopenteno-(*d*)pyrimidinethione (**11**) in 70% yield in one step from 1-amino-2-cyanocyclopentene (**10**). Similarly, 3-amino-4-cyanopyrazole (**12**) has been converted in a single step to 4(5*H*)-pyrazolo(3,4-*d*)pyrimidinethione (**13**) in 97% yield (96). The reaction sequence has been applied with equal success to a variety of aromatic and heterocyclic *o*-aminonitriles, and appears to be the method of choice for the preparation of fused 4(3*H*)-pyrimidinethiones, provided only that the substrate is stable to the alkaline conditions employed (96).

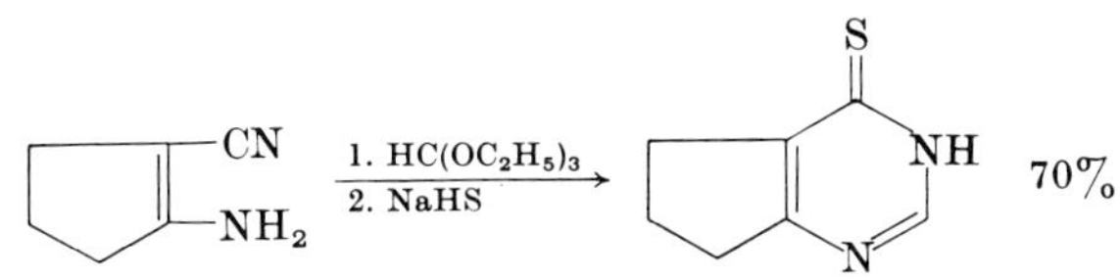

(**10**) (**11**)

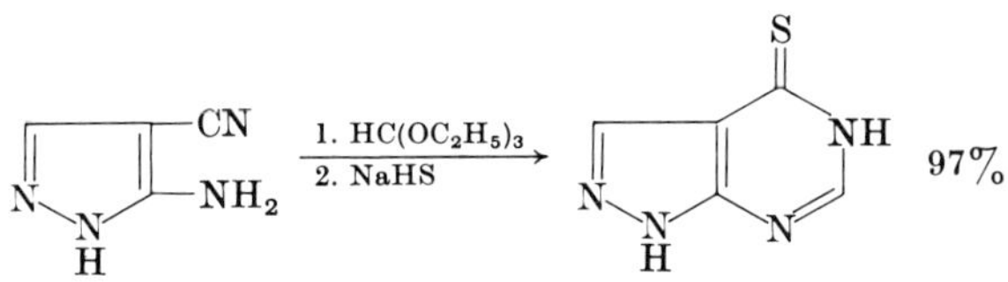

(**12**) (**13**)

2. *By Reaction of o-Aminonitriles with Thioamides and Acid*

A general synthesis of thioamides from nitriles is by hydrogen sulfide transfer from an added thioamide in dimethylformamide solution saturated with dry hydrogen chloride (464). This thioamide synthesis was postulated to take place by an initial protonation of the nitrile followed by a nucleophilic attack upon the protonated nitrile by the sulfur atom of the added thioamide (usually thioacetamide) to give a protonated imino sulfide intermediate (**1**). This latter intermediate was presumed to be in equilibrium both with the starting nitrile and thioacetamide (by cleavage of carbon–sulfur bond *a*) and with acetonitrile and a new thioamide derived by addition of hydrogen sulfide to the initial nitrile (by cleavage of carbon–sulfur bond *b*). Under the experimental conditions employed, the equilibrium is shifted to the right by elimination of the relatively volatile acetonitrile.

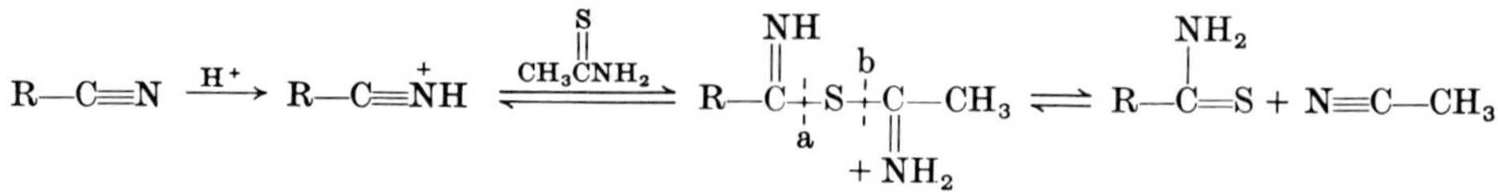

(1)

An extrapolation of this thioamide exchange reaction to *o*-aminonitriles offered the possibility that the intermediate imino sulfide (corresponding to **1**) might be trapped by an intramolecular displacement on carbon by the *o*-situated amino grouping, and that this reaction might take precedence over the thioamide exchange process observed with simple nitriles. It has been found (18) that a convenient, one-step synthesis of fused 4(3*H*)-pyrimidinethiones results from the reaction of *o*-aminonitriles with thioamides in ethanol solution saturated with dry hydrogen chloride, followed by evaporation and treatment of the reaction mixture with alkali (an important step, as discussed below). Variation

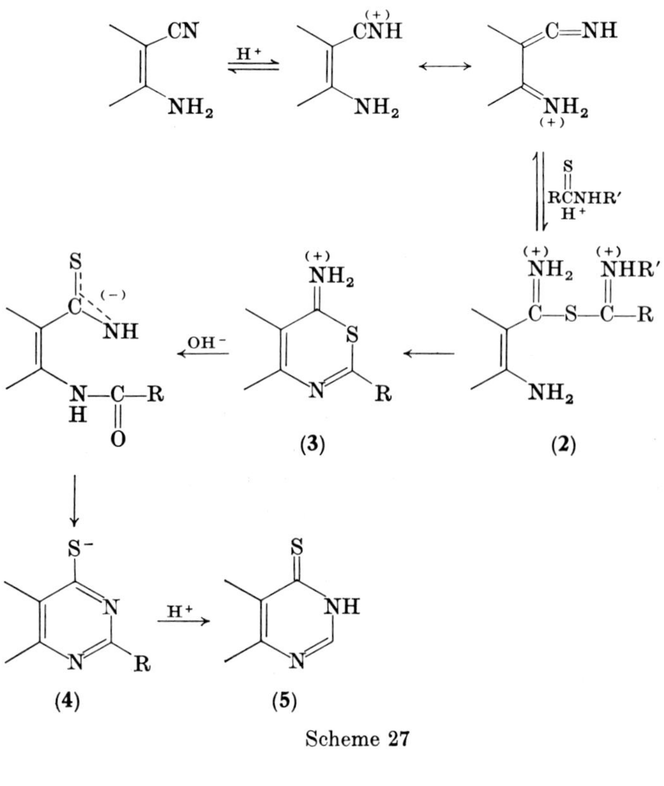

Scheme 27

of the thioamide allows the introduction of hydrogen, alkyl, or aryl groups into the 2-position of the condensed pyrimidine ring.

It was suggested that this reaction is initiated, in analogy with the thioamide exchange process described above, by addition of the sulfur atom of the thioamide to the protonated *o*-aminonitrile to give an intermediate imino sulfide **2** which then cyclizes by loss of ammonia (or an amine) to the *m*-thiazine **3**. Addition of alkali to **3**, which need not be isolated, initiates the ring-opening, ring-closure sequence characteristic of *m*-thiazines to give the stable, aromatic anion (**4**) of the fused 4(3*H*)-pyrimidinethione (**5**) (18) (Scheme 27).

A minor side reaction observed in the above synthesis of fused 4(3*H*)-pyrimidinethiones is dissociation of the intermediate **2** to the thioamide derived from the *o*-aminonitrile, and a new nitrile derived from the initially added thioamide (i.e., the reaction observed with simple nitriles). Thus, reaction of 2-amino-5-methoxybenzonitrile and thioformanilide in glacial acetic acid saturated with hydrogen bromide gave both 6-methoxyquinazoline-4(3*H*)-thione (the normal product) and 2-methyl-6-methoxy-

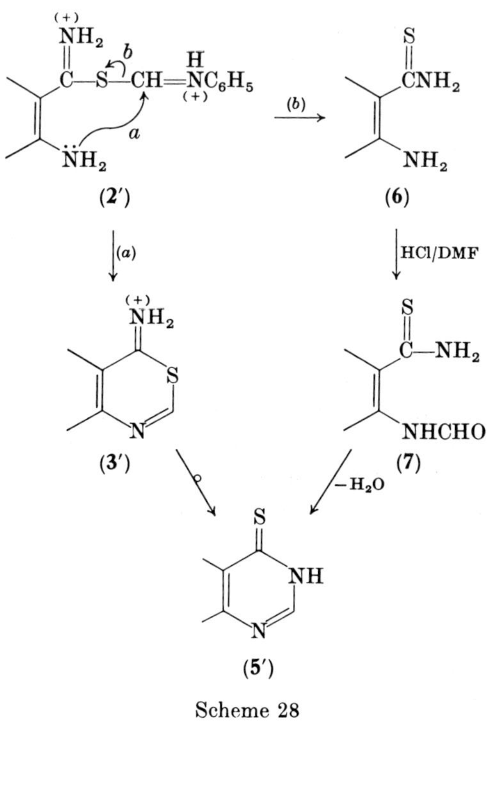

Scheme 28

quinazoline-4(3*H*)-thione. The latter compound arises by the intermediate formation of 2-amino-5-methoxythiobenzamide by the exchange reaction, followed by acetylation by the solvent and subsequent ring closure (538). This side reaction can actually be utilized to advantage in the synthesis of 2-unsubstituted 4(3*H*)-pyrimidinethiones from *o*-aminonitriles and thioformanilide by employing dimethylformamide–hydrogen chloride as the reaction medium. The thioamide **6** derived from the *o*-aminonitrile by the exchange reaction (via **2′**) is formylated by the dimethylformamide–hydrogen chloride mixture (a variant of the Vilsmeier-Haack procedure) and the resulting *o*-formylaminothioamide **7** then cyclizes under the reaction conditions to the same 4(3*H*)-pyrimidinethione (**5′**) obtained via the *m*-thiazine pathway (via **3′**). What is otherwise an undesired side reaction is thus diverted to give the desired product, and improved yields are generally obtained (18) (Scheme 28).

The product can, in fact, be controlled to give either thioamide or the fused 4(3*H*)-pyrimidinethione by changing the formylating power of the reaction medium. Thus, although 3-amino-4-cyano-1,2,2-trimethyl-3-pyrroline (**8**) reacts with thioacetamide in dimethylformamide (made 6*N* in hydrogen chloride) to give the thioamide **9**, reaction under otherwise identical conditions in dimethylformamide saturated with hydrogen chloride gives the pyrrolinopyrimidinethione **10** (16).

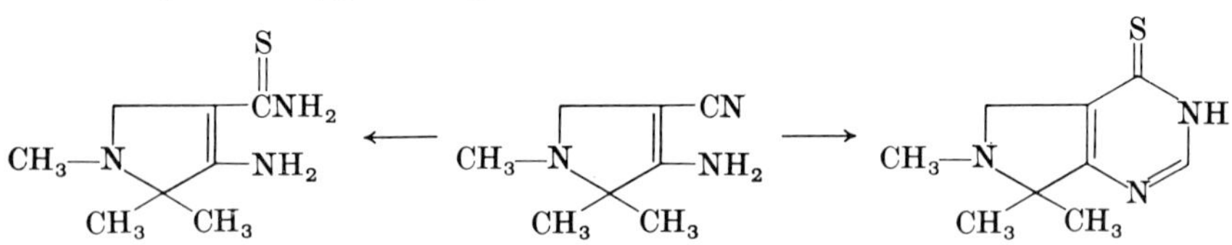

(**9**) (**8**) (**10**)

In one instance, a sulfur-free product was isolated from the thioamide reaction. Reaction of 2-amino-5-nitrobenzonitrile with thioacetamide in glacial acetic acid saturated with hydrogen bromide gave only 2-methyl-6-nitroquinazolin-4(3*H*)-one and no quinazoline-4(3*H*)-thione. Since the same product could be obtained under the same reaction conditions but in the absence of the thioacetamide, it is presumably formed by simple acetylation of the aminonitrile by the solvent followed by cyclization (538).

3. *By Reaction of o-Aminonitriles with Isothiocyanates and Acid*

As described above on p. 297, 2-aminobenzonitrile (**1**) reacts with phenyl isothiocyanate in the absence of solvent to give 4-anilino-2(1*H*)-quinazolinethione (**4**) via the sequential formation of *N*-phenyl-*N′*-(*o*-

cyanophenyl)thiourea (**2**) and 3-phenyl-4(3*H*)-imino-2(1*H*)-quinazolinethione (**3**) (for formulas see p. 296). Since it is known that nitriles in acidic solution suffer nucleophilic attack by sulfur in reaction with thioamides (18,464) (the thioamide exchange process described above on p. 301), the reaction of 2-aminobenzonitrile (**1**) with phenyl isothiocyanate in acidic solution is of particular interest, since an analogous reaction course would have led to 2-anilino-4(3*H*)-quinazolinethione, an isomer of **4** with the position of substituents reversed. However, the actual reaction takes a surprisingly different course. Thus, treatment of **1** with phenyl isothiocyanate in dimethylformamide solution saturated with dry hydrogen chloride leads to 4(3*H*)-quinazolinethione (**8**) in 79% yield (95,241). Other *o*-aminonitriles react similarly, and this process constitutes an alternative one-step conversion of *o*-aminonitriles to fused 4(3*H*)-pyrimidinethiones. For example, 1-methyl-4-cyano-5-aminopyrazole is converted to 1-methyl-4(5*H*)-pyrazolo(3,4-*d*)pyrimidinethione in 67% yield; the corresponding 1-phenyl derivative is similarly prepared in 48% yield.

The course of this unexpected conversion was uncovered when the reaction between 2-aminobenzonitrile and phenyl isothiocyanate was carried out using $H^{14}CON(CH_3)_2$. The 4(3*H*)-quinazolinethione (**8**) formed contained 97.3% of the calculated amount of ^{14}C, showing that the C_2 atom of the quinazoline had originated from the dimethylformamide employed as solvent (and not from the phenyl isothiocyanate). On the basis of this result, it was suggested (95) that the reaction sequence involves an initial nucleophilic addition of phenyl isothiocyanate to the protonated *o*-aminonitrile (a step analogous to the initial addition of thioamides to protonated nitriles and *o*-aminonitriles discussed above) to give (via **5**) an *o*-aminothioamide (**6**) which then is formylated by the dimethylformamide–hydrogen chloride mixture (the Vilsmeier-Haack

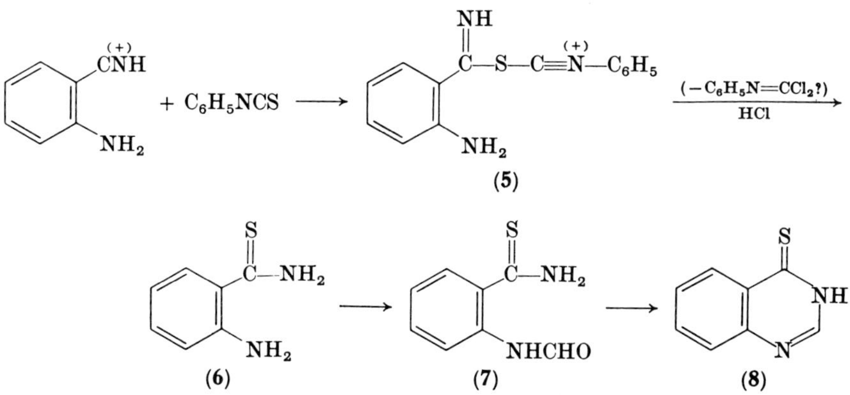

formylation) to give **7**. Subsequent dehydrative cyclization then gives the observed fused 4(3*H*)-pyrimidinethione (**8**).

I. SYNTHESIS OF FUSED 2,4(1*H*,3*H*)-PYRIMIDINEDITHIONES (2,4-DIMERCAPTOPYRIMIDINES) BY REACTION OF *o*-AMINONITRILES WITH CARBON DISULFIDE

The *m*-thiazine → pyrimidine rearrangement which is implicated in the synthesis of fused 4(3*H*)-pyrimidinethiones from *o*-cyanoethoxymethyleneamino compounds and sodium hydrosulfide as discussed on pp. 299–301 has been exploited further in a facile, one-step synthesis of fused 2,4(1*H*,3*H*)-pyrimidinedithiones (**3**) directly from *o*-aminonitriles (93,94,459,535). This conversion involves the initial formation of a dithiocarbamate salt (**1**), cyclization to a *m*-thiazine (**2**) (which has actually been isolated and characterized in the reaction of 3-amino-4-cyanopyrazole with carbon disulfide) followed by a ring-opening, ring-reclosure sequence, probably initiated by the solvent pyridine acting as the requisite base, leading rapidly and irreversibly to the observed pyrimidinedithione (**3**). Although this reaction has been applied to both aromatic and heterocyclic *o*-aminonitriles, its success is dependent both upon the basicity of the amino grouping and upon the ability of the *o*-situated nitrile to act as an electrophile. This reaction sequence from *o*-aminonitriles makes 2,4(1*H*,3*H*)-pyrimidinedithiones (**3**) readily available, and they should see considerable utilization as intermediates for heterocyclic synthesis.

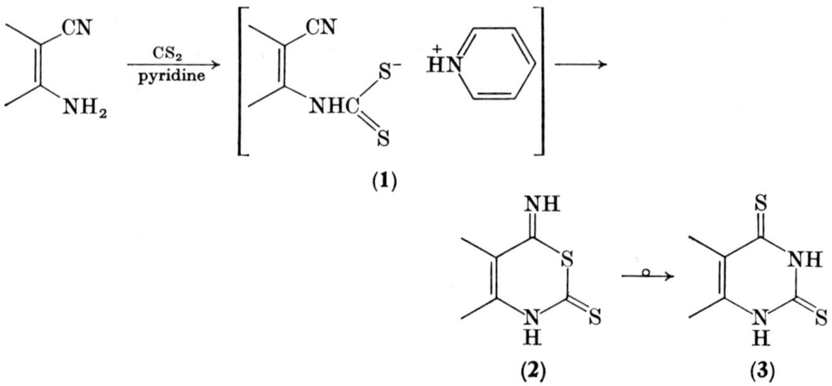

(**1**) (**2**) (**3**)

J. MICHAEL ADDITIONS OF *o*-AMINONITRILES

The amino group of 2-aminobenzonitrile and of some 5-substituted derivatives is sufficiently basic to add Michael-fashion to dimethyl acetylenedicarboxylate to give enamines (**1**); these cyclize upon heating in diphenyl ether to 2-carbomethoxy-8-cyano-4(1*H*)-quinolinones (**2**) (468).

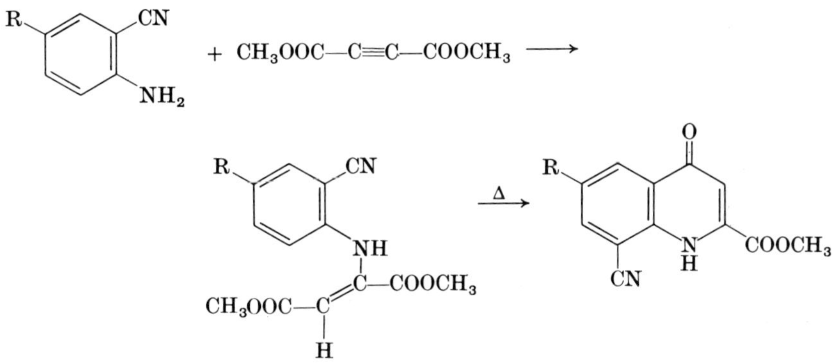

(**1**) (**2**)

TABLE XXX

Cyclic Enaminonitriles and *o*-Aminonitriles

Empirical formula	Compound	Preparation	References to use as synthetic intermediate	Other
$C_3H_3N_5$	4-Cyano-5-amino-1,2,3-triazole	344		
$C_4H_2ClN_3S$	3-Chloro-4-cyano-5-aminoisothiazole	173		
$C_4H_2N_2S_3$	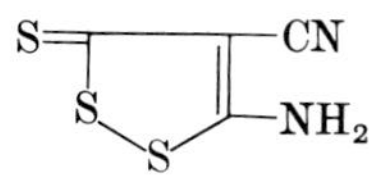	73, 356, 358	560	216
$C_4H_3N_3O$	4-Cyano-5-aminoisoxazole	176	11, 176	
$C_4H_3N_3O$	4-Cyano-5-aminooxazole	32, 335	32	
$C_4H_3N_3S$	4-Cyano-5-aminothiazole	114	114	
$C_4H_4N_4$	4-Amino-5-cyanoimidazole	32, 335, 336	32, 335, 336, 484	
$C_4H_4N_4$	3-Amino-4-cyanopyrazole	4, 154, 395	4, 17, 18, 27, 93–96, 168, 459, 467	
$C_4H_4N_4S$	2,4-Diamino-5-cyanothiazole	473		
$C_4H_5N_5$.	3,5-Diamino-4-cyanopyrazole	105		
$C_5H_2Cl_2N_4$	2-Amino-3-cyano-5,6-dichloropyrazine	562	562	
$C_5H_3BrN_4$	2-Amino-3-cyano-5-bromopyrazine	562	562	
$C_5H_3ClN_4$	2-Amino-3-cyano-5-chloropyrazine	562	562	
$C_5H_3F_3N_4$	3-Trifluoromethyl-4-cyano-5-aminopyrazole	349		
$C_5H_3N_5$	3,4-Dicyano-5-aminopyrazole	232	92, 232	
$C_5H_4ClN_5$	2,6-Diamino-3-cyano-5-chloropyrazine	562	562	
$C_5H_4ClN_5$	2-Chloro-4,6-diamino-5-cyanopyrimidine	561		
$C_5H_4N_2S$	2-Amino-3-cyanothiophene	74		
$C_5H_4N_4$	2-Amino-3-cyanopyrazine	23, 185	178, 562	

$C_5H_4N_4$	4-Amino-5-cyanopyrimidine	19, 23, 36, 44, 50, 165, 166, 259, 444, 544, 556	19, 27, 93–96, 130, 143, 161, 163, 254, 444, 459, 467	58, 146
$C_5H_4N_4O$	2-Hydroxy-4-amino-5-cyanopyrimidine	53, 161, 556	161, 163	
$C_5H_4N_4S$	2-Mercapto-4-amino-5-cyanopyrimidine	53, 161, 246, 556	161	
$C_5H_5N_3O$	3-Methyl-4-cyano-5-aminoisoxazole	11	11	
$C_5H_5N_3O$	2-Methyl-4-cyano-5-aminooxazole	32, 335		
$C_5H_5N_3OS$	3-Methylmercapto-4-cyano-5-amino-isoxazole	266		
$C_5H_5N_3S$	3-Methyl-4-cyano-5-aminoisothiazole	156, 283	156	
$C_5H_5N_3S$	2-Methyl-4-cyano-5-aminothiazole	114	114	
$C_5H_5N_3S_2$	2-Methylmercapto-4-cyano-5-aminothiazole	114	114	
$C_5H_5N_3S_2$	3-Amino-4-cyano-5-methylmercapto-isothiazole	156		
$C_5H_5N_5$	2,4-Diamino-5-cyanopyrimidine	2, 75, 556	2, 46, 122, 143, 161, 163, 467	
$C_5H_5N_5$	4,6-Diamino-5-cyanopyrimidine	458	458	
$C_5H_5N_5O$	1-Carbamoyl-4-cyano-5-aminopyrazole	232	232	
$C_5H_6N_4$	1-Methyl-4-amino-5-cyanoimidazole	24, 98, 136	20, 98, 136	
$C_5H_6N_4$	1-Methyl-4-cyano-5-aminoimidazole	114	114	
$C_5H_6N_4$	1-Methyl-4-cyano-5-aminopyrazole	3, 149, 405, 532	3, 17, 20, 93–96, 149, 168, 169, 405, 532	
$C_5H_6N_4$	3-Methyl-4-cyano-5-aminopyrazole	3, 124	3, 124, 467	
$C_5H_6N_4$	1-Methyl-3-amino-4-cyanopyrazole	97	97	
$C_5H_6N_4O_2S$	3-Methylsulfonyl-4-cyano-5-aminopyrazole	232, 406		
$C_5H_6N_4S$	3-Methylmercapto-4-cyano-5-amino-pyrazole	232, 266, 406	232, 266	
$C_5H_6N_6$	2-Hydrazino-4-amino-5-cyanopyrimidine	411, 412	400, 411, 412	
$C_6H_3BrN_4$	2-Amino-5-bromo-3,4-dicyanopyrrole	104	550	

(continued)

TABLE XXX (*continued*)

Empirical formula	Compound	Preparation	References to use as synthetic intermediate	Other
$C_6H_3F_3N_4$	2-Trifluoromethyl-4-amino-5-cyano-pyrimidine	346	346	
$C_6H_3N_5S$	2-Thiocyanato-4-amino-5-cyanopyrimidine	352		352
$C_6H_4BrN_3$	3-Amino-5-bromopicolinonitrile	78		
$C_6H_4ClN_3$	3-Amino-5-chloropicolinonitrile	78		
$C_6H_4N_4O_3S$	5-Amino-3,4-dicyano-2-pyrrolesulfonic acid	104		
$C_6H_4N_4S$	2,5-Diamino-3,4-dicyanothiophene	104, 374, 394, 398	10, 91, 104, 341	
$C_6H_4N_4S$	2-Amino-3,4-dicyano-5-mercaptopyrrole	104, 394	10, 104, 494	
$C_6H_4N_4S_2$	2-Cyanomethylmercapto-4-amino-5-cyanothiazole	107		
$C_6H_4N_6O$	1-Carbamoyl-3,4-dicyano-5-aminopyrazole	232	232	
$C_6H_4N_6S$	1-Thiocarbamoyl-3,4-dicyano-5-aminopyrazole	232		
$C_6H_5BrN_4$	2,4-Diamino-3(or 5)-cyano-6-bromopyridine	84		
$C_6H_5ClN_4$	2-Methyl-4-amino-5-cyano-6-chloropyrimidine	63	63	
$C_6H_5ClN_4O$	2-Amino-3-cyano-5-methyl-6-chloropyrazine 1-oxide	537	537	
$C_6H_5N_3$	2-Aminonicotinonitrile	23, 30	17, 27	
$C_6H_5N_3$	4-Aminonicotinonitrile	23, 536		
$C_6H_5N_3$	3-Aminopicolinonitrile	78		
$C_6H_5N_3O_2$	2,6-Dihydroxy-3-cyano-4-aminopyridine	508, 521		531
$C_6H_5N_5$	3-Cyanomethyl-4-cyano-5-aminopyrazole	26	26	
$C_6H_5N_5$	1-Methyl-3,4-dicyano-5-aminopyrazole	232		

$C_6H_5N_5O_2$	2-Amino-3-cyano-5-oximinomethylpyrazine 1-oxide	537	537	
$C_6H_6N_2O$	2-Amino-3-cyano-4-methylfuran	258	258	
$C_6H_6N_2S$	2-Amino-3-cyano-4-methylthiophene	74, 91, 334, 341	91, 334, 341	
$C_6H_6N_4$	4-Amino-5-cyano-6-methylpyrimidine	22, 444	19, 22, 27, 444	
$C_6H_6N_4$	2-Methyl-4-amino-5-cyanopyrimidine	42, 49, 50, 63, 161, 196, 203, 375, 376, 378, 380, 384, 387, 389, 411, 556	27, 42, 46, 49, 55, 161, 163, 196, 203, 362, 372, 392, 393, 444, 467	46, 57, 203, 255, 329, 360, 377, 379, 381, 382, 384–386, 388, 390, 391, 396, 397, 399, 401, 422, 462, 474
$C_6H_6N_4$	2-Amino-3-cyano-5-methylpyrazine	562	562	
$C_6H_6N_4O$	2-Methoxy-4-amino-5-cyanopyrimidine	22, 53	22, 53	
$C_6H_6N_4O$	3-Methyl-5-cyanocytosine	43		
$C_6H_6N_4O$	2-Amino-3-cyano-5-methylpyrazine 1-oxide	551	551	
$C_6H_6N_4S$	2-Methylmercapto-4-amino-5-cyanopyrimidine	147, 246, 404	147, 369, 404, 411–413	368
$C_6H_6N_6$	1,2,5-Triamino-3,4-dicyanopyrrole	365, 402	365	
$C_6H_7ClN_4$	1-(β-Chloroethyl)-4-cyano-5-aminopyrazole	407		368
$C_6H_7N_3$	2-Amino-3-cyano-4-methylpyrrole	124, 175	10	
$C_6H_7N_3$	1,2,5,6-Tetrahydro-3-cyano-4-aminopyridine	219		
$C_6H_7N_3O$	3-Ethyl-4-cyano-5-aminoisoxazole	11	11, 459	
$C_6H_7N_3O$	2-Ethyl-4-cyano-5-aminooxazole	32, 335		
$C_6H_7N_3O$	1-Methyl-2-oxo-3-amino-4-cyano-3-pyrroline	539		

(continued)

TABLE XXX (*continued*)

Empirical formula	Compound	Preparation	References to use as synthetic intermediate	Other
$C_6H_7N_3OS$	3-Amino-4-cyano-5-ethoxyisothiazole	156		
$C_6H_7N_3O_2$	3-Ethoxy-4-cyano-5-aminoisoxazole	105		
$C_6H_7N_3O_3$	3-(β-Hydroxyethoxy)-4-cyano-5-aminoisoxazole	105		
$C_6H_7N_5$	2,4-Diamino-5-cyano-6-methylpyrimidine	53, 122, 467		
$C_6H_7N_5$	2-Methyl-4,6-diamino-5-cyanopyrimidine	63, 338, 366	338	
$C_6H_7N_5$	2-Methylamino-4-amino-5-cyanopyrimidine	143, 163	143, 163	
$C_6H_7N_5O$	1-Carbamoyl-3-methyl-4-cyano-5-aminopyrazole	232		
$C_6H_7N_5OS$	1-Carbamoyl-3-methylmercapto-4-cyano-5-aminopyrazole	232, 406		
$C_6H_7N_5O_3S$	1-Carbamoyl-3-methylsulfonyl-4-cyano-5-aminopyrazole	232, 406		
$C_6H_7N_5S$	2,4-Diamino-5-cyano-6-methylmercaptopyrimidine		141	
$C_6H_8N_2$	1-Amino-2-cyanocyclopentene	59, 99, 194, 195, 233, 495	59, 95, 96, 179, 195, 233, 284, 288, 535	34, 66, 99, 236–239, 281, 284
$C_6H_8N_4$	1,3-Dimethyl-4-cyano-5-aminopyrazole	3	3, 25	
$C_6H_8N_4$	1,2-Dimethyl-4-cyano-5-aminoimidazole	114	114	
$C_6H_8N_4$	1-Ethyl-4-cyano-5-aminopyrazole	532	532	
$C_6H_8N_4O$	1-(β-Hydroxyethyl)-4-cyano-5-aminopyrazole	3, 150, 407	3, 150, 160, 168	368
$C_6H_8N_4O$	3-Ethoxy-4-cyano-5-aminopyrazole	105		
$C_6H_8N_4O$	1-Nitroso-3-cyano-4-amino-1,2,5,6-tetrahydropyridine	536	536	

$C_6H_8N_4O_2$	3-(β-Hydroxyethoxy)-4-cyano-5-aminopyrazole	105	105	
$C_6H_8N_4S$	1-Methyl-2-methylmercapto-4-amino-5-cyanoimidazole	339	339	
$C_6H_9N_3$	3-Cyano-4-amino-1,2,5,6-tetrahydropyridine	536	536	
$C_7H_2Br_2N_4$	2-Amino-4,6-dibromo-3,5-dicyanopyridine	106		
$C_7H_2F_4N_2$	2-Amino-3,4,5,6-tetrafluorobenzonitrile	546	546	
$C_7H_2F_6N_4$	2,4-bis(Trifluoromethyl)-5-cyano-6-aminopyrimidine	345		
$C_7H_3BrN_4$	2-Amino-6-bromo-3,5-dicyanopyridine	106		
$C_7H_3ClN_4$	2-Amino-3,5-dicyano-6-chloropyridine	106	133	
$C_7H_3Cl_3N_2$	2-Amino-3,4,5-trichlorobenzonitrile			302
$C_7H_3F_5N_4$	2-Pentafluoroethyl-4-amino-5-cyanopyrimidine	347	347	
$C_7H_4BrN_5$	2,4-Diamino-6-bromo-3,5-dicyanopyridine	106		
$C_7H_4BrN_5O$	1-Bromoacetyl-3,4-dicyano-5-aminopyrazole	232		
$C_7H_4Br_2N_2$	2-Amino-3,5-dibromobenzonitrile	79	79	302
$C_7H_4ClN_3O_2$	2-Amino-3-chloro-5-nitrobenzonitrile			295, 297, 298
$C_7H_4ClN_5$	2,4-Diamino-6-chloro-3,5-dicyanopyridine	106		
$C_7H_4Cl_2N_2$	2-Amino-4,5-dichlorobenzonitrile	563	563	
$C_7H_4IN_5$	2,4-Diamino-6-iodo-3,5-dicyanopyridine	106		
$C_7H_4N_4$	2-Amino-3,5-dicyanopyridine	133, 251	133	
$C_7H_4N_4O_4$	2-Amino-3,5-dinitrobenzonitrile	315		296, 299, 301, 315
$C_7H_5BrN_2$	2-Amino-5-bromobenzonitrile	17, 320, 468	17, 93, 94, 126, 320	
$C_7H_5BrN_4$	2-Amino-5-bromo-3,4-dicyano-1-methylpyrrole	104		

(continued)

TABLE XXX (*continued*)

Empirical formula	Compound	Preparation	References to use as synthetic intermediate	Other
$C_7H_5ClN_2$	2-Amino-4-chlorobenzonitrile	132, 563	129, 132, 563	248, 302, 306, 307
$C_7H_5ClN_2$	2-Amino-5-chlorobenzonitrile	1, 67, 433, 563	67, 563	
$C_7H_5ClN_2$	2-Amino-6-chlorobenzonitrile	312, 319	319	312, 319
$C_7H_5FN_2$	2-Amino-5-fluorobenzonitrile	320	320	
$C_7H_5N_3O_2$	2-Amino-3-nitrobenzonitrile	23		
$C_7H_5N_3O_2$	2-Amino-5-nitrobenzonitrile	1, 23, 37, 186	17, 21, 538, 554	295, 296, 300, 309, 310
$C_7H_5N_5$	2,6-Diamino-3,5-dicyanopyridine	106		
$C_7H_5N_5O$	1-Acetyl-3,4-dicyano-5-aminopyrazole	232		
$C_7H_6N_2$	2-Aminobenzonitrile	1, 39, 40, 62, 67, 123, 128, 137, 183, 197, 240, 253, 256, 264, 305, 317, 318, 332, 333, 433, 475, 493, 547, 553	15, 17, 27, 40, 45, 61, 67, 93–96, 118, 119, 123, 127, 128, 137, 142, 158, 244, 247, 249, 250, 252, 264, 320, 332, 340, 442, 459, 538, 563	129, 134, 183, 213, 214, 240, 256, 263, 308, 311, 314, 321, 322, 324, 326–328, 330, 331, 333
$C_7H_6N_2(^{14}CN)$	2-Aminobenzonitrile (^{14}CN)	554	554	
$C_7H_6N_4$	2-Vinyl-4-amino-5-cyanopyrimidine	51		
$C_7H_6N_4O_2S$	2-Carboxymethylmercapto-4-amino-5-cyanopyrimidine		413	
$C_7H_6N_4S$	2-Amino-3,4-dicyano-5-methylmercapto-pyrrole	104		
$C_7H_6N_6$	2-Amino-6-hydrazino-3,5-dicyanopyridine	125	125	

$C_7H_6N_6$	3,6-Diamino-5-cyanopyrazolo[3,4-*b*]-pyridine	125		
$C_7H_6N_6O$	1-*N*-Methylcarbamoyl-3,4-dicyano-5-aminopyrazole	232		
$C_7H_6N_6O$	1-Carbamoyl-3-cyanomethyl-4-cyano-5-aminopyrazole	232		
$C_7H_6N_6O$	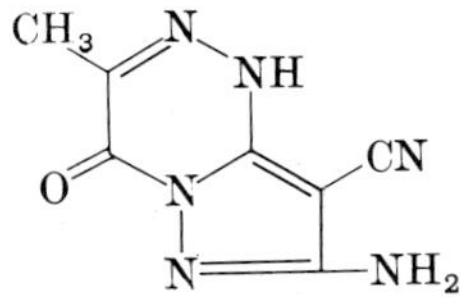	507		
$C_7H_7ClN_4$	2-(β-Chloroethyl)-4-amino-5-cyanopyrimidine	51		
$C_7H_7N_3$	2,5-Diaminobenzonitrile	553		
$C_7H_7N_3$	2-Amino-3-cyano-6-methylpyridine	87	87	
$C_7H_7N_5$	1-Methyl-3-cyanomethyl-4-cyano-5-aminopyrazole	25	25	
$C_7H_8ClN_5$	2-Amino-3-cyano-5-chloro-6-dimethylaminopyrazine	562	562	
$C_7H_8ClN_5$	2-Amino-3-cyano-5-chloro-6-ethylaminopyrazine	562	562	
$C_7H_8N_2O$	2-Amino-3-cyano-4,5-dimethylfuran	258	258	
$C_7H_8N_2S$	2,3-Dimethyl-4-cyano-5-aminothiophene	74, 113, 334	91, 334	
$C_7H_8N_4$	2,6-Dimethyl-4-amino-5-cyanopyrimidine	36, 48, 75, 122, 367, 467	48, 122, 367	364, 399
$C_7H_8N_4$	2-Ethyl-4-amino-5-cyanopyrimidine	75	55, 254	326
$C_7H_8N_4O$	3-Ethyl-5-cyanocytosine	43		

(*continued*)

TABLE XXX (*continued*)

Empirical formula	Compound	Preparation	References to use as synthetic intermediate	Other
$C_7H_8N_4O$	2-(β-Hydroxyethyl)-4-amino-5-cyano-pyrimidine	51		
$C_7H_8N_4OS$	2-Amino-3(or 4)-carbamoyl-4(or 3)-cyano-5-methylmercaptopyrrole	104		
$C_7H_8N_4O_2$	3-Carbethoxy-4-cyano-5-aminopyrazole	457	457	
$C_7H_8N_4S$	2-Methylmercapto-4-amino-5-cyano-6-methylpyrimidine	22, 416	22	
$C_7H_8N_4S$	2-Ethylmercapto-4-amino-5-cyano-pyrimidine	53, 75, 161, 246, 363	143, 161, 163, 363	
$C_7H_8N_4S_2$	2,6-bis(Methylmercapto)-4-amino-5-cyanopyrimidine	105		
$C_7H_8N_6$	1-Methylamino-2,5-diamino-3,4-dicyanopyrrole	365		
$C_7H_8N_6O_2S$	1-*N*,*N*-Dimethylaminosulphonyl-3,4-dicyano-5-aminopyrazole	232		
$C_7H_9N_3$	2-Amino-3-cyano-4,5-dimethylpyrrole	175		
$C_7H_9N_3O$	1-Acetyl-3-amino-4-cyano-3-pyrroline	76	76, 479	
$C_7H_9N_5$	2,4-Diamino-5-cyano-6-ethylpyrimidine	122	122	
$C_7H_9N_5$	2-Dimethylamino-4-amino-5-cyano-pyrimidine	143, 147, 163, 413, 414, 416	139, 143, 147, 163	
$C_7H_9N_5O$	2,4-Diamino-5-cyano-6-ethoxypyrimidine	105, 561		
$C_7H_9N_5O$	4,6-Diamino-5-cyano-2-ethoxypyrimidine	561		
$C_7H_9N_5O_2$	2,4-Diamino-5-cyano-6-(β-hydroxyethoxy)-pyrimidine	105		

$C_7H_9N_5O_2$	1-Carbamoyl-3-ethoxy-4-cyano-5-aminopyrazole	232		
$C_7H_{10}N_2$	1-Amino-2-cyanocyclohexene	89, 101, 495, 542	95, 96, 101, 459, 535, 542	89
$C_7H_{10}N_4$	1-Ethyl-2-methyl-4-amino-5-cyanoimidazole	98	98	
$C_7H_{10}N_4$	1-Methyl-3-ethyl-4-cyano-5-aminopyrazole	532	532	
$C_7H_{10}N_4$	1-*n*-Propyl-4-cyano-5-aminopyrazole	532	532	
$C_7H_{10}N_4$	1-*iso*-Propyl-4-cyano-5-aminopyrazole	148, 149, 373, 403	115, 140, 148, 149, 373, 403	
$C_7H_{10}N_4O$	1-(2-Methoxyethyl)-4-cyano-5-aminopyrazole	532	532	
$C_7H_{10}N_6$	2,4-Diamino-5-cyano-6-dimethylaminopyrimidine	141	141	
$C_7H_{11}N_3$	1,2-Dimethyl-3-amino-4-cyano-3-pyrroline	88	13, 88, 116	16, 88
$C_7H_{11}N_3$	1,2,5,6-Tetrahydro-3-cyano-4-amino-1-methylpyridine	88, 112, 117, 441	112, 117	88, 441
$C_8H_2BrN_5$	2-Amino-6-bromo-3,4,5-tricyanopyridine	106		
$C_8H_2ClN_5$	2-Amino-6-chloro-3,4,5-tricyanopyridine	106		
$C_8H_3F_7N_4$	2-*n*-Heptafluoropropyl-4-amino-5-cyanopyrimidine	347	347	
$C_8H_3N_5O_5$	2,4-Dicyano-3,5-dinitro-6-hydroxyaniline	83		
$C_8H_4ClN_3$	2-Amino-3-cyano-5-chlorobenzonitrile			303
$C_8H_4ClN_3$	2-Amino-3-chloro-5-cyanobenzonitrile			303
$C_8H_4N_4S_2$	2,5-Dicyano-3,4-diaminothieno[2,3-*b*]-thiophene	171		
$C_8H_4N_6$	2,6-Diamino-3,4,5-tricyanopyridine	106		

(*continued*)

TABLE XXX (*continued*)

Empirical formula	Compound	Preparation	References to use as synthetic intermediate	Other
$C_8H_4N_6$	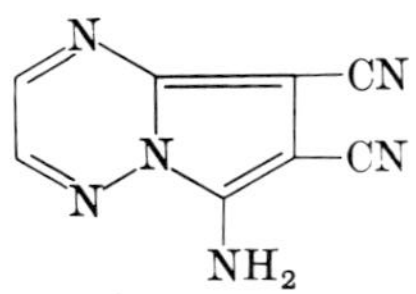	365		
$C_8H_5ClN_4$	2-Methyl-3-amino-4,5-dicyano-6-chloropyridine	77, 86	77, 86	
$C_8H_5N_3$	2,4-Dicyanoaniline	278		303, 309
$C_8H_6N_2O$	2-Amino-3-cyanotropone	9		
$C_8H_6N_2O_2$	2-Amino-3-cyanobenzoic acid	1		
$C_8H_6N_4O$	2-Amino-3,5-dicyano-6-methoxypyridine	246		
$C_8H_6N_4O_2S \cdot H_2O$	2-Amino-5-carboxymethylmercapto-3,4-dicyanopyrrole monohydrate	104		
$C_8H_6N_4O_3$	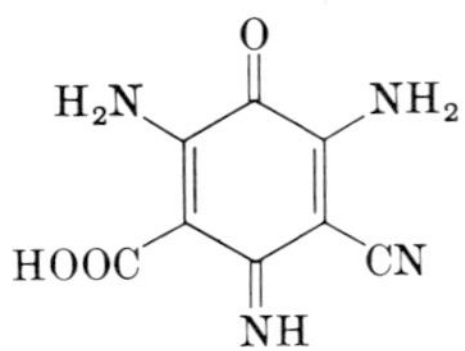	83		
$C_8H_6N_4S$	5-Amino-3,4-dicyanopyrrolo[2,1-*b*]-thiazolidine	104		
$C_8H_6N_4S$	2-Amino-3,5-dicyano-6-methylmercapto-pyridine	246		

$C_8H_6N_5O_3K$	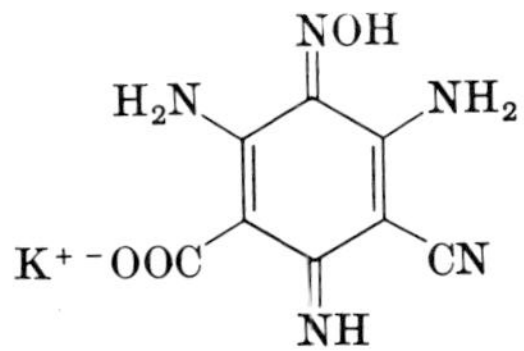	83		
$C_8H_6N_6$	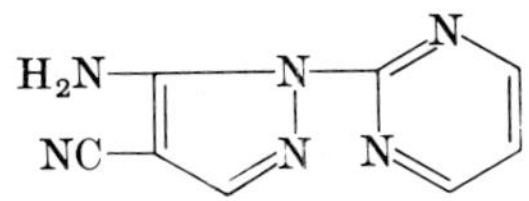	419		
$C_8H_7ClN_2$	2-Amino-4-chloro-6-methylbenzonitrile		304	304
$C_8H_7N_5O_2$	1-Carbethoxy-3,4-dicyano-5-aminopyrazole	232		
$C_8H_8ClN_5$	2-Amino-3-cyano-5-chloro-6-allylamino-pyrazine	562	562	
$C_8H_8N_2$	2-Amino-3-methylbenzonitrile	274		271, 275
$C_8H_8N_2$	2-Amino-4-methylbenzonitrile	198, 199, 291, 293, 294, 563	198, 199, 291–294, 320, 563	
$C_8H_8N_2$	2-Amino-5-methylbenzonitrile	1, 174, 184, 320, 433, 563	93, 94, 174, 184, 320, 538, 563	
$C_8H_8N_2$	2-Amino-6-methylbenzonitrile	272, 274, 320, 563	272, 274, 313, 563	
$C_8H_8N_2O$	2-Amino-4-methoxybenzonitrile	131, 240, 250, 563	131, 538, 563	240
$C_8H_8N_2O$	2-Amino-5-methoxybenzonitrile	131, 240, 318, 468	93, 94, 131, 318, 538	240
$C_8H_8N_2O$	2-Amino-6-methoxybenzonitrile	121, 245, 563	245, 563	
$C_8H_8N_2O_2$	2-Amino-3-cyano-4-acetyl-5-methylfuran	276		276
$C_8H_8N_4$	1,4-diamino-2,5-dicyanocyclohexa-1,4-diene	440		440
$C_8H_8N_4$	2-Amino-3-cyano-5-cyclopropylpyrazine	562	562	
$C_8H_8N_4O$	3-Allyl-5-cyanocytosine	43		

(*continued*)

TABLE XXX (*continued*)

Empirical formula	Compound	Preparation	References to use as synthetic intermediate	Other
$C_8H_8N_4O_3$	2,4,6-Triamino-3-cyano-5-hydroxybenzoic acid	83		
$C_8H_8N_4OS$	2-Amino-3,4-dicyano-5-(β-hydroxyethylmercapto)pyrrole	104		
$C_8H_8N_4S$	1-Methyl-2-amino-3,4-dicyano-5-methylmercaptopyrrole	494	494	
$C_8H_8N_4S$	2-Amino-3,4-dicyano-5-ethylmercaptopyrrole	10	10	
$C_8H_8N_6$	1-Methyl-3,6-diamino-5-cyanopyrazolo-[3,4-*b*]pyridine	125		
$C_8H_8N_6O$	1-*N*-Ethylcarbamoyl-3,4-dicyano-5-aminopyrazole	232		
$C_8H_8N_6O$	1-*N*,*N*-Dimethylcarbamoyl-3,4-dicyano-5-aminopyrazole	232		
$C_8H_9N_3$	2-Amino-3-cyano-4,6-dimethylpyridine	80, 108	80	
$C_8H_9N_3$	2-Amino-3-cyano-5,6-dimethylpyridine(?)	108		
$C_8H_9N_5$	2-Allylamino-4-amino-5-cyanopyrimidine	22, 143	22, 143	
$C_8H_9N_5O_2$	2,4,6-Triamino-3-hydroxy-5-carboxamidobenzonitrile	83		
$C_8H_9N_5O_2$	2,4,5,6-Tetraamino-3-cyanobenzoic acid	83		
$C_8H_{10}ClN_5$	2-Amino-3-cyano-5-chloro-6-isopropylaminopyrazine	562	562	

$C_8H_{10}ClN_5$	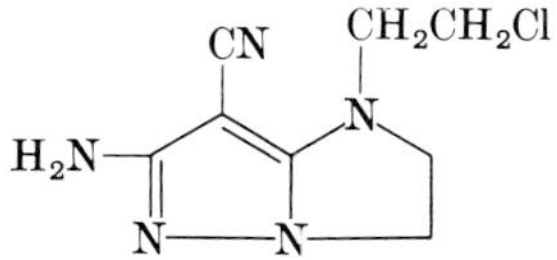	543	
$C_8H_{10}N_2$	*cis*-2-Cyano-3-aminobicyclo[3.2.0]-heptene-2	273	
$C_8H_{10}N_2O_2$	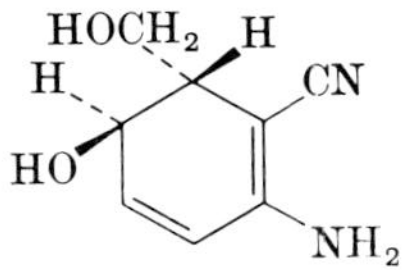	470	
$C_8H_{10}N_2S$	2-Amino-3-cyano-4-ethyl-5-methyl-thiophene	74	
$C_8H_{10}N_2S$	2-Amino-3-cyano-4-isopropylthiophene	47	
$C_8H_{10}N_4$	2-*n*-Propyl-4-amino-5-cyanopyrimidine	163, 409, 415	163
$C_8H_{10}N_4$	2-*iso*-Propyl-4-amino-5-cyanopyrimidine	415	
$C_8H_{10}N_4O$	3-*n*-Propyl-5-cyanocytosine	43	
$C_8H_{10}N_4O$	3-*iso*-Propyl-5-cyanocytosine	43	
$C_8H_{10}N_4O$	2-Methyl-4-amino-5-cyano-6-ethoxypyrimidine	63, 164	
$C_8H_{10}N_4O$	2-Amino-3-cyano-5-*n*-propylpyrazine 1-oxide	537	537
$C_8H_{10}N_4OS$	1-(β-Methoxyethyl)-2-thio-5-cyano-6-aminopyrimidine	35	
$C_8H_{10}N_4O_2$	3-(2-Methoxyethyl)-5-cyanocytosine	43	
$C_8H_{10}N_4O_2$	4-Amino-5-cyano-6-(β-hydroxyethoxy)-2-methylpyrimidine	105	

(continued)

TABLE XXX (*continued*)

Empirical formula	Compound	Preparation	References to use as synthetic intermediate	Other
$C_8H_{10}N_4O_2S$	2-Methylmercapto-4-amino-5-cyano-6-(β-hydroxyethoxy)pyrimidine	105		
$C_8H_{10}N_4S$	1-*iso*-Propyl-2-thio-5-cyano-6-amino-pyrimidine	35		
$C_8H_{10}N_4S$	2-Ethylmercapto-4-amino-5-cyano-6-methylpyrimidine	52		
$C_8H_{11}Cl_2N_5$	3-Amino-4-cyano-5-bis(2-chloroethyl)-aminopyrazole	543		
$C_8H_{11}N_3O$	1-Acetyl-2-methyl-3-amino-4-cyano-3-pyrroline	76	76, 479	
$C_8H_{11}N_3O$	1-Acetyl-3-cyano-4-amino-1,2,5,6-tetra-hydropyridine	536	536	
$C_8H_{11}N_5$	2-Dimethylamino-4-amino-5-cyano-6-methylpyrimidine	413, 416		
$C_8H_{11}N_5$	2-*n*-Propylamino-4-amino-5-cyano-pyrimidine	22	22	
$C_8H_{12}N_2$	1-Cyano-2-amino-3-methylcyclohexene	101	101	
$C_8H_{12}N_2$	1-Amino-2-cyano-5,5-dimethylcyclopentene	281, 495	281	281
$C_8H_{12}N_2$	1-Amino-2-cyanocycloheptene	190, 200, 202, 495, 500	187, 190, 200, 202, 270	
$C_8H_{12}N_4$	1-*sec*-Butyl-4-cyano-5-aminopyrazole	144, 155, 408	144, 155	
$C_8H_{12}N_4$	1-*n*-Butyl-4-cyano-5-aminopyrazole	170	170	

$C_8H_{13}N_2P$	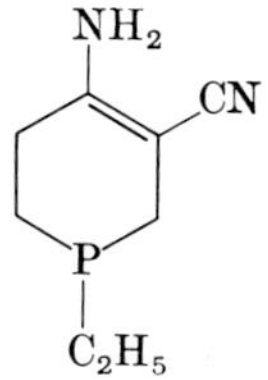	231, 417	231	
$C_8H_{13}N_3$	1-Ethyl-1,2,5,6-tetrahydro-3-cyano-4-aminopyridine	117, 441	117	
$C_8H_{13}N_3$	1,2,2-Trimethyl-3-amino-4-cyano-3-pyrroline	88	13, 16, 28, 88, 116	28
$C_8H_{13}N_3$	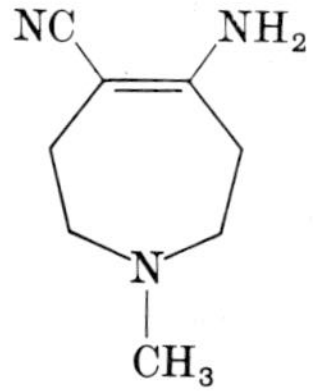	432		
$C_8H_{13}N_3$	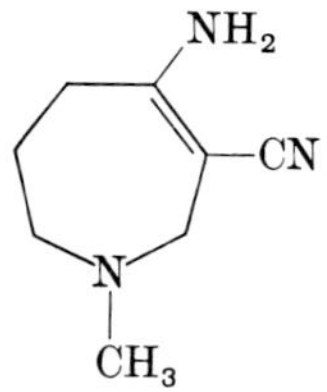	432		
$C_8H_{13}N_5$	1-(β-Dimethylaminoethyl)-4-cyano-5-aminopyrazole	160	160	

(continued)

TABLE XXX (*continued*)

Empirical formula	Compound	Preparation	References to use as synthetic intermediate	Other
$C_9H_2Cl_2N_4$	2,4,6-Tricyano-3,5-dichloroaniline	481, 482		
$C_9H_4BrN_5$	2-Amino-3,5-dicyano-4-cyanomethyl-6-bromopyridine	555		
$C_9H_4ClN_5$	2,4,6-Tricyano-3-chloro-5-aminoaniline	481, 482		
$C_9H_4ClN_5$	2-Amino-3,5-dicyano-4-cyanomethyl-6-chloropyridine	555	555	
$C_9H_6ClN_5$	1-*p*-Chlorophenyl-4-cyano-5-amino-1,2,3-triazole	343		
$C_9H_6Cl_6N_4O_2$	2,4-bis-(2,2,2-Trichloroethoxy)-5-cyano-6-aminopyrimidine	72		
$C_9H_6N_4$	2-Amino-3-cyanopyrido[2,3-*b*]pyridine	541	541	
$C_9H_6N_6$	2,4,6-Tricyano-1,3,5-triaminobenzene	481, 482		
$C_9H_6N_6$	2-Amino-6-methylamino-3,4,5-tricyanopyridine	106		
$C_9H_6N_6$	2,4-Diamino-3,5-dicyano-6-cyanomethylpyridine			364
$C_9H_6N_6$	2,4-bis(Cyanomethyl)-5-cyano-6-aminopyrimidine	342		342, 364
$C_9H_6N_6$	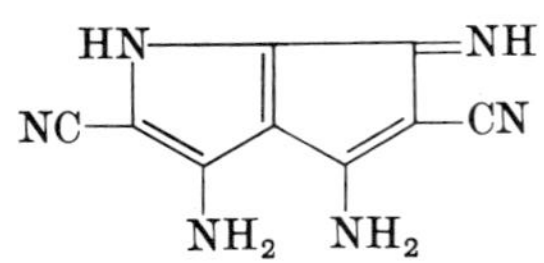	548		

$C_9H_6N_6$	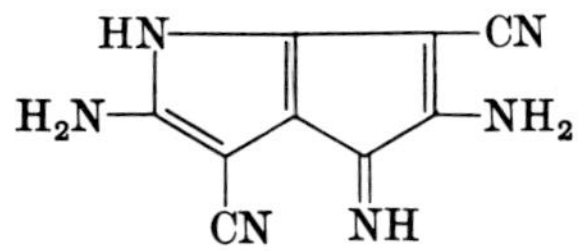	548	
$C_9H_6N_6O$	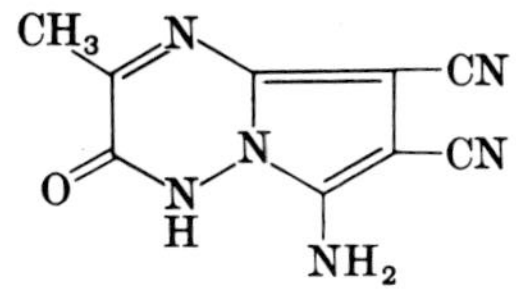	365	
$C_9H_6N_6O_2$	1-*p*-Nitrophenyl-4-cyano-5-amino-1,2,3-triazole	343	
$C_9H_6N_8$	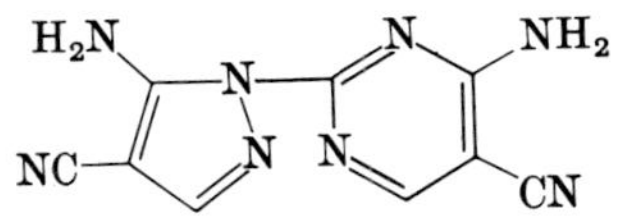	411, 412, 419	
$C_9H_7ClN_4O$	2-Amino-6-chloro-3,5-dicyano-4-ethoxypyridine	106	
$C_9H_7N_3$	2-Amino-3-methyl-5-cyanobenzonitrile		303
$C_9H_7N_5$	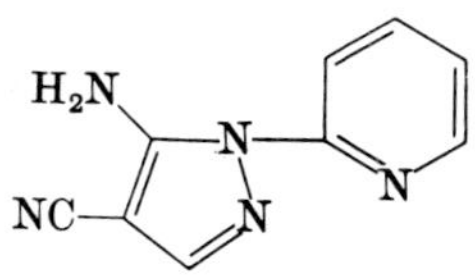	419	

(continued)

TABLE XXX (*continued*)

Empirical formula	Compound	Preparation	References to use as synthetic intermediate	Other
$C_9H_7N_5$	1-Phenyl-4-cyano-5-amino-1,2,3-triazole	343		
$C_9H_7N_5O$	7-Amino-6-cyano-5-methylpyrido[2,3-*d*]-pyrimidin-4(3*H*)-one	11		
$C_9H_7N_5S$	2,6-Diamino-4-(2-thienyl)-5-cyano-pyrimidine		141	
$C_9H_7N_5S$	2-(2-Thienyl)-4,6-diamino-5-cyano-pyrimidine		141	
$C_9H_8ClN_5$	2-Amino-6-chloro-3,5-dicyano-4-dimethylaminopyridine	106		
$C_9H_8N_2S_2$	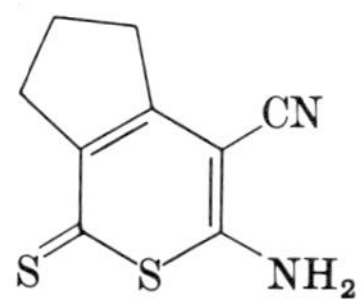	357		
$C_9H_8N_4O$	2-Amino-3,5-dicyano-6-ethoxypyridine	106		
$C_9H_8N_4S$	2-Amino-3,5-dicyano-6-ethylmercapto-pyridine	246	133	
$C_9H_8N_6$	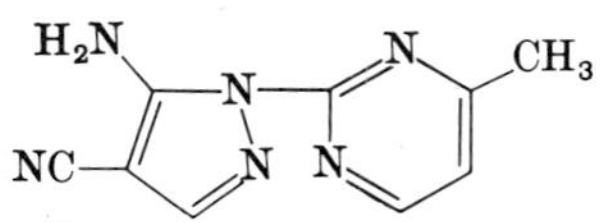	400, 419		
$C_9H_8N_6O$	1-Methyl-3-*N*-formamido-6-amino-5-cyanopyrazolo[3,4-*b*]pyridine	125		

$C_{10}H_6BrN_3O$	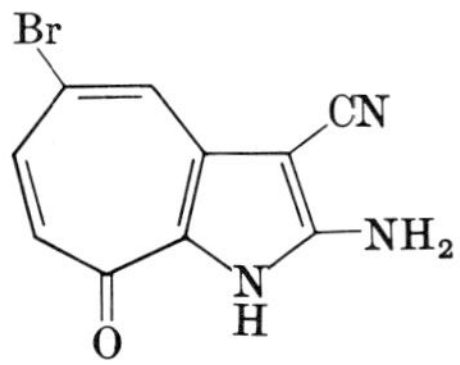	421		
$C_{10}H_6ClN_3O$	3-*p*-Chlorophenyl-4-cyano-5-aminoisoxazole	177		
$C_{10}H_6N_2O_3$	3-Amino-4-cyano-5-methylphthalic anhydride	258		
$C_{10}H_7BrN_4$	1-*p*-Bromophenyl-4-cyano-5-aminopyrazole	3	3, 168	
$C_{10}H_7ClN_4$	1-*p*-Chlorophenyl-4-cyano-5-aminopyrazole	3, 157	3, 157, 168, 169	
$C_{10}H_7ClN_4$	1-*o*-Chlorophenyl-4-cyano-5-aminopyrazole	3, 157	3, 157, 168	
$C_{10}H_7N_3$	2-Amino-3-cyanoquinoline	31, 172, 260, 261	14, 27, 31, 93, 94, 172	
$C_{10}H_7N_3O$	3-Phenyl-4-cyano-5-aminoisoxazole	6, 177	11, 177	
$C_{10}H_7N_3O$	2-Phenyl-4-cyano-5-aminooxazole	335, 337	459	350
$C_{10}H_7N_3O$	2-Amino-3-cyano-8-hydroxyquinoline	420		
$C_{10}H_7N_3O$	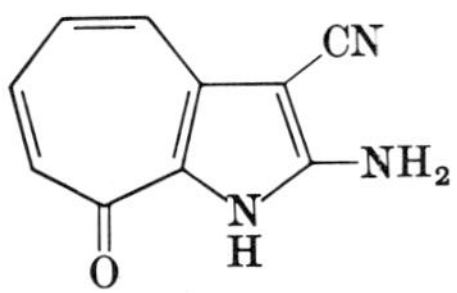	421		
$C_{10}H_7N_3O$	2-Amino-3-cyanoquinoline 1-oxide	465		
$C_{10}H_7N_3S$	2-Phenyl-4-amino-5-cyanothiazole	473		

(*continued*)

TABLE XXX (*continued*)

Empirical formula	Compound	Preparation	References to use as synthetic intermediate	Other
$C_{10}H_7N_3S_2$	3-Phenyl-4-amino-5-cyanothiazoline-2-thione	38		
$C_{10}H_7N_5O$	1-Amino-2,4-dicyano-5-hydroxy-7-methylpyrazolo[2,3-*a*]pyridine, or its 5-methyl-7-hydroxy isomer	26		
$C_{10}H_7N_5O$	2-Amino-6-ethoxy-3,4,5-tricyanopyridine	106		
$C_{10}H_7N_5O_2$	1-*p*-Nitrophenyl-4-cyano-5-aminopyrazole	3	3, 168	
$C_{10}H_7N_5O_2$	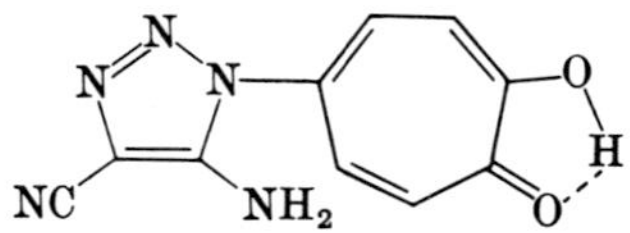	529		
$C_{10}H_8BrN_5$	"Bromodiaminocyanomethylnaphthyridine"	106		
$C_{10}H_8ClN_5$	"Chlorocyanodiaminomethylnaphthyridine"	106		
$C_{10}H_8N_2$	2-Amino-3-cyanoindene	64, 191, 454, 455	64, 191, 449, 450, 452, 453	449, 450, 454, 455
$C_{10}H_8N_4$	1-Phenyl-4-cyano-5-aminopyrazole	3, 145, 157	3, 17, 93–96, 145, 157, 168, 169	351
$C_{10}H_8N_4$	3-Phenyl-4-cyano-5-aminopyrazole	33		
$C_{10}H_8N_6$	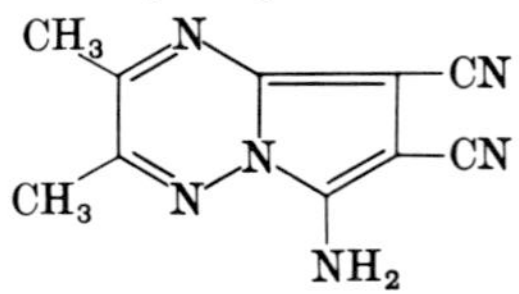	365		

$C_{10}H_8N_6O$	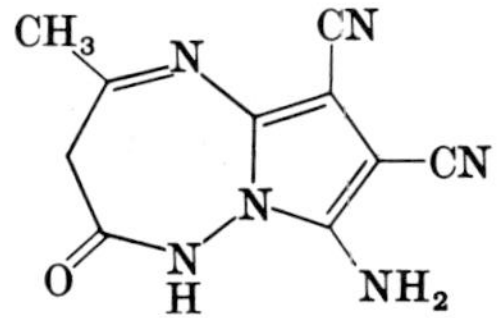	365		
$C_{10}H_9N_3$	2-Amino-3,6-dimethyl-5-cyanobenzonitrile			303
$C_{10}H_9N_3$	2-Amino-3-cyano-4,6-dimethylbenzonitrile			328
$C_{10}H_9N_5$	1-Benzyl-4-cyano-5-amino-1,2,3-triazole	343, 344	344	343
$C_{10}H_9N_5O_2$	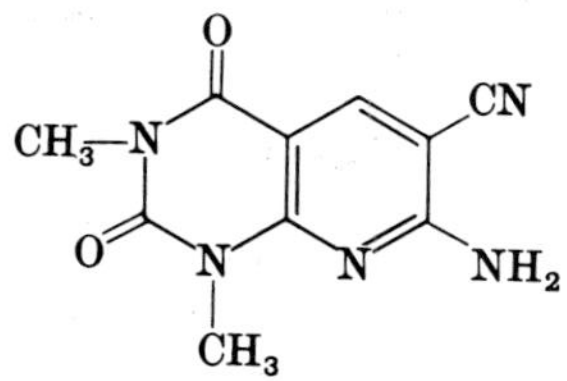	557		
$C_{10}H_{10}N_2$	4-Cyano-5-aminoindane	313	313	
$C_{10}H_{10}N_2O_2$	3-Amino-2-cyanobenzyl acetate	470		
$C_{10}H_{10}N_2S_2$	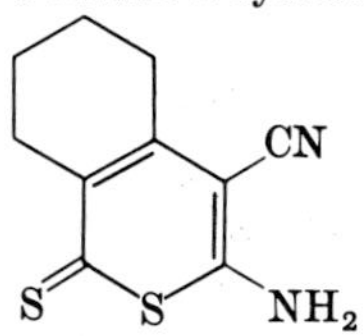	357		
$C_{10}H_{10}N_6$	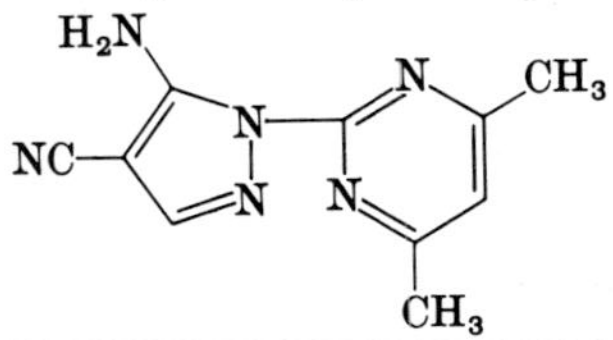	400, 419		

(continued)

TABLE XXX (*continued*)

Empirical formula	Compound	Preparation	References to use as synthetic intermediate	Other
$C_{10}H_{10}N_6$	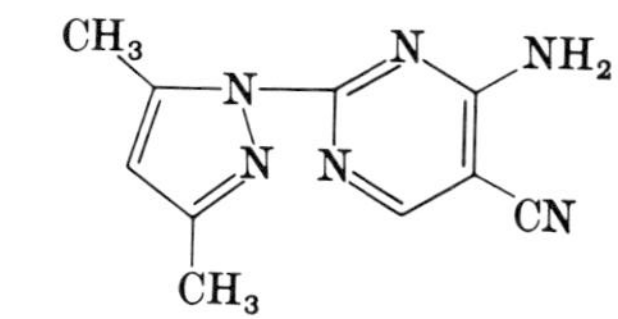	400, 411, 412, 419		
$C_{10}H_{10}N_6O$	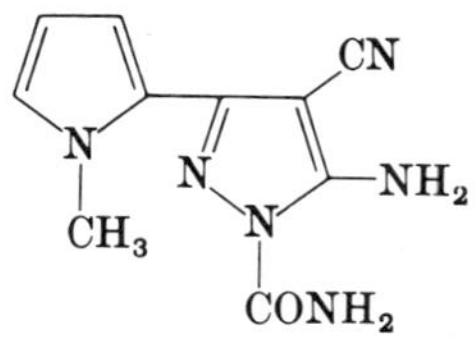	232		
$C_{10}H_{10}N_6O$	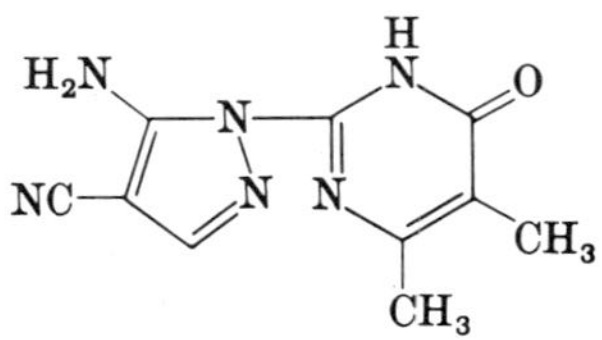	411, 412		
$C_{10}H_{10}N_6O_3$	1-*N*-Carbethoxymethylcarbamoyl-3,4-dicyano-5-aminopyrazole	232		

$C_{10}H_{10}N_6S$	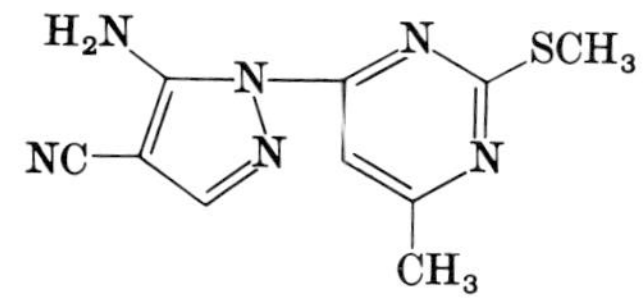	419	
$C_{10}H_{11}N_3$	2-Amino-3-cyano-5,6,7,8-tetrahydroquinoline	108	
$C_{10}H_{11}N_3$	3-Amino-4-cyano-5,6,7,8-tetrahydroisoquinoline	509	509
$C_{10}H_{12}ClN_5$	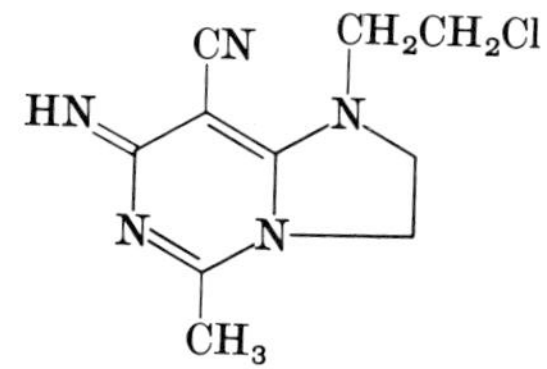	543	
$C_{10}H_{12}N_2$	*trans*-2-Amino-3-cyano-4,7,8,9-tetrahydroindene	273	273
$C_{10}H_{12}N_2$	*cis*-2-Amino-3-cyano-4,7,8,9-tetrahydroindene	273	273
$C_{10}H_{12}N_2OS$	2-Amino-3-cyano-6-methoxy-4,5,6,7-tetrahydrothianaphthene	91, 341	91, 341
$C_{10}H_{12}N_4$	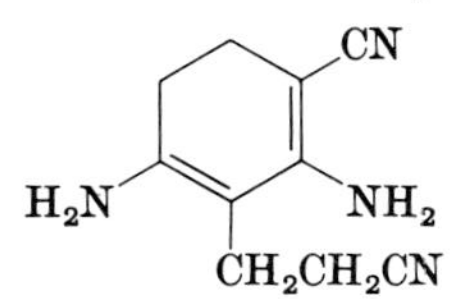	440	

(*continued*)

TABLE XXX (*continued*)

Empirical formula	Compound	Preparation	References to use as synthetic intermediate	Other
$C_{10}H_{12}N_6O$	1-*n*-Butylcarbamoyl-3,4-dicyano-5-aminopyrazole	232		
$C_{10}H_{12}N_6O$	1-*N*,*N*-Diethylcarbamoyl-3,4-dicyano-5-aminopyrazole	232		
$C_{10}H_{13}N_5$	2-Piperidyl-4-amino-5-cyanopyrimidine	22, 143, 416	22, 143	
$C_{10}H_{14}N_2$	1-Amino-2-cyano-3,5,5-trimethylcyclohexa-1,3-diene			89
$C_{10}H_{14}N_4$	1-Cyclohexyl-4-cyano-5-aminopyrazole	144, 149, 155, 170, 408	144, 149, 155, 170	
$C_{10}H_{14}N_4$	2-*n*-Pentyl-4-amino-5-cyanopyrimidine	415		
$C_{10}H_{14}N_4O$	3-*iso*-Pentyl-5-cyanocytosine	43		
$C_{10}H_{14}N_4O$	3-*n*-Pentyl-5-cyanocytosine	43, 361		
$C_{10}H_{14}N_4OS$	1-(β-Methoxyethyl)-2-thio-4-ethyl-5-cyano-6-aminopyrimidine	35		
$C_{10}H_{14}N_4S$	1-*n*-Butyl-2-thio-4-methyl-5-cyano-6-aminopyrimidine	35		
$C_{10}H_{14}N_4S$	1-*iso*-Propyl-2-thio-4-ethyl-5-cyano-6-aminopyrimidine	35		
$C_{10}H_{14}N_6$	2-*N*-Methylpiperazyl-4-amino-5-cyano-pyrimidine	22, 416	22	
$C_{10}H_{15}N_5O$	1-(β-Morpholinoethyl)-4-cyano-5-aminopyrazole	160	160	
$C_{10}H_{16}N_2$	1-Amino-2-cyano-5,5-diethylcyclopentene	281	281	
$C_{10}H_{16}N_2$	1-Amino-2-cyano-7,7-dimethylcycloheptene	495		

$C_{10}H_{16}N_2O$	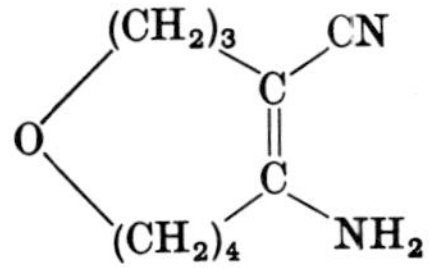	189		
$C_{10}H_{16}N_4$	1-*n*-Hexyl-4-cyano-5-aminopyrazole	170	170	
$C_{10}H_{17}N_3$	1-*n*-Butyl-2-methyl-3-amino-4-cyano-3-pyrroline	88	88	
$C_{10}H_{17}N_3O$	2-Phenyl-4-cyano-5-aminooxazole	32		
$C_{10}H_{17}N_5$	1-(β-Diethylaminoethyl)-4-cyano-5-aminopyrazole	160	160	
$C_{11}H_3N_7$	2-Amino-6-dicyanomethyl-3,4,5-tricyanopyridine	106		89
$C_{11}H_6N_6$	1-Amino-2,4,4,6,6-pentacyanocyclohexene	204		66, 89
$C_{11}H_6N_6O_2$	1-*p*-Nitrophenyl-3,4-dicyano-5-aminopyrazole	232		
$C_{11}H_7ClN_4$	2-Phenyl-4-amino-5-cyano-6-chloropyrimidine	458	458	
$C_{11}H_7ClN_4O$	3-*p*-Chlorophenyl-5-cyanocytosine	43		
$C_{11}H_7N_5$	1-Phenyl-3,4-dicyano-5-aminopyrazole	232		
$C_{11}H_7N_7S$	2-(2-Thienyl)-4,7-diamino-6-cyanopteridine	439		
$C_{11}H_7N_7S$	2-(3-Thienyl)-4,7-diamino-6-cyanopteridine	439		
$C_{11}H_8ClN_5$	2,4-Diamino-5-cyano-6-*p*-chlorophenylpyrimidine	33		
$C_{11}H_8Cl_2N_6$	2,4-Diamino-5-cyano-6-(3,4-dichlorophenyl)pyrimidine	348		
$C_{11}H_8N_2$	2-Amino-3-cyanonaphthalene	65		
$C_{11}H_8N_2$	1-Amino-2-cyanonaphthalene	68, 70, 280		68, 70, 280

(continued)

TABLE XXX (*continued*)

Empirical formula	Compound	Preparation	References to use as synthetic intermediate	Other
$C_{11}H_8N_2$	1-Cyano-2-aminonaphthalene	152, 212, 267, 289, 340, 549	152, 267, 289, 340	
$C_{11}H_8N_2O_3$	3-Amino-4-cyano-5,6-dimethylphthalic anhydride	258		
$C_{11}H_8N_2S$	2-Amino-3-cyano-4-phenylthiophene	74		
$C_{11}H_8N_4$	2-Phenyl-4-amino-5-cyanopyrimidine	75		
$C_{11}H_8N_4$	4-Phenyl-5-cyano-6-aminopyrimidine	122	122	
$C_{11}H_8N_4O$	3-Phenyl-5-cyanocytosine	43		
$C_{11}H_8N_4O$	2-Amino-3-cyano-5-phenylpyrazine 1-oxide	537	537	
$C_{11}H_8N_4OS$	2-Benzoylmercapto-4-amino-5-cyano-pyrimidine	53		
$C_{11}H_8N_4S$	1-Phenyl-2-thio-5-cyano-6-aminopyrimidine	35		
$C_{11}H_8N_6O_3$	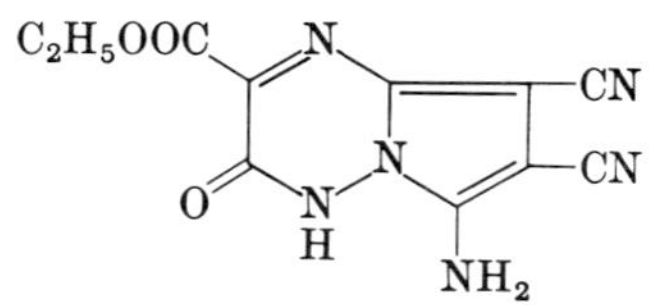	365		
$C_{11}H_9ClN_4O_2S$	1-*p*-Toluenesulfonyl-3-chloro-4-cyano-5-aminopyrazole	232, 406		
$C_{11}H_9BrN_6$	2,4-Diamino-5-cyano-6-*p*-bromophenyl-pyrimidine	348		
$C_{11}H_9IN_6$	2,4-Diamino-5-cyano-6-*p*-iodophenyl-pyrimidine	348		
$C_{11}H_9N_3$	1-Phenyl-2-amino-3-cyanopyrrole	370		

$C_{11}H_9N_3$	2-Amino-3-cyano-4-phenylpyrrole	175	10	
$C_{11}H_9N_3S$	2,5-Diamino-3-cyano-4-phenylthiophene	103, 398		
$C_{11}H_9N_3S$	3-Amino-4-cyano-5-benzylisothiazole	156		
$C_{11}H_9N_5$	2-Anilino-4-amino-5-cyanopyrimidine	22, 143	22, 143	
$C_{11}H_9N_5$	2,4-Diamino-5-cyano-6-phenylpyrimidine	33, 122	122, 141, 180	
$C_{11}H_9N_5$	2-Phenyl-4,6-diamino-5-cyanopyrimidine	105, 122, 366, 458	141, 458	
$C_{11}H_9N_5O$	2-Amino-6-iso-propoxy-3,4,5-tricyano-pyridine	106		
$C_{11}H_9N_5O$	1-Carbamoyl-3-phenyl-4-cyano-5-amino-pyrazole	232		
$C_{11}H_{10}BrN_5$	Aminobromocyanomethylmethylamino-naphthyridine	106		
$C_{11}H_{10}N_2$	1-Cyano-2-amino-3,4-dihydronaphthalene	549	549	549
$C_{11}H_{10}N_4$	1-*p*-Tolyl-4-cyano-5-aminopyrazole	3	3, 168	
$C_{11}H_{10}N_4$	1-Phenyl-3-methyl-4-cyano-5-amino-pyrazole	3	3, 25, 95, 96, 459	
$C_{11}H_{10}N_4$	1-Benzyl-4-amino-5-cyanoimidazole	71	71, 472	
$C_{11}H_{10}N_4$	1-Benzyl-4-cyano-5-aminopyrazole	170, 371	170	
$C_{11}H_{10}N_4S$	1-Phenyl-3-methylmercapto-4-cyano-5-aminopyrazole	266		
$C_{11}H_{10}N_6$	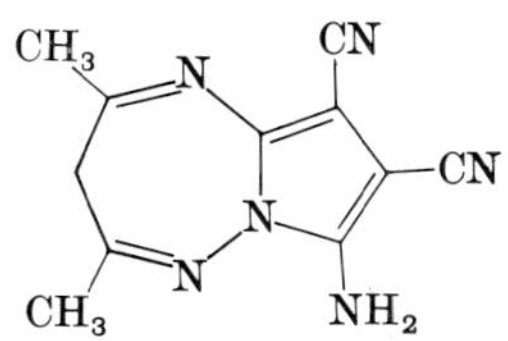	365		

(continued)

TABLE XXX (*continued*)

Empirical formula	Compound	Preparation	References to use as synthetic intermediate	Other
$C_{11}H_{10}N_6O$	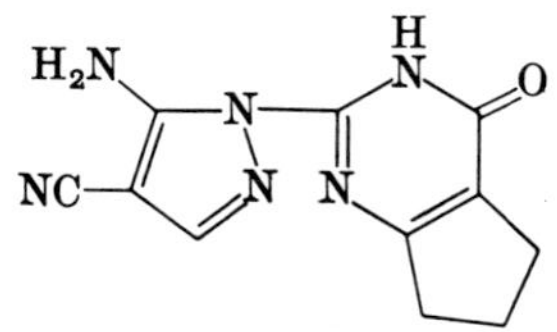	400, 419		
$C_{11}H_{10}N_6O_3$	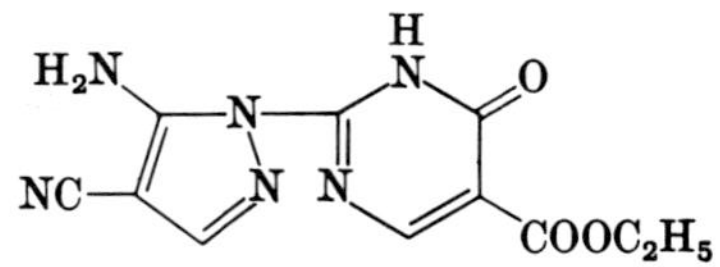	411, 419		
$C_{11}H_{11}N_7O_2$	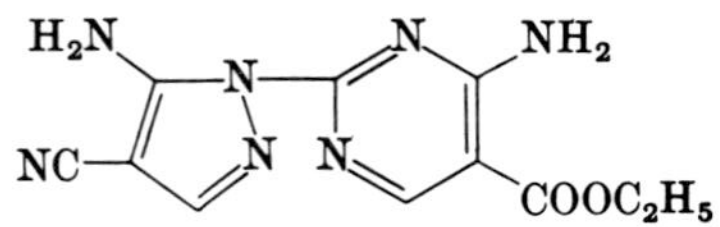	400, 412		
$C_{11}H_{11}N_7O_2$	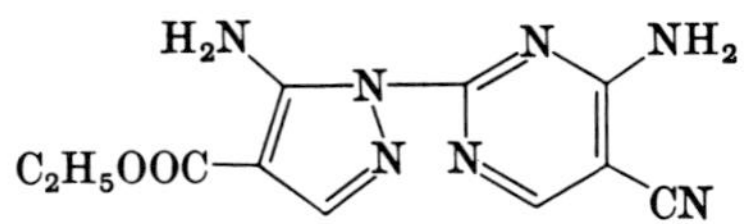	400, 411, 412, 419		

$C_{11}H_{12}N_2$	1-Cyano-2-amino-5,6,7,8-tetrahydro-naphthalene	313	313	
$C_{11}H_{12}N_2O_5$	2-Amino-3-cyano-4,5-dicarbethoxyfuran	5		
$C_{11}H_{14}N_2$	2-Amino-3-*t*-butylbenzonitrile	257		256, 257
$C_{11}H_{14}N_2$	*cis*-2-Amino-3-cyano-4,7,8,9-tetrahydro-8-methylindene	273	273	
$C_{11}H_{15}N_5$	2-Cyclohexylamino-4-amino-5-cyano-pyrimidine	22	22	
$C_{11}H_{16}N_2$	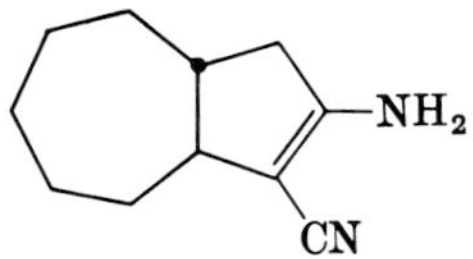	511	511	
$C_{11}H_{16}N_4O$	3-*n*-Hexyl-5-cyanocytosine	43		
$C_{11}H_{17}N_3$	1-Cyclohexyl-3-amino-4-cyano-3-pyrroline	100		100
$C_{11}H_{17}N_5$	2-*n*-Hexylamino-4-amino-5-cyano-pyrimidine	22, 413	22	
$C_{11}H_{17}N_5$	1-(β-Piperidinoethyl)-4-cyano-5-amino-pyrazole	160	160	
$C_{11}H_{18}N_2$	1-Amino-2-cyano-6,6-diethylcyclohexene	542	542	
$C_{11}H_{18}N_4$	1-(2-*n*-Heptyl)-4-cyano-5-aminopyrazole	408		
$C_{11}H_{18}N_6$	2-(2-*N*,*N*-Diethylaminoethylamino)-4-amino-5-cyanopyrimidine	413, 414		
$C_{11}H_{19}N_3$	1-*n*-Pentyl-2-methyl-3-amino-4-cyano-3-pyrroline	88	88	
$C_{12}H_6N_5S_2$	bis-(5-Amino-3,4-dicyano-2-pyrryl)-disulfide	104		

(continued)

TABLE XXX (*continued*)

Empirical formula	Compound	Preparation	References to use as synthetic intermediate	Other
$C_{12}H_6N_6$	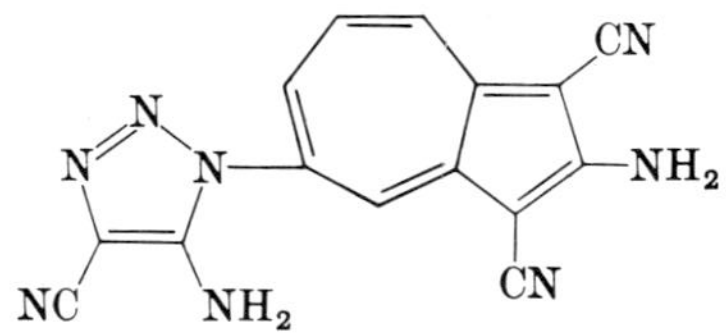	529		
$C_{12}H_7N_5O$	1-Benzoyl-3,4-dicyano-5-aminopyrazole	232		
$C_{12}H_8ClN_5O$	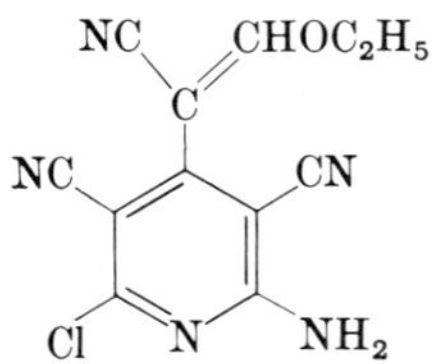	555		
$C_{12}H_8N_2S_2$	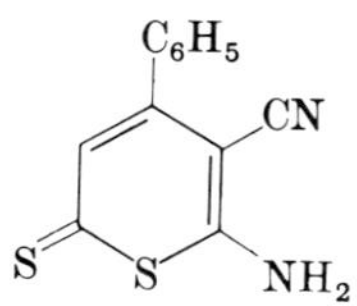	357		

$C_{12}H_8N_2S_3$	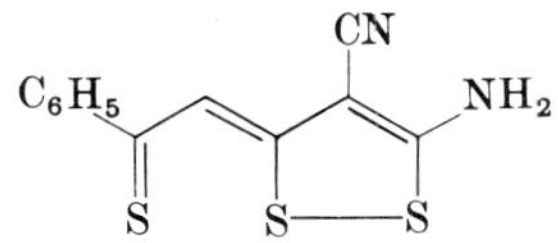	471, 489	
$C_{12}H_8N_4S$	2-Amino-3,4-dicyano-5-phenylmercapto-pyrrole	104	
$C_{12}H_8N_6O$	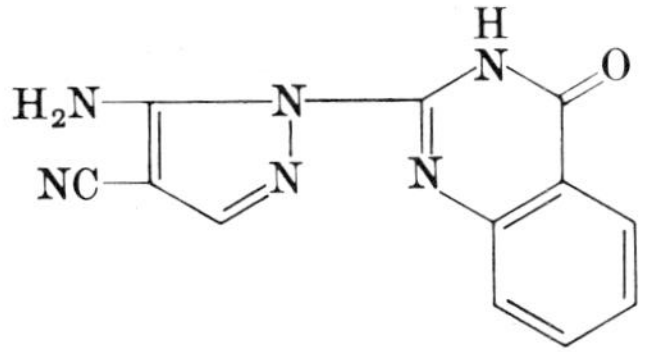	419, 423	
$C_{12}H_8N_6O$	1-*N*-Phenylcarbamoyl-3,4-dicyano-5-aminopyrazole	232	
$C_{12}H_8N_6O$	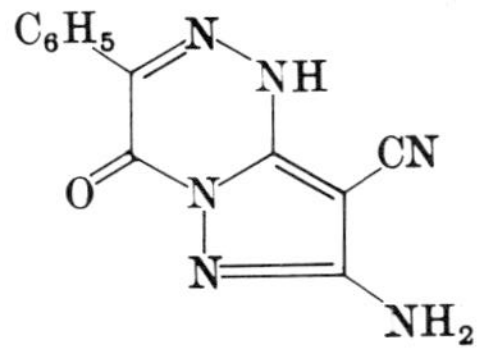	507	
$C_{12}H_9ClN_4$	2-Methyl-4-amino-5-cyano-6-*p*-chloro-phenylpyrimidine	33	
$C_{12}H_9N_5$	1-Phenyl-3-cyanomethyl-4-cyano-5-aminopyrazole	25, 84	25
$C_{12}H_9N_5O_2S$	1-*p*-Toluenesulphonyl-3,4-dicyano-5-aminopyrazole	232	

(*continued*)

TABLE XXX (*continued*)

Empirical formula	Compound	Preparation	References to use as synthetic intermediate	Other
$C_{12}H_{10}ClN_5$	2-*o*-Chlorobenzylamino-4-amino-5-cyano-pyrimidine	22, 143	22, 143	
$C_{12}H_{10}N_2$	1-Amino-2-cyano-3-methylnaphthalene	435		
$C_{12}H_{10}N_2$	1-Amino-2-cyano-4-methylnaphthalene	435		
$C_{12}H_{10}N_2O$	1-Cyano-2-amino-6-methoxynaphthalene	290, 340	290, 340	
$C_{12}H_{10}N_4$	2-Methyl-4-amino-5-cyano-6-phenyl-pyrimidine	33		
$C_{12}H_{10}N_4$	2-Benzyl-4-amino-5-cyanopyrimidine	50, 378	55	
$C_{12}H_{10}N_4O$	3-Benzyl-5-cyanocytosine	43		
$C_{12}H_{10}N_4O$	3-*p*-Tolyl-5-cyanocytosine	43		
$C_{12}H_{10}N_4O_2$	3-*p*-Methoxyphenyl-5-cyanocytosine	43		
$C_{12}H_{10}N_4O_2$	1-Phenyl-3-carboxymethyl-4-cyano-5-aminopyrazole	25	25	
$C_{12}H_{10}N_4S$	2-Benzylmercapto-4-amino-5-cyano-pyrimidine	53, 246		
$C_{12}H_{10}N_4S$	2-Methylmercapto-4-amino-5-cyano-6-phenylpyrimidine	141	141	
$C_{12}H_{10}N_4S$	2-Phenyl-4-amino-5-cyano-6-methyl-mercaptopyrimidine	105(?), 122	122	
$C_{12}H_{10}N_4S_4$	bis-[5-Amino-4-cyano-3-thienyl]disulfide	74		
$C_{12}H_{11}N_3$	2-Amino-3-cyano-4-phenyl-5-methylpyrrole	175		
$C_{12}H_{11}N_5$	2-Benzylamino-4-amino-5-cyanopyrimidine	22, 143	22, 143	
$C_{12}H_{12}N_2$	1-Amino-2-cyano-5-phenylcyclopentene	281		
$C_{12}H_{12}N_4$	1-Amino-2,6,6-tricyano-3,5,5-trimethyl-cyclohexa-1,3-diene	89, 242		89, 242

$C_{12}H_{12}N_6O$	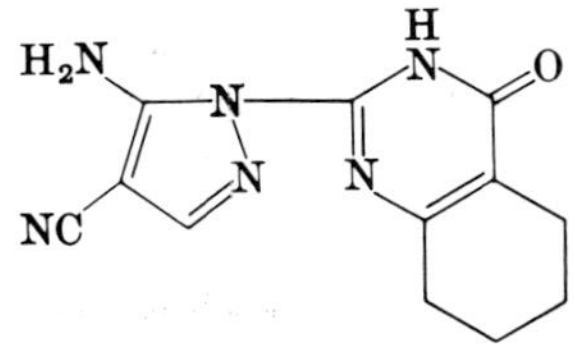	411, 412, 419		
$C_{12}H_{12}N_6O_2$	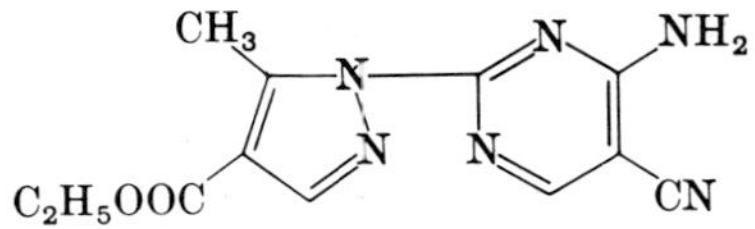	400, 411, 419		
$C_{12}H_{12}N_6O_2$	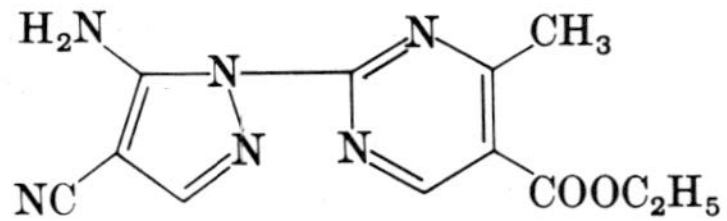	411, 412, 419		
$C_{12}H_{13}N_2As$	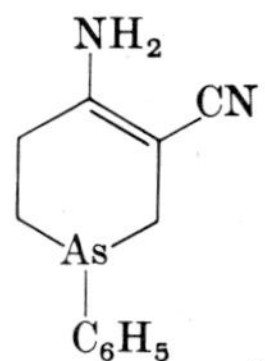	231, 417		111
$C_{12}H_{13}N_2P$	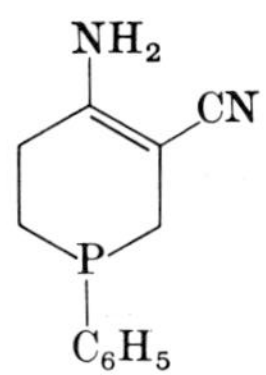	231, 417, 441	111	109, 111

(continued)

TABLE XXX (*continued*)

Empirical formula	Compound	Preparation	References to use as synthetic intermediate	Other
$C_{12}H_{13}N_3$	1,2,5,6-Tetrahydro-3-cyano-4-amino-1-phenylpyridine	111, 117, 441	117	111
$C_{12}H_{13}N_3$	1-Methyl-2-phenyl-3-amino-4-cyano-3-pyrroline	13	13	
$C_{12}H_{13}N_3$	1-*p*-Tolyl-3-amino-4-cyano-3-pyrroline	100		100
$C_{12}H_{13}N_5$	3-*p*-Dimethylaminophenyl-4-cyano-5-aminopyrazole	232		
$C_{12}H_{14}N_2O_4S_2$	2-Carbethoxy-3-amino-4-cyano-5-carbethoxymethylmercaptothiophene	171	171	
$C_{12}H_{15}N_3$	2-Amino-5-piperidinobenzonitrile	93, 94	93, 94	
$C_{12}H_{15}N_3O_2$	1-Amino-2,4-dicyano-3,5-dimethyl-4-carbethoxycyclopentene	491, 515		
$C_{12}H_{15}N_5$	1,3-Diamino-2,6,6-tricyano-3,5,5-trimethylcyclohexene	89		89
$C_{12}H_{16}N_4S$	1-Cyclohexyl-2-thio-4-methyl-5-cyano-6-aminopyrimidine	35		
$C_{12}H_{18}N_4O$	3-*n*-Heptyl-5-cyanocytosine	43		
$C_{12}H_{19}N_5$	1-(β-Hexamethyleneiminoethyl)-4-cyano-5-aminopyrazole	160	160	
$C_{12}H_{20}N_6$	2-(3-*N*,*N*-Diethylaminopropylamino)-4-amino-5-cyanopyrimidine	414		343
$C_{13}H_7ClN_4$	2-Amino-3,5-dicyano-4-phenyl-6-chloropyridine	106		
$C_{13}H_8BrN_3O_2$	2-Amino-3-nitro-6-(*p*-bromophenyl)-benzonitrile	478		

$C_{13}H_8BrN_7$	2-(*m*-Bromophenyl)-4,7-diamino-6-cyanopteridine	439	
$C_{13}H_8ClN_7$	2-(*p*-Chlorophenyl)-4,7-diamino-6-cyanopteridine	439	
$C_{13}H_8ClN_7$	2-(*o*-Chlorophenyl)-4,7-diamino-6-cyanopteridine	439	
$C_{13}H_8N_4O_2$	2,6-di(2-Furyl)-4-amino-5-cyanopyrimidine	36	
$C_{13}H_8N_4S$	2-Amino-3,4-dicyano-5-benzylideneaminothiophene	365	
$C_{13}H_8N_8O_3$	2-(3-Nitro-4-hydroxyphenyl)-4,7-diamino-6-cyanopteridine	439	
$C_{13}H_9ClN_8$	2-(2-Amino-4-chlorophenyl)-4,7-diamino-6-cyanopteridine	439	
$C_{13}H_9N_3S_2$	2-Cyano-3-amino-4-phenyl-5-cyanomethylmercaptothiophene	171	171
$C_{13}H_9N_5$	2,4-Diamino-3-cyanopyrido[2,3-*b*]quinoline	172	172
$C_{13}H_9N_7$	2-Phenyl-4,7-diamino-6-cyanopteridine	439	
$C_{13}H_{10}N_2O_2$	2-Amino-3-cyano-4-acetyl-5-phenylfuran	5	
$C_{13}H_{10}N_2O_3$	1,2,3,4-Tetrahydro-5-cyano-6-amino-7,8-naphthalene dicarboxylic acid anhydride	258	
$C_{13}H_{10}N_2S$	2-Amino-3-cyano-4,5-dihydrobenzo[*g*]thianaphthene	91, 341	91, 341
$C_{13}H_{10}N_4O$	2-Amino-3-cyano-5-styrylpyrazine 1-oxide	537	537
$C_{13}H_{10}N_4S$	2-Amino-5-benzylmercapto-3,4-dicyanopyrrole	104	
$C_{13}H_{10}N_8$	2-(3-Aminophenyl)-4,7-diamino-6-cyanopteridine	439	

(continued)

TABLE XXX (*continued*)

Empirical formula	Compound	Preparation	References to use as synthetic intermediate	Other
$C_{13}H_{12}BrN_3O$	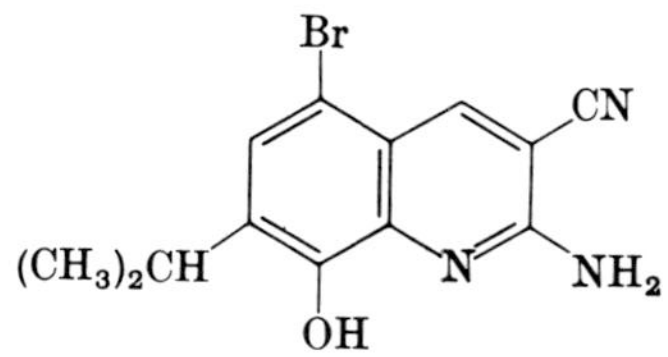	418		
$C_{13}H_{12}N_4O$	3-α-Phenylethyl-5-cyanocytosine	43		
$C_{13}H_{12}N_4O$	3-(2,6-Dimethylphenyl)-5-cyanocytosine	43		
$C_{13}H_{12}N_4O_2$	4-Amino-5-cyano-6-(β-hydroxyethoxy)-2-phenylpyrimidine	105		
$C_{13}H_{12}N_4S$	2-Ethylmercapto-4-amino-5-cyano-6-phenylpyrimidine	52		
$C_{13}H_{12}N_4S$	2-Benzylmercapto-4-amino-5-cyano-6-methylpyrimidine	53		
$C_{13}H_{12}N_6$	2-Amino-6-piperidino-3,4,5-tricyano-pyridine	106		
$C_{13}H_{13}N_3O$	1-Benzoyl-2-methyl-3-amino-4-cyano-3-pyrroline	76	76	
$C_{13}H_{13}N_3O$	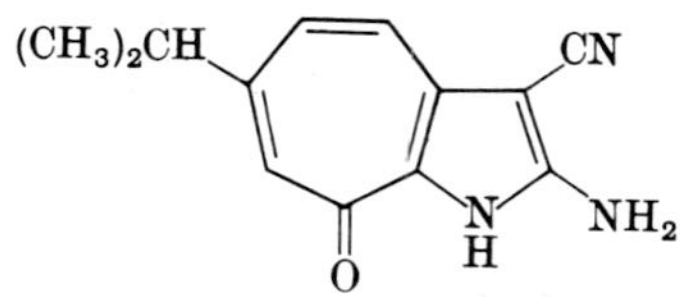	421		

$C_{13}H_{13}N_3OS$	2-Amino-3(or 4)-cyano-4(or 3)-phenyl-5-(β-hydroxyethylmercapto)pyrrole	103		
$C_{13}H_{13}N_3S$	2-Amino-3(or 4)-cyano-4(or 3)-phenyl-5-ethylmercaptopyrrole	103		
$C_{13}H_{13}N_5$	2-Dimethylamino-4-amino-5-cyano-6-phenylpyrimidine		141	
$C_{13}H_{14}N_2$	1-Cyano-2-amino-3-phenyl-3-methylcyclopentene	182	182	
$C_{13}H_{14}N_2$	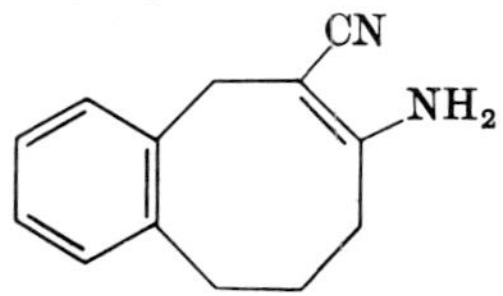	262	262	
$C_{13}H_{14}N_2$	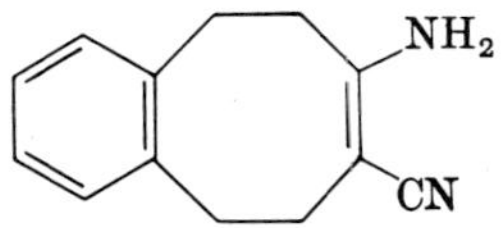	262	262	
$C_{13}H_{14}N_6O$	1-Carbamoyl-3-p-dimethylaminophenyl-4-cyano-5-aminopyrazole	232		
$C_{13}H_{15}N_3$	1-Benzyl-2-methyl-3-amino-4-cyano-3-pyrroline	88	13	88
$C_{13}H_{15}N_3$	N-Benzyl-3-cyano-4-amino-1,2,5,6-tetrahydropyridine	95, 96, 459	95, 96, 459, 535	
$C_{13}H_{15}N_3$	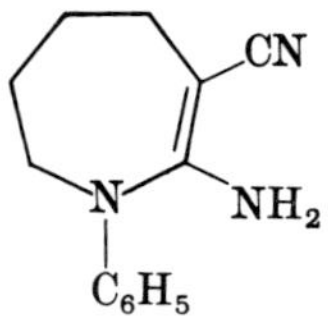	431, 447		

(continued)

TABLE XXX (*continued*)

Empirical formula	Compound	Preparation	References to use as synthetic intermediate	Other
$C_{13}H_{15}N_7$	2-Amino-3,5-dicyano-4-cyanomethyl-6-(2-dimethylaminoethylamino)pyridine	555		
$C_{13}H_{16}N_4$	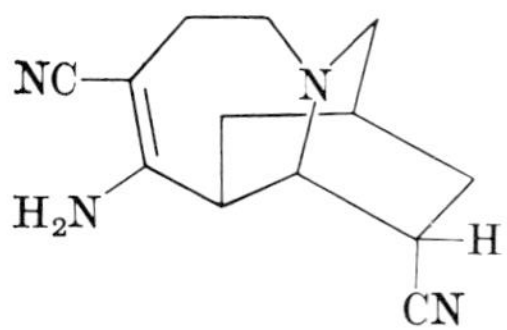	558	558	
$C_{13}H_{16}N_4O$	1-Amino-2,6,6-tricyano-3,5,5-trimethyl-3-methoxycyclohexene	89		89
$C_{13}H_{20}N_4O$	3-*n*-Octyl-5-cyanocytosine	43		
$C_{13}H_{20}N_4S$	1-*n*-Octyl-2-thio-5-cyano-6-aminopyrimidine	35		
$C_{13}H_{22}N_2$	1-Amino-2-cyano-5-*t*-butylcyclooctene	559	559	
$C_{14}H_7ClN_6$	2-Amino-6-*p*-chloroanilino-3,4,5-tricyanopyridine	106		
$C_{14}H_7N_5O_2S$	2-Amino-6-benzenesulphonyl-3,4,5-tricyanopyridine	106		
$C_{14}H_8N_6$	2-Amino-6-anilino-3,4,5-tricyanopyridine	106		
$C_{14}H_9ClN_6$	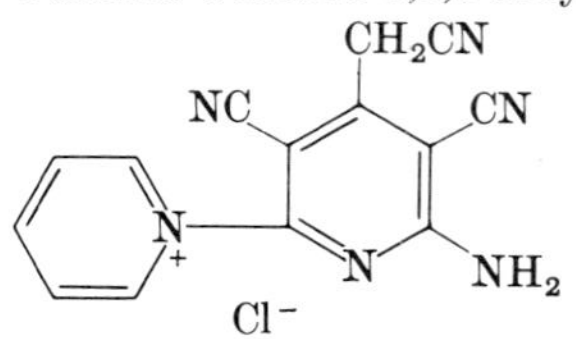	555		

$C_{14}H_9N_5$	4′-Amino-2,3′-dicyanoazobenzene	553	
$C_{14}H_9N_5$	2′-Amino-2,3′-dicyanoazobenzene	553	
$C_{14}H_{10}BrN_7O$	2-(3-Bromo-4-methoxyphenyl)-4,7-diamino-6-cyanopteridine	439	
$C_{14}H_{10}N_4O$	2-Amino-6-methoxy-3,5-dicyano-4-phenylpyridine	106	
$C_{14}H_{10}N_4S$	2-Amino-3,5-dicyano-6-benzylmercaptopyridine	246	
$C_{14}H_{10}N_6O$	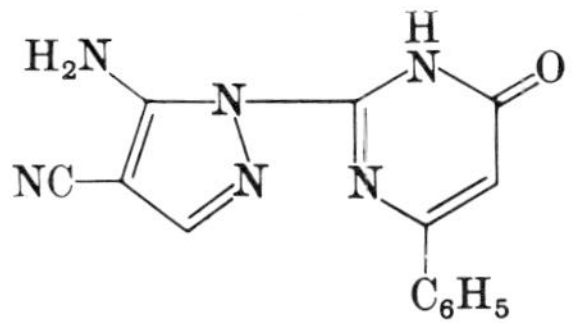	400, 411, 412, 419	
$C_{14}H_{11}Cl_3N_4O_3$	2-(2,2,2-Trichloroethoxy)-4-(*p*-methoxyphenoxy)-5-cyano-6-aminopyrimidine	72	
$C_{14}H_{11}N_5$	4′-Amino-2,3′-dicyanohydrazobenzene	553	
$C_{14}H_{11}N_7$	2-(*p*-Tolyl)-4,7-diamino-6-cyanopteridine	439	
$C_{14}H_{11}N_7$	2-(*m*-Tolyl)-4,7-diamino-6-cyanopteridine	439	
$C_{14}H_{11}N_7$	2-Phenyl-4-methylamino-6-cyano-7-aminopteridine	439	
$C_{14}H_{12}N_2O_3$	2-Amino-3-cyano-4-carbethoxy-5-phenylfuran	5	
$C_{14}H_{12}N_4S$	2-Amino-5-benzylmercapto-3,4-dicyano-1-methylpyrrole	104	
$C_{14}H_{13}N_3$	1-Amino-2,4-dicyano-4-phenylcyclohexene	230, 488	230
$C_{14}H_{14}N_4O_2$	1-Phenyl-3-carbethoxymethyl-4-cyano-5-aminopyrazole	25	25

(continued)

TABLE XXX (*continued*)

Empirical formula	Compound	Preparation	References to use as synthetic intermediate	Other
$C_{14}H_{14}N_8$	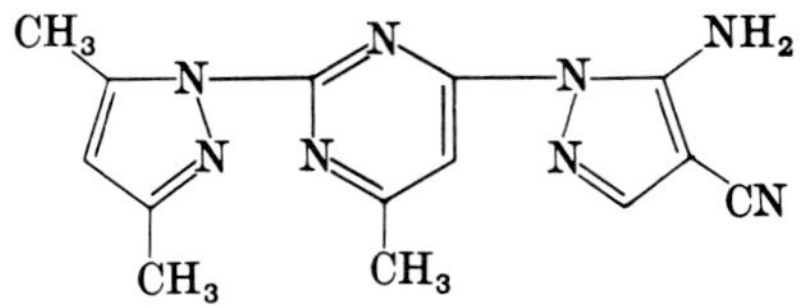	419		
$C_{14}H_{15}Cl_2N_5$	3-Amino-4-cyano-5-[*p*-{bis-(β-chloroethyl)amino}phenyl]pyrazole	477		
$C_{14}H_{15}N_3O$	1-Acetyl-2-benzyl-3-amino-4-cyano-3-pyrroline	76	76	
$C_{14}H_{16}N_2$	1-Amino-2-cyano-5-phenyl-5-ethylcyclopentene	269, 281, 486	269, 281, 486	
$C_{14}H_{16}N_2$	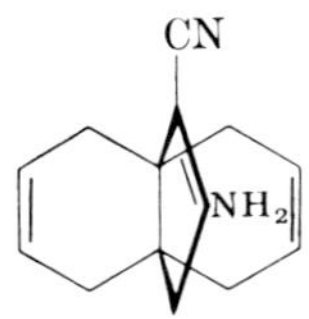	7, 12, 534	7, 12	7, 12
$C_{14}H_{16}N_4$	1-Amino-2,6,6-tricyano-3,5-diethyl-5-methylcyclohexa-1,3-diene	242		
$C_{14}H_{17}N_3$	1-Benzyl-2,2-dimethyl-3-amino-4-cyano-3-pyrroline	28	13, 28	
$C_{14}H_{17}N_3$	1-β-Phenylethyl-2-methyl-3-amino-4-cyano-3-pyrroline	88	88	
$C_{14}H_{18}N_4O$	1-Amino-2,6,6-tricyano-3,5,5-trimethyl-3-ethoxycyclohexene	89		89

$C_{14}H_{20}N_2$	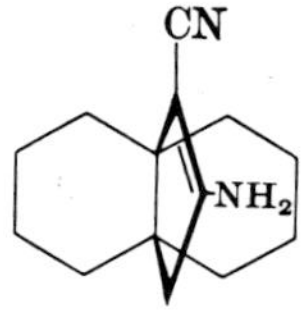	7	7	7
$C_{14}H_{24}N_2O_2$	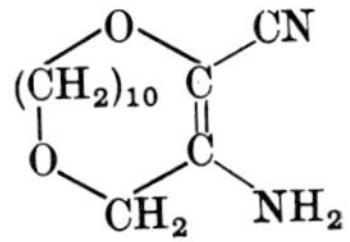	445		
$C_{14}H_{25}N_2P$	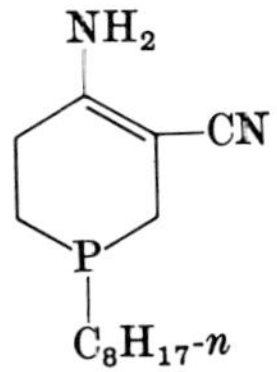	417		
$C_{15}H_8ClN_5S$	2-Amino-3,5-dicyano-4-cyanomethyl-6-(p-chlorophenylthio)pyridine	555		
$C_{15}H_8N_2O_2$	1-Cyano-2-aminoanthraquinone	325	153	
$C_{15}H_8N_4$	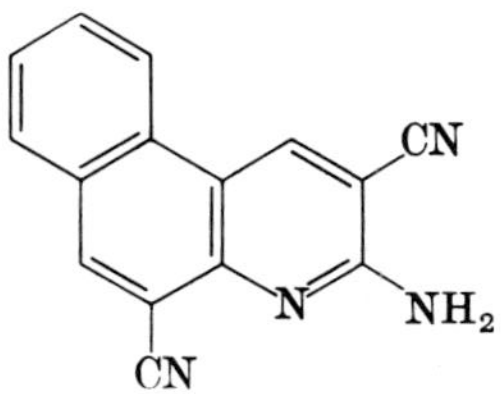	469		

(continued)

TABLE XXX (*continued*)

Empirical formula	Compound	Preparation	References to use as synthetic intermediate	Other
$C_{15}H_8N_4$	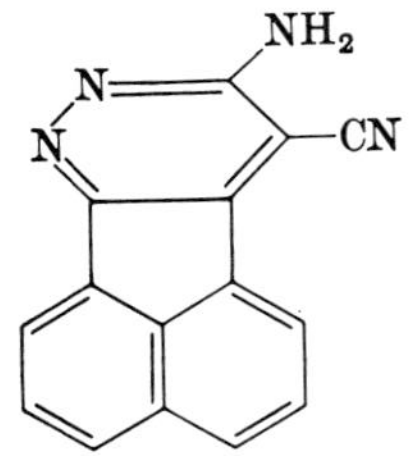	487		
$C_{15}H_8N_8$	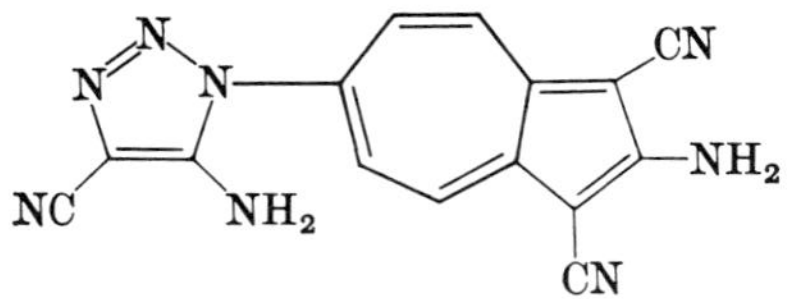	529		
$C_{15}H_8N_8$	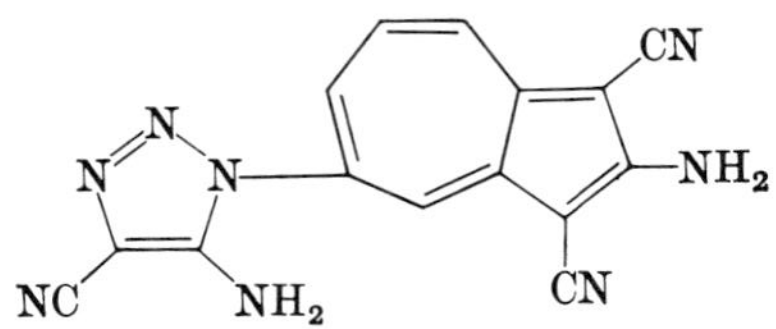	529		
$C_{15}H_9N_5S$	2-Amino-3,5-dicyano-4-cyanomethyl-6-phenylthiopyridine	555		

$C_{15}H_{10}N_4$	2-Amino-3-cyano-7-phenylpyrido-[2,3-*b*]pyridine	162	162
$C_{15}H_{11}N_3O_2$	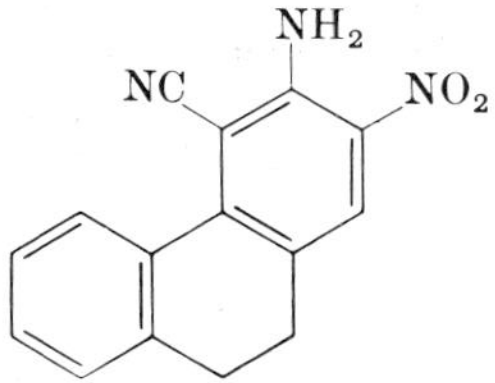	478	
$C_{15}H_{11}N_5O$	7-Amino-6-cyano-3-methyl-5-phenyl-pyrido[2,3-*d*]pyrimidin-4(3*H*)-one	11	
$C_{15}H_{12}ClN_5$	2-Amino-6-chloro-3,5-dicyano-4-*p*-dimethylaminophenylpyridine	106	
$C_{15}H_{12}N_4O$	2-Amino-6-ethoxy-4-phenyl-3,5-dicyano-pyridine	106	
$C_{15}H_{13}N_3O_2$	1-Amino-2,4-dicyano-4-(3′,4′-methylene-dioxyphenyl)cyclohexene	230	230
$C_{15}H_{13}N_7$	2-Phenyl-4-dimethylamino-6-cyano-7-aminopteridine	439	
$C_{15}H_{13}N_7$	2-(*p*-Ethylphenyl)-4,7-diamino-6-cyano-pteridine	439	
$C_{15}H_{13}N_7$	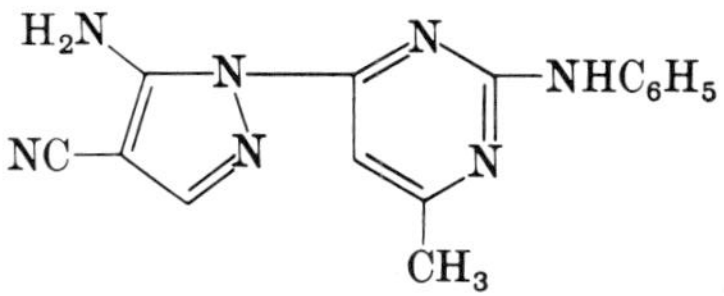	419	

(continued)

TABLE XXX (*continued*)

Empirical formula	Compound	Preparation	References to use as synthetic intermediate	Other
$C_{15}H_{14}IN_5O$	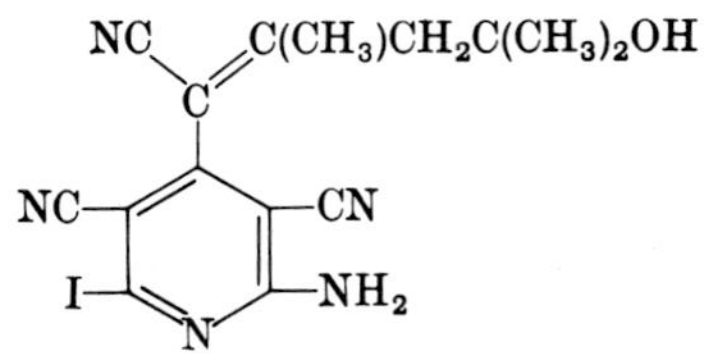	555		
$C_{15}H_{14}N_2$	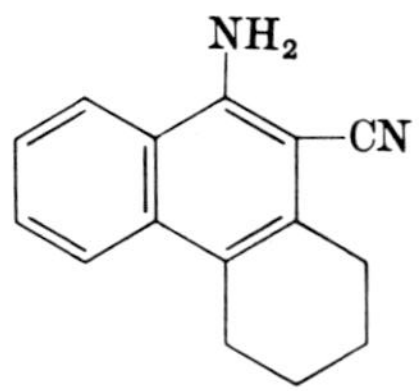	435		
$C_{15}H_{14}N_6$	1-Amino-2,6,6-tricyano-3,5,5-trimethyl-3-dicyanomethylcyclohexene	89		89
$C_{15}H_{16}Cl_2N_6$	2,4-Diamino-5-cyano-6[*p*-{bis-(*β*-chloroethyl)amino}phenyl]pyrimidine	477		
$C_{15}H_{17}Cl_2N_5$	1-Methyl-3-[*p*-{bis-(*β*-chloroethyl)amino}-phenyl]-4-cyano-5-aminopyrazole	477		

$C_{15}H_{17}Cl_2N_7$	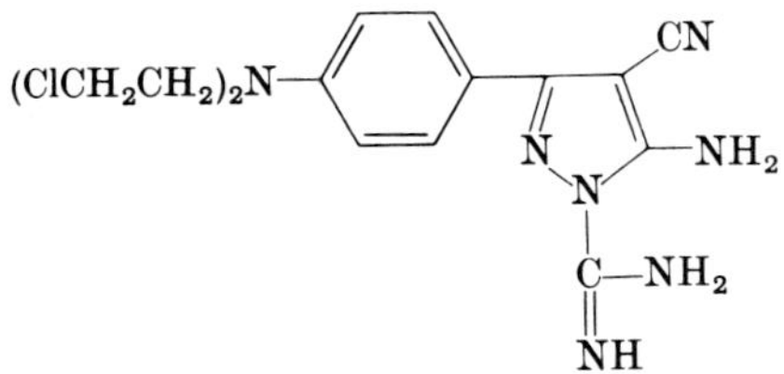	477	
$C_{15}H_{18}N_2$	1-Cyano-2-amino-3-phenyl-3-*n*-propylcyclopentene	138, 486	138, 486
$C_{15}H_{18}N_2$	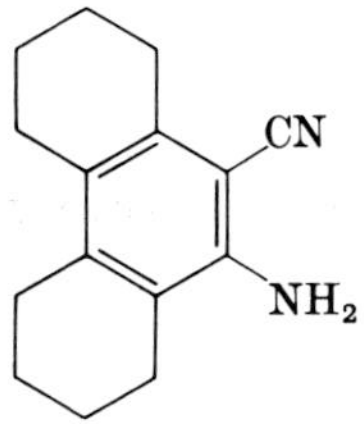	434	
$C_{15}H_{18}N_2$	1-Amino-2-cyano-6-ethyl-6-phenylcyclohexene	542	542
$C_{15}H_{18}N_2$	1-Amino-2-cyano-5-phenylcyclooctene	167, 181	167, 181
$C_{15}H_{18}N_2O_2$	2-Cyano-2-(2′,3′-dimethoxyphenyl)-cyclohexanoneimine	222	222
$C_{15}H_{18}N_4O_7$	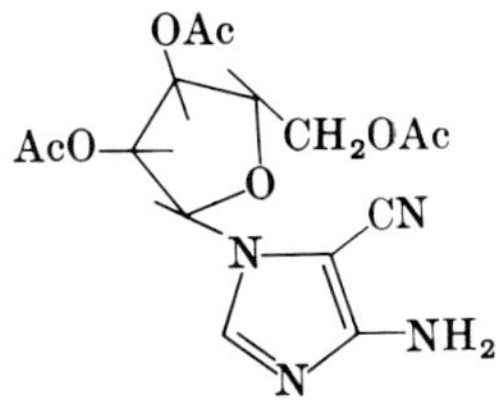	355, 552	355, 552

(*continued*)

TABLE XXX (*continued*)

Empirical formula	Compound	Preparation	References to use as synthetic intermediate	Other
$C_{15}H_{19}N_3$	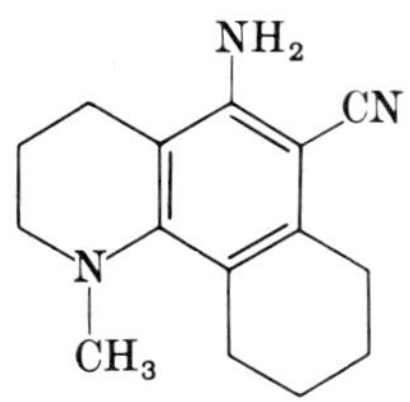	490		
$C_{15}H_{22}N_2$	2-Amino-3,5-di-*t*-butylbenzonitrile	547		
$C_{15}H_{24}N_4S$	1-*n*-Octyl-2-thio-4-ethyl-5-cyano-6-aminopyrimidine	35		
$C_{15}H_{26}N_2$	1-Amino-2-cyanocyclotetradecene	190, 201, 202	190, 201, 202	
$C_{16}H_{10}N_4$	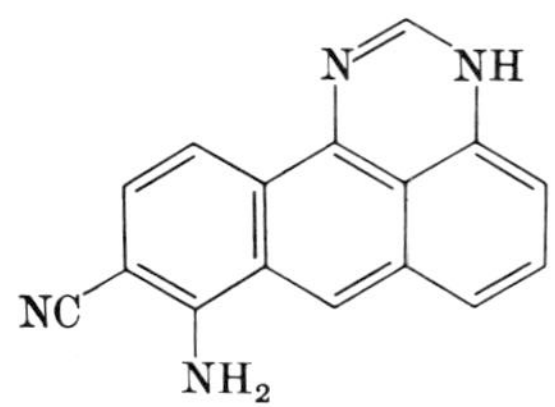			153

$C_{16}H_{10}N_4O_4$	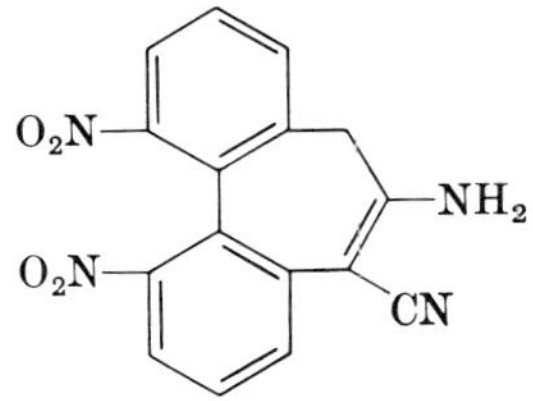 (+) (−)	205, 206	205	66, 207, 208
$C_{16}H_{12}N_2$	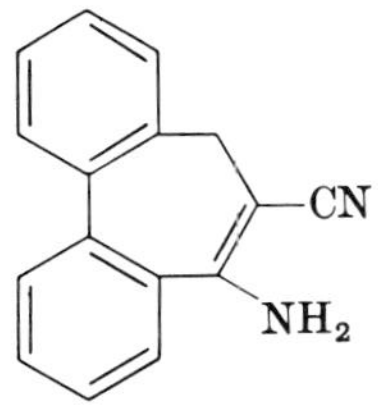	29, 224, 229	29, 224, 229	
$C_{16}H_{12}N_2$	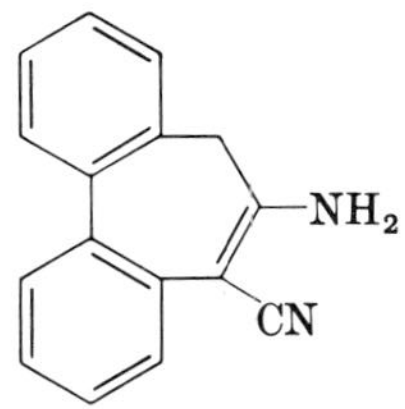	29, 192, 223, 224, 265	29, 192, 223, 224, 265	
$C_{16}H_{12}N_4$	1,3-Diphenyl-4-cyano-5-aminopyrazole	85, 232	85	
$C_{16}H_{12}N_4S$	2-(1-Naphthylmethylmercapto)-4-amino-5-cyanopyrimidine			53

(continued)

TABLE XXX (*continued*)

Empirical formula	Compound	Preparation	References to use as synthetic intermediate	Other
$C_{16}H_{12}N_4S$	2-Phenylimino-3-phenyl-4-amino-5-cyano-4-thiazoline	473		
$C_{16}H_{12}N_6O$	2-Amino-3,5-dicyano-4-cyanomethyl-6-*p*-anisidinopyridine	555	555	
$C_{16}H_{13}N_2P$	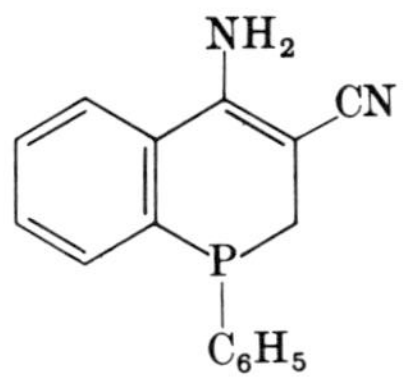	110	110	110
$C_{16}H_{13}N_5O_3S$	*N*-Methylquinolinium 5-amino-3,4-dicyano-2-pyrrolesulfonate	104		
$C_{16}H_{15}N_7$	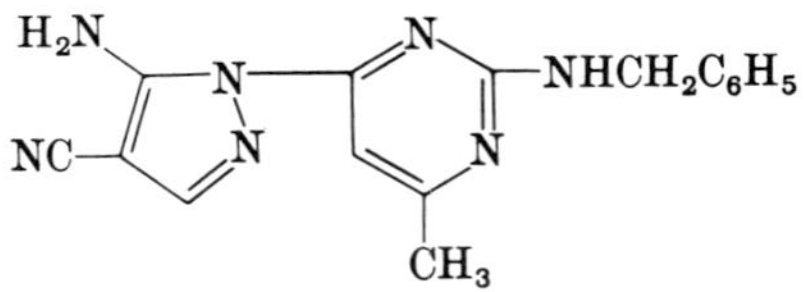	419		
$C_{16}H_{16}N_2O$	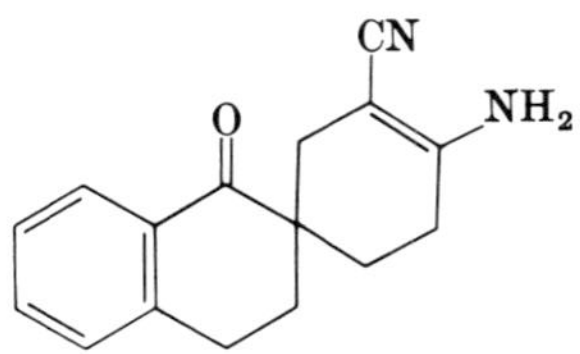			66

$C_{16}H_{16}N_8$	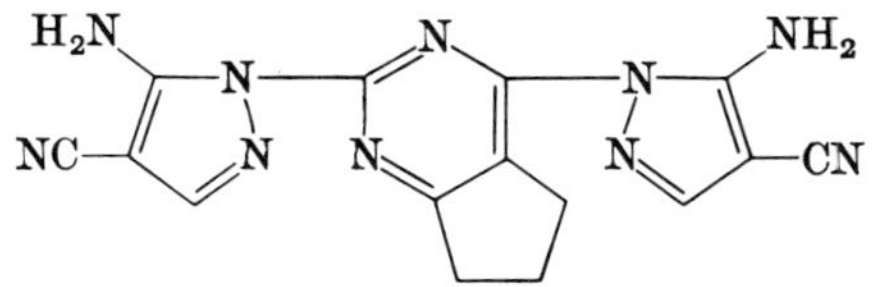	419	
$C_{16}H_{17}Cl_2N_5$	2-Methyl-4-amino-5-cyano-6-[*p*-{bis-(β-chloroethyl)amino}phenyl]pyrimidine	477	
$C_{16}H_{18}N_2O$	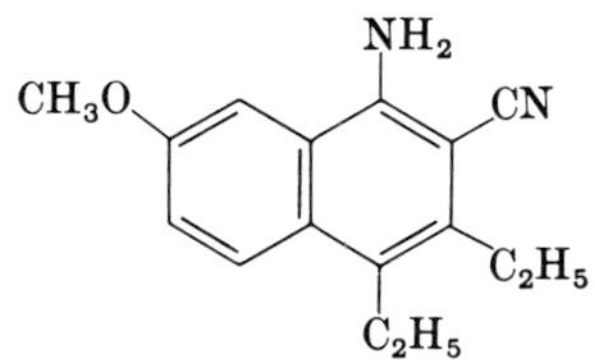	435	
$C_{16}H_{19}Cl_2N_5$	1-Ethyl-3-[*p*-{bis(β-chloroethyl)amino}-phenyl]-4-cyano-5-aminopyrazole	477	
$C_{16}H_{20}N_2$	1-Cyano-2-amino-3-phenyl-3-*n*-butyl-cyclopentene	138, 486	138, 486
$C_{16}H_{20}N_4$	1-Amino-2,6,6-tricyano-3,5,5-triethyl-4-methylcyclohexa-1,3-diene	242	
$C_{16}H_{28}N_2$	1-Amino-2-cyanocyclopentadecene	187, 200, 202, 243	187, 200, 202
$C_{17}H_9ClN_6$	*N*-(2-[6-Amino-3,4,5-tricyanopyridyl])-quinolinium chloride	106	
$C_{17}H_{10}Cl_2N_4$	2,6-bis(*p*-Chlorophenyl)-4-amino-5-cyanopyrimidine	359	
$C_{17}H_{11}BrN_2O$	2-Amino-3-cyano-4-bromophenyl-5-phenylfuran	492	492

(*continued*)

TABLE XXX (*continued*)

Empirical formula	Compound	Preparation	References to use as synthetic intermediate	Other
$C_{17}H_{11}ClN_2O$	2-Amino-3-cyano-4-chlorophenyl-5-phenylfuran	492	492	
$C_{17}H_{11}ClN_4O_2$	2-Phenoxy-4-*p*-chlorophenoxy-5-cyano-6-aminopyrimidine	72		
$C_{17}H_{12}N_2$	1-Benzylidene-2-amino-3-cyanoindene	450	450	
$C_{17}H_{12}N_2O$	2-Amino-3-cyano-4,5-diphenylfuran	5, 258, 492	258, 492	
$C_{17}H_{12}N_2O$	1-Salicylidene-2-amino-3-cyanoindene	450	450, 451	
$C_{17}H_{12}N_4$	2-Amino-3-cyano-5,6-diphenylpyrazine	8, 23		
$C_{17}H_{12}N_4$	2,4-Diphenyl-5-cyano-6-aminopyrimidine	36, 122, 354, 359	122, 359	
$C_{17}H_{14}N_2$	1-Benzyl-2-amino-3-cyanoindene	450		
$C_{17}H_{14}N_6O$	2-Amino-3,5-dicyano-4-cyanomethyl-6-(*N*-methyl-*p*-anisidino)pyridine	555	555	
$C_{17}H_{14}N_6O_2$	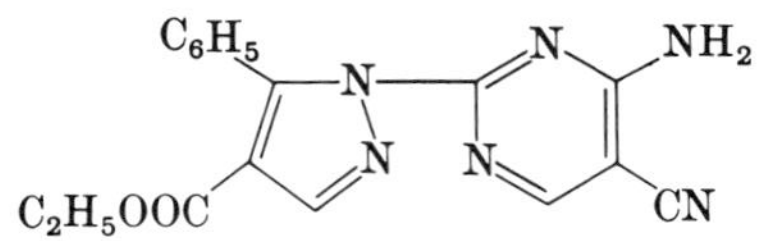	411, 412, 419		
$C_{17}H_{15}N_3O_2$	3-Amino-4-cyano-7-benzoyloxy-5,6,7,8-tetrahydroisoquinoline	509		

$C_{17}H_{16}N_6O_2$	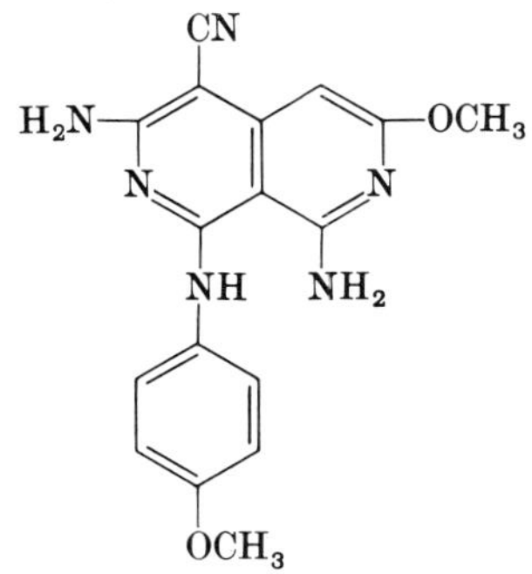	555	
$C_{17}H_{17}N_7O$	2-(*o*-Butoxyphenyl)-4,7-diamino-6-cyano-pteridine	439	
$C_{17}H_{18}N_8$	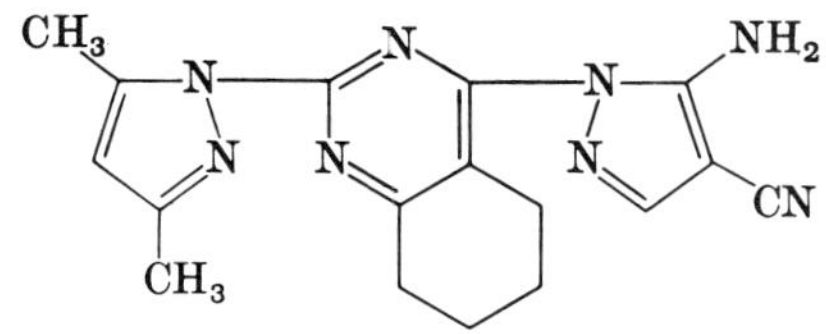	419	
$C_{17}H_{20}N_4O_2$	2,4-bis-(2,4-Dimethylphenoxy)-5-cyano-6-aminopyrimidine	72	
$C_{17}H_{22}N_2$	1-Cyano-2-amino-3-phenyl-3-*n*-pentyl-cyclopentene	138, 486	138, 486

(*continued*)

TABLE XXX (*continued*)

Empirical formula	Compound	Preparation	References to use as synthetic intermediate	Other
$C_{17}H_{22}N_2$	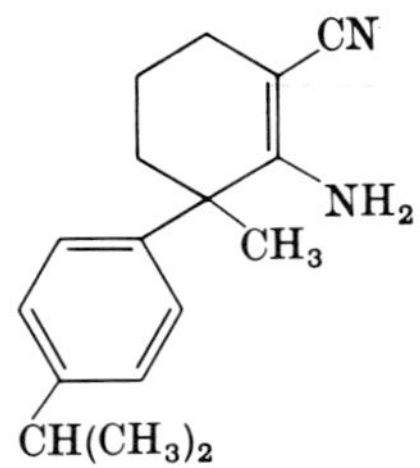	226	226	
$C_{17}H_{22}N_2O_2$	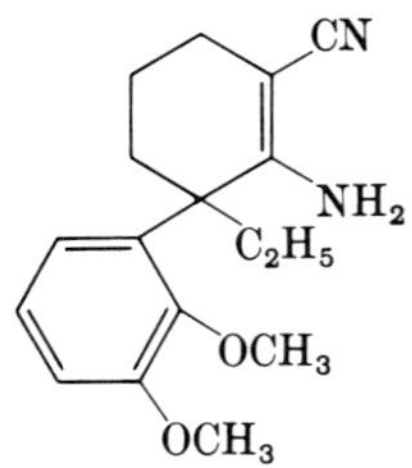	227, 228	227, 228	
$C_{17}H_{22}N_6$	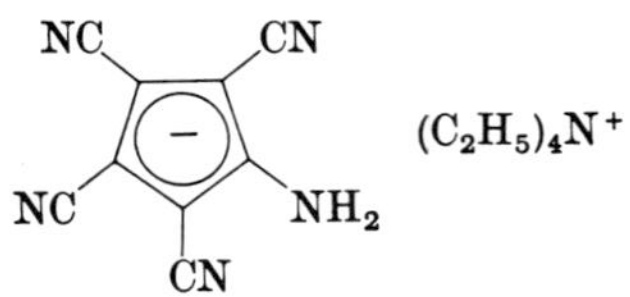	483		

$C_{17}H_{30}N_2$	1-Amino-2-cyanocyclohexadecene	201, 202	201, 202	
$C_{18}H_{12}N_2O_2$	1-(*o*-Carboxybenzylidene)-2-amino-3-cyanoindene	450	450	
$C_{18}H_{12}N_2O_2$	1-[Phthalidyl-(1)]-2-amino-3-cyanoindene	453	453	
$C_{18}H_{12}N_2O_2$	2-Amino-3-cyano-4-benzoyl-5-phenylfuran	5		
$C_{18}H_{13}N_3$	2-Amino-3-cyano-4,6-diphenylpyridine	354		
$C_{18}H_{14}N_2$	1-Phenyl-3-imino-4-cyano-4-phenylcyclopentene	533		
$C_{18}H_{14}N_2$	1-Amino-2-cyano-3-benzylnaphthalene	286	286	
$C_{18}H_{14}N_2O$	2-Amino-3-cyano-4-phenyl-5-tolylfuran	492	492	
$C_{18}H_{14}N_2O$	2-Amino-3-cyano-4-tolyl-5-phenylfuran	492	492	
$C_{18}H_{14}N_2O_2$	2-Amino-3-cyano-4-phenyl-5-anisylfuran	492	492	
$C_{18}H_{14}N_2O_2$	2-Amino-3-cyano-4-anisyl-5-phenylfuran	492	492	
$C_{18}H_{14}N_6O_2$	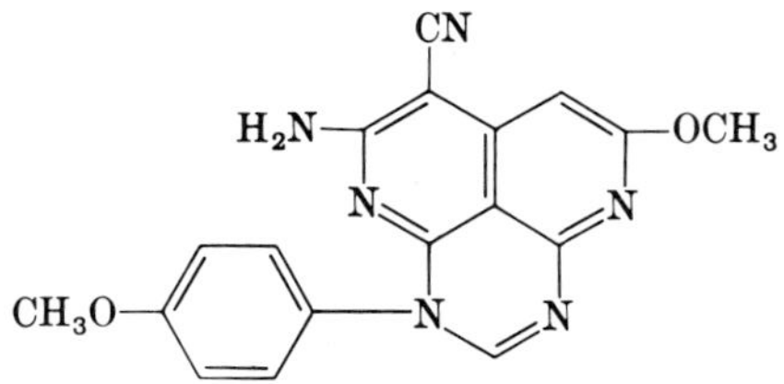	555		
$C_{18}H_{16}N_2$	1-Amino-2-cyano-5,5-diphenylcyclopentene	221, 268, 277, 282, 285, 287, 486	221, 268, 486	281
$C_{18}H_{16}N_2$	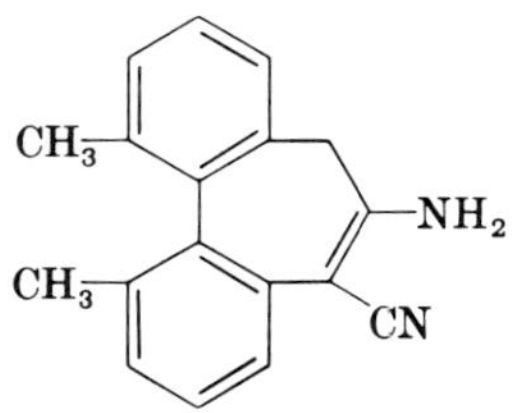	205		66

(*continued*)

TABLE XXX (*continued*)

Empirical formula	Compound	Preparation	References to use as synthetic intermediate	Other
$C_{18}H_{18}N_{12}O_2$	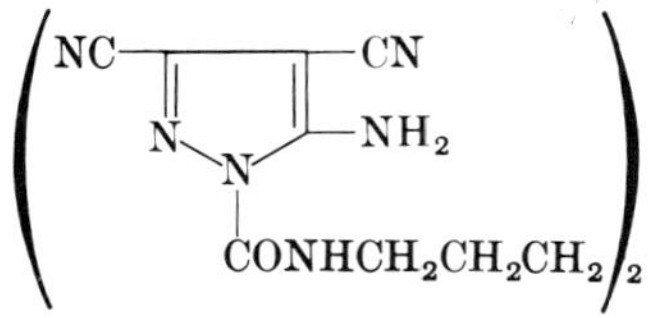	232		
$C_{18}H_{19}N_7O$	2-(2-Butyl-3-methoxyphenyl)-4,7-diamino-6-cyanopteridine	439		
$C_{18}H_{20}N_4$	Cyclohexylidenemalononitrile dimer	69, 242		69
$C_{18}H_{22}Cl_3N_5O$	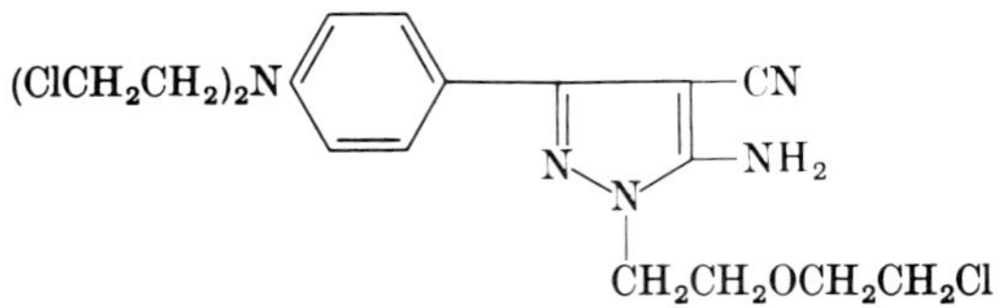	477		
$C_{18}H_{22}N_2$	1-Cyano-2-amino-3-phenyl-3-cyclohexylcyclopentene	138, 486	138, 486	
$C_{18}H_{32}N_2$	1-Amino-2-cyanocycloheptadecene	187, 188	187, 188	
$C_{19}H_{14}N_2$	1-Cinnamylidene-2-amino-3-cyanoindene	450	450	
$C_{19}H_{14}N_4O_3$	2-Phenoxy-4-*p*-acetylphenoxy-5-cyano-6-aminopyrimidine	72		
$C_{19}H_{14}N_4O_3$	2-*p*-Acetylphenoxy-4-phenoxy-5-cyano-6-aminopyrimidine	72		

$C_{19}H_{14}N_4O_4$	2-*o*-Carbomethoxyphenoxy-4-phenoxy-5-cyano-6-aminopyrimidine	72		
$C_{19}H_{16}N_2$	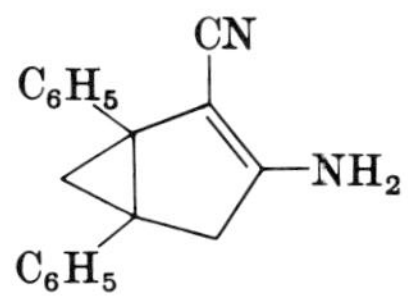	545	545	
$C_{19}H_{16}N_2$	4-Aminospiro[Δ^3-cyclohexene-1,9′-fluorene]-3-carbonitrile	81, 210, 211	81, 210, 211	66
$C_{19}H_{16}N_2O_2$	2-Amino-3-cyano-4-phenyl-5-(*p*-ethoxyphenyl)furan	492	492	
$C_{19}H_{16}N_8$	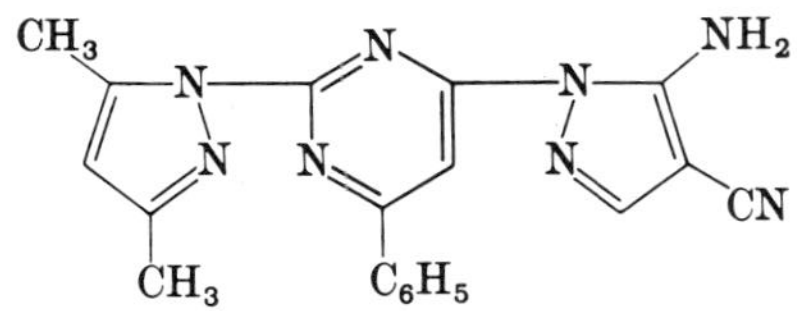	419		
$C_{19}H_{17}N_3O$	1-Benzoyl-2-benzyl-3-amino-4-cyano-3-pyrroline	76	76	
$C_{19}H_{18}N_2$	1-(3-Phenylpropyl)-2-amino-3-cyanoindene	450		
$C_{19}H_{18}N_2$	1-Cyano-2-amino-3-phenyl-3-benzylcyclopentene	138, 486	138, 486	
$C_{19}H_{18}N_2$	1-Cyano-2-amino-3,3-diphenyl-4-methylcyclopentene	220, 485	220, 485	
$C_{19}H_{18}N_2$	1-Amino-2-cyano-3-methyl-5,5-diphenylcyclopentene	220	220	
$C_{19}H_{18}N_2$	1-Amino-2-cyano-6,6-diphenylcyclohexene	542	542	

(continued)

TABLE XXX (*continued*)

Empirical formula	Compound	Preparation	References to use as synthetic intermediate	Other
$C_{19}H_{26}N_2O$	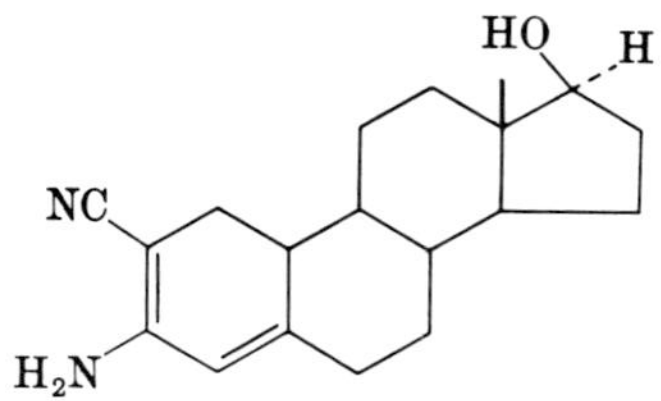	438	438	
$C_{19}H_{26}N_2O_3$	1-Cyano-2-amino-3-(β-ethoxyethyl)-3-(2′,3′-dimethoxyphenyl)cyclohexene	225	225	
$C_{19}H_{28}N_2O$	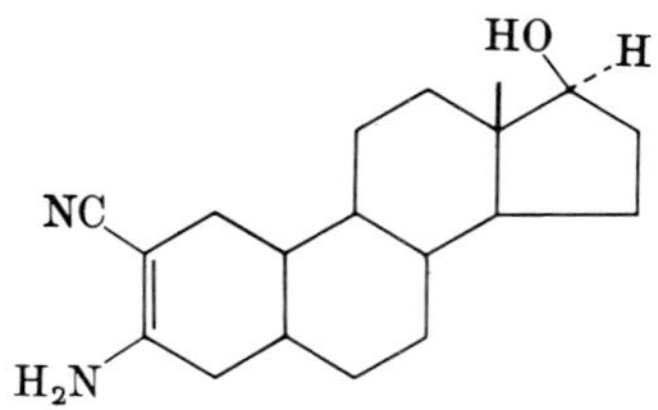	438	438	
$C_{20}H_{12}N_6$	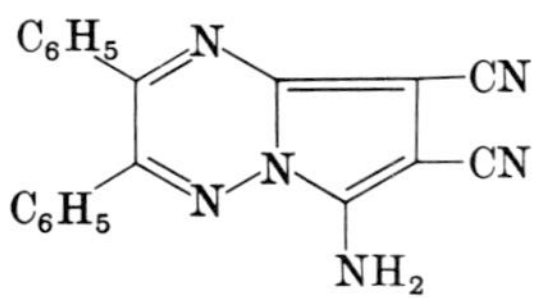	365		

$C_{20}H_{12}N_8O_4$	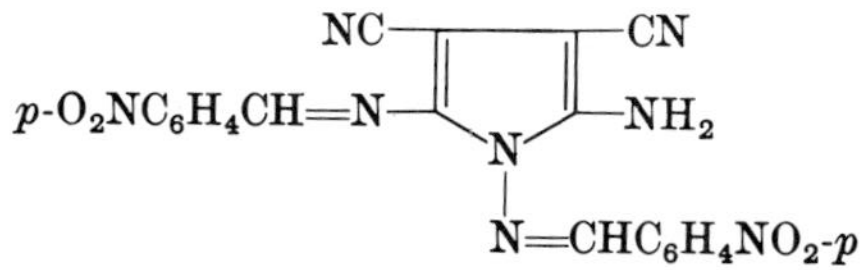	365	
$C_{20}H_{13}N_5O_2$	2-(5-Quinolyloxy)-4-phenoxy-5-cyano-6-aminopyrimidine	72	
$C_{20}H_{13}N_5O_2$	2-Phenoxy-4-(5-quinolyloxy)-5-cyano-6-aminopyrimidine	72	
$C_{20}H_{14}N_6$	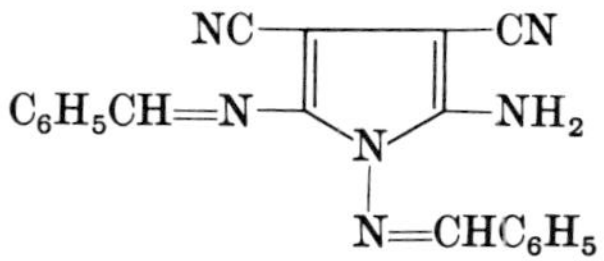	365	
$C_{20}H_{19}Cl_2N_5$	1-Phenyl-3-[*p*-{bis(*β*-chloroethyl)-amino}-phenyl]-4-cyano-5-aminopyrazole	477	
$C_{20}H_{19}Cl_2N_5$	2-Phenyl-4-amino-5-cyano-6-[*p*-{bis-(*β*-chloroethyl)amino}phenyl]pyrimidine	477	
$C_{20}H_{20}N_2$	1-Cyano-2-amino-3-phenyl-3-*β*-phenylethylcyclopentene	138, 486	138, 486
$C_{20}H_{20}N_2O_4$	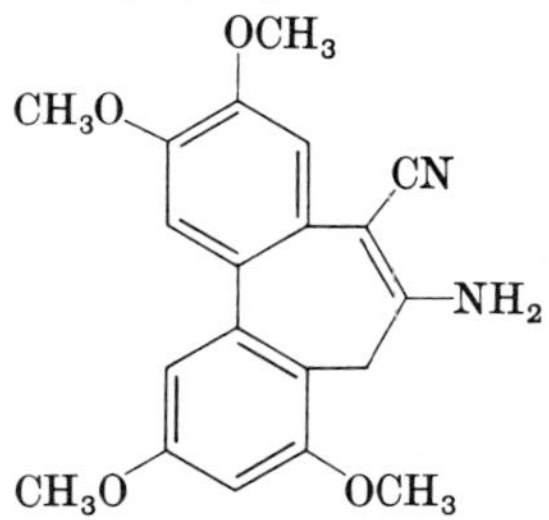	476	476
	or		

(continued)

TABLE XXX (*continued*)

Empirical formula	Compound	Preparation	References to use as synthetic intermediate	Other
	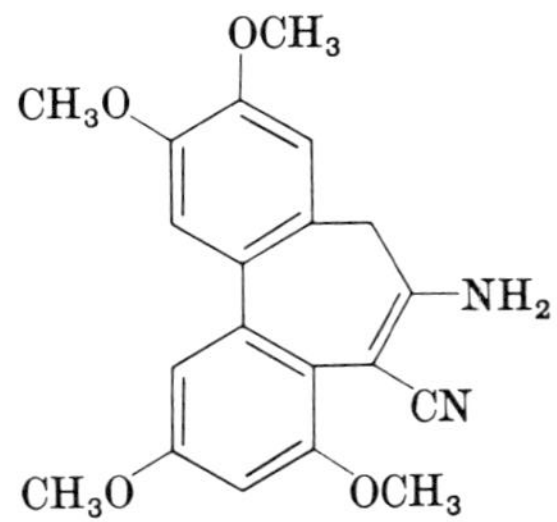			
$C_{20}H_{28}N_2O$	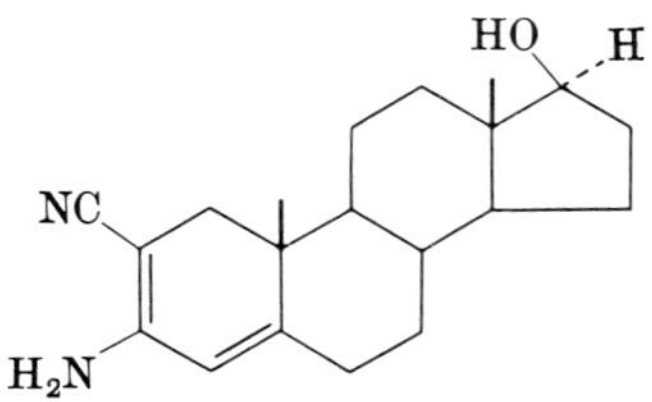	438	438	
$C_{20}H_{28}N_2O_2$	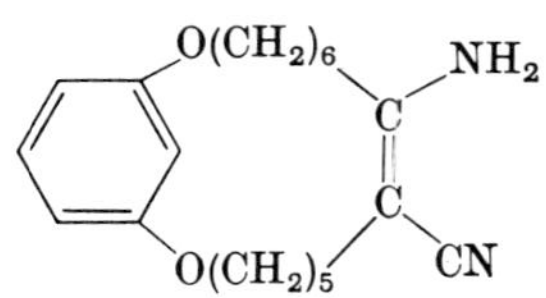	193	193	

$C_{20}H_{28}N_2O_2$	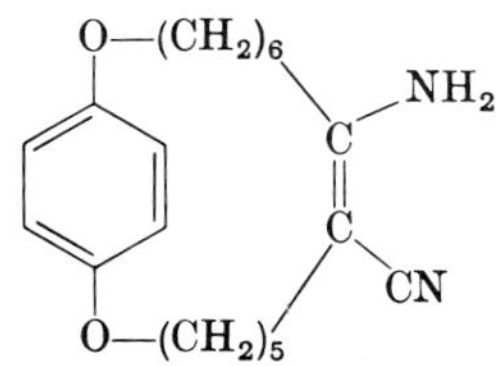	193	193
$C_{20}H_{30}N_2O$	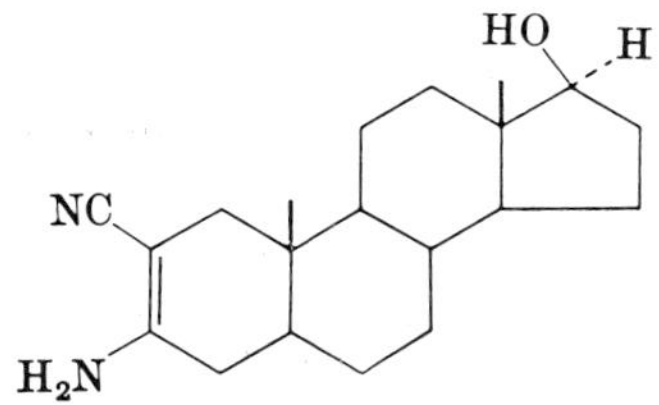	438	438
$C_{20}H_{32}N_4$	1,10-Diamino-2,11-dicyanocyclo-octadecadiene-1,10	190	190
$C_{21}H_{12}N_2O_3$	3-Amino-4-cyano-5,6-diphenylphthalic anhydride	258	
$C_{21}H_{25}N_7$	2-Phenyl-4-dibutylamino-6-cyano-7-aminopteridine	439	
$C_{21}H_{28}N_2O_2$	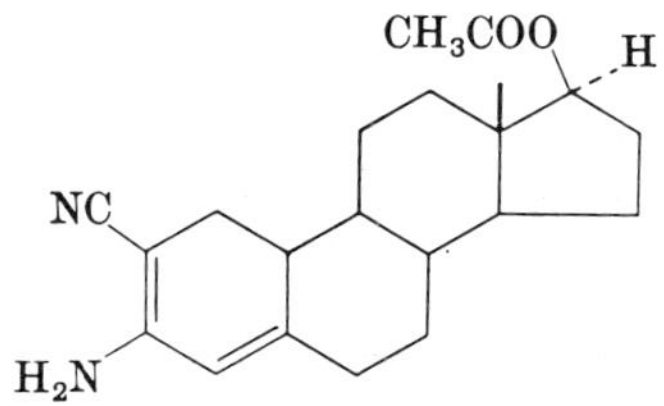	438	438

(continued)

TABLE XXX (*continued*)

Empirical formula	Compound	Preparation	Reference to use as synthetic intermediate	Other
$C_{21}H_{30}N_2O_2$	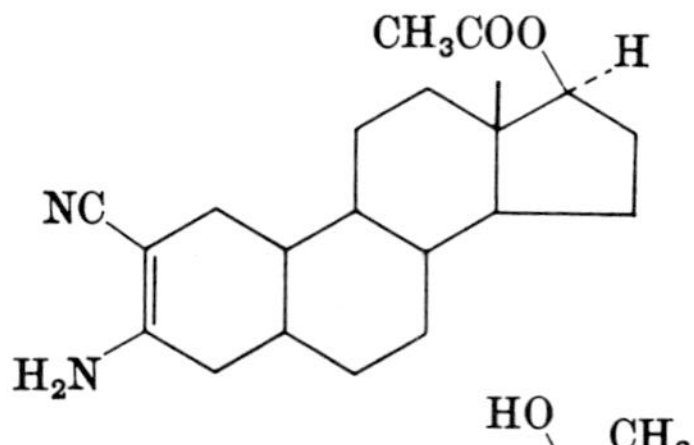	438	438	
$C_{21}H_{32}N_2O$	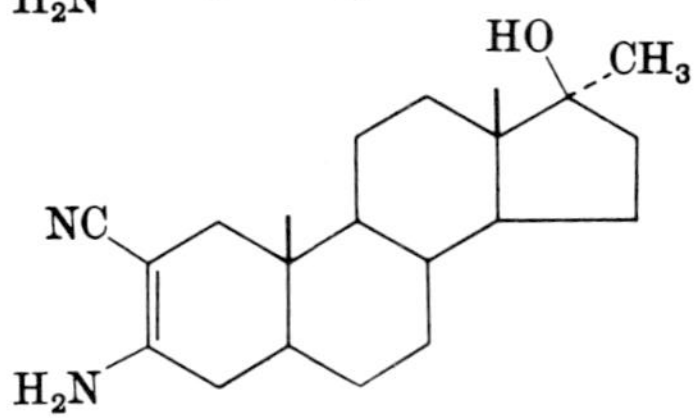	438	438	
$C_{22}H_{21}Cl_2N_5S$	2-Benzylmercapto-4-amino-5-cyano-6-[*p*-{bis(β-chloroethyl)amino}phenyl]-pyrimidine	477		
$C_{22}H_{32}N_2O_2$	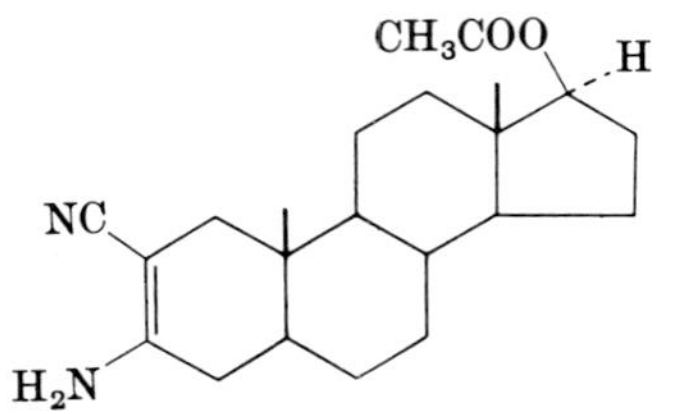	438	438	

$C_{22}H_{33}ClN_6$	2-(6-Amino-3,4,5-tricyanopyridyl)-dimethyldodecylammonium chloride	106		
$C_{22}H_{36}N_4$	1,11-Diamino-2,12-dicyanoeicosane-1,11-diene	190	190	
$C_{23}H_{15}N_3$	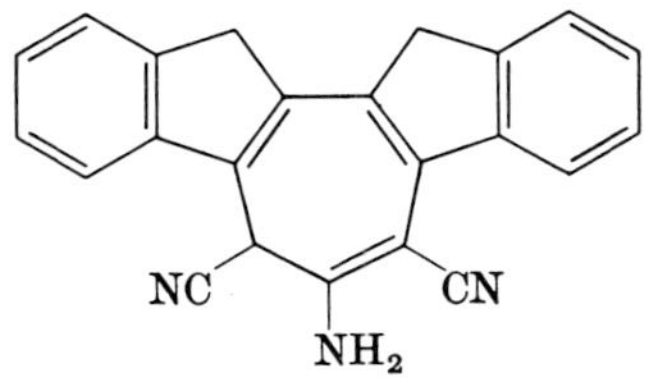	342		
$C_{23}H_{16}N_2O$	2-Amino-3-cyano-4-phenyl-5-(p-biphenylyl)furan	492	492	
$C_{23}H_{16}N_2O$	2-Amino-3-cyano-4-(p-biphenylyl)-5-phenylfuran	492	492	
$C_{24}H_{16}N_2$	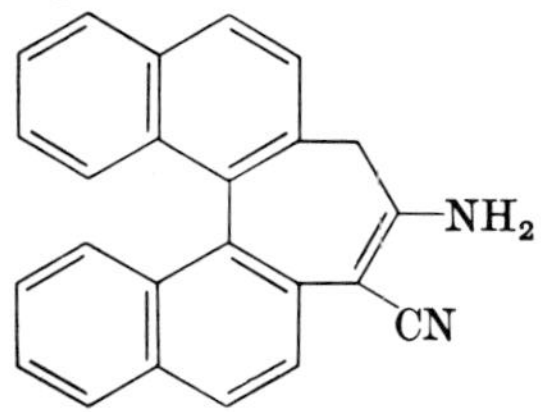 (±) (−)	209	209	66
$C_{24}H_{24}N_4$	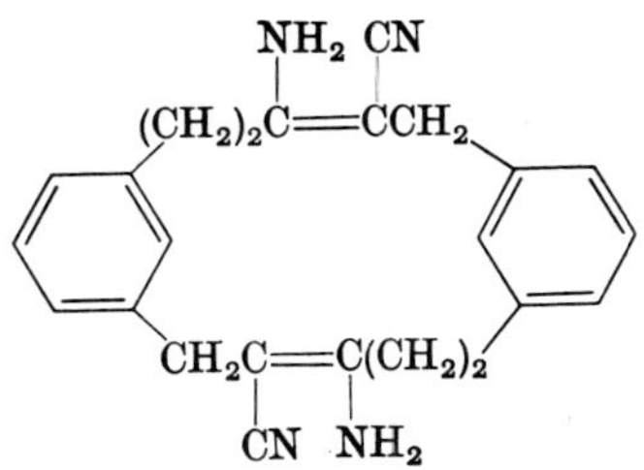	193	193	

(continued)

TABLE XXX (*continued*)

Empirical formula	Compound	Preparation	References to use as synthetic intermediate	Other
$C_{24}H_{40}N_4$	1,12-Diamino-2,13-dicyanodocosane-1,12-diene	190	190	
$C_{28}H_{46}N_2$	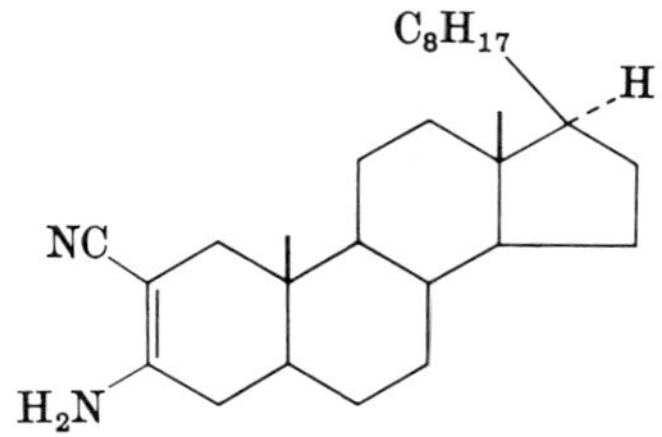	438	438	
$C_{32}H_{56}N_6$	2,17-Diamino-1,16-dicyanocyclotriaconta-1,16-diene	187, 200, 202	187, 200, 202	
$C_{36}H_{32}N_4$	1,6-Diamino-2,7-dicyano-5,5,10,10-tetraphenylcyclodeca-1,6-diene	282		
$C_{69}H_{125}N_3O_2$	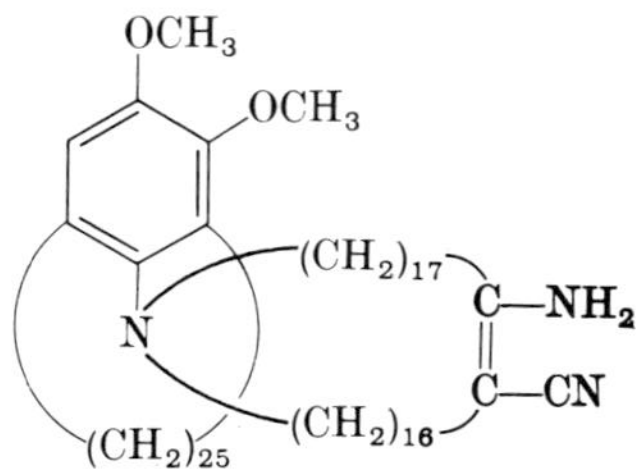	530	530	

REFERENCES

1. G. R. Bedford and M. W. Partridge, *J. Chem. Soc.*, **1959**, 1633.
2. W. Huber, *J. Am. Chem. Soc.*, **65**, 2222 (1943).
3. C. C. Cheng and R. K. Robins, *J. Org. Chem.*, **21**, 1240 (1956).
4. R. K. Robins, *J. Am. Chem. Soc.*, **78**, 784 (1956).
5. T. I. Temnikova and Y. A. Sharanin, *Zh. Org. Khim.*, **2** (11), 2018 (1966); *Index Chemicus*, 76168.
6. G. Desimoni, P. Grünanger, and P. Vita Finzi, *Tetrahedron*, **23**, 687 (1967).
7. J. Altman, E. Babad, J. Itzchaki, and D. Ginsburg, *Tetrahedron*, Suppl. 8, Part I, 279 (1966).
8. E. C. Taylor and W. W. Paudler, *Chem. Ind.*, **1955**, 1061.
9. T. Asao and M. Kobayashi, *J. Chem. Soc. Japan*, **39**, 2538 (1966).
10. E. C. Taylor and R. W. Hendess, *J. Am. Chem. Soc.*, **87**, 1995 (1965).
11. E. C. Taylor and E. E. Garcia, *J. Org. Chem.*, **29**, 2116 (1964).
12. J. J. Bloomfield and A. Mitra, *Chem. Ind.*, **1966**, 2012.
13. J. F. Cavalla and J. A. D. Willis, *J. Chem. Soc.* (*C*), **1967**, 693.
14. E. C. Taylor and N. W. Kalenda, *J. Am. Chem. Soc.*, **78**, 5108 (1956).
15. E. C. Taylor and R. V. Ravindranathan, *J. Org. Chem.*, **27**, 2622 (1962).
16. J. F. Cavalla, N. E. Webb, and J. A. D. Willis, *J. Chem. Soc.* (*C*), **1967**, 698.
17. E. C. Taylor and A. L. Borror, *J. Org. Chem.*, **26**, 4967 (1961).
18. E. C. Taylor and J. A. Zoltewicz, *J. Am. Chem. Soc.*, **83**, 248 (1961).
19. E. C. Taylor and W. A. Ehrhart, *J. Am. Chem. Soc.*, **82**, 3138 (1960).
20. E. C. Taylor and P. K. Loeffler, *J. Am. Chem. Soc.*, **82**, 3147 (1960).
21. E. C. Taylor, R. J. Knopf, and A. L. Borror, *J. Am. Chem. Soc.*, **82**, 3152 (1960).
22. E. C. Taylor, R. J. Knopf, R. F. Meyer, A. Holmes, and M. L. Hoefle, *J. Am. Chem. Soc.*, **82**, 5711 (1960).
23. E. C. Taylor, R. J. Knopf, J. A. Cogliano, J. W. Barton, and W. Pfleiderer, *J. Am. Chem. Soc.*, **82**, 6058 (1960).
24. E. C. Taylor and P. K. Loeffler, *J. Org. Chem.*, **24**, 2035 (1959).
25. E. C. Taylor and K. S. Hartke, *J. Am. Chem. Soc.*, **81**, 2456 (1959).
26. E. C. Taylor and K. S. Hartke, *J. Am. Chem. Soc.*, **81**, 2452 (1959).
27. E. C. Taylor, A. J. Crovetti, and R. J. Knopf, *J. Am. Chem. Soc.*, **80**, 427 (1958).
28. J. F. Cavalla, A. R. Katritzky, M. J. Sewell, and G. R. Bedford, *J. Chem. Soc.*, **1965**, 4546.
29. B. Eistert and H. Minas, *Chem. Ber.*, **97**, 2479 (1964).
30. E. C. Taylor and A. J. Crovetti, *J. Org. Chem.*, **19**, 1633 (1954).
31. E. C. Taylor and N. W. Kalenda, *J. Org. Chem.*, **18**, 1755 (1953).
32. J. P. Ferris and L. E. Orgel, *J. Am. Chem. Soc.*, **88**, 3829 (1966).
33. A. Dornow and E. Schleese, *Chem. Ber.*, **91**, 1830 (1958).
34. C. F. Hammer and R. A. Hines, *J. Am. Chem. Soc.*, **77**, 3649 (1955).
35. C. W. Whitehead and J. J. Traverso, *J. Am. Chem. Soc.*, **78**, 5294 (1956).
36. G. W. Kenner, B. Lythgoe, A. R. Todd, and A. Topham, *J. Chem. Soc.*, **1943**, 388.
37. H. Ph. Baudet, *Rec. Trav. Chim.*, **43**, 707 (1924).
38. K. Gewald, *J. prakt. Chem.*, **32**, 26 (1966).
39. W. Borsche, H. Weussmann, and A. Fritzsche, *Chem. Ber.*, **57**, 1149 (1924).
40. J. Pinnow and C. Sämann, *Chem. Ber.*, **29**, 623 (1896).

41. C. W. Jefford, Ph.D. Thesis, Princeton University, Princeton, N.J., 1962.
42. F. Korte and H. Weithamp, *Ann.*, **622**, 121 (1959).
43. C. W. Whitehead and J. J. Traverso, *J. Am. Chem. Soc.*, **77**, 5867 (1955).
44. J. Baddiley, B. Lythgoe, and A. R. Todd, *J. Chem. Soc.*, **1943**, 386.
45. M. T. Bogert, H. C. Breneman, and W. F. Hand, *J. Am. Chem. Soc.*, **25**, 372 (1903).
46. W. Huber, *J. Am. Chem. Soc.*, **66**, 876 (1944).
47. K. Gewald and E. Schinke, *Chem. Ber.*, **99**, 2712 (1966).
48. E. Ochiai, K. Yanai, and T. Naito, *J. Pharm. Soc. Japan*, **63**, 25 (1943); *Chem. Abstr.*, **45**, 609fgh (1951).
49. Y. Sawa, F. Osawa, and H. Kaneko, *J. Pharm. Soc. Japan*, **67**, 204 (1947); *Chem. Abstr.*, **45**, 9063g (1951).
50. M. Ohta, *J. Pharm. Soc. Japan*, **67**, 161 (1947); *Chem. Abstr.*, **45**, 9545def (1951).
51. T. Matsukawa and S. Yurugi, *J. Pharm. Soc. Japan*, **72**, 1585 (1952); *Chem. Abstr.*, **47**, 9330g (1953).
52. R. G. Jarque and C. Vallmitjana Sala, *Anales Real Soc. Espan. Fis. Quim. (Madrid)*, **42**, 349 (1946); *Chem. Abstr.*, **41**, 4797hi (1947).
53. H. Suter and E. Habicht, U.S. Pat. 2,698,326 (Dec. 28, 1954); *Chem. Abstr.*, **50**, 1093e (1956).
54. H. Andersag and K. Westphal, U.S. Pat. 2,377,395 (June 5, 1945); *Chem. Abstr.*, **40**, 98, 100 (1946).
55. T. Matsukawa, *J. Pharm. Soc. Japan*, **62**, 417 (1942); *Chem. Abstr.*, **45**, 4723i, 4724a (1951).
56. M. E. Kreling and A. F. McKay, *Can. J. Chem.*, **40**, 143 (1962).
57. M. M. Delépine, *Bull. Soc. Chim. France*, **1938**, 1539.
58. D. J. Brown and L. N. Short, *J. Chem. Soc.*, **1953**, 331.
59. H. E. Schroeder and G. W. Rigby, *J. Am. Chem. Soc.*, **71**, 2205 (1949).
60. L. F. Fieser and M. M. Pechet, *J. Am. Chem. Soc.*, **68**, 2577 (1946).
61. K. W. Breukink and P. E. Verkade, *Rec. Trav. Chim.*, **79**, 443 (1960).
62. G. Bargellini and C. J. Turi, *Gazz. Chim. Ital.*, **84**, 157 (1954).
63. Z. Budesinsky and J. Kopecky, *Collection Czech. Chem. Commun.*, **20**, 52 (1955).
64. D. L. Garmaise and S. Gelblum, *Can. J. Chem.*, **38**, 1639 (1960).
65. A. Etienne and A. Staehelin, *Bull. Soc. Chim. France*, **1954**, 743.
66. S. Baldwin, *J. Org. Chem.*, **26**, 3288 (1961).
67. K. W. Breukink, L. H. Krol, P. E. Verkade, and B. M. Wepster, *Rec. Trav. Chim.*, **76**, 401 (1957).
68. H. Rupe and A. Metzger, *Helv. Chim. Acta*, **8**, 838 (1925).
69. M. R. S. Weir and J. B. Hyne, *Can. J. Chem.*, **41**, 2905 (1963).
70. P. Friedländer and S. Littner, *Chem. Ber.*, **48**, 328 (1915).
71. N. J. Leonard, K. L. Carraway, and J. P. Helgeson, *J. Heterocyclic Chem.*, **2**, 291 (1965).
72. E. Grigat and R. Putter, *Angew. Chem.*, **77**, 913 (1965).
73. K. Gewald, *Angew. Chem.*, **77**, 916 (1965); *Intern. Ed. (English)*, **4**, 881 (1965).
74. K. Gewald, *Chem. Ber.*, **98**, 3571 (1965).
75. W. Huber and H. A. Hölscher, *Chem. Ber.*, **71**, 87 (1938).
76. T. Sheradsky and P. L. Southwick, *J. Org. Chem.*, **30**, 194 (1965).
77. J. H. Mowat, F. J. Pilgrim, and G. H. Carlson, *J. Am. Chem. Soc.*, **65**, 954 (1943).

78. A. H. Berrie, G. T. Newbold, and F. S. Spring, *J. Chem. Soc.*, **1952**, 2042.
79. M. T. Bogert and W. F. Hand, *J. Am. Chem. Soc.*, **25**, 935 (1903).
80. P. J. Vanderhorst and C. S. Hamilton, *J. Am. Chem. Soc.*, **75**, 656 (1953).
81. S. Baldwin, *J. Org. Chem.*, **26**, 3280 (1961).
82. R. Pschorr and G. Hoppe, *Chem. Ber.*, **43**, 2543 (1910).
83. R. Nietzki and W. Petri, *Chem. Ber.*, **33**, 1788 (1900).
84. R. A. Carboni, D. D. Coffman, and E. G. Howard, *J. Am. Chem. Soc.*, **80**, 2838 (1958).
85. R. Justoni and R. Fusco, *Gazz. Chim. Ital.*, **68**, 59 (1938).
86. A. Ichiba, S. Emoto, and M. Nagai, *J. Sci. Res. Inst.* (*Tokyo*), **43**, 23 (1948); *Chem. Abstr.*, **43**, 4673 (1949).
87. A. Lespagnol, E. Cuingnet, and H. Beerens, *Bull. Soc. Pharm. Lille*, No. 1, 60 (1955); *Chem. Abstr.*, **50**, 3438 (1956).
88. J. F. Cavalla, *J. Chem. Soc.*, **1962**, 4664.
89. J. K. Williams, *J. Org. Chem.*, **28**, 1054 (1963).
90. E. D. Bergmann and Z. Pelchowicz, *J. Am. Chem. Soc.*, **75**, 4281 (1953).
91. E. C. Taylor and J. G. Berger, *Angew. Chem.*, **78**, 144 (1966); *Intern. Ed.* (*English*), **5**, 131 (1966).
92. E. C. Taylor and A. Abul-Husn, *J. Org. Chem.*, **31**, 342 (1966).
93. E. C. Taylor, R. N. Warrener, and A. McKillop, *Angew. Chem.*, **78**, 333 (1966); *Intern. Ed.* (*English*), **5**, 309 (1966).
94. E. C. Taylor, A. McKillop, and R. N. Warrener, *Tetrahedron*, **23**, 891 (1967).
95. E. C. Taylor, S. Vromen, A. McKillop, and R. V. Ravindranathan, *Angew. Chem.*, **78**, 332 (1966); *Intern. Ed.* (*English*), **5**, 308 (1966).
96. E. C. Taylor, A. McKillop, and S. Vromen, *Tetrahedron*, **23**, 885 (1967).
97. P. Schmidt, K. Eichenberger, M. Wilhelm, and J. Druey, *Helv. Chim. Acta*, **42**, 763 (1959).
98. F. Montequi, *Anales Real Soc. Espan. Fis. Quim.* (*Madrid*), **25**, 182 (1927).
99. Q. E. Thompson, *J. Am. Chem. Soc.*, **80**, 5483 (1958).
100. K. Dimroth, D. Holzner, and H. G. Aurieh, *Chem. Ber.*, **98**, 3907 (1965).
101. K. v. Auwers, T. Bahr, and E. Frese, *Ann.*, **441**, 68 (1925).
102. T. L. Cairns, R. A. Carboni, D. D. Coffman, V. A. Engelhardt, R. E. Heckert, E. L. Little, E. G. McGeer, B. C. McKusick, W. J. Middleton, R. M. Scribner, C. W. Theobald, and H. E. Winberg, *J. Am. Chem. Soc.*, **80**, 2775 (1958).
103. G. N. Sausen, V. A. Engelhardt, and W. J. Middleton, *J. Am. Chem. Soc.*, **80**, 2815 (1958).
104. W. J. Middleton, V. A. Engelhardt, and B. S. Fisher, *J. Am. Chem. Soc.*, **80**, 2822 (1958).
105. W. J. Middleton and V. A. Engelhardt, *J. Am. Chem. Soc.*, **80**, 2829 (1958).
106. E. L. Little, Jr., W. J. Middleton, D. D. Coffman, V. A. Engelhardt, and G. N. Sausen, *J. Am. Chem. Soc.*, **80**, 2832 (1958).
107. K. Gewald, P. Blauschmidt, and R. Mayer, *J. prakt. Chem.*, **35**, 97 (1967).
108. A. Dornow and E. Neuse, *Arch. Pharm.*, **288**, 174 (1955).
109. G. Märkl, *Angew. Chem.*, **77**, 1109 (1965); *Intern. Ed.* (*English*), **4**, 1023 (1965).
110. M. J. Gallagher, E. C. Kirby, and F. G. Mann, *J. Chem. Soc.*, **1963**, 4846.
111. M. J. Gallagher and F. G. Mann, *J. Chem. Soc.*, **1962**, 5110.
112. A. H. Cook and K. J. Reed, *J. Chem. Soc.*, **1945**, 399.
113. K. Gewald, E. Schinke, and H. Böttcher, *Chem. Ber.*, **99**, 94 (1966).
114. G. Shaw and D. N. Butler, *J. Chem. Soc.*, **1959**, 4040.

115. P. Schmidt, K. Eichenberger, M. Wilhelm, and C. A. Burckhardt, *Ann. Chim.*, **53**, 62 (1963).
116. J. F. Cavalla, *Tetrahedron Letters*, **1964**, 2807.
117. J. Colonge, G. Descotes, and A. Frenay, *Bull. Soc. Chim. France*, **1963**, 2264.
118. F. C. Cooper and M. W. Partridge, *J. Chem. Soc.*, **1954**, 3429.
119. M. A. Aron and J. A. Elvidge, *Chem. Ind.*, **1958**, 1234.
120. F. C. Cooper, *J. Chem Soc* , **1958**, 4212.
121. B. R. Baker, R. E. Schaub, J. P. Joseph, F. J. McEvoy, and J. H. Williams, *J. Org. Chem.*, **17**, 141 (1952).
122. H. Graboyes, G. E. Jaffe, I. J. Pachter, J. P. Rosenbloom, A. J. Villani, J. W. Wilson, and J. Weinstock, *J. Med. Chem.*, **11**, 568 (1968).
123. M. T. Bogert and W. F. Hand, *J. Am. Chem. Soc.*, **24**, 1031 (1902).
124. G. M. Badger and R. P. Rao, *Austr. J. Chem.*, **18**, 1267 (1965).
125. S. G. Cottis, P. B. Clarke, and H. Tieckelmann, *J. Heterocyclic Chem.*, **2**, 192 (1965).
126. M. T. Bogert and W. F. Hand, *J. Am. Chem. Soc.*, **28**, 94 (1906).
127. M. T. Bogert and Y. G. Chen, *J. Am. Chem. Soc.*, **44**, 2352 (1922).
128. S. Gabriel, *Chem. Ber.*, **36**, 800 (1903).
129. C. Grundmann and H. Ulrich, *J. Org. Chem.*, **24**, 272 (1959).
130. H. G. Mautner *J. Org. Chem.*, **23**, 1450 (1958).
131. R. L. McKee, M. K. McKee, and R. W. Bost, *J.Am. Chem. Soc.*, **68**, 1902 (1946).
132. R. L. McKee, M. K. McKee, and R. W. Bost, *J. Am. Chem. Soc.*, **69**, 940 (1947).
133. D. M. Mulvey, S. G. Cottis, and H. Tieckelmann, *J. Org. Chem.*, **29**, 2903 (1964).
134. E. W. Parnell, *J. Chem. Soc.*, **1961**, 4930.
135. H. Adkins and G. M. Whitman, *J. Am. Chem. Soc.*, **64**, 150 (1942).
136. R. N. Prasad and R. K. Robins, *J. Am. Chem. Soc.*, **79**, 6401 (1957).
137. A. Reissert and F. Grube, *Chem. Ber.*, **42**, 3710 (1909).
138. F. S.-Legagneur and J. Rabadeux, *Compt. Rend.*, **261**, 5524 (1965).
139. J. Druey, P. Schmidt, K. Eichenberger, and M. Wilhelm, U.S. Pat. 3,055,900 (Sept. 25, 1962); *Chem. Abstr.*, **59**, 8763 (1963).
140. P. Schmidt, K. Eichenberger, and M. Wilhelm, Ger. Pat. 1,147,234 (April 18, 1963); *Chem. Abstr.*, **59**, 11529 (1963).
141. J. Weinstock and V. D. Wiebelhaus, Fr. Pat. 1,335,354 (Aug. 16, 1963); *Chem. Abstr.*, **60**, 2975 (1964).
142. W. Zerweck and W. Kunze, Ger. Pat. 737,931 (July 1, 1943); *Chem. Abstr.*, **38**, 3993 (1944).
143. M. L. Hoefle and R. F. Meyer, U.S. Pat. 2,949,466 (Aug. 16, 1960); *Chem. Abstr.*, **55**, 589 (1961).
144. J. Druey, P. Schmidt, and K. Eichenberger, Ger. Pat. 1,089,388 (Sept. 22, 1960) (Swiss Pat. 368,499 (Nov. 26, 1957)); *Chem. Abstr.*, **57**, 4681 (1962).
145. J. Druey and P. Schmidt, U.S. Pat. 2,965,643 (Dec. 20, 1960); *Chem. Abstr.*, **57**, 11211 (1962).
146. D. J. Brown and M. N. Paddon-Row, *J. Chem. Soc.*, **1966**, 164.
147. Brit. Pat. 893,235 (April 4, 1962); *Chem. Abstr.*, **58**, 12581 (1963).
148. J. Druey and P. Schmidt, U.S. Pat. 2,980,677 (April 18, 1961); *Chem. Abstr.*, **55**, 18784 (1961).

149. J. Druey, P. Schmidt, and K. Eichenberger, U.S. Pat. 3,098,075 (July 16, 1963); *Chem. Abstr.*, **61**, 10689 (1964).
150. J. Druey, P. Schmidt, and K. Eichenberger, U.S. Pat. 2,989,537 (June 20, 1961); *Chem. Abstr.*, **55**, 25990 (1961).
151. S. G. Cottis, Ph.D. Thesis, University of Buffalo, Buffalo, N.Y., 1962.
152. W. Dymek and D. Sybistowicz, *Roczniki Chem.*, **37**, 547 (1963); *Chem. Abstr.*, **59**, 10040 (1963).
153. Brit. Pat. 483,585 (April 22, 1938); *Chem. Abstr.*, **32**, 7283 (1938).
154. Brit. Pat. 798,662 (July 23, 1958); *Chem. Abstr.*, **53**, 1382 (1959).
155. J. Druey, P. Schmidt, and K. Eichenberger, Swiss Pat. 373,761 (Jan. 31, 1964); *Chem. Abstr.*, **61**, 5667 (1964).
156. K. Hartke and L. Peshkar, *Angew. Chem.*, **79**, 56 (1967); *Intern. Ed.* (*English*), **6**, 83 (1967).
157. J. Druey and P. Schmidt, Ger. Pat. 1,058,519 (June 4, 1959); *Chem. Abstr.*, **55**, 13459 (1961). (Same patent: U.S. Pat. 2,925,418 (Feb. 16, 1960); *Chem. Abstr.*, **54**, 9971 (1960).)
158. M. S. Puar, H. S. Sachdev, and N. K. Ralhan, *Indian J. Chem.*, **2**, 285 (1964).
159. J. Druey, P. Schmidt, K. Eichenberger, and M. Wilhelm, Ger. Pat. 1,092,922, appl. Nov. 7, 1958; *Chem. Abstr.*, **56**, 5979 (1962).
160. Brit. Pat. 937,879 (Sept. 25, 1963); *Chem. Abstr.*, **60**, 2944 (1964).
161. S. K. Chatterjee and N. Anand, *J. Sci. Ind. Res.* (*India*), **17B**, 63 (1958); *Chem. Abstr.*, **52**, 20188 (1958).
162. E. M. Hawes and D. G. Wibberley, *J. Chem. Soc.*, **1966**, 315.
163. S. K. Chatterji and N. Anand, *J. Sci. Ind. Res.* (*India*), **18B**, 272 (1959); *Chem. Abstr.*, **54**, 6745 (1960).
164. H. F. Mower and C. L. Dickinson, *J. Am. Chem. Soc.*, **81**, 4011 (1959).
165. F. C. Schaefer, K. R. Huffman, and G. A. Peters, *J. Org. Chem.*, **27**, 548 (1962).
166. K. R. Huffman, F. C. Schaefer, and G. A. Peters, *J. Org. Chem.*, **27**, 551 (1962).
167. A. C. Cope and R. B. Kinnel, *J. Am. Chem. Soc.*, **88**, 752 (1966).
168. C. C. Cheng and R. K. Robins, *J. Org. Chem.*, **23**, 191 (1958).
169. C. C. Cheng and R. K. Robins, *J. Org. Chem.*, **23**, 852 (1958).
170. E. Y. Sutcliffe, K. Y. Zee-Cheng, C. C. Cheng, and R. K. Robins, *J. Med. Pharm. Chem.*, **5**, 588 (1962).
171. R. Gompper, E. Kutter, and W. Töpfl, *Ann.*, **659**, 90 (1962).
172. H. Junek, *Monatsh. Chem.*, **94**, 890 (1963).
173. W. R. Hatchard, *J. Org. Chem.*, **29**, 660 (1964).
174. F. Ehrlich, *Chem. Ber.*, **34**, 3366 (1901).
175. K. Gewald, *Z. Chem.*, **1**, 349 (1961).
176. H. Kanō, Y. Makisumi, and K. Ogata, *Chem. Pharm. Bull.* (*Tokyo*), **6**, 105 (1958).
177. A. Dornow and H. Teckenburg, *Chem. Ber.*, **93**, 1103 (1960).
178. A. Albert, D. J. Brown, and G. Cheeseman, *J. Chem. Soc.*, **1951**, 474.
179. G. de Stevens, A. Halamandaris, P. Wenk, R. A. Mull, and E. Schlittler, *Arch. Biochem. Biophys.*, **83**, 141 (1959).
180. J. Weinstock and V. D. Wiebelhaus, U.S. Pat. 3,134,778 (May 26, 1964); *Chem. Abstr.*, **61**, 4372 (1964).
181. A. C. Cope and R. J. Cotter, *J. Org. Chem.*, **29**, 3467 (1964).
182. M. S. Newman and R. D. Closson, *J. Am. Chem. Soc.*, **66**, 1553 (1944).
183. P. Grammaticakis, *Bull. Soc. Chim. France*, **1953**, 207.

184. W. Findeklee, *Chem. Ber.*, **38**, 3542 (1905).
185. R. C. Ellingson, R. L. Henry, and F. G. McDonald, *J. Am. Chem. Soc.*, **67**, 1711 (1945).
186. H. M. A. Hartmans, *Rec. Trav. Chim.*, **65**, 468 (1946).
187. K. Ziegler, H. Eberle, and H. Ohlinger, *Ann.*, **504**, 94 (1933).
188. K. Ziegler and W. Hechelhammer, *Ann.*, **528**, 114 (1937).
189. K. Ziegler and H. Hole, *Ann.*, **528**, 143 (1937).
190. K. Ziegler and R. Aurnhammer, *Ann.*, **513**, 43 (1934).
191. C. W. Moore and J. F. Thorpe, *J. Chem. Soc.*, **1908**, 165.
192. J. Kenner and E. G. Turner, *J. Chem. Soc.*, **1911**, 2101.
193. K. Ziegler and A. Lüttringhaus, *Ann.*, **511**, 1 (1934).
194. S. R. Best and J. F. Thorpe, *J. Chem. Soc.*, **1909**, 685.
195. J F. Thorpe, *J. Chem. Soc.*, **1909**, 1901.
196. A. R. Todd and F. Bergel, *J. Chem. Soc.*, **1937**, 364.
197. J. Pinnow and E. Müller, *Chem. Ber.*, **28**, 149 (1895).
198. S. von Niementowski, *J. prakt. Chem.*, **40**, 1 (1889).
199. S. von Niementowski, *Chem. Ber.*, **21**, 1534 (1888).
200. K. Ziegler, H. Ohlinger, and H. Eberle, Ger. Pat. 591,269 (Jan. 19, 1934); *Chem. Abstr.*, **28**, 2364 (1934).
201. K. Ziegler, Ger. Pat. 620,904 (Oct. 30, 1935); *Chem. Abstr.*, **30**, 735 (1936).
202. K. Ziegler, U.S. Pat. 2,068,854 (Jan. 19, 1937); *Chem. Abstr.*, **31**, 1820 (1937).
203. R. Grewe, *Z. Physiol. Chem.*, **242**, 89 (1936).
204. J. C. Westfahl and T. L. Gresham, *J. Org. Chem.*, **21**, 319 (1956).
205. P. Newman, P. Rutkin, and K. Mislow, *J. Am. Chem. Soc.*, **80**, 465 (1958).
206. D. D. Fitts, M. Siegel, and K. Mislow, *J. Am. Chem. Soc.*, **80**, 480 (1958).
207. K. Mislow and P. Newman, *J. Am. Chem. Soc.*, **79**, 1769 (1957) (prelim. comm. of ref. 205).
208. K. Mislow, P. Rutkin, and A. K. Lazarus, *J. Am. Chem. Soc.*, **79**, 2974 (1957) (prelim. comm. of ref. 205,6).
209. K. Mislow and F. A. McGinn, *J. Am. Chem. Soc.*, **80**, 6036 (1958).
210. D. A. Stauffer and O. E. Fancher, *J. Org. Chem.*, **25**, 935 (1960).
211. D. A. Stauffer and O. E. Fancher, U.S. Pat. 2,647,896 (Aug. 4, 1953); *Chem. Abstr.*, **48**, 9405 (1954) (patent for ref. 210).
212. L. H. Krol, P. E. Verkade, and B. M. Wepster, *Rec. Trav. Chim.*, **71**, 545 (1952).
213. C. Haslinger, *Chem. Ber.*, **41**, 1444 (1908).
214. L. Alessandri, *Atti Accad. Nazl. Lincei, Mem. Classe. Sci. Fis. Mat. Nat. Sez. II*, **22**, 150–155, 227–234 (1913); *Chem. Abstr.*, **8**, 337 (1914).
215. D. J. Brown, *The Pyrimidines* (*Chemistry of Heterocyclic Compounds*, vol. 16, A. Weissberger, Ed.) Interscience, New York, 1962.
216. E. Söderbäck, *Acta Chem. Scand.*, **17**, 362 (1963).
217. K. Gewald, *Angew. Chem.*, **73**, 114 (1961).
218. F. M. Dean and K. Manunapichu, *J. Chem. Soc.*, **1957**, 3112.
219. G. B. Bachman and R. S. Barker, *J. Am. Chem. Soc.*, **69**, 1535 (1947).
220. N. R. Easton, H. E. Reiff, G. Svamas, and V. B. Fish, *J. Am. Chem. Soc.*, **74**, 260 (1952).
221. N. R. Easton and S. J. Nelson, *J. Am. Chem. Soc.*, **75**, 640 (1953).
222. E. C. Horning, M. G. Horning, and E. J. Platt, *J. Am. Chem. Soc.*, **69**, 2929 (1947).

223. T. Sakan and M. Nakazaki, *J. Inst. Polytech., Osaka City Univ. Japan*, **1**, No. 2, 23–29 (1950); *Chem. Abstr.*, **46**, 5036 (1952).
224. C. W. Muth, W-L. Sung, and Z. B. Papanastassiou, *J. Am. Chem. Soc.*, **77**, 3393 (1955).
225. E. C. Horning, M. G. Horning, and E. J. Platt, *J. Am. Chem. Soc.*, **70**, 2072 (1948).
226. E. C. Horning, M. G. Horning, and E. J. Platt, *J. Am. Chem. Soc.*, **72**, 2731 (1950).
227. M. S. Newman and W. L. Mosby, *J. Am. Chem. Soc.*, **73**, 3738 (1951).
228. E. C. Horning and A. F. Finelli, *J. Am. Chem. Soc.*, **73**, 3741 (1951).
229. H. Rapoport and A. R. Williams, *J. Am. Chem. Soc.*, **71**, 1774 (1949).
230. H. Irie, Y. Tsuda, and S. Uyeo, *J. Chem. Soc.*, **1959**, 1446.
231. R. P. Welcher, G. A. Johnson, and V. P. Wystrach, *J. Am. Chem. Soc.*, **82**, 4437 (1960).
232. C. L. Dickinson, J. K. Williams, and B. C. McKusick, *J. Org. Chem.*, **29**, 1915 (1964).
233. M. Lamant, *Ann. Chim. (France)*, **4**, 87 (1959).
234. K. Gewald, *Z. Chem.*, **2**, 305 (1962).
235. K. Ziegler and K. Weber, *Ann.*, **512**, 164 (1934).
236. A. Compère, *Bull. Soc. Chim. (Belg.)*, **44**, 523 (1935).
237. W. A. Lazier and B. W. Howk, U.S. Pat. 2,292,949 (Aug. 11, 1943); *Chem. Abstr.*, **37**, 889 (1943).
238. R. H. Halliwell, U.S. Pat. 2,768,132 (Oct. 23, 1956); *Chem. Abstr.*, **51**, 5815 (1957).
239. Brit. Pat. 728,522 (Apr. 20, 1955); *Chem. Abstr.*, **50**, 5729 (1956).
240. M. F. Bartlett, D. F. Dickel, and W. I. Taylor, *J. Am. Chem. Soc.*, **80**, 126 (1958).
241. R. V. Ravindranathan, Ph.D. Thesis, Princeton University, Princeton, N.J., 1962.
242. M. R. S. Weir and J. B. Hyne, *Can. J. Chem.*, **42**, 1440 (1964).
243. K. Ziegler, U.S. Pat. 2,068,586 (Jan. 19, 1937); *Chem. Abstr.*, **31**, 1820 (1937).
244. C. E. Kwartler and P. Lucas, *J. Am. Chem. Soc.*, **65**, 1804 (1943).
245. P. Friedländer, S. Bruckner, and G. Deutsch, *Ann.*, **388**, 23 (1912).
246. S. G. Cottis and H. Tieckelmann, *J. Org. Chem.*, **26**, 79 (1961).
247. M. T. Bogert and W. F. Hand, *J. Am. Chem. Soc.*, **27**, 1476 (1905).
248. E. B. Hunn, *J. Am. Chem. Soc.*, **45**, 1024 (1923).
249. A. H. Cook, I. M. Heilbron, and K. J. Reed, *J. Chem. Soc.*, **1945**, 182.
250. A. H. Cook, I. M. Heilbron, K. J. Reed, and M. N. Strachan, *J. Chem. Soc.*, **1945**, 861.
251. S. Trofimenko, *J. Org. Chem.*, **28**, 2755 (1963).
252. H. C. Carrington, *J. Chem. Soc.*, **1955**, 2527.
253. Ger. Pat. 212,207 (Dec. 6, 1907); *Chem. Abstr.*, **3**, 2756 (1909).
254. P. Karrer, R. Schwyzer, and K. Kostić, *Helv. Chim. Acta*, **33**, 1482 (1950).
255. K. Miyatake and M. Tsunoo, *J. Pharm. Soc. Japan*, **72**, 630 (1952); *Chem. Abstr.*, **47**, 2177 (1953).
256. H. Musso and H. Schröder, *Chem. Ber.*, **98**, 1562 (1965).
257. H. Musso and H. Schröder, *Chem. Ber.*, **98**, 1577 (1965).
258. K. Gewald, *Chem. Ber.*, **99**, 1002 (1966).
259. A. Kreutzberger and C. Grundmann, *J. Org. Chem.*, **26**, 1121 (1961).

260. L. Horner and K. Klüpfel, *Ann.*, **591**, 69 (1955).
261. B. C. McKusick, R. E. Heckert, T. L. Cairns, D. D. Coffman, and H. F. Mower, *J. Am. Chem. Soc.*, **80**, 2806 (1958).
262. E. M. Fry and L. F. Fieser, *J. Am. Chem. Soc.*, **62**, 3489 (1940).
263. H. Hübner, *Chem. Ber.*, **10**, 1697 (1877).
264. P. Friedländer, *Monatsh. Chem.*, **19**, 627 (1898).
265. R. Weitzenböck, *Monatsh. Chem.*, **34**, 193 (1913).
266. R. Gompper and W. Töpfl, *Chem. Ber.*, **95**, 2881 (1962).
267. H. Bretschneider and K. Hohenlohe-Oehringen, *Monatsh. Chem.*, **89**, 358 (1958).
268. F. S.-Legagneur and C. Neven, *Bull. Soc. Chim. France*, **1956**, 929.
269. P. Nedenskov, W. Taub, and D. Ginsburg, *Acta Chem. Scand.*, **12**, 1405 (1958).
270. C. Iwanoff, *Chem. Ber.*, **87**, 1600 (1954).
271. V. Jürgens, *Chem. Ber.*, **40**, 4409 (1907).
272. J. Kenner and E. Witham, *J. Chem. Soc.*, **1921**, 1452.
273. J. J. Bloomfield and P. V. Fennessey, *Tetrahedron Letters*, **1964**, 2273.
274. S. Gabriel and A. Thieme, *Chem. Ber.*, **52**, 1079 (1919).
275. M. Mayer, *J. prakt. Chem.*, **92**, 137 (1915).
276. G. Westöö, *Acta Chem. Scand.*, **13**, 692 (1959).
277. F. S.-Legagneur, *Bull. Soc. Chim. France*, **1956**, 411.
278. Fr. Pat. 828,202 (May 12, 1938); *Chem. Abstr.*, **33**, 176 (1939).
279. R. L. Perkins and A. J. Sweet, U.S. Pat. 2,044,015 (June 16, 1933); *Chem. Abstr.*, **30**, 5427 (1936).
280. H. Schröder, U. Schwabe, and H. Musso, *Chem. Ber.*, **98**, 2556 (1965).
281. S. S. Kulp, V. B. Fish, and N. R. Easton, *Can. J. Chem.*, **43**, 2512 (1965).
282. S. S. Kulp, V. B. Fish, and N. R. Easton, *J. Med. Chem.*, **6**, 516 (1963).
283. J. Goerdeler and H. W. Pohland, *Chem. Ber.*, **96**, 526 (1963).
284. O. Riobé and L. Gouin, *Compt. Rend.*, **234**, 1889 (1952).
285. F. S.-Legagneur and C. Neven, *Compt. Rend.*, **239**, 1809 (1954).
286. C. Dufraisse, A. Etienne, and H. V. de Pradenne, *Compt. Rend.*, **239**, 1744 (1954).
287. F. S.-Legagneur and C. Neven, *Compt. Rend.*, **237**, 64 (1953).
288. M. Lamant, *Compt. Rend.*, **238**, 1591 (1954).
289. A. Rosowsky and E. J. Modest, *J. Org. Chem.*, **31**, 2607 (1966).
290. E. C. Taylor, A. McKillop, Y. Shvo, and G. H. Hawks, *Tetrahedron*, **23**, 2081 (1967).
291. G. Glock, *Chem. Ber.*, **21**, 2659 (1888).
292. Z. v. Jakubowski, *Chem. Ber.*, **43**, 3026 (1910).
293. M. T. Bogert and A. Hoffman, *J. Am. Chem. Soc.*, **27**, 1293 (1905).
294. G. T. Morgan and E. A. Coulson, *J. Chem. Soc.*, **1929**, 2551.
295. B. J. Heywood and A. H. Knight, Brit. Pat. 594,218 (Nov. 5, 1947); *Chem. Abstr.*, **42**, 3579 (1948).
296. J. B. Dickey, U.S. Pat. 2,436,100 (Feb. 17, 1948); *Chem. Abstr.*, **42**, 3578 (1948).
297. Swiss Pat. 216,157 (Nov. 17, 1941); *Chem. Abstr.*, **42**, 6127 (1948).
298. B. J. Heywood, U.S. Pat. 2,474,737 (June 28, 1949); *Chem. Abstr.*, **43**, 8155 (1949).
299. Swiss Pat. 235,462 (April 3, 1945); *Chem. Abstr.*, **43**, 9459 (1949).
300. Swiss Pat. 227,512 (Sept. 1, 1943); *Chem. Abstr.*, **43**, 4474 (1949).

301. J. B. Dickey, U.S. Pat. 2,516,302 (July 25, 1950); *Chem. Abstr.*, **45**, 8254 (1951).
302. W. Kruckenberg, Ger. Pat. 928,902 (June 13, 1955); *Chem. Abstr.*, **50**, 11677 (1956).
303. Ger. Pat. 942,221 (April 26, 1956); *Chem. Abstr.*, **50**, 11678 (1956).
304. E. Hoffa and H. Heyna, U.S. Pat. 1,774,650 (Sept. 2, 1930); *Chem. Abstr.*, **24**, 5044 (1930).
305. O. v. Schickh, Ger. Pat. 681,639 (Sept. 7, 1939); *Chem. Abstr.*, **36**, 1953 (1942).
306. W. Kruckenberg, Ger. Pat. 953,548 (Dec. 6, 1956); *Chem. Abstr.*, **51**, 10068 (1957).
307. W. Kruckenberg, Ger. Pat. 960,752 (March 28, 1957); *Chem. Abstr.*, **51**, 12498 (1957).
308. K. Watanabe, *Nippon Kagaku Zasshi*, **76**, 391 (1955); *Chem. Abstr.*, **51**, 17815 (1957).
309. W. Kruckenberg and K. Weis, U.S. Pat. 2,945,849 (July 19, 1960); *Chem. Abstr.*, **55**, 3998 (1961).
310. Brit. Pat. 849,994 (Sept. 28, 1960); *Chem. Abstr.*, **55**, 7852 (1961).
311. K. Akanuma, H. Amamiya, T. Hayashi, K. Watanabe, and K. Hata, *Nippon Kagaku Zasshi*, **81**, 333 (1960); *Chem. Abstr.*, **56**, 406 (1962).
312. J. Yates, Brit. Pat. 901,977 (July 25, 1962); *Chem. Abstr.*, **57**, 13696 (1962).
313. G. H. Hitchings, E. A. Falco, and K. W. Ledig, Ger. Pat. 1,125,939 (March 22, 1962); *Chem. Abstr.*, **57**, 16633 (1962).
314. I. K. Fel'dman, N. N. Bel'tsova, and V. K. Grishkova, *Mechenye Biol. Aktivn. Veshchestva, Sb. Statei*, **1962**, 87; *Chem. Abstr.*, **59**, 7419 (1963).
315. Fr. Pat. 1,369,628 (Aug. 14, 1964); *Chem. Abstr.*, **61**, 16021 (1964).
316. L. J. Sargent, *J. Org. Chem.*, **19**, 599 (1954).
317. E. A. Braude, R. P. Linstead, and K. R. H. Wooldridge, *J. Chem. Soc.*, **1954**, 3586 (R. P. Linstead and E. A. Braude, Brit. Pat. 705,919 (March 24, 1954); *Chem. Abstr.*, **50**, 1079 (1956)).
318. E. D. Bergmann and M. Bentov, *J. Org. Chem.*, **20**, 1654 (1955).
319. H. Koopman, *Rec. Trav. Chim.*, **80**, 1075 (1961).
320. M. W. Partridge, H. J. Vipond, and J. A. Waite, *J. Chem. Soc.*, **1962**, 2549.
321. K. Butler and M. W. Partridge, *J. Chem. Soc.*, **1959**, 2396.
322. P. J. Krueger, *Can. J. Chem.*, **40**, 2300 (1962).
323. T. R. Govindachari, S. Rajappa, and V. Sudarsanam, *Ind. J. Chem.*, **1**, 247 (1963).
324. A. N. Hambly and B. V. O'Grady, *Austral. J. Chem.*, **15**, 626 (1962).
325. A. Schaarschmidt, *Ann.*, **405**, 95 (1914).
326. M. Delépine and K. A. Jensen, *Bull. Soc. Chim. France*, **6**, 1663 (1939).
327. M. St. C. Flett, *Spectrochim. Acta*, **18**, 1537 (1962).
328. P. J. Krueger and H. W. Thompson, *Proc. Roy. Soc.* (*London*), **250**, 22 (1959).
329. S. Mizukami and E. Hirai, *J. Org. Chem.*, **31**, 1199 (1966).
330. S. Läufer and F. Lingens, *Z. Anal. Chem.*, **181**, 494 (1961).
331. G. Jacini, *Gazz. Chim. Ital.*, **71**, 532 (1941).
332. G. Jacini, *Gazz. Chim. Ital.*, **77**, 308 (1947).
333. P. Sensi and G. G. Gallo, *Gazz. Chim. Ital.*, **85**, 235 (1955).
334. A. M. Chacko, Ph.D. Dissertation, University of North Carolina at Chapel Hill, 1965; Univ. Microfilms, 65-14320, Ann Arbor, Michigan.
335. J. P. Ferris and L. E. Orgel, *J. Am. Chem. Soc.*, **87**, 4976 (1965).

336. J. P. Ferris and L. E. Orgel, *J. Am. Chem. Soc.*, **88**, 1074 (1966).
337. W. R. Boon, H. C. Carrington, J. S. H. Davies, P. Gaubert, W. G. M. Jones, G. R. Ramage, and W. S. Waring, cited in *The Chemistry of Penicillin*, H. T. Clarke, J. R. Johnson, and Sir R. Robinson, Eds., Princeton University Press, Princeton, N.J., 1949, pp. 702, 729.
338. C. D. May and P. Sykes, *J. Chem. Soc.*, **1966**, 649.
339. R. Gompper, M. Gäng, and F. Saygin, *Tetrahedron Letters*, **1966**, 1885.
340. E. C. Taylor and Y. Shvo, *J. Org. Chem.*, **33**, 1719 (1968).
341. E. C. Taylor and J. G. Berger, *J. Org. Chem.*, **32**, 2376 (1967).
342. D. M. W. Anderson, F. Bell, and J. L. Duncan, *J. Chem. Soc.*, **1961**, 4705.
343. A. Dornow and J. Helberg, *Chem. Ber.*, **93**, 2001 (1960).
344. J. R. E. Hoover and A. R. Day, *J. Am. Chem. Soc.*, **78**, 5832 (1956).
345. J. A. Barone, *J. Med. Chem.*, **6**, 39 (1963).
346. J. A. Barone, E. Peters, and H. Tieckelmann, *J. Org. Chem.*, **24**, 198 (1959).
347. J. A. Barone and H. Tieckelmann, *J. Org. Chem.*, **26**, 598 (1961).
348. D. E. O'Brien, F. Baiocchi, R. K. Robins, and C. C. Cheng, *J. Med. Chem.*, **6**, 467 (1963).
349. A. D. Josey, *J. Org. Chem.*, **29**, 707 (1964).
350. A. B. A. Jansen and M. Szelke, *J. Chem. Soc.*, **1961**, 405.
351. J. Kinugawa, M. Ochiai, C. Matsumura, and H. Yamamoto, *Chem. Pharm. Bull. Japan*, **12**, 182 (1964).
352. J. Kinugawa, M. Ochiai, and H. Yamamoto, *J. Pharm. Soc. Japan*, **83**, 1086 (1963).
353. D. M. Mulvey, *Dissertation Abstr.*, **26**, 5043 (1966); State University of New York at Buffalo, 1965; Univ. Microfilm, Order No. 65-10,167, Ann Arbor, Michigan.
354. R. R. Schmidt, *Chem. Ber.*, **98**, 3892 (1965).
355. R. J. Rousseau, L. B. Townsend, and R. K. Robins, *Chem. Commun.*, **1966**, 265.
356. K. Gewald, *Z. Chem.*, **3**, 26 (1963).
357. K. Gewald, *J. prakt. Chem.*, **31**, 205 (1966).
358. K. Gewald, *J. prakt. Chem.*, **31**, 214 (1966).
359. R. R. Schmidt, *Chem. Ber.*, **98**, 346 (1965).
360. M. M. Stimson, *J. Am. Chem. Soc.*, **71**, 1470 (1949).
361. C. W. Whitehead, *J. Am. Chem. Soc.*, **75**, 671 (1953).
362. A. Dornow and G. Petsch, *Chem. Ber.*, **86**, 1404 (1953).
363. A. Dornow and G. Petsch, *Ann.*, **588**, 45 (1954).
364. O. E. Polansky and M. A. Grassberger, *Monatsh. Chem.*, **94**, 662 (1963).
365. C. L. Dickinson, W. J. Middleton, and V. A. Engelhardt, *J. Org. Chem.*, **27**, 2470 (1962).
366. S. Trofimenko, E. L. Little, Jr., and H. F. Mower, *J. Org. Chem.*, **27**, 433 (1962).
367. J. Biggs and P. Sykes, *J. Chem. Soc.*, **1961**, 2595.
368. P. Schmidt, K. Eichenberger, and M. Wilhelm, *Angew. Chem.*, **73**, 15 (1961).
369. J. G. Nairn and H. Tieckelmann, *J. Org. Chem.*, **25**, 1127 (1960).
370. R. Huisgen and E. Laschtuvka, *Chem. Ber.*, **93**, 65 (1960).
371. P. Schmidt, K. Eichenberger, M. Wilhelm, and J. Druey, *Helv. Chim. Acta*, **42**, 349 (1959).
372. A. Dornow and E. Hinz, *Chem. Ber.*, **91**, 1834 (1958).
373. P. Schmidt, K. Eichenberger, and J. Druey, *Helv. Chim. Acta*, **41**, 1052 (1958).

374. W. J. Middleton, *Org. Syn.*, Coll. Vol. 4, p. 243.
375. Swiss Pats. 195,951 and 195,952 (May 16, 1938); *Chem. Abstr.*, **32**, 7216 (1938).
376. Brit. Pat. 486,414 (June 2, 1938); *Chem. Abstr.*, **32**, 7928 (1938).
377. M. Delépine, *Compt. Rend.*, **206**, 865 (1938).
378. O. Hromatka, Ger. Pat. 667,990 (Nov. 24, 1938); *Chem. Abstr.*, **33**, 2909 (1939). (This patent is U.S. 2,235,638 (March 18, 1941); *Chem. Abstr.*, **35**, 4041 (1941).)
379. T. Matukawa, Japan. Pat. 133,464 (Nov. 24, 1939); *Chem. Abstr.*, **35**, 4041 (1941).
380. K. Yano, Y. Ikeda, Y. Sawa, F. Osawa, and H. Kaneko, Japan. Pat. 178,360 (March 31, 1949); *Chem. Abstr.*, **46**, 541 (1952).
381. Y. Tanaka and H. Matsuoka, *J. Agr. Chem. Soc. Japan*, **24**, 74 (1950); *Chem. Abstr.*, **47**, 1507 (1953).
382. K. Miyatake and M. Tsunoo, *J. Pharm. Soc. Japan*, **72**, 630 (1952); *Chem. Abstr.*, **47**, 2177 (1953).
383. T. Iwatsu, *J. Pharm. Soc. Japan*, **72**, 354 (1952); *Chem. Abstr.*, **47**, 2178 (1953).
384. A. Ito, *J. Soc. Org. Syn. Chem* (*Japan*), **11**, 252 (1953); *Chem. Abstr.*, **47**, 12056 (1953).
385. S. Murahashi and A. Nishio, *J. Pharm. Soc. Japan*, **73**, 977 (1953); *Chem. Abstr.*, **48**, 11426 (1954).
386. J. Ishikawa and K. Akita, Japan. Pat. 1839 (April 7, 1954); *Chem. Abstr.*, **49**, 11726 (1955).
387. I. A. Rubtsov, M. V. Balyakina, E. V. Zaïtseva, and N. A. Preobrazhenskiĭ, *Tr. Vses. Nauchn. Issle. Vitamin. Inst.*, **4**, 20 (1953); *Chem. Abstr.*, **50**, 4156 (1956).
388. H. Hirayama, *Ann. Repts. Shionogi Research Lab.*, **1953**, 41; *Chem. Abstr.*, **50**, 14766 (1956).
389. Swiss Pat. 215,659 (Oct. 1, 1941); *Chem. Abstr.*, **42**, 7346 (1948).
390. K. Heyns and E. Tauber, *Z. Physiol. Chem.*, **282**, 31 (1945).
391. M. Sekiya, *J. Pharm. Soc. Japan*, **70**, 62 (1950); *Chem. Abstr.*, **44**, 5368 (1950).
392. M. Sekiya, *J. Pharm. Soc. Japan*, **70**, 524 (1950); *Chem. Abstr.*, **45**, 5640 (1951).
393. K. Washimi, *J. Pharm. Soc. Japan*, **66**, 62 (1946); *Chem. Abstr.*, **45**, 6207 (1951).
394. W. J. Middleton, U.S. Pat. 2,779,766 (Jan. 29, 1957); *Chem. Abstr.*, **51**, 10586 (1957).
395. G. H. Hitchings and E. A. Falco, U.S. Pat. 2,759,949 (Aug. 21, 1956); *Chem. Abstr.*, **51**, 11391 (1957).
396. H. Nakayama, *Vitamins* (Kyoto), **10**, 356 (1956); *Chem. Abstr.*, **51**, 16678 (1957).
397. K. Sano, *Takamine Kenkyujo Nempo*, **8**, 36 (1956); *Chem. Abstr.*, **52**, 395 (1958).
398. W. J. Middleton, U.S. Pat. 2,801,908 (Aug. 6, 1957); *Chem. Abstr.*, **52**, 1261 (1958).
399. M. Narisada, T. Nakagawa, T. Kubota, and I. Tanaka, *Shionogi Kenkyusho Nempo*, **8**, 141 (1958); *Chem. Abstr.*, **53**, 5870 (1959).
400. K. Shirakawa, Japan. Pat. 3032 (April 4, 1960); *Chem. Abstr.*, **55**, 1669 (1961).
401. J. Ishikawa, *Nippon Kagaku Zasshi*, **81**, 1489 (1960); *Chem. Abstr.*, **55**, 7003 (1961).
402. W. J. Middleton, U.S. Pat. 2,961,447 (Nov. 22, 1960); *Chem. Abstr.*, **55**, 8874 (1961).
403. J. Druey and P. Schmidt, Ger. Pat. 1,056,613 (May 6, 1959); *Chem. Abstr.*, **55**, 13457 (1961).

404. Brit. Pat. 859,716 (Jan. 25, 1961); *Chem. Abstr.*, **55**, 17669 (1961).
405. J. Druey and P. Schmidt, Ger. Pat. 1,065,421 (Sept. 17, 1959); *Chem. Abstr.*, **55**, 18786 (1961).
406. C. L. Dickinson, Jr., and B. C. McKusick, U.S. Pat. 2,998,426 (Nov. 2, 1959); *Chem. Abstr.*, **56**, 5974 (1962).
407. Brit. Pat. 880,256 (Nov. 20, 1959); *Chem. Abstr.*, **56**, 5976 (1962).
408. Brit. Pat. 869,552 (May 31, 1961); *Chem. Abstr.*, **56**, 1459 (1962).
409. E. F. Rogers and R. L. Clark, U.S. Pat. 3,030,364 (April 17, 1962); *Chem. Abstr.*, **57**, 9863 (1962).
410. Brit. Pat. 889,146 (Feb. 7, 1962); *Chem. Abstr.*, **57**, 15126 (1962).
411. F. J. Meyer, Ger. Pat. 1,135,913 (Sept. 6, 1962); *Chem. Abstr.*, **57**, 16634 (1962).
412. K. Sirakawa, U.S. Pat. 3,040,047 (June 19, 1962); *Chem. Abstr.*, **58**, 533 (1963).
413. Swiss Pat. 358,426 (Jan. 15, 1962); *Chem. Abstr.*, **58**, 3443 (1963).
414. E. Habicht, Swiss Pat. 358,424 (Jan. 15, 1962); *Chem. Abstr.*, **58**, 3444 (1963).
415. Brit. Pat. 911,551 (Nov. 28, 1962); *Chem. Abstr.*, **58**, 10214 (1963).
416. Brit. Pat. 901,749 (July 25, 1962); *Chem. Abstr.*, **59**, 1660 (1963).
417. R. P. Welcher, G. A. Johnson, and V. P. Wystrach, Ger. Pat. 1,149,004 (May 22, 1963); *Chem. Abstr.*, **59**, 14023 (1963).
418. Y. Sato, *Sankyo Kenkyusho Nempo*, **15**, 47 (1963); *Chem. Abstr.*, **60**, 11979 (1964).
419. K. Shirakawa and T. Tsujikawa, *Takeda Kenkyusho Nempo*, **22**, 19, 27 (1963); *Chem. Abstr.*, **60**, 12009 (1964).
420. G. Sunagawa and N. Soma, Japan. Pat. 3843 (April 8, 1964); *Chem. Abstr.*, **61**, 5618 (1964).
421. G. Sunagawa and N. Soma, Japan. Pat. 8543 (May 25, 1964); *Chem. Abstr.*, **61**, 11973 (1964).
422. L. M. Jampolsky, J. Kiss, B. Recherer, and J. J. Plati, Belg. Pat. 638,097 (April 2, 1964); *Chem. Abstr.*, **62**, 7648 (1965).
423. T. Tsujikawa, Japan. Pat. 23,409 (Oct. 20, 1964); *Chem. Abstr.*, **62**, 10451 (1965).
424. N. L. Allinger and S. Greenberg, *J. Am. Chem. Soc.*, **84**, 2394 (1962).
425. N. L. Allinger and S. Greenberg, *J. Am. Chem. Soc.*, **81**, 5733 (1959).
426. N. L. Allinger and Shih-En Hu, *J. Am. Chem. Soc.*, **83**, 1664 (1961).
427. N. L. Allinger, M. Nakazaki, and V. Zalkow, *J. Am. Chem. Soc.*, **81**, 4074 (1959).
428. A. C. Cope and A. S. Mehta, *J. Am. Chem. Soc.*, **86**, 5626 (1964).
429. H. Tiefenthaler, W. Dörscheln, H. Göth, and H. Schmid, *Tetrahedron Letters*, **1964**, 2999.
430. H. E. Schroeder, Ph.D. Thesis, Harvard University, Cambridge, Mass., 1938.
431. G. Faerber, Dissertation, T.U., Berlin, 1951; (*Angew. Chem.*, **63**, 491 (1951)).
432. J. Diamond, Ph.D. Thesis, Temple University, Philadelphia, Penn., 1955.
433. Badische Anilin- u. Soda-Fabrik A.-G., Neth. Appl. 6,505,782 (Dec. 21, 1965); Ger. Appl. (June 20, 1964); *Chem. Abstr.*, **64**, 17500 (1966).
434. H. Jäger, *Chem. Ber.*, **95**, 242 (1962).
435. E. Campaigne, D. R. Maulding, and W. L. Roelofs, *J. Org. Chem.*, **29**, 1543 (1964).
436. W. L. Roelofs, *Mass. Inst. Technol. Seminars*, **1965**, 274.
437. D. H. Hunneman, *Mass. Inst. Technol. Seminars*, **1964**, 43.
438. P. de Ruggieri, C. Gandolfi, and U. Guzzi, *Gazz. Chim. Ital.*, **96**, 152 (1966).

439. E. C. Taylor, U.S. Pat. 2,963,479 (Dec. 6, 1960); *Chem. Abstr.*, **57**, 11214 (1962).
440. J. Décombe and C. Verry, *Compt. Rend.*, **256**, 5156 (1963).
441. J. Colonge, G. Descotes, and G. Fresnay, *Compt. Rend.*, **256**, 2638 (1963).
442. P. Grammaticakis, *Compt. Rend.*, **235**, 546 (1952).
443. M. E. Kuehne, *J. Am. Chem. Soc.*, **81**, 5400 (1959).
444. W. A. Ehrhart, Ph.D. Thesis, Princeton University, Princeton, N.J., 1960.
445. C. F. H. Allen and J. A. VanAllen, *J. Org. Chem.*, **14**, 754 (1949).
446. H.-J. Nitzschke and H. Budka, *Chem. Ber.*, **88**, 264 (1955).
447. H.-J. Nitzschke and G. Faerber, *Chem. Ber.*, **87**, 1635 (1954).
448. K. Ziegler, *Chem. Ber.*, **67A**, 139 (1934).
449. W. Treibs and W. Schroth, *Angew. Chem.*, **71**, 578 (1959).
450. W. Schroth and W. Treibs, *Ann.*, **639**, 214 (1961).
451. W. Treibs and W. Schroth, *Ann.*, **642**, 82 (1961).
452. W. Treibs, W. Schroth, H. Lichtmann, and G. Fischer, *Ann.*, **642**, 97 (1961).
453. W. Schroth and W. Treibs, *Ann.*, **642**, 108 (1961).
454. E. Eidebenz, *Chem. Ber.*, **74**, 1798 (1941).
455. J. v. Braun, O. Kruber, and E. Danziger, *Chem. Ber.*, **49**, 2642 (1916).
456. E. C. Taylor and A. McKillop, unpublished results.
457. J. A. Zoltewicz, Ph.D. Thesis, Princeton University, Princeton, N.J., 1960.
458. C. D. S. Tomlin, Ph.D. Thesis, Princeton University, Princeton, N.J., 1964.
459. E. C. Taylor and S. Vromen, unpublished results.
460. C. J. Cavallito, C. M. Martini, and F. C. Nachod, *J. Am. Chem. Soc.*, **73**, 2544 (1951).
461. M. R. Atkinson, G. Shaw, K. Schaffner, and R. N. Warrener, *J. Chem. Soc.*, 3847 (1956).
462. T. S. Gardner, E. Wenis, and J. Lee, *J. Org. Chem.*, **19**, 753 (1954).
463. I. Heilbron, *J. Chem. Soc.*, **1949**, 2099.
464. E. C. Taylor and J. A. Zoltewicz, *J. Am. Chem. Soc.*, **82**, 2656 (1960).
465. E. C. Taylor and J. Bartulin, *Tetrahedron Letters*, No. 25, **1967**, 2337.
466. K.-H. Wünsch and A. J. Boulton, in *Advances in Heterocyclic Chemistry*, Vol. 8, A. R. Katritzky and A. J. Boulton, Eds., Academic Press, New York, pp. 303–342.
467. M. Ochiai and K. Morita, *Tetrahedron Letters*, No. 25, **1967**, 2349.
468. N. D. Heindel, T. A. Brodof, and J. E. Kogelschatz, *J. Heterocyclic Chem.*, **3**, 222 (1966).
469. C. F. Huebner, P. L. Strachan, E. M. Donoghue, N. Cahoon, L. Dorfman, R. Margerison, and E. Wenkert, *J. Org. Chem.*, **32**, 1126 (1967).
470. F. Johnson and J. P. Heeschen, *J. Org. Chem.*, **29**, 3252 (1964).
471. E. Klingsberg, *J. Org. Chem.*, **31**, 3489 (1966).
472. N. J. Leonard and K. L. Carraway, *J. Heterocyclic Chem.*, **3**, 485 (1966).
473. W. J. Middleton, *J. Org. Chem.*, **31**, 3731 (1966).
474. S. Mizukami and E. Hirai, *Chem. Pharm. Bull. Japan*, **14**, 1321 (1966).
475. R. Huisgen, F. Wimmer, and K. Bast, cited in a review by R. Huisgen, *Angew. Chem.*, **72**, 371 (1960).
476. B. Pecherer and A. Brossi, *J. Org. Chem.*, **32**, 1053 (1967).
477. W. Schulze, H. Willitzer, and H. Fritzsche, *Chem. Ber.*, **99**, 3492 (1966).
478. T. Severin, B. Brück, and P. Adhikary, *Chem. Ber.*, **99**, 3097 (1966).
479. T. Sheradsky and P. L. Southwick, *J. Heterocyclic Chem.*, **4**, 1 (1967).

480. E. C. Taylor and S. Vromen, *Israel J. Chem.*, **2**, 310 (1964).
481. K. Wallenfels, F. Witzler, and K. Friedrich, *Tetrahedron*, **23**, 1845 (1967).
482. K. Wallenfels, *Chimia*, **20**, 303 (1966).
483. O. W. Webster, *J. Am. Chem. Soc.*, **88**, 4055 (1966).
484. Fr. Pat. 1,437,213 (April 29, 1966); *Chem. Abstr.*, **65**, 20136g (1966).
485. F. S.-Legagneur, L. Picquet, and C. Neveu, *Bull. Soc. Chim. France*, **1966**, 2025.
486. F. S.-Legagneur and J. Rabadeux, *Bull. Soc. Chim. France*, **1967**, 1310.
487. H. Junek, H. Hamböck, and B. Hornischer, *Monatsh. Chem.*, **98**, 315 (1967).
488. C. F. Koelsch, *J. Org. Chem.*, **25**, 164 (1960).
489. H. Behringer and R. Wiedenmann, *Tetrahedron Letters*, No. 41, **1965**, 3705.
490. R. Gompper, W. Elser, and H.-J. Müller, *Angew. Chem.*, **79**, 473 (1967); *Intern. Ed.* (*English*), **6**, 453 (1967).
491. M. A. G. Nonell, *Rev. Real Acad. Cienc. Exact. Fis. Nat. Madrid*, **60** (1), 35 (1966); *Chem. Abstr.*, **66**, 10630a (1967).
492. T. I. Temnikova, Y. A. Sharanin, and V. S. Karavan, *Zh. Org. Khim.*, **3** (4), 681 (1967); *Index Chemicus*, 82579.
493. M. F. G. Stevens, *J. Chem. Soc.* (*C*), **1967**, 1096.
494. R. L. Tolman, R. K. Robins, and L. B. Townsend, *J. Heterocyclic Chem.*, **4**, 230 (1967).
495. D. W. Karle, Ph.D. Thesis, University of Southern California, Los Angeles, Calif., 1965; University Microfilms, 65-7233.
496. K. von Auwers and W. Susemihil, *Chem. Ber.*, **63**, 1072 (1930).
497. K. von Auwers and H. Wunderling, *Chem. Ber.*, **64**, 2758 (1931).
498. J. J. Conn and A. Taurins, *Can. J. Chem.*, **31**, 1211 (1953).
499. R. E. Meyer, *Helv. Chim. Acta.*, **16**, 1291 (1933).
500. C. Krüger, *J. Organometal. Chem.*, **9**, 125 (1967).
501. W. Pfleiderer, *Angew. Chem.*, **75**, 993 (1963); *Intern. Ed.* (*English*), **3**, 114 (1964) gives a review of the Isay reaction in the synthesis of pteridines.
502. R. G. W. Spickett and G. M. Timmis, *J. Chem. Soc.*, **1954**, 2887.
503. G. M. Timmis, *Nature*, **164**, 133 (1949).
504. G. M. Timmis, D. G. I. Felton, and T. S. Osdene, in *Chemistry and Biology of Pteridines*, a Ciba Symposium, G. E. W. Wolstenholme and M. P. Cameron, Eds., J. and A. Churchill, Ltd., London, 1954, p. 93.
505. T. S. Osdene, in *Pteridine Chemistry*, W. Pfleiderer and E. C. Taylor, Eds., Pergamon Press, London, 1964, p. 65.
506. F. Johnson and R. Madroñero, in *Advances in Heterocyclic Chemistry*, Vol. 6, A. R. Katritzky and A. J. Boulton, Eds., Academic Press, New York, 1966, p. 95.
507. A. Dornow and H. Pietsch, *Chem. Ber.*, **100**, 2585 (1967).
508. H. Junek and A. Schmidt, *Monatsh. Chem.*, **98**, 1097 (1967).
509. E. C. Taylor and J. Klug, unpublished results.
510. N. L. Allinger and V. B. Zalkow, *J. Am. Chem. Soc.*, **83**, 1144 (1961).
511. D. C. Ayres and R. A. Raphael, *J. Chem. Soc.*, **1958**, 1779.
512. H. O. House, P. P. Wickham, and H. C. Müller, *J. Am. Chem. Soc.*, **84**, 3139 (1962).
513. W. E. Truce, W. W. Bannister, and R. H. Knospe, *J. Org. Chem.*, **27**, 2821 (1962).
514. H. C. H. Carpenter and W. H. Perkin, Jr., *J. Chem. Soc.*, **75**, 921 (1899).

515. A. Alberola, M. A. Gunther, M. Lora-Tamayo, and J. L. Soto, *An. Soc. Esp. Fis Quim.*, **B63** (6), 691 (1967); *Index Chemicus*, 83329.
516. H. C. Brown, J. H. Brewster, and H. Schechter, *J. Am. Chem. Soc.*, **76**, 467 (1954).
517. K. Ziegler, in *Methoden der Organischen Chemie* (Houben-Weyl), Allegemeine Chemische Methoden, Band IV, Teil 2, Georg Thieme Verlag, Stuttgart, 1955, p. 729 ff.
518. J. P. Schaefer and J. J. Bloomfield, in *Organic Reactions*, Vol. 15, Wiley, New York, 1967, pp. 28–30, 186–196.
519. P. A. S. Smith, *The Chemistry of Open-Chain Nitrogen Compounds*, Vol. 1, Benjamin, New York, 1965, pp. 177–184.
520. D. J. Brown and M. N. Paddon-Row, *J. Chem. Soc.* (*C*), **1967**, 903 and preceding papers in this series.
521. H. Baron, F. H. P. Remfry, and J. F. Thorpe, *J. Chem. Soc.*, **1904**, 1726.
522. G. Schill and A. Lüttringhaus, *Angew. Chem.*, **76**, 567 (1964); *Intern. Ed.* (*English*), **3**, 546 (1964).
523. G. Schill, *Chem. Ber.*, **98**, 2906 (1965).
524. J. Mohr, *J. Chem. Soc.*, **81**, 100 (1902).
525. N. K. Kochetkov and S. D. Sokolov, in *Advances in Heterocyclic Chemistry*, Vol. 2, A. R. Katritzky, Ed., Academic Press, New York, 1963, pp. 398–410.
526. E. C. Taylor and W. A. Ehrhart, *J. Org. Chem.*, **28**, 1108 (1963).
527. R. J. Schnitzer and F. Hawking, Eds., *Experimental Chemotherapy*, Vol. IV, Part I, Academic Press, New York, 1966.
528. L. F. Larionov, *Cancer Chemotherapy*, Pergamon Press, Oxford, 1965.
529. T. Nozoe, H. Horino, and T. Toda, *Tetrahedron Letters*, No. 52, **1967**, 5349.
530. A. Lüttringhaus and G. Isele, *Angew. Chem.*, **79**, 945 (1967); *Intern. Ed.* (*English*), **11**, 956 (1967).
531. H. Sterk and H. Junek, *Monatsh. Chem.*, **98**, 1763 (1967).
532. H. A. Burch, *J. Med. Chem.*, **11**, 79 (1968).
533. W. A. Nasutavicus, S. W. Tobey, and F. Johnson, *J. Org. Chem.*, **32**, 3325 (1967).
534. W. B. Scott and R. E. Pincock, *J. Org. Chem.*, **32**, 3374 (1967).
535. A. Aviram and S. Vromen, *Chem. Ind.*, **1967**, 1452.
536. D. Taub, C. H. Kuo, and N. L. Wendler, *J. Chem. Soc.* (*C*), **1967**, 1558.
537. E. C. Taylor, K. Lenard, I. P. Sword, J. V. Berrier, and P. Jacobi, unpublished observations.
538. J. A. Zoltewicz and T. W. Sharpless, *J. Org. Chem.*, **32**, 2681 (1967).
539. M. D. Nair and P. A. Malik, *Indian J. Chem.*, **5**, 603 (1967).
540. M. D. Nair, S. R. Mehta, and S. M. Kalbag, *Indian J. Chem.*, **5**, 464 (1967).
541. E. M. Hawes and D. G. Wibberley, *J. Chem. Soc.* (*C*), **1967**, 1564.
542. S. S. Kulp, *Can. J. Chem.*, **45**, 1981 (1967).
543. W. Schulze and H. Willitzer, *Chem. Ber.*, **100**, 3460 (1967).
544. H. Bredereck, G. Simchen, and H. Traut, *Chem. Ber.*, **100**, 3664 (1967).
545. W. Broser and D. Rahn, *Chem. Ber.*, **100**, 3472 (1967).
546. L. J. Belf, M. W. Buxton, and J. F. Tilney-Bassett, *Tetrahedron*, **23**, 4719 (1967).
547. H. Tiefenthaler, W. Dörscheln, H. Göth, and H. Schmid, *Helv. Chim. Acta*, **50**, 2244 (1967).
548. H. Junek and H. Sterk, *Z. Naturforsch.*, **22b**, 732 (1967).

549. E. P. Burrows, A. Rosowsky, and E. J. Modest, *J. Org. Chem.*, **32**, 4090 (1967).
550. R. L. Tolman, R. K. Robins, and L. B. Townsend, *J. Am. Chem. Soc.*, **90**, 524 (1968).
551. E. C. Taylor and K. Lenard, *J. Am. Chem. Soc.*, **90**, 2424 (1968).
552. R. J. Rousseau, R. K. Robins, and L. B. Townsend, *J. Am. Chem. Soc.*, **90**, 2661 (1968).
553. M. F. G. Stevens, *J. Chem. Soc.* (*C*), **1968**, 348.
554. H. G. Dean, R. J. Grout, M. W. Partridge, and H. J. Vipond, *J. Chem. Soc.* (*C*), **1968**, 142.
555. J. D. Atkinson and M. C. Johnson, *J. Chem. Soc.* (*C*), **1968**, 1252.
556. T. Seiyaku Co., Lt., Brit. Pat. 1,087,505 (Oct. 18, 1967); *Chem. Abstr.*, **68**, 114628v (1968).
557. H. Bredereck, G. Simchen, R. Wahl, and F. Effenberger, *Chem. Ber.*, **101**, 512 (1968).
558. Y. Ban, T. Wakamatsu, Y. Fujimoto, and T. Oishi, *Tetrahedron Letters*, No. 30, **1968**, 3383.
559. N. L. Allinger and W. Szkrybalo, *Tetrahedron*, **24**, 4699 (1968).
560. H. Behringer, D. Bender, J. Falkenberg, and R. Wiedenmann, *Chem. Ber.*, **101**, 1428 (1968).
561. E. Allenstein and R. Fuchs, *Chem. Ber.*, **101**, 1244 (1968).
562. J. H. Jones and E. J. Cragoe, Jr., *J. Med. Chem.*, **11**, 322 (1968).
563. A. Rosowsky, personal communication.
564. W. Pfleiderer and E. C. Taylor, Eds., *Pteridine Chemistry*, Pergamon Press, Ltd., London, 1964.
565. R. C. Elderfield and A. C. Mehta, in *Heterocyclic Compounds*, Vol. 9, R. C. Elderfield, Ed., Wiley, New York, 1967, pp. 1–117.
566. J. P. Ferris and L. E. Orgel, *J. Am. Chem. Soc.*, **88**, 3829 (1966).
567. E. C. Taylor, O. Vogl, and C. C. Cheng, *J. Am. Chem. Soc.*, **81**, 2442 (1959).
568. K. Hartke, Diplomarbeit, University of Marburg, 1958.
569. R. Huisgen, private communication (K. Bast, Dissertation, München, 1962).
570. R. Huisgen, private communication (F. Wimmer, Dissertation, München, 1959).

AUTHOR INDEX

Abul-Husn, A., 257, 308(**92**)
Adhikary, P., 20, 41, 42, 346, 355(**478**)
Adkins, H., 4, 11(**135**)
Akanuma, K., 190, 314(**311**)
Akita, K., 224, 311(**386**)
Alberola, A., 35, 346(**515**)
Albert, A., 231, 308(**178**)
Alessandri, L., 314(**214**)
Allen, C. F. H., 23, 54, 353(**445**)
Allenstein, E., 308, 316(**561**)
Allinger, N. L., 68(**424**); 68(**425**); 68(**426**); 65(**427**); 65, 66(**510**); 20, 45, 68, 350(**559**)
Altman, J., 11, 14, 37, 63, 352, 353(**7**)
Amamiya, H., 190, 314(**311**)
Anand, N., 109, 113, 114, 231, 309, 311, 316(**161**); 111, 112, 282, 283, 309, 311, 312, 316, 321(**163**)
Anderson, D. M. W., 109, 126, 324, 373(**342**)
Aron, M. A., 225, 314(**119**)
Asao, T., 195, 318(**9**)
Atkinson, J. D., 161, 324, 342, 350, 353, 354, 356, 360, 362, 365(**555**)
Atkinson, M. R., 299(**461**)
Aurieh, H. G., 49, 73, 341, 346(**100**)
Aurnhammer, R., 1, 20, 42, 45, 46, 47, 65, 68, 69, 71, 322, 330, 358, 371, 373, 374(**190**)
von Auwers, K., 60, 64, 187, 188, 317, 322(**101**); 11(**496**); 11(**497**)
Aviram, A., 306, 312, 317, 349(**535**)
Ayres, D. C., 36, 63, 169, 341(**511**)

Babad, E., 11, 14, 37, 63, 352, 353(**7**)
Bachman, G. B., 22, 50, 311(**219**)
Baddiley, J., 104, 108, 309(**44**)
Badger, G. M., 90, 231, 295, 309, 311(**124**)
Bahr, T., 60, 64, 187, 188, 317, 322(**101**)
Baiocchi, F., 195, 337, 338(**348**)
Baker, B. R., 199, 319(**121**)
Baldwin, S., 4, 8, 11, 312, 337, 359, 360, 365, 367, 373(**66**); 11, 19, 41, 65, 367(**81**)
Balyakina, M. V., 109, 224, 311(**387**)
Ban, Y., 29, 55, 76, 350(**558**)
Bannister, W. W., 62, 64(**513**)
Bargellini, G., 204, 314(**62**)
Barker, R. S., 22, 50, 311(**219**)
Baron, H., 162, 310(**521**)
Barone, J. A., 108, 313(**345**); 110, 224, 310(**346**); 110, 111, 224, 313, 317(**347**)
Bartlett, M. F., 203, 205, 206, 207, 226, 314, 319(**240**)
Barton, J. W., 209, 210, 212, 214, 215, 216, 308, 309, 310, 314, 362(**23**)
Bartulin, J., 167, 171, 331(**465**)
Bast, K., 217, 314(**475**)
Baudet, H. Ph., 190, 314(**37**)
Bedford, G. R., 203, 204, 205, 314, 318, 319(**1**); 50, 78, 323, 352(**28**)
Beerens, H., 193, 224, 315(**87**)
Behringer, H., 157, 343(**489**); 125, 194, 308(**560**)
Belf, L. J., 191, 313(**546**)
Bell, F., 109, 126, 324, 373(**342**)
Bel'tsova, N. N., 198, 220, 314(**314**)
Bender, D., 125, 194, 308(**560**)
Bentov, M., 196, 220, 314, 319(**318**)
Bergel, F., 183, 194, 224, 311(**196**)
Berger, J. G., 138, 139, 241, 242, 249, 250, 251, 310, 311, 315, 328, 335, 347(**91**); 137, 138, 140, 241, 242, 249, 250, 251, 310, 311, 335, 347(**341**)
Bergmann, E. D., 16(**90**); 196, 220, 314, 319(**318**)
Berrie, A. H., 193, 219, 310(**78**)
Berrier, J. V., 172, 173, 174, 176, 222, 293, 294, 310, 311, 321, 327, 328, 329, 338, 347(**537**)
Best, S. R., 12, 31, 62, 312, 329(**194**)
Biggs, J., 104, 108, 224, 315(**367**)
Blauschmidt, P., 26, 56, 310(**107**)
Bloomfield, J. J., 11, 15, 37, 63, 352(**12**); 11, 13, 14, 35, 36, 321, 335, 341(**273**); 2(**518**)
Bogert, M. T., 227, 228, 231, 314(**45**); 227, 294, 313(**79**); 197, 227, 314(**123**); 227, 313(**126**); 228, 314(**127**); 226, 314(**247**)

199, 227, 319**(293)**
Boon, W. R., 146, 147, 330, 331**(337)**
Borror, A. L., 200, 237, 308, 309, 310, 313, 314, 332**(17)**; 190, 236, 237, 314**(21)**
Borsche, W., 204, 314**(39)**
Bost, R. W., 119, 227, 319**(131)**; 119, 227, 314**(132)**
Böttcher, H., 137, 138, 139, 315, 328**(113)**
Boulton, A. J., 170**(466)**
Braude, E. A., 198, 314**(317)**
v. Braun, J., 16, 332**(455)**
Bredereck, H., 193, 309**(544)**; 169, 333**(557)**
Breneman, H. C., 227, 228, 231, 314**(45)**
Bretschneider, H., 191, 338**(267)**
Breukink, K. W., 228, 229, 298, 314**(61)**; 190, 197, 228, 314**(67)**
Brewster, J. H., 11**(516)**
Brodof, T. A., 305, 307, 313, 319**(468)**
Broser, W., 35, 63, 367**(545)**
Brossi, A., 20, 44, 67, 369**(476)**
Brown, D. J., 309**(58)**; 309**(146)**; 231, 308**(178)**; 103, 177**(215)**; 240**(520)**
Brown, H. C., 11**(516)**
Brück, B., 20, 41, 42, 346, 355**(478)**
Bruckner, S., 199, 319**(245)**
Budesinsky, Z., 194, 224, 310, 311, 312, 321**(63)**
Budka, H., 24**(446)**
Burch, H. A., 82, 83, 91, 227, 237, 309, 312, 317**(532)**
Burckhardt, C. A., 231, 317**(115)**
Burrows, E. P., 11, 20, 41, 65, 338, 339**(549)**
Butler, D. N., 149, 182, 259, 308, 312**(114)**
Butler, K., 196, 314**(321)**
Buxton, M. W., 191, 313**(546)**

Cahoon, N., 26, 168, 171, 353**(469)**
Cairns, T. L., 142, 207, 214, 242, 299**(102)**; 166, 170, 331**(261)**
Campaigne, E., 177, 178, 179, 344, 356, 361**(435)**
Carboni, R. A., 1, 81, 95, 157, 158, 310, 343**(84)**; 142, 207, 214, 242, 299**(102)**
Carlson, G. H., 200, 318**(77)**
Carpenter, H. C. H., 31, 329**(514)**
Carraway, K. L., 201, 261, 339**(71)**; 262, 339**(472)**
Carrington, H. C., 222, 223, 314**(252)**; 146, 147, 330, 331**(337)**
Cavalla, J. F., 22, 49, 78, 273, 288, 295, 317, 323, 346, 349, 352**(13)**; 304, 317, 323**(16)**; 50, 78, 323, 352**(28)**; 22, 48, 49, 50, 51, 73, 78, 317, 323, 337, 341, 349, 352**(88)**; 78, 273, 288, 295, 317, 323**(116)**
Cavallito, C. J., 299**(460)**
Chacko, A. M., 137, 138, 139, 274, 275, 287, 295, 311, 315, 328**(334)**
Chatterjee, S. K., 109, 113, 114, 231, 309, 311, 316**(161)**
Chatterji, S. K., 111, 112, 282, 283, 309, 311, 312, 316, 321**(163)**
Cheeseman, G., 231, 308**(178)**
Chen, Y. G., 228, 314**(127)**
Cheng, C. C., 82, 83, 85, 86, 90, 91, 231, 275, 276, 278, 279, 309, 312, 331, 332, 339**(3)**; 82, 83, 85, 86, 227, 308, 309, 312, 331, 332, 339**(168)**; 82, 85, 295, 309, 331, 332**(169)**; 84, 85, 86, 277, 278, 322, 336, 337, 339**(170)**; 195, 337, 338**(348)**; 177**(567)**
Clark, R. L., 111, 321**(409)**
Clarke, P. B., 220, 284, 314, 315, 320, 326**(125)**
Closson, R. D., 32, 62, 349**(182)**
Coffman, D. D., 1, 81, 95, 157, 158, 310, 343**(84)**; 142, 207, 214, 242, 299**(102)**; 159, 160, 161, 164, 193, 313, 314, 317, 324, 325, 326, 332, 337, 339, 346, 348, 350, 351, 355, 361, 373**(106)**; 166, 170, 331**(261)**
Cogliano, J. A., 209, 210, 212, 214, 215, 216, 308, 309, 310, 314, 362**(23)**
Colonge, J., 22, 51, 317, 323, 346**(117)**; 22, 51, 317, 323, 345, 346**(441)**
Compère, A., 312**(236)**
Conn, J. J., 11**(498)**
Cook, A. H., 22, 51, 61, 73, 317**(112)**; 227, 314**(249)**; 119, 226, 314, 319**(250)**
Cooper, F. C., 233, 314**(118)**; 202, 225**(120)**
Cope, A. C., 20, 45, 68, 229, 357**(167)**; 11, 20, 45, 68, 357**(181)**; 65**(428)**
Cotter, R. J., 11, 20, 45, 68, 357**(181)**
Cottis, S. G., 220, 284, 314, 315, 320, 326**(125)**; 220, 281, 282, 313, 326**(133)**; 164, 193, 220, 221**(151)**; 113, 114, 159, 164, 309, 311, 316, 318, 326, 344, 351**(246)**
Coulson, E. A., 199, 225, 319**(294)**
Cragoe, E. J., Jr., 185, 186, 224, 272, 292,

293, 294, 308, 311, 315, 319, 320, 329(**562**)
Crovetti, A. J., 235, 237, 308, 309, 310, 311, 314, 331(**27**); 192, 310(**30**)
Cuingnet, E., 193, 224, 315(**87**)

Danziger, E., 16, 332(**455**)
Davies, J. S. H., 146, 147, 330, 331(**337**)
Day, A. R., 151, 308, 333(**344**)
Dean, F. M., 19(**218**)
Dean, H. G., 197, 314(**554**)
Décombe, J., 319, 335(**440**)
Delépine, M., 224, 311(**57**); 314, 315(**326**); 311(**377**)
Descotes, G., 22, 51, 317, 323, 346(**117**); 22, 51, 317, 323, 345, 346(**441**)
Desimoni, G., 143, 331(**6**)
Deutsch, G., 199, 319(**245**)
Diamond, J., 323(**432**)
Dickel, D. F., 203, 205, 206, 207, 226, 314, 319(**240**)
Dickey, J. B., 225, 313, 314(**296**); 225, 313(**301**)
Dickinson, C. L., 121, 122, 321(**164**); 81, 86, 91, 92, 93, 94, 95, 96, 97, 98, 103, 225, 308, 309, 310, 313, 314, 315, 316, 317, 319, 320, 327, 334, 336, 337, 338, 339, 342, 343, 346, 349, 359(**232**); 133, 136, 311, 316, 318, 325, 328, 329, 332, 333, 338, 339, 347, 368, 369(**365**)
Dickinson, C. L., Jr., 93, 94, 309, 312, 338(**406**)
Dimroth, K., 49, 73, 341, 346(**100**)
Donoghue, E. M., 26, 168, 171, 353(**469**)
Dorfman, L., 26, 168, 171, 353(**469**)
Dornow, A., 91, 118, 119, 332, 337, 339, 343, 344(**33**); 158, 163, 164, 166, 169, 320, 335(**108**); 143, 144, 219, 331(**177**); 151, 183, 324, 325, 326, 333, 346(**343**); 224, 311(**362**); 194, 316(**363**); 231, 311(**372**); 26, 102, 219, 315, 343(**507**)
Dörscheln, W., 146(**429**); 217, 314, 358(**547**)
Druey, J., 86, 276, 309(**97**); 283, 316(**139**); 84, 85, 225, 231, 322, 330, 333(**144**); 85, 231, 332(**145**); 84, 231, 317(**148**); 82, 84, 85, 309, 317, 330, 336(**149**); 83, 312(**150**); 84, 85, 276, 277, 278, 322, 329, 330, 336(**155**); 85, 231, 331, 332(**157**); 84, 85, 277, 329, 330(**159**); 86, 339(**371**); 84, 317(**373**); 84, 317(**403**); 82, 231, 309(**405**)
Dufraisse, C., 177, 178, 365(**286**)
Duncan, J. L., 109, 126, 324, 373(**342**)
Dymek, W., 191, 227, 338(**152**)

Easton, N. R., 12, 34, 35, 63, 367(**220**); 34, 62, 365(**221**); 11, 31, 32, 62, 312, 322, 336, 344, 352, 365(**281**); 21, 34, 47, 365, 374(**282**)
Eberle, H., 1, 46, 47, 61, 65, 68, 69, 71, 72, 322, 361, 366, 374(**187**); 1, 20, 42, 46, 47, 65, 68, 69, 71, 72, 322, 361, 374(**200**)
Effenberger, F. 169, 333(**557**)
Ehrhart, W. A., 267, 268, 290, 309, 311(**19**); 169, 224, 267, 269, 290, 294, 309, 311(**444**); 241(**526**)
Ehrlich, F., 199, 220, 319(**174**)
Eichenberger, K., 86, 276, 309(**97**); 231, 317(**115**); 283, 316(**139**); 227, 317(**140**); 84, 85, 225, 231, 322, 330, 333(**144**); 82, 84, 85, 309, 317, 330, 336(**149**); 83, 312(**150**); 84, 85, 276, 278, 322, 329, 330, 336(**155**); 84, 85, 227, 329, 330(**159**); 219, 311, 312(**368**); 86, 339(**371**); 84, 317(**373**)
Eidebenz, E., 16, 332, 365(**354**)
Eistert, B., 20, 43, 66, 359(**29**)
Elderfield, R. C., 171(**565**)
Ellingson, R. C., 184, 219, 308(**185**)
Elser, W., 180, 358(**490**)
Elvidge, J. A., 225, 314(**119**)
Emoto, S., 200, 318(**86**)
Engelhardt, V. A., 142, 207, 214, 242, 299(**102**); 132, 133, 135, 136, 141, 339, 349(**103**); 129, 131, 132, 133, 134, 135, 140, 142, 207, 214, 219, 242, 299, 309, 310, 313, 314, 316, 318, 320, 341, 343, 347, 351, 360(**104**); 80, 81, 92, 93, 105, 121, 122, 143, 144, 308, 312, 313, 316, 321, 332, 339, 344, 348(**105**); 159, 160, 161, 164, 193, 313, 314, 317, 324, 325, 326, 332, 337, 339, 346, 348, 350, 351, 355, 361, 373(**106**); 133, 136, 311, 316, 318, 325, 328, 329, 332, 333, 338, 339, 347, 368, 369(**365**)
Etienne, A., 205, 337(**65**); 177, 178, 365(**286**)

Faerber, G., 23, 52, 349(**431**); 22, 52, 103, 349(**447**)

Falco, E. A., 192, 204, 205, 286, 319, 328, 333, 341**(313)**; 82, 308**(395)**
Falkenberg, J., 125, 194, 308**(560)**
Fancher, O. E., 19, 41, 65, 367**(210)**; 367**(211)**
Fel'dman, I. K., 198, 220, 314**(314)**
Felton, D. G. I., 76**(504)**
Fennessey, P. V., 11, 13, 14, 35, 36, 321, 335, 341**(273)**
Ferris, J. P., 145, 146, 183, 219, 268, 289, 308, 309, 311, 337**(32)**; 145, 146, 183, 289, 308, 309, 311, 331**(335)**; 145, 268, 289, 308**(336)**; 176**(566)**
Fieser, L. F., 16**(60)**; 20, 46, 68, 69, 349**(262)**
Findeklee, W., 198, 225, 319**(184)**
Finelli, A. F., 40, 61, 64, 364**(228)**
Finzi, P. Vita, 143, 331**(6)**
Fischer, G., 16, 332**(452)**
Fish, V. B., 12, 34, 35, 63, 367**(220)**; 11, 31, 32, 62, 312, 322, 336, 344, 352, 365**(281)**; 21, 34, 47, 365, 374**(282)**
Fisher, B. S., 129, 131, 132, 133, 134, 135, 140, 142, 207, 214, 219, 242, 299, 309, 310, 313, 314, 316, 318, 320, 341, 343, 347, 351, 360**(104)**
Fitts, D. D., 20, 44, 359**(206)**
Flett, M. St. C., 314**(327)**
Frenay, A., 22, 51, 317, 323, 346**(117)**
Frese, E., 60, 64, 187, 188, 317, 322**(101)**
Fresnay, G., 22, 51, 317, 323, 345, 346**(441)**
Friedländer, P., 200, 220, 337**(70)**; 199, 319**(245)**; 197, 225, 314**(264)**
Friedrich, K., 190, 191, 203, 324**(481)**
Fritzsche, A., 204, 314**(39)**
Fritzsche, H., 98, 99, 100, 103, 107, 125, 352, 356, 357, 361, 366, 369, 372**(477)**
Fry, E. M., 20, 46, 68, 69, 349**(262)**
Fuchs, R., 308, 316**(561)**
Fujimoto, Y., 29, 55, 76, 350**(558)**
Fusco, R., 101, 231, 294, 359**(85)**

Gabriel, S., 181, 225, 314**(128)**; 198, 199, 220, 319**(274)**
Gallagher, M. J., 54, 74, 360**(110)**; 11, 22, 51, 53, 73, 345, 346**(111)**
Gallo, G. G., 197, 314**(333)**
Gandolfi, C., 187, 189, 264, 265, 280, 281, 295, 368, 370, 371, 372, 374**(438)**
Gäng, M., 24, 55, 279, 313**(339)**
Garcia, E. E., 143, 144, 252, 253, 254, 308, 309, 311, 326, 331, 355**(11)**
Gardner, T. S., 224, 299, 311**(462)**
Garmaise, D. L., 35, 77, 332**(64)**
Gaubert, P., 146, 147, 330, 331**(337)**
Gelblum, S., 35, 77, 332**(64)**
Gewald, K., 147, 148, 332**(38)**; 137, 141, 321, 329**(47)**; 147, 308**(73)**; 137, 138, 139, 141, 142, 308, 311, 315, 321, 328, 338, 344**(74)**; 26, 56, 310**(107)**; 137, 138, 139, 315, 328**(113)**; 129, 130, 311, 316, 339, 344**(175)**; 137**(217)**; 137, 139, 328**(234)**; 127, 128, 213, 216, 271, 273, 274, 311, 315, 328, 331, 338, 347, 362, 371**(258)**; 308**(356)**; 154, 155, 222, 326, 328, 333, 342**(357)**; 153, 155, 308**(358)**
Ginsburg, D., 11, 14, 37, 63, 352, 353**(7)**; 32, 62, 352**(269)**
Glock, G., 199, 225, 319**(291)**
Goerdeler, J., 183, 309**(283)**
Gompper, R., 25, 26, 55, 56, 252, 317, 346, 347**(171)**; 81, 93, 144, 258, 309, 339**(266)**; 24, 55, 279, 313**(339)**; 180, 358**(490)**
Göth, H., 146**(429)**; 217, 314, 358**(547)**
Gouin, L., 312**(284)**
Govindachari, T. R., 199, 328**(323)**
Graboyes, H., 105, 108, 116, 118, 121, 219, 226, 272, 285, 290, 292, 294, 309, 312, 315, 316, 338, 339, 344, 362**(122)**
Grammaticakis, P., 196, 198, 314**(183)**; 224, 314**(442)**
Grassberger, M. A., 315, 324**(364)**
Greenberg, S., 68**(424)**; 68**(425)**
Gresham, T. L., 57, 337**(204)**
Grewe, R., 39, 109, 224, 311**(203)**
Grigat, E., 107, 108, 109, 324, 351, 362, 363, 366, 367, 369**(72)**
Grishkova, V. K., 198, 220, 314**(314)**
Grout, R. J., 197, 314**(554)**
Grube, F., 197, 226, 314**(137)**
Grünanger, P., 143, 331**(6)**
Grundmann, C., 223, 314**(129)**; 214, 309**(259)**
Gunther, M. A., 35, 346**(515)**
Guzzi, U., 187, 189, 264, 265, 280, 281, 295, 368, 370, 371, 372, 374**(438)**

Habicht, E., 113, 114, 115, 116, 224, 309, 311, 312, 316, 338, 344, 348, 359**(53)**; 112, 113, 316, 330, 341, 346**(414)**

Halamandaris, A., 273, 295, 312(**179**)
Halliwell, R. H., 312(**238**)
Hambly, A. N., 314(**324**)
Hamböck, H., 354(**487**)
Hamilton, C. S., 193, 320(**80**)
Hammer, C. F., 11, 312(**34**)
Hand, W. F., 227, 228, 231, 314(**45**); 227, 294, 313(**79**); 197, 227, 314(**123**); 227, 313(**126**); 226, 314(**247**)
Hartke, K. S., 81, 90, 91, 95, 243, 257, 258, 279, 295, 312, 315, 339, 343, 344, 351(**25**); 1, 81, 94, 310, 332(**26**); 150, 242, 254, 309, 312, 339(**156**); 101(**568**)
Hartmans, H. M. A., 190, 314(**186**)
Haslinger, C., 181, 314(**213**)
Hata, K., 190, 314(**311**)
Hatchard, W. R., 308(**173**)
Hawes, E. M., 166, 170, 226, 355(**162**); 219, 324(**541**)
Hawking, F., 272(**527**)
Hawks, G. H., 266, 280, 297, 298, 344(**290**)
Hayashi, T., 190, 314(**311**)
Hechelhammer, W., 1, 47, 69, 70, 366(**188**)
Heckert, R. E., 142, 207, 214, 242, 299(**102**); 166, 170, 331(**261**)
Heeschen, J. P., 218, 321, 333(**470**)
Heilbron, I., 299(**463**)
Heilbron, I. M., 227, 314(**249**); 119, 226, 314, 319(**250**)
Heindel, N. D., 205, 307, 313, 319(**468**)
Helberg, J., 151, 183, 324, 325, 326, 333, 346(**343**)
Helgeson, J. P., 201, 261, 339(**71**)
Hendess, R. W., 214, 241, 242, 245, 246, 247, 248, 288, 310, 311, 320, 339(**10**)
Henry, R. L., 184, 219, 308(**185**)
Heyna, H., 200, 319(**304**)
Heyns, K., 311(**390**)
Heywood, B. J., 225, 313, 314(**295**); 225, 313(**298**)
Hines, R. A., 11, 312(**34**)
Hinz, E., 231, 311(**372**)
Hirai, E., 311(**329**); 311(**474**)
Hirayama, H., 224, 311(**388**)
Hitchings, G. H., 192, 204, 205, 286, 319, 328, 333, 341(**313**); 82, 308(**395**)
Hoefle, M. L., 113, 116, 272, 282, 283, 290, 291, 294, 311, 316, 320, 322, 330, 336, 339, 341, 344(**22**); 112, 309, 312, 316, 320, 329, 330, 336, 339, 344(**143**)
Hoffa, E., 200, 319(**304**)
Hoffman, A., 199, 227, 319(**293**)
Hohenlohe-Oehringen, K., 191, 338(**267**)
Hole, H., 1, 22, 54, 74, 337(**189**)
Holmes, A., 113, 116, 272, 282, 283, 290, 291, 294, 311, 316, 320, 322, 330, 336, 339, 341, 344(**22**)
Hölscher, H. A., 110, 111, 114, 116, 309, 315, 316, 338(**75**)
Holzner, D., 49, 73, 341, 346(**100**)
Hoover, J. R. E., 151, 308, 333(**344**)
Horino, H., 151, 152, 332, 342, 354(**529**)
Horner, L., 166, 170, 331(**260**)
Horning, E. C., 18, 39, 64, 291, 357(**222**); 40, 64, 368(**225**); 40, 64, 364(**226**); 40, 61, 64, 364(**228**)
Horning, M. G., 18, 39, 64, 291, 357(**222**); 40, 64, 368(**225**); 40, 64, 364(**226**)
Hornischer, B., 354(**487**)
House, H. O., 31, 62, 329(**512**)
Howard, E. G., 1, 81, 95, 157, 158, 310, 343(**84**)
Howk, B. W., 77, 312(**237**)
Hromatka, O., 123, 311, 344(**378**)
Hu, Shih-En, 68(**426**)
Huber, W., 111, 224, 309(**2**); 224, 309, 311(**46**); 110, 111, 114, 116, 309, 315, 316, 338(**75**)
Hübner, H., 181, 197, 314(**263**)
Huebner, C. F., 26, 168, 171, 353(**469**)
Huffman, K. R., 208, 209, 214, 309(**165**); 208, 214, 309(**166**)
Huisgen, R., 180, 338(**370**); 217, 314(**475**); 197, 217, 233(**569**); 233(**570**)
Hunn, E. B., 314(**248**)
Hunneman, D. H., 16(**437**)
Hyne, J. B., 57, 366(**69**); 60, 344, 352, 361, 366(**242**)

Ichiba, A., 200, 318(**86**)
Ikeda, Y., 110, 311(**380**)
Irie, H., 18, 39, 65, 351, 355(**230**)
Isele, G., 75, 374(**530**)
Ishikawa, J., 224, 311(**386**); 311(**401**)
Ito, A., 109, 224, 311(**384**)
Itzchaki, J., 11, 14, 37, 63, 352, 353(**7**)
Iwanoff, C., 322(**270**)
Iwatsu, T., 111, 224, 329(**383**)

Jacini, G., 182, 314(**331**); 197, 227, 314(**332**)

Jacobi, P., 172, 173, 174, 176, 222, 293, 294, 310, 311, 321, 327, 328, 329, 338, 347**(537)**
Jaffe, G. E., 105, 108, 116, 118, 121, 219, 226, 272, 285, 290, 292, 294, 309, 312, 315, 316, 338, 339, 344, 362**(122)**
Jäger, H., 177, 179, 357**(434)**
v. Jakubowski, Z., 319**(292)**
Jampolsky, L. M., 311**(422)**
Jansen, A. B. A., 331**(350)**
Jarque, R. G., 116, 118, 322, 348**(52)**
Jefford, C. W., 194, 202, 224**(41)**
Jensen, K. A., 314, 315**(326)**
Johnson, F., 218, 321, 333**(470)**; 16**(506)**; 17, 38, 365**(533)**
Johnson, G. A., 11, 22, 53, 73, 323, 345**(231)**; 53, 323, 345, 353**(417)**
Johnson, M. C., 161, 324, 342, 350, 353, 354, 356, 360, 362, 365**(555)**
Jones, J. H., 185, 186, 224, 272, 292, 293, 294, 308, 311, 315, 319, 320, 329**(562)**
Jones, W. G. M., 146, 147, 330, 331**(337)**
Joseph, J. P., 119, 319**(121)**
Josey, A. D., 93, 308**(349)**
Junek, H., 167, 170, 226, 331, 347**(172)**; 354**(487)**; 162, 310**(508)**; 30**(531)**; 134, 324, 325**(548)**
Jürgens, V., 319**(271)**
Justoni, R., 101, 231, 294, 359**(85)**

Kalbag, S. M., 81**(540)**
Kalenda, N. W., 166, 170, 282, 295, 331**(14)**; 167, 170, 331**(31)**
Kaneko, H., 105, 110, 224, 311**(49)**; 110, 311**(380)**
Kanō, H., 143, 221, 308**(176)**
Karavan, V. S., 127, 128, 129, 249, 361, 362, 365, 367, 373**(492)**
Karle, D. W., 2, 4, 8, 11, 20, 31, 38, 42, 45, 312, 317, 322, 330, 336**(495)**
Karrer, P., 224, 309, 315**(254)**
Katritzky, A. R., 50, 78, 323, 352**(28)**
Kenner, G. W., 104, 108, 123, 309, 315, 347, 362**(36)**
Kenner, J., 20, 43, 359**(192)**; 198, 225, 319**(272)**
Kinnel, R. B., 20, 45, 68, 229, 357**(167)**
Kinugawa, J., 332**(351)**; 310**(352)**
Kirby, E. C., 54, 74, 360**(110)**
Kiss, J., 311**(422)**
Klingsberg, E., 156, 157, 343**(471)**
Klug, J., 168, 170, 266, 335, 362**(509)**
Klüpfel, K., 166, 170, 331**(260)**
Knight, A. H., 225, 313, 314**(295)**
Knopf, R. J., 190, 236, 237, 314**(21)**; 113, 116, 272, 282, 283, 290, 291, 294, 311, 316, 320, 322, 330, 336, 339, 341, 344**(22)**; 209, 210, 212, 214, 215, 216, 308, 309, 310, 314, 362**(23)**; 235, 237, 308, 309, 310, 311, 314, 331**(27)**
Knospe, R. H., 62, 64**(513)**
Kobayashi, M., 195, 318**(9)**
Kochetkov, N. K., 221**(525)**
Koelsch, C. F., 39**(488)**
Kogelschatz, J. E., 305, 307, 313, 319**(468)**
Koopman, H., 199, 314**(319)**
Kopecky, J., 194, 224, 310, 311, 312, 321**(63)**
Korte, F., 109, 224, 311**(42)**
Kostić, K., 224, 309, 315**(254)**
Kreutzberger, A., 214, 309**(259)**
Krol, L. H., 190, 197, 228, 314**(67)**; 191, 338**(212)**
Kruber, O., 16, 332**(455)**
Kruckenberg, W., 225, 313, 314**(302)**; 225, 314**(306)**; 225, 314**(307)**; 225, 314, 318**(309)**
Krueger, P. J., 314**(322)**; 314, 333**(328)**
Krüger, C., 3, 11, 20, 322**(500)**
Kubota, T., 311, 315**(399)**
Kuehne, M. E., 60**(443)**
Kulp, S. S., 11, 31, 32, 62, 312, 322, 336, 344, 352, 365**(281)**; 21, 34, 47, 365, 374**(282)**; 11, 38, 39, 64, 317, 330, 341, 357, 367**(542)**
Kunze, W., 286, 314**(142)**
Kuo, C. H., 22, 50, 73, 165, 310, 312, 313, 322**(536)**
Kutter, E., 25, 26, 55, 56, 252, 317, 346, 347**(171)**
Kwartler, C. E., 225, 314**(244)**

Lamant, M., 60, 62, 312**(233)**; 312**(288)**
Larionov, L. F., 272**(528)**
Laschtuvka, E., 180, 338**(370)**
Läufer, S., 314**(330)**
Lazarus, A. K., 359**(208)**
Lazier, W. A., 77, 312**(237)**
Ledig, K. W., 192, 204, 205, 286, 319, 328, 333, 341**(313)**

Lee, J., 224, 299, 311(**462**)
Legagneur, F. S.-, 33, 34, 62, 63, 357, 361, 363, 366, 367, 369(**138**); 34, 77, 365(**268**); 34, 365(**277**); 12, 34, 365(**285**), 12, 34, 365(**287**); 34, 63, 367(**485**); 32, 33, 34, 62, 63, 352, 357, 361, 363, 365, 366, 367, 369(**486**)
Lenard, K., 172, 173, 174, 176, 222, 293, 294, 310, 311, 321, 327, 328, 329, 338, 347(**537**); 172, 176, 293, 294, 311(**551**)
Leonard, N. J., 201, 261, 339(**71**); 262, 339(**472**)
Lespagnol, A., 193, 224, 315(**87**)
Lichtmann, H., 16, 332(**452**)
Lingens, F., 314(**330**)
Linstead, R. P., 198, 314(**317**)
Little, E. L., 142, 207, 214, 242, 299(**102**)
Little, E. L., Jr., 159, 160, 161, 164, 193, 313, 314, 317, 324, 325, 326, 332, 337, 339, 346, 348, 350, 351, 355, 361, 373(**106**); 106, 312, 339(**366**)
Littner, S., 200, 220, 337(**70**)
Loeffler, P. K., 82, 238, 239, 254, 255, 256, 259, 260, 261, 309(**20**); 201, 309(**24**)
Lora-Tamayo, M., 35, 346(**515**)
Lucas, P., 225, 314(**244**)
Lüttringhaus, A., 1, 20, 22, 48, 54, 55, 70, 74, 370, 371, 373(**193**); 29, 70(**522**); 75, 374(**530**)
Lythgoe, B., 104, 108, 123, 309, 315, 347, 362(**36**); 104, 108, 309(**44**)

McDonald, F. G., 184, 219, 308(**185**)
McEvoy, F. J., 199, 319(**121**)
McGeer, E. G., 142, 207, 214, 242, 299(**102**)
McGinn, F. A., 45, 68, 373(**209**)
McKee, M. K., 199, 227, 319(**131**); 199, 227, 314(**132**)
McKee, R. L., 199, 227, 319(**131**); 199, 227, 314(**132**)
McKillop, A., 200, 306, 308, 309, 313, 314, 319, 331, 332, 346(**93**); 200, 306, 308, 309, 313, 314, 319, 331, 332, 346(**94**); 22, 51, 224, 299, 305, 308, 309, 312, 314, 317, 332, 339, 349(**95**); 22, 51, 224, 299, 301, 308, 309, 312, 314, 317, 332, 339, 349(**96**); 266, 280, 297, 298, 344(**290**); 232, 233(**456**)
McKusick, B. C., 142, 207, 214, 242, 299(**102**); 81, 86, 91, 92, 93, 94, 95, 96, 97, 98, 103, 225, 308, 309, 310, 313, 314, 315, 316, 317, 319, 320, 327, 334, 336, 337, 338, 339, 342, 343, 346, 349, 359(**232**); 166, 170, 331(**261**); 93, 94, 309, 312, 338(**406**)
Madroñero, R., 16(**506**)
Makisumi, Y., 143, 221, 308(**176**)
Malik, P. A., 118, 311, 330(**539**)
Mann, F. G., 54, 74, 360(**110**); 11, 22, 51, 53, 73, 345, 346(**111**)
Manunapichu, K., 19(**218**)
Margerison, R., 26, 168, 171, 353(**469**)
Märkl, G., 345(**109**)
Martini, C. M., 299(**460**)
Matsukawa, T., 110, 224, 314, 315, 316, 329(**51**); 224, 311, 315, 344, 363(**55**)
Matsumura, C., 332(**351**)
Matsuoka, H., 224, 311(**381**)
Matukawa, T., 224, 311(**379**)
Maudling, D. R., 177, 178, 179, 344, 356, 361(**435**)
Mautner, H. G., 224, 231, 309(**130**)
May, C. D., 194, 224, 312(**338**)
Mayer, M., 198, 319(**275**)
Mayer, R., 26, 56, 310(**107**)
Mehta, A. C., 171(**565**)
Mehta, A. S., 65(**428**)
Mehta, S. R., 81(**540**)
Metzger, A., 200, 337(**68**)
Meyer, F. J., 87, 88, 124, 309, 311, 325, 327, 334, 340, 345, 351, 362(**411**)
Meyer, R. E., 60(**499**)
Meyer, R. F., 113, 116, 272, 282, 283, 290, 291, 294, 311, 316, 320, 322, 330, 336, 339, 341, 344(**22**); 112, 309, 312, 316, 320, 329, 330, 336, 339, 344(**143**)
Middleton, W. J., 142, 207, 214, 242, 299(**102**); 132, 133, 135, 136, 141, 339, 349(**103**); 129, 131, 132, 133, 134, 135, 140, 142, 207, 214, 219, 242, 299, 309, 310, 313, 314, 316, 318, 320, 341, 343, 347, 351, 360(**104**); 80, 81, 92, 93, 105, 121, 122, 143, 144, 308, 312, 313, 316, 321, 332, 339, 344, 348(**105**); 159, 160, 161, 164, 193, 313, 314, 317, 324, 325, 326, 332, 337, 339, 346, 348, 350, 351, 355, 361, 373(**106**); 133, 136, 311, 316, 318, 325, 328, 329, 332, 333, 338, 339, 347, 368, 369(**365**); 140, 310(**374**); 214, 310(**394**); 141, 301, 339(**398**); 133,

311(**402**); 148, 149, 308, 331, 360(**473**)
Minas, H., 20, 43, 66, 359(**29**)
Mislow, K., 20, 43, 44, 67, 359, 365(**205**); 20, 44, 359(**206**); 359(**208**); 45, 68, 373(**209**)
Mitra, A., 11, 15, 37, 63, 352(**12**)
Miyatake, K., 224, 311(**255**); 224, 311(**382**)
Mizukami, S., 311(**329**); 311(**474**)
Modest, E. J., 191, 280, 287, 294, 295, 338(**289**); 11, 20, 41, 65, 338, 339(**549**)
Mohr, J., 1(**524**)
Montequi, F., 201, 231, 294, 309, 317(**98**)
Moore, C. W., 1, 16, 35, 63, 77, 332(**191**)
Morgan, G. T., 199, 225, 319(**294**)
Morita, K., 105, 237, 269, 289, 308, 309, 311, 312, 315(**467**)
Mosby, W. L., 40, 364(**227**)
Mowat, J. H., 200, 318(**77**)
Mower, H. F., 121, 122, 321(**164**); 166, 170, 331(**261**); 106, 312, 339(**366**)
Mull, R. A., 273, 295, 312(**179**)
Müller, E., 197, 314(**197**)
Müller, H. C., 31, 62, 329(**512**)
Müller, H.-J., 180, 358(**490**)
Mulvey, D. M., 220, 281, 282, 313, 326(**133**); 220(**353**)
Murahashi, S., 224, 311(**385**)
Musso, H., 198, 199, 202, 314, 341(**256**); 199, 341(**257**); 200, 337(**280**)
Muth, C. W., 3, 20, 43, 66, 359(**224**)

Nachod, F. C., 299(**460**)
Nagai, M., 200, 318(**86**)
Nair, M. D., 118, 311, 330(**539**); 81(**540**)
Nairn, J. G., 311(**369**)
Naito, T., 194, 224, 315(**48**)
Nakagawa, T., 311, 315(**399**)
Nakayama, H., 311(**396**)
Nakazaki, M., 3, 20, 43, 359(**223**); 65(**427**)
Narisada, M., 311, 315(**399**)
Nasutavicus, W. A., 17, 38, 365(**533**)
Nedenskov, P., 32, 62, 352(**269**)
Nelson, S. J., 34, 62, 365(**221**)
Neuse, E., 158, 163, 164, 166, 169, 320, 335(**108**)
Neven, C., 34, 77, 365(**268**); 12, 34, 365(**285**); 12, 34, 365(**287**); 34, 63, 367(**485**)
Newbold, G. T., 193, 219, 310(**78**)
Newman, M. S., 32, 62, 349(**182**); 40, 364(**227**)
Newman, P., 20, 43, 44, 67, 359, 365(**205**)
von Niementowski, S., 199, 319(**198**); 199, 319(**199**)
Nietzki, R., 317, 318, 319, 320(**83**)
Nishio, A., 224, 311(**385**)
Nitzschke, H.-J., 24(**446**); 22, 52, 103, 349(**447**)
Nonell, M. A. G., 35, 346(**491**)
Nozoe, T., 151, 152, 332, 342, 354(**529**)

O'Brien, D. E., 195, 337, 338(**348**)
Ochiai, E., 194, 224, 315(**48**)
Ochiai, M., 332(**351**); 310(**352**); 105, 237, 269, 289, 308, 309, 311, 312, 315(**467**)
Ogata, K., 143, 221, 308(**176**)
O'Grady, B. V., 314(**324**)
Ohlinger, H., 1, 46, 47, 61, 65, 68, 69, 71, 72, 322, 361, 366, 374(**187**); 1, 20, 42, 46, 47, 65, 68, 69, 71, 72, 322, 361, 374(**200**)
Ohta, M., 105, 123, 309, 311, 344(**50**)
Oishi, T., 29, 55, 76, 350(**558**)
Orgel, L. E., 145, 146, 183, 219, 268, 289, 308, 309, 311, 337(**32**); 145, 146, 183, 289, 308, 309, 311, 331(**335**); 145, 268, 289, 308(**336**); 176(**566**)
Osawa, F., 105, 110, 224, 311(**49**); 110, 311(**380**)
Osdene, T. S., 76(**504**); 176(**505**)

Pachter, I. J., 105, 108, 116, 118, 121, 219, 226, 272, 285, 290, 292, 294, 309, 312, 315, 316, 338, 339, 344, 362(**122**)
Paddon-Row, M. N., 309(**146**); 240(**520**)
Papanastassiou, Z. B., 3, 20, 43, 66, 359(**224**)
Parnell, E. W., 217, 219, 222, 223, 314(**134**)
Partridge, M. W., 203, 204, 205, 314, 318, 319(**1**); 233, 314(**118**); 204, 205, 233, 313, 314, 319(**320**); 196, 314(**321**); 197, 314(**554**)
Paudler, W. W., 195, 362(**8**)
Pecherer, B., 20, 44, 67, 369(**476**)
Pechet, M. M., 16(**60**)
Pelchowicz, Z., 16(**90**)
Perkin, W. H., Jr., 31, 329(**514**)
Perkins, R. L., 199, 328(**279**)
Peshkar, L., 150, 242, 254, 309, 312, 339(**156**)
Peters, E., 110, 224, 310(**346**)
Peters, G. A., 208, 209, 214, 309(**165**);

208, 214, 309(**166**)
Petri, W., 317, 318, 319, 320(**83**)
Petsch, G., 224, 311(**362**); 194, 316(**363**)
Pfleiderer, W., 209, 210, 212, 214, 215, 216, 308, 309, 310, 314, 362(**23**); 176(**501**); 171, 294(**564**)
Picquet, L., 34, 63, 367(**485**)
Pietsch, H., 26, 102, 219, 315, 343(**507**)
Pilgrim, F. J., 200, 318(**77**)
Pincock, R. E., 15, 37, 352(**534**)
Pinnow, J., 197, 222, 225, 314(**40**); 197, 314(**197**)
Plati, J. J., 311(**422**)
Platt, E. J., 18, 39, 64, 291, 357(**222**); 40, 64, 368(**225**); 40, 64, 364(**226**)
Pohland, H. W., 183, 309(**283**)
Polansky, O. E., 315, 324(**364**)
de Pradenne, H. V., 177, 178, 365(**286**)
Prasad, R. N., 201, 279, 309(**136**)
Preobrazhenskiǐ, N. A., 109, 224, 311(**387**)
Puar, M. S., 298, 314(**158**)
Putter, R., 107, 108, 109, 324, 351, 362, 363, 366, 367, 369(**72**)

Rabadeux, J., 33, 34, 62, 63, 357, 361, 363, 366, 367, 369(**138**); 32, 33, 34, 62, 63, 352, 357, 361, 363, 365, 366, 367, 369(**486**)
Rahn, D., 35, 63, 367(**545**)
Rajappa, S., 199, 328(**323**)
Ralhan, N. K., 298, 314(**158**)
Ramage, G. R., 146, 147, 330, 331(**337**)
Rao, R. P., 90, 231, 295, 309, 311(**124**)
Raphael, R. A., 36, 63, 169, 341(**511**)
Rapoport, H., 43, 66, 359(**229**)
Ravindranathan, R. V., 204, 297, 298, 314(**15**); 22, 51, 224, 299, 305, 308, 309, 312, 314, 317, 332, 339, 349(**95**); 181, 305(**241**)
Recherer, B., 311(**422**)
Reed, K. J., 22, 51, 61, 73, 317(**112**); 227, 314(**249**); 119, 226, 314, 319(**250**)
Reiff, H. E., 12, 34, 35, 63, 367(**220**)
Reissert, A., 197, 226, 314(**137**)
Remfry, F. H. P., 162, 310(**521**)
Rigby, G. W., 123, 232, 312(**59**)
Riobé, O., 312(**284**)
Robins, R. K., 82, 83, 85, 86, 90, 91, 231, 275, 276, 278, 279, 309, 312, 331, 332, 339(**3**); 82, 275, 295, 308(**4**); 201, 279, 309(**136**); 82, 83, 85, 86, 227, 308, 309, 312, 331, 332, 339(**168**); 82, 85, 295, 309, 331, 332(**169**); 84, 85, 86, 277, 278, 322, 336, 337, 339(**170**); 195, 337, 338(**348**); 245, 246, 310, 320(**494**); 289, 309(**550**); 201, 241, 262, 357(**552**); 201, 262, 357(**355**)
Roelofs, W. L., 177, 178, 179, 344, 356, 361(**435**); 177(**436**)
Rogers, E. F., 111, 321(**409**)
Rosenbloom, J. P., 105, 108, 116, 118, 121, 219, 226, 272, 285, 290, 292, 294, 309, 312, 315, 316, 338, 339, 344, 362(**122**)
Rosowsky, A., 191, 280, 287, 294, 295, 338(**289**); 11, 20, 41, 65, 338, 339(**549**); 190, 198, 199, 204, 286, 287, 294, 313, 314, 319, 328(**563**)
Rousseau, R. J., 201, 241, 262, 357(**552**); 201, 262, 357(**355**)
Rubtsov, I. A., 109, 224, 311(**387**)
de Ruggieri, P., 187, 189, 264, 265, 280, 281, 295, 368, 370, 371, 372, 374(**438**)
Rupe, H., 200, 337(**68**)
Rutkin, P., 20, 43, 44, 67, 359, 365(**205**); 359(**208**)

Sachdev, H. S., 298, 314(**158**)
Sakan, R., 3, 20, 43, 359(**223**)
Sala, C. Vallmitjana, 116, 118, 322, 348(**52**)
Sämann, C., 197, 222, 225, 314(**40**)
Sano, K., 224, 311(**397**)
Sargent, L. J., 199, 328(**316**)
Sato, Y., 137, 167, 348(**418**)
Sausen, G. N., 132, 133, 135, 136, 141, 339, 349(**103**); 159, 160, 161, 164, 193, 313, 314, 317, 324, 325, 326, 332, 337, 339, 346, 348, 350, 351, 355, 361, 373(**106**)
Sawa, Y., 105, 110, 224, 311(**49**); 110, 311(**380**)
Saygin, F., 24, 55, 279, 313(**339**)
Schaarschmidt, A., 192, 353(**325**)
Schaefer, F. C., 208, 209, 214, 309(**165**); 208, 214, 309(**166**)
Schaefer, J. P., 2(**518**)
Schaffner, K., 299(**461**)
Schaub, R. E., 119, 319(**121**)
Schechter, H., 11(**516**)
v. Schickh, O., 197, 314(**305**)
Schill, G., 29, 70(**522**); 75(**523**)
Schinke, E., 137, 141, 321, 329(**47**); 137,

138, 139, 315, 328(**113**)
Schleese, E., 91, 118, 119, 332, 337, 339, 343, 344(**33**)
Schlittler, E., 273, 295, 312(**179**)
Schmid, H., 146(**429**); 217, 314, 358(**547**)
Schmidt, A., 162, 310(**508**)
Schmidt, P., 86, 276, 309(**97**); 231, 317(**115**); 283, 316(**139**); 227, 317(**140**); 84, 85, 225, 231, 322, 330, 333(**144**); 85, 231, 332(**145**); 84, 231, 317(**148**); 82, 84, 85, 309, 317, 330, 336(**149**); 83, 312(**150**); 84, 85, 276, 277, 278, 322, 329, 330, 336(**155**); 85, 231, 331, 332(**157**); 84, 85, 277, 329, 330(**159**); 219, 311, 312(**368**); 86, 339(**371**); 84, 317(**373**); 84, 317(**403**); 82, 231, 309(**405**)
Schmidt, R. R., 193, 362(**354**); 195, 283, 361, 362(**359**)
Schnitzer, R. J., 272(**527**)
Schröder, H., 198, 199, 202, 314, 341(**256**); 199, 341(**257**); 200, 337(**280**)
Schroeder, H. E., 123, 232, 312(**59**); 69(**430**)
Schroth, W., 332(**449**); 11, 16, 332, 362, 365, 366, 367(**450**); 16(**451**); 16, 332(**452**); 16, 332, 365(**453**)
Schulze, W., 98, 99, 100, 103, 107, 125, 352, 356, 357, 361, 366, 369, 372(**477**); 26, 101, 125, 321, 322, 329, 335(**543**)
Schwabe, U., 200, 337(**280**)
Schwyzer, R., 224, 309, 315(**254**)
Scott, W. B., 15, 37, 352(**534**)
Scribner, R. M., 142, 207, 214, 242, 299(**102**)
Sekiya, M., 311(**391**); 224, 311(**392**)
Sensi, P., 197, 314(**333**)
Severin, T., 20, 41, 42, 346, 355(**478**)
Sewell, M. J., 50, 78, 323, 352(**28**)
Sharanin, Y. A., 127, 128, 129, 328, 341, 347, 351, 362, 365(**5**); 127, 128, 129, 249, 361, 362, 365, 367, 373(**492**)
Sharpless, T. W., 304, 314, 319(**538**)
Shaw, G., 149, 182, 259, 308, 309, 312(**114**); 299(**461**)
Sheradsky, T., 187, 188, 196, 244, 245, 288, 316, 322, 348, 352, 367(**76**); 300, 316, 322(**479**)
Shirakawa, K., 87, 88, 309, 326, 327, 333, 334, 340, 345, 351(**400**); 87, 88, 89, 90, 319, 325, 326, 327, 328, 333, 334, 335, 340, 343, 345, 351, 352, 355, 360, 361, 362, 363, 367(**419**)
Short, L. N., 309(**58**)
Shvo, Y., 266, 280, 297, 298, 344(**290**); 191, 192, 229, 314, 338, 344(**340**)
Siegel, M., 20, 44, 359(**206**)
Simchen, G., 193, 309(**544**); 169, 333(**557**)
Sirakawa, K., 309, 311, 325, 327, 334, 340, 345, 351, 362(**412**)
Smith, P. A. S., 238(**519**)
Söderbäck, E., 153, 308(**216**)
Sokolov, S. D., 221(**525**)
Soma, N., 137, 167, 331(**420**); 130, 131, 137, 331, 348(**421**)
Soto, J. L., 35, 346(**515**)
Southwick, P. L., 187, 188, 196, 224, 245, 288, 316, 322, 348, 352, 367(**76**); 300, 316, 322(**479**)
Spickett, R. G. W., 176(**502**)
Spring, F. S., 193, 219, 310(**78**)
Staehelin, A., 205, 337(**65**)
Stauffer, D. A., 19, 41, 65, 367(**210**); 367(**211**)
Sterk, H., 310(**531**); 134, 324, 325(**548**)
Stevens, M. F. G., 202, 314(**493**); 202, 314, 315, 351(**553**)
de Stevens, G., 273, 295, 312(**179**)
Stimson, M. M., 311(**360**)
Strachan, M. N., 119, 226, 314, 319(**250**)
Strachan, P. L., 26, 168, 171, 353(**469**)
Sudarsanam, V., 199, 328(**323**)
Sunagawa, G., 137, 167, 331(**420**); 130, 131, 137, 331, 348(**421**)
Sung, W-L., 3, 20, 43, 66, 359(**224**)
Susemihil, W., 11(**496**)
Sutcliffe, E. Y., 84, 85, 86, 277, 278, 322, 336, 337, 339(**170**)
Suter, H., 113, 114, 115, 116, 224, 309, 311, 312, 316, 338, 344, 348, 359(**53**)
Svamas, G., 12, 34, 35, 63, 367(**220**)
Sweet, A. J., 199, 328(**279**)
Sword, I. P., 172, 173, 174, 176, 222, 293, 294, 310, 311, 321, 327, 328, 329, 338, 347(**537**)
Sybistowicz, D., 91, 227, 338(**152**)
Sykes, P., 194, 224, 312(**338**); 104, 108, 224, 315(**367**)
Szelke, M., 331(**350**)
Szkrybalo, W., 20, 45, 68, 350(**559**)

Tanaka, I., 311, 315(**399**)
Tanaka, Y., 224, 311(**381**)
Taub, D., 22, 50, 73, 165, 310, 312, 313, 322(**536**)
Taub, W., 32, 62, 352(**269**)
Tauber, E., 311(**390**)
Taurins, A., 11(**498**)
Taylor, E. C., 195, 362(**8**); 214, 241, 242, 245, 246, 247, 248, 288, 310, 311, 320, 339(**10**); 143, 144, 252, 253, 254, 308, 309, 311, 326, 331, 355(**11**); 166, 170, 282, 295, 331(**14**); 204, 297, 298, 314(**15**); 200, 237, 308, 309, 310, 313, 314, 332(**17**); 82, 302, 303, 304, 305, 308(**18**); 267, 268, 290, 309, 311(**19**); 82, 238, 239, 254, 255, 256, 259, 260, 261, 309(**20**); 190, 236, 237, 314(**21**); 113, 116, 272, 282, 283, 290, 291, 294, 311, 316, 320, 322, 330, 336, 339, 341, 344(**22**); 209, 210, 212, 214, 215, 216, 308, 309, 310, 314, 362(**23**); 201, 309(**24**); 81, 90, 91, 95, 243, 257, 258, 279, 295, 312, 315, 339, 343, 344, 351(**25**); 1, 81, 94, 310, 332(**26**); 235, 237, 308, 309, 310, 311, 314, 331(**27**); 192, 310(**30**); 167, 170, 331(**31**); 138, 139, 241, 242, 249, 250, 251, 310, 311, 315, 328, 335, 347(**91**); 257, 308(**92**); 200, 306, 308, 309, 313, 314, 319, 331, 332, 346(**93**); 200, 306, 308, 309, 313, 314, 319, 331, 332, 346(**94**); 22, 51, 224, 299, 305, 308, 309, 312, 314, 317, 332, 339, 349(**95**); 22, 51, 224, 299, 301, 308, 309, 312, 314, 317, 332, 339, 349(**96**); 266, 280, 297, 298, 344(**290**); 191, 192, 229, 314, 338, 344(**340**); 137, 138, 140, 241, 242, 249, 250, 251, 310, 311, 335, 347(**341**); 175, 176, 337, 347, 351, 355, 363, 366, 371(**439**); 232, 233(**456**); 262, 263, 264, 297, 300, 306, 308, 309, 311, 314, 317, 331, 339, 349(**459**); 301, 305(**464**); 167, 171, 331(**465**); 300(**480**); 168, 170, 266, 335, 362(**509**); 241(**526**); 172, 173, 174, 176, 222, 293, 294, 310, 311, 321, 327, 328, 329, 338, 347(**537**); 172, 176, 293, 294, 311(**551**); 171, 294(**564**); 177(**567**)
Taylor, W. I., 203, 205, 206, 207, 226, 314, 319(**240**)
Teckenburg, H., 143, 144, 219, 331(**177**)
Temnikova, T. I., 127, 128, 129, 328, 341, 347, 351, 362, 365(**5**); 127, 128, 129, 249, 361, 362, 365, 367, 373(**492**)
Theobald, C. W., 142, 207, 214, 242, 299(**102**)
Thieme, A., 198, 199, 220, 319(**274**)
Thompson, H. W., 314, 333(**328**)
Thompson, Q. E., 11, 26, 31, 312(**99**)
Thorpe, J. F., 1, 16, 35, 63, 77, 332(**191**); 12, 31, 62, 312, 329(**194**); 1, 62, 312(**195**); 162, 310(**521**)
Tieckelmann, H., 220, 284, 314, 315, 320, 326(**125**); 220, 281, 282, 313, 326(**133**); 113, 114, 159, 164, 309, 311, 316, 318, 326, 344, 351(**246**); 110, 224, 310(**346**); 110, 111, 224, 313, 317(**347**); 311(**369**)
Tiefenthaler, H., 146(**429**); 217, 314, 358(**547**)
Tilney-Bassett, J. F., 191, 313(**546**)
Timmis, G. M., 176(**502**); 176(**503**); 176(**504**)
Tobey, S. W., 17, 38, 365(**533**)
Toda, T., 151, 152, 332, 342, 354(**529**)
Todd, A. R., 104, 108, 123, 309, 315, 347, 362(**36**); 104, 108, 309(**44**); 183, 194, 224, 311(**196**)
Tolman, R. L., 245, 246, 310, 320(**494**); 289, 309(**550**)
Tomlin, C. D. S., 180, 184, 194, 213, 267, 272, 283, 284, 290, 294, 309, 337, 339(**458**)
Töpfl, W., 25, 26, 55, 56, 252, 317, 346, 347(**171**); 81, 93, 144, 258, 309, 339(**266**)
Topham, A., 104, 108, 123, 309, 315, 347, 362(**36**)
Townsend, L. B., 201, 262, 357(**355**); 245, 246, 310, 320(**494**); 289, 309(**550**); 201, 241, 262, 357(**552**)
Traut, H., 193, 309(**544**)
Traverso, J. J., 114, 115, 117, 321, 322, 329, 330, 336, 338, 346, 350, 358(**35**); 107, 115, 119, 120, 311, 315, 319, 321, 329, 336, 337, 338, 341, 344, 346, 348, 350(**43**)
Treibs, W., 332(**449**); 11, 16, 332, 362, 365, 366, 367(**450**); 16(**451**); 16, 332(**452**); 16, 332, 365(**453**)
Trofimenko, S., 159, 220, 313(**251**); 106, 312, 339(**366**)
Truce, W. E., 62, 64(**513**)
Tsuda, Y., 18, 39, 65, 351, 355(**230**)

Tsujikawa, T., 87, 88, 89, 90, 319, 325, 326, 327, 328, 333, 334, 335, 340, 343, 345, 351, 352, 355, 360, 361, 362, 363, 367**(419)**; 88, 343**(423)**
Tsunoo, M., 224, 311**(255)**; 224, 311**(382)**
Turi, C. J., 204, 314**(62)**
Turner, E. G., 20, 43, 359**(192)**

Ulrich, H., 223, 314**(129)**
Uyeo, S., 18, 39, 65, 351, 355**(230)**

VanAllen, J. A., 23, 54, 353**(445)**
Vanderhorst, P. J., 193, 320**(80)**
Verkade, P. E., 228, 229, 298, 314**(61)**; 190, 197, 228, 314**(67)**; 191, 338**(212)**
Verry, C., 319, 335**(440)**
Villani, A. J., 105, 108, 116, 118, 121, 219, 226, 272, 285, 290, 292, 294, 309, 312, 315, 316, 338, 339, 344, 362**(122)**
Vipond, H. J., 204, 205, 233, 313, 314, 319**(320)**; 197, 314**(554)**
Vogl, O., 177**(567)**
Vromen, S., 22, 51, 224, 299, 305, 308, 309, 312, 314, 317, 332, 339, 349**(95)**; 22, 51, 224, 299, 301, 308, 309, 312, 314, 317, 332, 339, 349**(96)**; 262, 263, 264, 297, 300, 306, 308, 309, 311, 314, 317, 331, 339, 349**(459)**; 300**(480)**; 306, 312, 317, 349**(535)**

Wahl, R., 169, 333**(557)**
Waite, J. A., 204, 205, 233, 313, 314, 319**(320)**
Wakamatsu, T., 29, 55, 76, 350**(558)**
Wallenfels, K., 190, 191, 203, 324**(481)**; 190, 191, 203, 324**(482)**
Waring, W. S., 146, 147, 330, 331**(337)**
Warrener, R. N., 200, 306, 308, 309, 313, 314, 319, 331, 332, 346**(93)**; 200, 306, 308, 309, 313, 314, 319, 331, 332, 346**(94)**; 299**(461)**
Washimi, K., 224, 311**(393)**
Watanabe, K., 198, 314**(308)**; 190, 314**(311)**
Webb, N. E., 304, 317, 323**(16)**
Weber, K., 1, 69**(235)**
Webster, O. W., 16, 37, 330, 364**(483)**
Weinstock, J., 105, 108, 116, 118, 121, 219, 226, 272, 285, 290, 292, 294, 309, 312, 315, 316, 338, 339, 344, 362**(122)**; 118, 269, 291, 294, 312, 317, 326, 339, 344, 349**(141)**; 339**(180)**
Weir, M. R. S., 57, 366**(69)**; 60, 344, 352, 361, 366**(242)**
Weis, K., 225, 314, 318**(309)**
Weithamp, H., 109, 224, 311**(42)**
Weitzenböck, R., 20, 43, 77, 359**(265)**
Welcher, R. P., 11, 22, 53, 73, 323, 345**(231)**; 53, 323, 345, 353**(417)**
Wendler, N. L., 22, 50, 73, 165, 310, 312, 313, 322**(536)**
Wenis, E., 224, 299, 311**(462)**
Wenk, P., 273, 295, 312**(179)**
Wenkert, E., 26, 168, 171, 353**(469)**
Wepster, B. M., 190, 197, 228, 314**(67)**; 191, 338**(212)**
Westfahl, J. C., 57, 337**(204)**
Westöö, G., 128, 129, 221, 319**(276)**
Weussmann, H., 204, 314**(39)**
Whitehead, C. W., 114, 115, 117, 321, 322, 329, 330, 336, 338, 346, 350, 358**(35)**; 107, 115, 119, 120, 311, 315, 319, 321, 329, 336, 337, 338, 341, 344, 346, 348, 350**(43)**; 119, 336**(361)**
Whitman, G. M., 4, 11**(135)**
Wibberley, D. G., 166, 170, 226, 355**(162)**; 219, 324**(541)**
Wickham, P. P., 31, 62, 329**(512)**
Wiebelhaus, V. D., 118, 269, 291, 294, 312, 317, 326, 339, 344, 349**(141)**; 339**(180)**
Wiedenmann, R., 157, 343**(489)**; 125, 194, 308**(560)**
Wilhelm, M., 86, 276, 309**(97)**; 231, 317**(115)**; 283, 316**(139)**; 227, 317**(140)**; 84, 85, 227, 329, 330**(159)**; 219, 311, 312**(368)**; 86, 339**(371)**
Williams, A. R., 43, 66, 359**(229)**
Williams, J. H., 199, 319**(121)**
Williams, J. K., 11, 17, 38, 57, 64, 317, 336, 337, 344, 346, 350, 352, 356**(89)**; 81, 86, 91, 92, 93, 94, 95, 96, 97, 98, 103, 225, 308, 309, 310, 313, 314, 315, 316, 317, 319, 320, 327, 334, 336, 337, 338, 339, 342, 343, 346, 349, 359**(232)**
Willis, J. A. D., 22, 49, 78, 273, 288, 295, 317, 323, 346, 349, 352**(13)**; 304, 317, 323**(16)**
Willitzer, H., 98, 99, 100, 103, 107, 125, 352, 356, 357, 361, 366, 369, 372**(477)**; 26, 101, 125, 321, 322, 329, 335**(543)**
Wilson, J. W., 105, 108, 116, 118, 121, 219,

226, 272, 285, 290, 292, 294, 309, 312, 315, 316, 338, 339, 344, 362(**122**)
Wimmer, F., 217, 314(**475**)
Winberg, H. E., 142, 207, 214, 242, 299(**102**)
Witham, E., 198, 225, 319(**272**)
Witzler, F., 190, 191, 203, 324(**481**)
Wooldridge, K. R. H., 198, 314(**317**)
Wunderling, H., 11(**497**)
Wünsch, K.-H., 170(**466**)
Wystrach, V. P., 11, 22, 53, 73, 323, 345(**231**); 53, 323, 345, 353(**417**)

Yamamoto, H., 332(**351**); 310(**352**)
Yanai, K., 194, 224, 315(**48**)
Yano, K., 110, 311(**380**)
Yates, J., 199, 314(**312**)
Yurugi, S., 110, 224, 314, 315, 316, 329(**51**)

Zaǐtseva, E. V., 109, 224, 311(**387**)
Zalkow, V., 65(**427**)
Zalkow, V. B., 65, 66(**510**)
Zee-Cheng, K. Y., 84, 85, 86, 277, 278, 322, 336, 337, 339(**170**)
Zerwech, W., 286, 314(**142**)
Ziegler, K., 1, 46, 47, 61, 65, 68, 69, 71, 72, 322, 361, 366, 374(**187**); 1, 47, 69, 70, 366(**188**); 1, 22, 54, 74, 337(**189**); 1, 20, 42, 45, 46, 47, 65, 68, 69, 71, 322, 330, 358, 371, 373, 374(**190**); 1, 20, 22, 48, 54, 55, 70, 74, 370, 371, 373(**193**); 1, 20, 42, 46, 47, 65, 68, 69, 71, 72, 322, 361, 374(**200**); 1, 46, 69, 358, 365(**201**); 1, 20, 42, 46, 47, 65, 68, 69, 71, 72, 322, 358, 361, 365, 374(**202**); 1, 69(**235**); 1, 46, 361(**243**); 1, 351(**448**); 2(**517**)
Zoltewicz, J. A., 82, 302, 303, 304, 305, 308(**18**); 110, 230, 275, 316(**457**); 301, 305(**464**); 304, 314, 319(**538**)

SUBJECT INDEX

Acetamidine, 149, 269
reaction with aminomethylenemalono-nitrile, 105
reaction with ethoxymethylenemalono-nitrile, 105
reaction with malononitrile, 104
Acetonitrile, dimerization of, 1, 10
reaction with 1,1-diamino-2,2-dicyano-ethylene, 106
Acetonitrile ethyl imino ether, reaction with aminomethylenemalononitrile, 105
Acetonitrile ethyl imino thioether, reaction with aminomethylenemalononitrile, 105
Acetylacetone, 164
2-Acetylaminobenzonitrile, reaction with sodamide, 232
reaction with sodium methoxide, 228
2-Acetylamino-1-cyanocyclopentene, reaction with sodamide, 232
o-Acylaminonitriles, conversion to fused pyridines, 231-233
conversion to fused pyrimidines, 226-231
reaction with alkali, 227
reaction with alkaline hydrogen peroxide, 227
reaction with alkoxides, 228-229
reaction with dry hydrogen chloride, 229
reaction with hydrogen sulfide, 227
reaction with sodamide, 232
Adenine, 268
Adipic acid, 2
Adiponitrile, intramolecular cyclization of, 1, 2, 11, 27
3-Alkoxy-4-cyano-5-aminoisoxazoles, 143
3-Alkoxy-4-cyano-5-aminopyrazoles, 80
Alkylamines, cyclization of bis-cyano-methylation, products of, 22
N-Alkyl-3-amino-4-cyano-3-pyrrolines, 22
β-Alkylaminopropionitriles, reaction with cyanohydrins, 21
Alkyl azides, reaction with malononitrile, 150
N-Alkyl-N-(cyanoethyl)-N-(cyanomethyl)-amines, cyclization of, 22
Alkyl hydrazines, reaction with ethoxy-methylenemalononitrile, 80
reaction with malononitrile, 81
reaction with tetracyanoethylene, 81
reaction with 1,1,2-tricyanostyrenes, 103
Alkylidenemalononitriles, acid-catalyzed cyclization of, 177-180
Table, 178-179
conversion to furan *o*-aminonitriles, 126
dimerization of, 57-60, 137, 152
Table, 58-59
reaction with ammonia, 169
reaction with carbon disulfide, 152
reaction with sulfur and triethylamine, 137
preparation, 57
N-Alkyl ureas, reaction with ethyl ortho-formate and malononitrile, 107
Allyl isothiocyanate, reaction with 2-amino-benzonitrile, 298
Amide, exchange, 270
Amidines, reaction with *o*-aminonitriles, 243, 268-270
Table, 286-293
reaction with ethoxymethylenemalono-nitrile, 105
Amidine exchange, 243, 268-270
1-Amino-2-aminomethylcyclopentane, 77
3-Aminoanthranil, 202
o-Aminobenzaldehyde, reaction with malononitrile, 170
2-Aminobenzamide, 202
2-Aminobenzamidine, 223
2-Aminobenzamidoxime, reduction of, 223
2-Aminobenzonitrile, 219, 223
dimerization of, 233-235
from reduction of 2-nitrobenzonitrile, 196, 202
reaction, with acid anhydrides and sodium sulfide, 228
with allyl isothiocyanate, 298
with dimethyl acetylenedicarboxylate, 306-307
with DMF/HCl, 230

with N,N′-diphenylformamidine, 297
with ethyl iodide, 225
with formic acid, 227
with hydrogen sulfide, 224
with hydroxylamine, 222
with methylmagnesium iodide, 224
with 4-nitrobenzonitrile, 237
with phenylmagnesium bromide, 224
with phenyl isocyanate, 298
with phenyl isothiocyanate, 297, 305
with thioacids, 228
4-Aminobenzo-1,2,3-triazine, 223
2-Aminobenzthioamide, 224
reaction with triethyl orthoformate, 224
2-Amino-5-bromo-3,4-dicyanopyrrole, 129
2-Amino-3-bromotropone, reaction with malononitrile, 136
4-Aminocarbostyril, 232
o-Aminocarboxylic acids, 220
decarboxylation of, 220
2-Amino-4-chlorobenzonitrile, 223
β-Aminocrotononitrile, 1, 10
2-Amino-3-cyano-4-acetyl-5-methylfuran, ring cleavage of, 221
1-Amino-2-cyano-3-benzylnaphthalene, 177
2-Amino-3-cyano-5-chloropyrazine,
reaction with methyl mercaptan, 224
1-Amino-2-cyanocycloheptene, hydrolysis of, 20, 61
1-Amino-2-cyanocyclohexene, 3, 17, 187, 232
spiro derivatives of, 18
1-Amino-2-cyanocyclooctene, 20
hydrolysis of, 61
1-Amino-2-cyanocyclopentene, 2, 11, 27
dimerization of, 27-28
hydrolysis of, 60
reaction with ethyl orthoformate/sodium hydrosulfide, 301
reduction of, 77
2-Amino-3-cyano-4,5-dimethylpyridine, 163
2-Amino-3-cyano-4,6-dimethylpyridine, 164
2-Amino-3-cyano-4,5-dimethylthiophene, 137
2-Amino-3-cyano-5,6-diphenylpyrazine, 209
2-Amino-3-cyanofurans, 126
Diels-Alder reactions of, 213
2-Amino-3-cyano-8-hydroxyquinoline, 136
4-Amino-5-cyanoimidazole,
from aminomalononitrile and formamidine, 146
from irradiation of 3-amino-4-cyanopyrazole, 144-146
from irradiation of 1,1-diamino-2,2-dicyanoethylene, 144, 146
from irradiation of HCN tetramer, 144, 146
reaction with formamidine, 268
2-Amino-3-cyanoindene, 16, 77
3-Amino-4-cyanoisothiazoles, 149, 150
3-Amino-4-cyano-5-methylisothiazole, 242
4-Amino-5-cyano-6-methylmercaptopyrrolo-(2,3-*d*)pyrimidine, 242
3-Amino-4-cyano-5-methylpyrazole, reaction with acetamidine, 269
4-Amino-5-cyano-6-methylpyrimidine, 224
2-Amino-5-cyanonicotinamide, 220, 221
4-Amino-5-cyano-3-phenyl-2-phenylimino-4-thiazoline, 149
2-Amino-3-cyano-7-phenylpyrido(2,3-*b*)-quinoline, reaction with nitrous acid, 226
2-Amino-3-cyanopyrazine, 209
2-Amino-3-cyanopyrazine 1-oxides, 176
reaction with guanidine, 176
3-Amino-4-cyanopyrazole, irradiation of, 144, 146
reaction with carbon disulfide, 306
reaction with ethyl orthoformate/sodium hydrosulfide, 301
4-Amino-5-cyanopyrimidine, 104, 208, 209, 212, 220
irradiation of, 105
reaction with N,N′-dialkylformamidines, 269
reaction with formamidine, 213, 268
reaction with hydrogen sulfide, 224
reduction of, 225
2-Amino-3-cyanopyrroles, 129
3-Amino-4-cyano-3-pyrrolines, reaction with urea, 295
2-Amino-3-cyanoquinoline, 170
reaction with nitrous acid, 226
reaction with urea, 295
2-Amino-3-cyanoquinoline 1-oxide, 171
trans-2-Amino-3-cyano-4,7,8,9-tetrahydroindene, 14
3-Amino-4-cyano-5,6,7,8-tetrahydroisoquinoline, 169
2-Amino-3-cyano-5,6,7,8-tetrahydroquinoline, 165

4-Amino-5-cyanothiazoline-2-thiones, 147
Aminocyanothioacetamide, 149
2-Amino-3-cyanothiophenes, 241, 242
3-Amino-4-cyano-1,2,2-trimethyl-3-pyrroline, reaction with thioacetamide, 304
2-Amino-3,4-dicyano-5-alkylthiopyrroles, 136
3-Amino-2,5-dicyanobenzo(*f*)quinoline, 26
2-Amino-3,5-dicyano-6-chloropyridine, 164
2-Amino-3,5-dicyano-6-ethoxypyridine, 164
2-Amino-3,4-dicyano-5-mercaptopyrrole, 242
 reaction with methyl orthoformate, 242
2-Amino-3,5-dicyano-6-methoxypyridine, 164
2-Amino-3,5-dicyano-6-methylthiopyridine, 164
2-Amino-3,5-dicyanopyridine, 220
4-Amino-6,7-dihydro-2(1*H*)-pyrindinone, 232
4-Amino-5-formylpyrimidine, 224
2-Aminofuran, as 1,3-diene, 213
3-Aminoindazole, 202, 225
α-Aminoketones, reaction with malononitrile, 129
Aminomalononitrile, reaction, with formamidine, 146
  with mixed formic-acetic anhydride, 146
  with α-oximinoketones, 176
3-Aminomethylene-2-butanone, reaction with malononitrile, 163
Aminomethylenemalononitrile, reaction, with acetamidine, 105
  with acetonitrile ethyl imino ether, 105
  with acetonitrile ethyl imino thioether, 105
  with thioacetamidine, 105
 reduction of, 220
 use in the preparation of *o*-aminonitriles, 79
2-Aminonicotinonitrile, 210
 dimerization of, 235-236
4-Aminonicotinonitrile, 210
o-Aminonitriles, 79-374
 acylation of, 226-233
 alkylation of, 225
 conversion, to *o*-aminoaldehydes, 224-225
  to *o*-aminoamidoximes, 223
  to *o*-aminocarboxamides, 219-221
  to *o*-aminoketones, 224
  to *o*-aminothioamides, 224-228, 231
  to *o*-cyanoamidines, 238
  to fused 4-alkoxypyrimidines, 228-229
  to fused 4-aminopyrimidines, 233-299
  to fused pyridines, 231-233
  to fused 2,4(1*H*,3*H*)-pyrimidinedithiones, 306
  to fused 4(3*H*)-pyrimidinethiones, 231, 299-306
  to fused 4(3*H*)-pyrimidinones, 226-231
 diazotization of, 225-226
 dimerization of, 233-237
 from the Bedford-Partridge Reaction, 203-207
  Table, 204-206
 from dehydration of amides (Table), 180-186
 from desulfurization of thioamides (Table), 180-186
 from nucleophilic displacement of Hal-, O- or S-, Table, 180, 188-195
 from pteridine-4(3*H*)-thiones, 209-213
 from rearrangement or ring cleavage reactions, Table, 214-217
 from reduction of *o*-nitronitriles, 196-202
  Table, 197-201
 from ring cleavage of thiophenes, 207
 from Thorpe-Ziegler reactions, 19-20
 from *s*-triazine, 208-209
 hydrolysis of, 219-222, 231
 Michael additions of, 306-307
 reaction with amidines, 243, 268-270
  Table, 286-293
 reaction with carbon disulfide, 306
 reaction with cyanamide, 272, 285, 294
  Table, 286-293
 reaction with dicyandiamide, 272, 285, 294
  Table, 286-293
 reaction with diethyl oxalate, 295
 reaction with DMF/HCl, 230-231
 reaction with ethyl chloroformate, 295
 reaction with formamide, 270-272
  Table, 273-284
 reaction with formamidine, 243
 reaction with guanidine, 272, 285, 294
  Table, 286-293
 reaction with isocyanates, 295-299
 reaction with isothiocyanates, 295-299
 reaction with isothiocyanates and acid, 304-306

reaction with nitriles, 233-237
reaction with orthoformate esters and amines, 238-243
Table, 244-267
reaction with orthoformate esters and hydrogen sulfide, 299-301
reaction with thioamides and acid, 301-304
reaction with thioformanilide, 303-304
reaction with thiourea, 294-295
reaction with urea, 294-295
reaction with urethane, 294
reduction of, 224-225
Table of all known examples, 308-374
o-Aminonitriles, heterocyclic, from the Thorpe-Ziegler reaction, Table, 55-56
o-Aminonitriles, miscellaneous, S-heterocyclic (Table), 154-156
2-Amino-5-nitrobenzonitrile, dimerization of, 236
reaction with 4-nitrobenzonitrile, 237
reaction with thioacetamide, 304
2-Amino-3-nitro-6-(*p*-bromophenyl)benzonitrile, 20
2-(2-Amino-5-nitrophenyl)-4-amino-6-nitroquinazoline, 236
4-Amino-5-nitrosopyrimidines, reaction with malononitrile, 176
1-Amino-2,3,4,5,5-pentacyanocyclopentadiene, 16
2-(*o*-Aminophenyl)-4-aminoquinazoline, 233
2-Aminopyridine, 220
2-(3-(2-Aminopyridyl))-4-aminopyrido-(2,3-*d*)pyrimidine, 235
4-Aminopyrimidine, 220
4-Aminopyrimido(4,5-*d*)pyrimidine, 268
cleavage of, 213
4-Amino-2(1*H*)-quinazolone, 229
1-Amino-2,3,4,5-tetracyanocyclopentadienide, tetraethylammonium salt of, 16
4-Aminothieno(2,3-*d*)pyrimidines, 242
desulfurization of, 242
2-Amino-1,1,3-tricyanopropene, 81; *see also* Malononitrile dimer
Ammonia, cyclization of bis-cyanoethylation product of, 22
reaction with malononitrile, 126
4-Anilinoquinazoline, 297
4-Anilino-2(1*H*)-quinazolinethione, 297
desulfurization of, 297
4-Anilino-2(1*H*)-quinazolone, 297, 298
Anthranil, reaction, with hydrazines, 170
with hydroxylamine, 170
with malononitrile, 171
ring-opening of, 170-171
Arsines, cyclization of bis-cyanoethylation products of, 22
Arylamines, cyclization of bis-cyanoethylation products of, 22
Aryl azides, reaction with malononitrile, 150
3-Aryl-5-cyanocytosines, 107
Aryl cyanates, reaction with malononitrile, 107
α-Arylcyclohexanones, from hydrolysis of enaminonitriles, 18
Aryl hydrazines, reaction, with ethoxymethylenemalononitrile, 80
with malononitrile, 81
with tetracyanoethylene, 81
with 1,1,2-tricyanostyrenes, 103
Azasteroids, 229
Azelonitrile, intramolecular cyclization of, 20

Beckman rearrangement, 203
Bedford-Partridge reaction, 203-207
Table, 204-206
Benzamide, 272
Benzamidine, reaction with malononitrile, 104, 269
Benzamidinium tricyanomethanide, 105
Benzamidomalondiamide, dehydration of, 147
Benzcyclooctanone, 20
Benzonitrile, reaction with 1,1-diamino-2,2-dicyanoethylene, 106
3-Benzoyl-2-phenylpropionitrile, Wittig reaction with, 16
2,2′-Bis(bromomethyl)biphenyl, reaction with potassium cyanide, 2, 20
cis-1,2-Bis(bromomethyl)cyclobutane, reaction with sodium cyanide, 13
1,2-Bis(cyanomethyl)benzene, intramolecular cyclization of, 16
1,3-Bis(2-cyanoethyl)benzene, dimerization of, 21
Bis-cyanoethylation, 18, 22
2,2′-Bis(cyanomethyl)biphenyl, intramolecular cyclization of, 2, 20
Bis-enaminonitriles, heterocyclic, 21, 77, 25

o-Bromoamines, reaction with cuprous cyanide, 180
4-*n*-Butylaminopyrimido(4,5-*d*)pyrimidine, 269

2-Carboethoxy-3-amino-4-cyano-5-carboethoxymethylthiophene, 26
5-Carboethoxy-1-amino-2-cyanocyclopentene, 12
1-Carboethoxy-1-cyanocyclopropane, condensation with ethyl cyanoacetate, 12
Carbon disulfide, reaction, with alkylidenemalononitriles, 152
  with *o*-aminonitriles, 306
Catalysts, for Thorpe-Ziegler reaction, 1, 3, 20, 23
Catenanes, 1, 29
Chloramine, 149
α-Chloroalkylidenemalononitriles, reaction with sodium hydrosulfide, 142
3-Chloro-4-cyano-5-aminopyrazole, 81
α-Chloro-β-diketones, reaction with malononitrile, 129
α-Chloro-β-keto esters, reaction with malononitrile, 129
α-Chloroketones, reaction, with cyanide ion, 60
  with malononitrile, 142
2-Chloronicotinonitrile, 235
Cuprous cyanide, reaction with *o*-bromoamines, 180
Cyanamide, 243
  reaction with *o*-aminonitriles, 272, 285, 294
    Table, 286-293
Cyanide ion, reaction with epichlorohydrin, 213, 218
1-Cyano-2-acetylaminonaphthalene, reaction with dry hydrogen chloride, 229
  reaction with sodium hydroxide, 227
o-Cyanoamidines, 238, 240, 243, 268
cis-2-Cyano-3-aminobicyclo[3.2.0]heptene-2, 13
1-Cyano-2-amino-3,4-dihydronaphthalene, 20, 294
  dehydrogenation of, 20
4-Cyano-5-amino-1,2-dithiol-3-thione, 153
1-Cyano-2-aminoindene, 294
4-Cyano-5-aminoisoxazole, ring cleavage of, 221
1-Cyano-2-aminonaphthalene, reaction with cyanamide, 294
4-Cyano-5-aminooxazole, 146
2-Cyano-3-amino-4-phenyl-5-cyanomethylthiophene, 25
3-Cyano-4-amino-3-piperideine, 22
  dehydrogenation of, 165
3-Cyano-4-aminopyridine, 165
3-Cyano-4-aminopyrrolo(2,3-*d*)pyrimidine, 241, 242
4-Cyano-5-aminothiazole, 149
5-Cyano-6-amino-2*H*-thiopyranethiones, 152
2-(4-Cyanobutyl)-4-amino-5,6-trimethylenepyrimidine, 28
3-Cyanocarbostyril, 226
2-Cyanocycloheptanone, 61
α-Cyanocyclohexanone, reaction with ammonium formate, 187
2-Cyanocyclooctanone, 61
2-Cyanocyclopentanone, 11, 60
5-Cyanocytosines, 106, 107
3-Cyano-1,2,6,7-dibenzo-4-azazulene, 16
1-(2-Cyanoethyl)-2-cyanomethylbenzene, 20
2-Cyano-6-ethyl-6-(2,3-dimethoxyphenyl)-cyclohexanone, 60
Cyanogen bromide, reaction with N-phenylpiperidine, 23
Cyanogen chloride, reaction with enamines, 60
Cyanohydrins, reaction with β-alkylaminopropionitriles, 21
α-Cyanoketones, cyclic, by hydrolysis of cyclic enaminonitriles, 60
    Table, 62-76
  reaction with ammonium formate, 187
2-Cyano(methoxycarbonyl)aminobenzene, reaction with sodium methoxide, 228
3-Cyanomethyl-4-cyano-5-aminopyrazole, 81
2-(Cyanomethylmercapto)-4-amino-5-cyanothiazole, 26
o-Cyanophenylurea, reaction, with aqueous base, 227
  with sodium methoxide, 229
Cyanothioacetamide, 153, 157
N-cycloalkyl ureas, reaction with ethyl orthoformate and malononitrile, 107
1,11-Cyclodocosanedione, 77
Cycloheptanones, by hydrolysis of cyclic enaminonitriles, Table, 65-68

Cyclohexanone, 169
Cyclohexanones, by hydrolysis of cyclic enaminonitriles, Table, 64-65
2-(Cyclohexylidene)cyclohexylidenemalononitrile, 177
Cyclohexylidenemalononitrile, 170, 180
Cyclopentanones, by hydrolysis of cyclic enaminonitriles, Table, 62-63
7(6*H*)-Cyclopenteno(*d*)pyrimidinethione, 301
Cyclooctanones, by hydrolysis of cyclic enaminonitriles, Table, 68
meta-Cyclophanes, 1, 21, 23
para-Cyclophanes, 23

Desulfurization, 242, 297
1,3-Diaminobenzo(*f*)quinazoline, 294
2,4-Diamino-5-cyano-6-dimethylamino-pyrimidine, reaction with benzamidine, 269
2,4-Diamino-5-cyano-6-methylpyrimidine, photochemical synthesis of, 105
3,5-Diamino-4-cyanopyrazole, 81
2,4-Diamino-3-cyanopyrido(2,3-*b*)quinoline, reaction with nitrous acid, 226
2,4-Diamino-5-cyanopyrimidine, irradiation of, 105
4,6-Diamino-5-cyanopyrimidine, 213
  reaction with formamide, 271
2,4-Diamino-5-cyanothiazole, 149
1,1-Diamino-2,2-dicyanoethylene, irradiation of, 144, 146
  reaction, with acetonitrile, 106
    with benzonitrile, 106
2,5-Diamino-3,4-dicyanothiophene, 142, 207, 241
  rearrangement of, 207, 242
2,5-Diamino-4-dimethylamino-7-phenyl-pyrimido(4,5-*d*)pyrimidine, 269
4,6-Diamino-5-formylpyrimidine, 213
(Diaminomethylene)sulfonium dicyano-methylid, 149
6,12-Diaminophenhomazine, 233, 234
2,4-Diaminopteridines, 272
2,4-Diaminopteridine 8-oxides, 176, 272
2,4-Diaminopyrimido(4,5-*d*)pyrimidines, 272
2,4-Diaryloxy-5-cyano-6-aminopyrimidines, 107
Dibenz(*a,c*)[1,3] cycloheptatrienes, 20
N,N′-Di(*n*-butyl)formamidine, 269
6,12-Dichlorophenhomazine, 233, 234
Dicyandiamide, reaction with *o*-aminonitriles, 272, 285, 294
  Table, 286-293
1,1-Dicyano-2-amino-2-(2-ethoxyethoxy)-ethylene, reaction with hydrazine, 81
1,10-Dicyanodecane, 61
1,1-Dicyano-2,2-dichloroethylene, reaction with hydrazine, 81
1,1-Dicyano-2,2-dimercaptoethylene, 153
1,1-Dicyano-2-hydrazinostyrenes, conversion to pyrazoles, 103
Dicyanoketene acetals, reaction, with hydrazine, 80
  with hydroxylamine, 143
Dicyanoketene dithioacetals, reaction with hydrazines, 81
1,1-Dicyano-2,2-bis(trifluoromethyl)ethylene oxide, 149
Dieckmann condensation, 2, 3
Diels-Alder reaction, 213, 218
Diethyl cyanomethylphosphonate, sodium salt, Wittig reaction with, 16
Diethyl oxalate, reaction with *o*-aminonitriles, 295
trans-3,4-Dihydro-2-cyano-3-hydroxymethyl-4-hydroxyaniline, 213, 218
Dimerization, of 2-aminobenzonitrile, 233-235
  of 2-aminonicotinonitrile, 235-236
  of *o*-aminonitriles, 233-237
  of 2-amino-5-nitrobenzonitrile, 236
Dimethyl acetylenedicarboxylate, reaction with 2-aminobenzonitrile, 306-307
2,6-Dimethyl-4-amino-5-cyanopyrimidine, photochemical synthesis of, 105
2,4-Dimethyl-5-cyano-6-aminopyrimidine, 104
Dimethylformamide/hydrogen chloride, reaction with *o*-aminonitriles, 230-231
1,6-Dimethyl-4(5*H*)-pyrazolo(3,4-*d*)pyrimidinone, 230
α,α-Dimethylsuberonitrile, intramolecular cyclization of, 20
Dimroth rearrangement, 239-242, 268, 298-299
α,ω-Dinitriles, intramolecular cyclization of, 1
Diphenylacetonitrile, alkylation of, 12
α,α-Diphenyladiponitrile, purported

synthesis of, 21
5,5-Diphenyl-1-amino-2-cyanocyclopentene, 12
2,4-Diphenyl-5-cyano-6-aminopyrimidine, 104
N,N′-Diphenylformamidine, reaction with 2-aminobenzonitrile, 297
1,3-Diphenylisopropylidenemalononitrile, 177
α,α-Diphenyl-β-methyladiponitrile, 12
6,7-Diphenyl-4(3*H*)-pteridinethione, cleavage of, 209
s-Diphenylthiourea, 149
Dipotassium cyanimidodithiocarbonate, 26
Disodium hexacyanobutenediide, acid cyclization of, 16

Enamines, reaction with cyanogen chloride, 60
Enaminonitriles, infrared spectra of, Table, 4-7
  ultraviolet spectra of, Table, 8-10
Enaminonitriles, cyclic, Table, 30-48
  5-membered rings, 2, 11-17
    Table, 30-38
  6-membered rings, 2, 17-20
    Table, 38-42
  7-membered rings, 2, 20
    Table, 42-45
  8-membered rings, 2, 20-21
    Table, 45-46
  10-membered and larger rings,
    Table, 45-48
  from α-cyanoketones and ammonium formate, 187
  hydrolysis of, 3, 11, 18, 29, 60-77
    Table, 62-76
  nmr spectra of, 11
  reaction with urea, 294
  reduction of, 77
  tautomerism of, 1, 3
Enaminonitriles, heterocyclic, 1, 21-29
    Table, 48-55
  reaction with urea, 294
Epichlorohydrin, reaction with cyanide ion, 213, 218
d,l-Epiibogamine, 29
Ethoxymethylenemalononitrile, reaction,
    with alkyl hydrazines, 80
    with amidines, 105
    with aryl hydrazines, 80
    with guanidines, 105
    with 2-hydroxyethylhydrazine, 220
    with hydroxylamine, 143
    with S-methylthiourea, 164
    with potassium hydroxide, 164
    with thioureas, 105
  use in the preparation of *o*-aminonitriles, 79
2-Ethyl-4-amino-5-formylpyrimidine, 225
Ethyl chloroformate, reaction with *o*-aminonitriles, 295
Ethyl cyanoacetate, condensation with 1-carboethoxy-1-cyanocyclopropane, 12
Ethyl cyanoacetimidate, reaction with *s*-triazine, 209
Ethyl orthoformate, 169, 238, 241, 242, 297, 300
  reaction with 2-aminobenzthioamide, 224
  reaction with *o*-aminocarboxamides, 231
  reaction with *o*-aminonitriles, (with amines), 238-243
    Table, 244-267
  reaction with *o*-aminonitriles and hydrogen sulfide, 299-301
  reaction with malononitrile and N-alkyl ureas, 107

Fluorene, Thorpe-Ziegler cyclization of, bis-cyanoethylation product of, 18
Formamide, reaction with *o*-aminonitriles, 270-272
    Table, 273-284
  reaction with 4,6-diamino-5-cyanopyrimidine, 271
  reaction with 2-phenyl-4,6-diamino-5-cyanopyrimidine, 272
Formamidine, reaction with aminomalononitrile, 146
  reaction with *o*-aminonitriles, 243, 268-269
  reaction with 4-amino-5-cyanopyrimidine, 213
  reaction with malononitrile, 104, 268
2-Formylaminobenzonitrile, reaction with sodium methoxide, 227, 228
Furan *o*-Aminonitriles, 126-129
    Table, 127-128

Guanidine, 242

reaction, with 2-amino-3-cyanopyrazine 1-oxides, 176
reaction with *o*-aminonitriles, 272, 285, 294
Table, 286-293
reaction with ethoxymethylenemalononitrile, 105
reaction with malononitrile, 103
reaction with 2-phenyl-4,6-diamino-5-cyanopyrimidine, 285
reaction with 1,1,2-tricyanostyrenes, 103

α-Haloketones, reaction with malononitrile, 129
o-Halonitriles, reaction with ammonia, 180
1*H*-1,3,4,6,7,9-Hexaazaphenalene, 271
Hofmann degradation, 180
Hydrazides, reaction with tetracyanoethylene, 81
Hydrazine, reaction, with anthranil, 170
with 1,1-dicyano-2-amino-2-(2-ethoxyethoxy)ethylene, 81
with 1,1-dicyano-2,2-dichloroethylene, 81
with dicyanoketene acetals, 80
with dicyanoketene dithioacetals, 81
with malononitrile, 81
with malononitrile dimer, 81
with tetracyanoethane, 136
with 1,1,2-tricyanostyrenes, 103
2-Hydrazinobenzamidoxime, 219
2-Hydrazinobenzonitrile, cyclization of, 202
Hydrogen cyanide tetramer, irradiation of, 144, 146
Hydrogen sulfide, reaction, with *o*-acylaminonitriles, 227
with 2-aminobenzonitrile, 224
with 4-amino-5-cyanopyrimidine, 224
with *o*-aminonitriles and orthoformate esters, 299-301
with 2-methyl-4-amino-5-cyanopyrimidine, 224
with nitriles, 301
with tetracyanoethylene, 241
Hydrolytic desulfurization, 297
cis-4-Hydroxycrotonitrile, Diels-Alder reaction with 2-aminofuran, 213
2-Hydroxyethylhydrazine, reaction with ethoxymethylenemalononitrile, 220
α-Hydroxyketones, reaction with malononitrile, 128
Hydroxylamine, reaction, with 2-aminobenzonitrile, 222
with anthranil, 170-171
with dicyanoketene acetals, 143
with ethoxymethylenemalononitrile, 143
Hydroxylamine-O-sulfonic acid, 149
α-Hydroxymethylenecyclohexanone, reaction with malononitrile and ammonia, 169

Imidazole *o*-aminonitriles, 144-147
Table, 154
Imino ethers, 238, 243
Iminonitriles, 'frozen' tautomers, 17, 18
tautomerism, 1, 3
1-Indanone, reaction with malononitrile, 126
Indazole, reaction with sodamide, 233
Indene, pyrolysis of adduct with tetracyanoethylene, 26
Irradiation, 105, 144-146
Isatin 3-oximes, pyrolysis of, 203
Isocyanates, reaction with *o*-aminonitriles, 295-299
Isoquinoline *o*-aminonitriles, 165-171
Table, 166-168
Isothiazole *o*-aminonitriles, 149-150
Table, 150
Isothiocyanates, reaction, with *o*-aminonitriles, 295-299, 304-306
with malononitrile and sulfur, 147
Isoxazole *o*-aminonitriles, 143-144
Table, 143-144
Isoxazoles, ring-opening of, 60

Ketene dithioacetals, 25-26
Ketones (cyclic), by hydrolysis of cyclic enaminonitriles, 60
Table, 62-76
medium and large-membered,
Table, 69-72
Ketones (heterocyclic), from hydrolysis of heterocyclic enaminonitriles, Table, 73-76

Malononitrile, 164
aldol condensations of, 60
dimerization of, 1
reaction, with acetamidine, 104
with alkyl azides, 150
with *o*-aminobenzaldehyde, 170

with 2-amino-3-bromotropone, 136
with α-aminoketones, 129
with 3-aminomethylene-2-butanone, 163
with 4-amino-5-nitrosopyrimidine, 176
with ammonia, 126
with anthranil, 171
with aryl azides, 150
with aryl cyanates, 107
with benzamidine, 104
with carbon disulfide, alkali and chloroacetonitrile, 25
with carbon disulfide and sulfur, 153
with α-chloro-β-diketones, 129
with α-chloro-β-keto esters, 129
with α-chloroketones, 142
with ethyl methyl ketone and sulfur, 137
with ethyl orthoformate and N-alkyl ureas, 107
with formamidine, 104, 268
with guanidine, 103
with α-haloketones, 129
with hydrazine, 81
with α-hydroxyketones, 128
with hydrogen bromide, 157
with α-hydroxymethylenecyclohexanone and ammonia, 169
with 1-indanone, 126
with isothiocyanates and sulfur, 147
with ketones and sulfur, 242
with α-mercaptoaldehydes, 137
with α-mercaptoketones, 137
with 5-phenyl-1,2-dithiol-3-one, 153-154
with 1-tetralone, 20
with thiourea, 103
with *s*-triazine, 208
with urea, 103
use in the preparation of *o*-aminonitriles, 79-180
Malononitrile dimer, 157, 163
formation of, 81
reaction with hydrazine, 81
α-Mercaptoaldehydes, reaction with malononitrile, 137
α-Mercaptoketones, reaction with malononitrile, 137
4-Methoxy-2-aminobenzonitrile, 207
4-Methoxyquinazoline, 228
4-Methoxy-2(1*H*)-quinazolone, 228
6-Methoxy-2,2-tetramethylene pseudoindoxyl oxime, 207
2-Methyl-4-amino-5-aminomethylpyrimidine, 224
2-Methyl-4-amino-5-cyanopyrimidine, 105
irradiation of, 105
reaction with hydrogen sulfide, 224
reduction of, 224
2-Methyl-4-amino-5-formylpyrimidine, 225
Methylation, by irradiation, 105
3-Methyl-1(2*H*)-benzo(*h*)quinazolone, 229
2-Methyl-4(3*H*)-benzo(*f*)quinazolone, 227
1-Methyl-4-cyano-5-aminopyrazole, reaction with ethyl orthoformate, 238
reaction with DMF/HCl, 230
reaction with phenyl isothiocyanate, 305
1-Methyl-3-cyano-4-piperidone, 61
2-Methyl-4,6-diamino-5-cyanopyrimidine, 106
3,4-Methylenedioxyphenylacetonitrile, Thorpe-Ziegler cyclization of biscyanoethylation, product of, 18
Methyl hydrazine, reaction with tetracyanoethane, 136
3-Methylmercapto-5-phenyl-1,2-dithiolium iodide, reaction with cyanothioacetamide, 157
2-Methyl-4-methoxyquinazoline, 228
1-Methyl-4-methylaminopyrazolo(3,4-*d*)-pyrimidine, 239
1-Methyl-2-methylthio-4-amino-5-cyanoimidazole, 24
2-Methyl-6-nitroquinazolin-4(3*H*)-one, 304
Methyl orthoformate, reaction with 5-amino-3,4-dicyano-2-mercaptopyrrole, 242
1-Methyl-4(5*H*)-pyrazolo(3,4-*d*)pyrimidinone, 230
1-Methyl-4(5*H*)-pyrazolo(3,4-*d*)pyrimidinethione, 305
2-Methylthio-4-amino-5-cyanopyrimidine, 164
S-Methylthiourea, reaction with ethoxymethylenemalononitrile, 164
Michael addition, 306
of *o*-aminonitriles, 306-307
to ethoxymethylenemalononitrile, 80
to malononitrile dimer, 81
to tetracyanoethylene, 81

Nitriles, reaction, with *o*-aminonitriles, 233-237
with hydrogen sulfide, 301

with thioamides, 301-304
3-Nitro-2-aminobenzonitrile, 212
5-Nitro-2-aminobenzonitrile, 212
2-Nitrobenzonitrile, reduction of, 196, 202
o-Nitrobenzylidenemalononitrile, reduction of, 170
1-Nitro-2-dimethylaminoethylene, 20
2-(4-Nitrophenyl)-4-amino-6-nitroquinazoline, 237
2-(4-Nitrophenyl)-4-aminoquinazoline, 237
6-Nitro-4(3*H*)-quinazolinethione, cleavage of, 212
8-Nitro-4(3*H*)-quinazolinethione, cleavage of, 212

m-Oxazines, rearrangement of, 230-231
Oxazole *o*-aminonitriles, 144-147
Table, 145-146
α-Oximinoketones, reaction with aminomalononitrile, 176
4(3*H*)-Oxo-2(1*H*)-quinazolinethione, 297
desulfurization of, 297

Phenyl acetonitrile, 25
2-Phenyl-4-amino-5-cyano-6(1*H*)-pyrimidone, 226
1-Phenyl-2-amino-3-cyanopyrrole, 180
2-Phenyl-4-amino-5-cyanothiazole, 149
1-Phenyl-2-carbamoyl-3-cyanopyrrole, Hofmann degradation of, 180
2-Phenyl-4-cyano-5-aminooxazole, 146, 147
N-Phenyl-N′-(*o*-cyanophenyl)thiourea, 297
N-Phenyl-N′-(*o*-cyanophenyl)urea, reaction with sodium methoxide, 229
2-Phenyl-4,6-diamino-5-cyanopyrimidine, 105, 106
reaction, with benzamide, 272
with formamide, 272
with guanidine, 285
with nitrous acid, 226
5-Phenyl-1,2-dithiol-3-one, reaction with malononitrile, 153
1*H*-2(or 5, or 8)-Phenyl-1,3,4,6,7,9-hexaazaphenalene, 271
3-Phenyl-4(3*H*)-imino-2(1*H*)-quinazolinethione, 297
rearrangement of, 297
3-Phenyl-4(3*H*)-imino-2(1*H*)-quinazolone, 298
Phenyl isocyanate, reaction with 2-aminobenzonitrile, 298
Phenyl isothiocyanate, reaction with 2-aminobenzonitrile, 297, 305
3-Phenyl-4(3*H*)-oxo-2(1*H*)-quinazolinethione, 297
N-Phenylpiperidine, cleavage with cyanogen bromide, 23
2-Phenyl-4,5,7-triaminopyrimido(4,5-*d*)-pyrimidine, 285
Phosphines, cyclization of bis-cyanoethylation, products of, 22
o-Phthaldehyde, 16
Pimelonitrile, intramolecular cyclization of, 17
[4.4.3]Propellane, 14
Pteridine *o*-aminonitriles, 171-177
Table, 175
4(3*H*)-Pteridinethione, cleavage of, 209
Pterins, 176
Purines, 237
Pyrazine *o*-aminonitriles, 171-177
Table, 172-174
Pyrazole *o*-aminonitriles, 80-103
Table, 82-102
reaction with urea, 295
Pyrazolo(2,3-*a*)imidazolidine, 220
Pyrazolo(3,4-*d*)pyrimidines, 237
4(5*H*)-Pyrazolo(3,4-*d*)pyrimidinethione, 301
Pyrazolo(3,4-*d*)-1,2,3-triazinones, 231
Pyridine *o*-aminonitriles, 157-165
Table, 158-162
Pyrido(2,3-*b*)pyridine *o*-aminonitriles, Table, 169
Pyrido(2,3-*d*)pyrimidine *o*-aminonitriles, Table, 169
4(3*H*)-Pyrido(2,3-*d*)pyrimidinethione, cleavage of, 210
4(3*H*)-Pyrido(4,3-*d*)pyrimidinethione, cleavage of, 210
Pyrimidine *o*-aminonitriles, 103-126
Table, 108-125
4(3*H*)-Pyrimidinethiones, fused, cleavage to *o*-aminonitriles, 209-213
synthesis of, 299-306
Pyrimido(4,5-*d*)pyrimidines, 231
4(3*H*)-Pyrimido(4,5-*d*)pyrimidinethione, 224
cleavage of, 212
Pyrrole *o*-aminonitriles, 129-137
Table, 130-134

2,4(1*H*,3*H*)-Quinazolinedione, 227
4(3*H*)-Quinazolinethione, 224, 228, 305
4(3*H*)-Quinazolinethiones, 227, 228
4(3*H*)-Quinazolone, 227, 229, 230, 297
Quinoline *o*-aminonitriles, 165-171
  Table, 166-168

Rueggli's high dilution technique, 1, 20, 22, 77

Sandmeyer reaction, 196, 225
Semicarbazides, reaction with tetracyanoethylene, 81
Suberonitrile, intramolecular cyclization of, 3, 20

Tetracyanoethane, reaction, with hydrazine, 136
    with hydrogen bromide, 129-130
    with mercaptans, 136
    with methyl hydrazine, 136
    with sodium hydrosulfide, 142
  use in the preparation of *o*-aminonitriles, 79
Tetracyanoethylene, reaction, with alkyl hydrazines, 81
    with aryl hydrazines, 81
    with hydrazides, 81
    with hydrogen bromide, 129
    with hydrogen sulfide, 142, 241
    with mercaptans, 136
    with semicarbazides, 81
    with thiosemicarbazides, 81
  use in the preparation of *o*-aminonitriles, 79
1,1,3,3-Tetracyanopropene, 164
1-Tetralone, reaction with malononitrile, 20
  Thorpe-Ziegler cyclization of bis-cyanoethylation, product of, 19
m-Thiazines, 299, 300
  rearrangement or, 303, 304, 306
Thiazole *o*-aminonitriles, 147-149
  Table, 148
Thioacetamide, reaction with *o*-aminonitriles, 301-304
Thioacetamidine, reaction with aminomethylenemalononitrile, 105
Thioamide exchange, 301-306
Thioamides, reaction with *o*-aminonitriles, 301-304
Thiobenzamide, 149
Thioformanilide, reaction with *o*-aminonitriles, 303-304
Thiophene *o*-aminonitriles, 137-142
  Table, 138-141
Thiosemicarbazides,
  reaction with tetracyanoethylene, 81
Thiothiophthene, 153, 157
Thiourea, reaction, with *o*-aminonitriles, 294, 295
    with ethoxymethylenemalononitrile, 105
    with malononitrile, 103
Thorpe-Ziegler reaction, 1-56
  acid-catalysis of, 16
Toyocamycin aglycone, 241, 242
1,2,5-Triamino-3,4-dicyanopyrrole, 136
1,2,3-Triazole *o*-aminonitriles, 150-151
  Table, 151-152
s-Triazine, reaction, with aminomethylenemalononitrile, 208
  with ethyl cyanoacetimidate, 209
  with malononitrile, 208
1,1,3-Tricyano-2-aminopropene, 1
Tricyanostyrene, reaction, with ethyl mercaptan, 135
    with 2-mercaptoethanol, 135
Tricyanostyrenes, reaction, with guanidine, 107
    with hydrazines, 103
2,4,6-Tricyano-1,3,5-triaminobenzene, 203
2,4,6-Tricyano-1,3,5-triazidobenzene, 203
2,4,6-Tricyano-1,3,5-trifluorobenzene, 203
2,4,6-Tricyano-1,3,5-triphthalimidobenzene, 203
Tricycloquinazoline, 233, 234
1*H*-2,5,8-Triphenyl-1,3,4,6,7,9-hexaazaphenalene, 272

Urea, reaction, with *o*-aminonitriles, 294-295
    with ethoxymethylenemalononitrile, 105
    with malononitrile, 103

Vilsmeier-Haack formylation, 230, 304-305

Wittig reaction, 16